THE SCHOOL OF HELLAS

THE SCHOOL OF HELLAS

Essays on
Greek History, Archaeology,
and Literature

A. E. RAUBITSCHEK

Edited by Dirk Obbink and Paul A. Vander Waerdt

New York Oxford
OXFORD UNIVERSITY PRESS
1991

Oxford University Press

Oxford New York Toronto
Delhi Bombay Calcutta Madras Karachi
Petaling Jaya Singapore Hong Kong Tokyo
Nairobi Dar es Salaam Cape Town
Melbourne Auckland

and associated companies in
Berlin Ibadan

Copyright © 1991 by Oxford University Press, Inc.

Published by Oxford University Press, Inc.,
200 Madison Avenue, New York, New York 10016

Oxford is a registered trademark of Oxford University Press

All rights reserved. No part of this publication may be reproduced,
stored in a retrieval system, or transmitted, in any form or by any means,
electronic, mechanical, photocopying, recording, or otherwise,
without prior permission of Oxford University Press.

Library of Congress Cataloging-in-Publication Data
Raubitschek, A. E. (Antony Erich), 1912–
[Selections. 1990]
The school of Hellas: Essays on Greek history, archaeology,
and literature / A. E. Raubitschek
edited by Dirk Obbink and P. A. Vander Waerdt.
p. cm. Bibliography: p.
Includes index.
ISBN 0-19-505691-4
1. Civilization, Classical. I. Obbink, D.
II. Vander Waerdt, Paul A. III. Title.
DE59.R32 1990 938—dc20
89-33003

1 3 5 7 9 8 6 4 2

Printed in the United States of America
on acid-free paper

Preface

This volume has been designed to make accessible between two covers the most important works of A. E. Raubitschek. Vienna born and educated, Raubitschek came to America shortly before the Second World War, during a period notable for new knowledge produced by the application of archaeology and epigraphy to the study of Greek history. With fifth-century Athens as the focus, a new school of historical research—led prominently by Benjamin Meritt and his associates—began to emerge, which combined the traditional interests of classical scholarship in texts and the theory of their study with an appreciation of the ancient historians themselves as inquirers and interpreters and as interested members of specific social, economic, and political groups. Their many publications stimulated a renewed interest in the history, archaeology, and epigraphy of the ancient world, as well as in classical politics and the interplay of poetry, art, and religion.

The selection of papers in the present volume has been arranged thematically (History; Institutions; Art, Monuments, and Inscriptions; Poets, Literature, and Historiography). It is to be regretted that no excerpt could be included from the author's magnum opus, *Dedications from the Athenian Acropolis* (Cambridge, Mass. 1949), co-authored by Raubitschek and the late Oxford historian Lilian H. Jeffery. The reader will note the omission here of the valuable introductory discussion in that volume, together with several other fundamental epigraphic publications (e.g., "Early Attic Votive Monuments," *Annual of the British School at Athens* 40 [1939] 17–37), the inclusion of which was prohibited primarily because of the extraordinarily large number of plates and figures. The present volume is likewise deficient insofar as it contains no representation of the detailed editorial work by the author that appeared in the monumental collaborative effort on the *Editio Tertia* of *Inscriptiones Graecae*, vol. 1, fasc. 1 (Berlin 1981). A version of the Raubitschek bibliography appeared in a Festschrift volume of essays (*The Greek Historians: Literature and History. Papers presented to A. E. Raubitschek*, ed. M. H. Jameson [Stanford 1985], pp. 10–23).

The editors wish to express thanks to A. E. Raubitschek, to the editorial and production departments at Oxford University Press, and to colleagues and associates who tirelessly provided advice and encouragement in the compilation of the present volume, especially H. Cherniss, T. Figueira, P. A. Hansen, A. Henrichs, M. H. Jameson, D. Lateiner, D. M. Lewis, Isabelle Raubitschek, and F. Solmsen.

New York City   D. O.
Durham, North Carolina   P. A. V. W.
November 1990

Contents

Introduction by A. E. Raubitschek, *xi*

I HISTORY

1 The Treaties between Persia and Athens, *3*
   Greek, Roman and Byzantine Studies 5 (1964) 151–159
2 The Covenant of Plataea, *11*
   Transactions of the American Philological Association 91 (1960) 178–183
3 The Peace Policy of Pericles, *16*
   American Journal of Archaeology 70 (1966) 37–41
4 Corinth and Athens before the Peloponnesian War, *23*
   Greece and the Eastern Mediterranean in Ancient and Prehistory: Studies Presented to F. Schachermeyer (Berlin, 1977), 266–269
5 Die Rükkehr des Aristeides, *26*
   Historia 8 (1959) 127–128
6 War Melos tributpflichtig?, *29*
   Historia 12 (1963) 78–83
7 The Heroes of Phyle, *35*
   Hesperia 10 (1941) 284–295
8 Review of A. Masaracchia, *Solone*, *45*
   Classical Philology 58 (1963) 137–140

II INSTITUTIONS

9 Ostracism: The Athenian Ostraca, *53*
   Actes du 2e Congrès International d'Épigraphie Grecque (Paris, 1953), 59–74
10 The Origin of Ostracism, *65*
   American Journal of Archaeology 55 (1951) 221–229
11 A Late Byzantine Account of Ostracism, *77*
   American Journal of Philology 93 (1972) 87–91 [with J. J. Keaney]

12 Theophrastos on Ostracism, *81*
   Classica et Mediaevalia 19 (1958) 73–109

13 The Ostracism of Xanthippos, *108*
   American Journal of Archaeology 51 (1947) 257–262

14 The Case against Alcibiades (Andocides IV), *116*
   Transactions of the American Philological Association 79 (1948) 191–210

15 Philochoros *Frag.* 30 (Jacoby), *132*
   Hermes 83 (1955) 119–120

16 A New Attic Club (Eranos), *134*
   J. Paul Getty Museum Journal 9 (1981) 93–98

17 Zur Frühgeschichte der Olympischen Spiele, *143*
   Lebendige Altertumswissenschaft: Festgabe zur Vollendung des 70. Lebenjahres von Hermann Vetters ((A. Hakkert), 1985), 64–65

18 Das Datislied, *146*
   Charites E. Langlotz: Studien zur Altertumswissenschaft (Bonn, 1957), 234–242

19 Die Gründungsorakel der Dionysien, *156*
   Pro Arte Antiqua: Festschrift für Hedwig Kenner, Österreichische Archaeologische Institut Sonderschriften 18 (Berlin, 1985), 2: 300–301

20 The Priestess of Pandrosos, *158*
   American Journal of Archaeology 49 (1945) 434–435

21 Review of F. Sartori, *La eterie nelle vita politica Ateniese del VI e V secolo A.C.*, *160*
   American Journal of Philology 80 (1959) 81–88

22 Review of M. P. Nilsson, *Die hellenistische Schule*, *167*
   Phoenix 12 (1958) 40–42

III ART, MONUMENTS, AND INSCRIPTIONS

23 Inschriften als Hilfsmittel der Geschichtsforschung, *173*
   Rivista Storica dell' Antichità 1 (1971) 177–195

24 Review of L. Jeffery, *The Local Scripts of Archaic Greece*, *187*
   Gnomon 34 (1962) 225–231

25 Leagros, *195*
   Hesperia 8 (1939) 155–164

26 Two Monuments Erected after the Victory of Marathon, *204*
   American Journal of Archaeology 44 (1940) 53–59

27 An Original Work of Endoios, *212*
   American Journal of Archaeology 46 (1942) 245–253

28 The New Homer, *220*
   Hesperia 23 (1954) 317–319

29 Demokratia, *223*
   Hesperia 31 (1962) 238–243

30 The Mission of Triptolemos, *229*
   Studies in Athenian Architecture, Sculpture and Topography Presented to Homer A. Thompson. Hesperia Supplement 20 (1982) 109–117 [with Isabelle Raubitschek]

31 The Eleusinian Spondai (*IG* I^3, 6, lines 8–47), *239*
   Philia Epē, Bibliothēkē tēs en Athēnais Archaiologikēs Hetaireias 103 (Athens 1986), 2: 263–265 [with Mariko Sakurai]

IV POETS, LITERATURE, AND HISTORIOGRAPHY

32 Das Denkmal-Epigramm, *245*
   Fondation Hardt, *Entretiens sur l'antiquité classique 14, L'Epigramme Grecque* (Vandoevres-Genève, 1969), 3–36

33 Review of F. Solmsen, *Hesiod and Aeschylus*, *266*
   Classical Weekly 45 (1951) 70–72

34 *Erga megala te kai thomasta*, *270*
   Revue des Etudes Anciennes 41 (1939) 217–222

35 Herodotus and the Inscriptions, *275*
   Bulletin of the Institute of Classical Studies 8 (1961) 59–61

36 The Speech of the Athenians at Sparta, *279*
   The Speeches in Thucydides, ed. P. A. Stadter (Chapel Hill, 1973), 32–48

37 Andocides and Thucydides, *292*
   Classical Contributions: Studies in Honor of M. F. McGregor (Locust Valley, N. Y., 1981), 121–123

38 Die sogenannten Interpolation in den ersten beiden Büchern von Xenophons griechischer Geschichte, *296*
   Vestigia 17 (1973) 315–325

39 Theopompos on Thucydides the Son of Melesias, *306*
   Phoenix 14 (1960) 81–95

40 Theopompos on Hyperbolos, *320*
   Phoenix 9 (1955) 122–126

41 Review of L. Pearson, *The Local Historians of Attica*, *325*
   American Journal of Philology 45 (1944) 294–297

42 Review of F. Jacoby, *Atthis: The Local Chronicles of Ancient Athens*, *328*
   Classical Weekly 44 (1951) 135

43 Die schamlose Ehefrau (Herodot 1, 8, 3), *330*
   Rheinisches Museum 100 (1957) 139–140

44 Damon, *332*
   Classica et Mediaevalia 16 (1955) 78–83

45 Phaidros and His Roman Pupils, *337*
   Hesperia 18 (1949) 96–103
46 Zu einigen Wiederholungen bei Lukrez, *345*
   American Journal of Philology 59 (1938) 218–223

Plates follow page *349*
Bibliography of A. E. Raubitschek, *351*
Subject Index, *364*
Index of Greek Words, *381*
Index of Latin Words, *384*

Introduction

Higher education in the United States has been expanded during the past fifty years, and so have the number of colleges and graduate students in the classics, the size and the numbers of our meetings and publications have increased, and more financial support is available in the form of fellowships and grants. If the quality of our performance and production has not proportionately improved, this is also true of other fields; it is the result of the equalizing and democratic tendencies present in our society—of which I approve and which I consider worth the price we are paying.

The subject matter and the method of classical studies have also changed somewhat, more in religion, ancient history, and archaeology than in classical philology. Here our favorite texts have remained the same, while the computer has opened up new possibilities and literary criticism has made the understanding of the originals more difficult. In general, there is a tendency to substitute "ancient" for "classical," and to use "culture" in the sense that anthropologists have given it. Emphasis is placed more and more on what people—the common people—thought, said, did, and experienced, rather than on universally accepted masterpieces and their creators. This puts the classics on the same level with other disciplines, but it does not remove the recognition and the esteem accorded these great works nor does it explain them. The core period and age have not changed, but we have learned much about Homer, his poems, and the age with which they deal, and also about Athens through the ages. Our knowledge of Athens has been greatly increased by the excavations conducted by the American School of Classical Studies in the Agora. I have been familiar with them since 1934, and I have been connected with them and with the American School of Classical Studies since I came to Princeton in 1938 to work with Ben Meritt at the Institute for Advanced Study.

My first task was to finish with L. H. Jeffery the work on the *Dedications from the Athenian Acropolis*, and then I engaged in the study of the Agora inscriptions under the direction of Ben Meritt. At the same time, I helped J. Hondius finish the tenth volume of the *Supplementum Epigraphicum Graecum*, and I began working on the Agora ostraka that T. L. Shear generously put at my disposal and that M. Lang is going to publish, using E. Vanderpool's notes.

At Yale, during 1942–1944 and 1945–1947, I devoted much of my time to the study of ancient history and ancient political thought and theory, both in support of my teaching, and I extended this study to include the history of the Balkans up to the present time, also in support of my teaching in the Army Specialized Training Progam. Back at Princeton, from 1947–1963, I continued my association with Ben Meritt at the Institute for Advanced Study, taught various language courses in the classics department, and conducted with Paul Ramsey the humanities course, which gave me the opportunity to study the humanities and to lecture on problems connected with them. After coming to Stanford in 1963, I taught language courses in the classics department including graduate seminars on the historians and on epigrams, as well as courses on humanities, classical politics, and classical athletics.

This summary is to explain the way I came to write most of my articles; except for those on inscriptions, they were byproducts of my preparation for courses and seminars. I never felt a conflict between teaching and scholarship. Teaching requires an understanding of the information one is to convey, and whenever the available information is not satisfactory, one is obliged to improve it by scholarship, study, research. If these efforts are successful, one should convey them to others in the form of publications. Of course, in most cases one's own efforts are not successful, or not satisfactory at once, and they must be put aside until one can improve them later.

But how much of one's own scholarly interests should be passed on to students and friends? As much as possible, I believe. Any research topics that one considers worthwhile should be passed on as well. The disappointment that sometimes follows should not discourage research. It is also part of scholarly education to convey to others that the search is more important than the personal pleasure or profit derived from conducting it.

One of the most difficult problems in classical studies concerns its general purpose. This was pointed out to me for the first time in 1938 when some young and very liberal doctors in London observed that what I was doing was pure luxury and did not contribute to the *salus generis humani* I had sworn an oath to serve when I received my doctorate. For many years I smarted under this charge, especially because of the social consciousness I found so highly developed in the United States. Then, led by Plato and Jefferson, I began to understand that culture itself in its best achievements was indeed luxury, but a luxury that made our life peculiarly and significantly human. In the end, I realized that the great contributions of the Greeks and the Romans were not merely the starting point of Western culture, closely connected to the political, social, economic, artistic, and intellectual history and tradition of Europe and of those parts of the world under European influence or control. These contributions also had a universal meaning, significance, and value. In art and architecture, in literature and philosophy, in government, law and administration, in science and technology, the Greeks and the Romans found and formulated answers to humankind's universal question of how to make human life better. While the rise of human civilization during the past three thousand years has been spectacular all over the globe, it is proper for some of us to renew again and again our acquaintance with and our knowledge of the way problems were first formulated and solved and, if possible, to add to this knowledge. This is essentially the task of the student of the classics, just as the student of the Bible seeks to preserve and to increase our knowledge of universal religion.

I would like to add some reminiscences in order to illustrate my gratitude to my teachers and older friends, and also to provide some more vivid pictures of especially attractive aspects of classical studies.

When I entered university in 1931, Wilamowitz had just died and his student Johannes Mewaldt devoted his first lecture (which was the very first lecture I attended) to a tribute to the great scholar and teacher, saying that he represented classical studies at their best. During the following hour, Ludwig Radermacher was to give his first lecture on Greek comedy. He began by saying that his aim had been to correct all the mistakes made by Herr von Wilamowitz in the text and in the interpretation of Greek comedy, and he added that since the man had died, there was no point in doing this anymore and that he would lecture to us on the Homeric Hymns.

On the same day I also attended the Greek proseminar on Lysias offered by Radermacher. Students were not eligible to take seminars in Greek or Latin until they had taken two proseminars each in Greek and Latin and two composition courses in each of the two languages. I was told to give the first report on Lysias with no other instruction than to "establish" the text and prepare a correct translation. I did the best I could, but of course I did not understand the *apparatus criticus*. Radermacher was pleased and praised me for accepting an ἐγώ in the text which had been rejected by other editors. "You should investigate Lysias' use of ἐγώ," he said in closing. I never thought of it again until the end of the semester when Radermacher distributed the certificates with the grades, beginning with the lowest. Finally he held one certificate in his hands, and I had not yet received mine. He looked at me and said, "Where is the paper on ἐγώ in Lysias?" I replied that I had thought it was a joke. He said, "We philologists do not joke," and tore the certificate up and threw the pieces in the wastepaper basket. I was never admitted to his seminar, but at the request of the students I was permitted to attend, and many years later he wrote me very friendly letters.

The only seminar I was permitted to attend was that in archaeology, but even there I was told by the assistant, Fr. Hedwig von Kenner (who later became professor of archaeology at Vienna, and who is a dear friend), that as a first-year student I could not give a report but only listen and study. Just before Christmas I was informed that so many of the older students had become ill that I would be allowed to prepare a report. I was given an outline drawing of a Greek vase (at that time only a few photographs were available) and a bibliography. When the time came, I sat next to the professor, Emil Reisch, and read my paper. He paid absolutely no attention to me but criticized what I had said very sharply. At the end of the semester, the custodian of the seminar presented me with a five-schilling bill (about one dollar), telling me that the *Herr Hofrat* had decided to give me the prize for the best paper. When I asked whether I could express my thanks in person, I was told that this would be quite inappropriate. Naturally I took Reisch's seminar again and again, and I received the prize every time, but I was never able to speak with him in person. He died soon afterwards, and his successor, Camillo Praschniker, was much more accessible.

After I finished my dissertation on Epicurus and Lucretius under Mewaldt in 1935, I was told that I had to get a certificate for teaching high-school Latin, Greek, and German if I wanted to be considered for a position at the university. The reason was that the faculty did not want to recommend the appointment of young scholars who could not or would not teach school if need be. I therefore took the additional examinations and taught Latin and Greek to twelve and thirteen year olds full-time for a whole year as an unpaid volunteer. At the end of it, in 1936, I was fortunate to receive a fellowship for work in Athens on the inscribed statue bases from the Acropolis.

This project had been undertaken many years earlier by Rudolf Heberday, who had abandoned it in favor of his more important work on the Poros sculpture. He still had his old notebooks and he invited me to visit him in Graz to discuss this matter. I was there three whole days, being examined in all fields of classical studies. At the end, Heberday gave me his notebooks and his blessing. Unfortunately, he did not live to see the publication of the first modest results of the work that he had sponsored. A list of joining fragments was submitted to the Vienna Academy by my teachers Camillo Praschniker and Adolf Wilhelm, and it was printed on one small page of the *Anzeiger der Akademie der Wissenschaften in Wien* for 1936.

Fifty years later, in 1985, I returned to Vienna and had the pleasure of lecturing in the same classroom in which I had sat as a student, beginning in 1931, and of studying in the same seminar (now called an "Institute") in which I used to study. Of course there are many changes, but one thing has remained the same: the spirit of cooperation between faculty and students. Both groups are using the same books in the same way for the same purpose, and they are anxious to help each other as much as possible. The departmental libraries in our universities are supposed to serve the same purpose, not only to make important books available, but especially to facilitate their use on the part of all members of the academic family.

It is often said that not enough is done for the humanities and for the classics especially, as if their prominence and even their survival depended on our effort and interest. I sometimes wonder whether it is not the humanities and especially the classics that provide *us* with preeminence and perhaps even cultural survival, and which we should foster and transmit to future generations for all our sakes. The classics give significance and satisfaction to us and make our efforts worthwhile by revealing the meaning of our existence.

At this point, the feeling uppermost in my mind is that of gratitude, to my parents, my teachers, and my colleagues, my wife, our children and their families, and to the many young people who studied with me, especially to the two who initiated the publication of this book and did all the work on it.

Stanford, California A. E. Raubitschek

I

HISTORY

1

The Treaties between Persia and Athens

According to the ancient tradition represented by Diodorus (12.26.2, derived [151] from Ephorus) and Demosthenes (15.29), there were two treaties (διτταὶ συνθῆκαι) between Persia and Athens (or Greece), the famous Peace of Kallias and the equally famous Peace of Antalkidas.[1] The genuineness and the historicity of the Peace of Kallias have been questioned by Theopompus (*FGrHist* 115 F 154) because its Attic copy was written in Ionic script (which came into official use not before the end of the fifth century) and by the moderns because Thucydides does not specifically refer to it. The date assigned to this treaty is the time immediately following the death of Kimon (Diodorus 12.4.4), but its absolute date and its connection with the so-called Congress Decree of Pericles (Plutarch, *Pericles* 17) will have to be argued separately.[2] The historicity of the Peace is further attested by Diodorus (9.10.5), who emphasizes the fact that the Athenians in signing the treaty with Artaxerxes were breaking the oath of everlasting enmity against the barbarian, which they had sworn on the battle field of Plataea.[3] Its validity, however, is made doubtful by the fact that the chief negotiator, Kallias, was accused of bribery, condemned, almost executed, and fined fifty talents.[4] Plutarch's account (*Cimon* 13) suggests that the document itself was available in the Athenian archives, and that the treaty was kept by the Persians while they were militarily weak and disregarded by the Athenians while they were militarily strong. In fact, Demosthenes referred to the two treaties with Persia as examples of the general | statement ἅπαντας [152]

---

1. See now H. Bengtson, *Die Staatsvertraege des Altertums* 2 (1962) nos. 152 and 242, where may be found all significant references.
2. See my tentative suggestion in "Kimons Zurückberufung," *Historia* 3 (1954/5) 379–380, that the death of Kimon be placed before 455 B.C.
3. See my remarks in "The Covenant of Plataea," *TAPA* 91 (1960) 181 with n. 6 (also on p. 182) [*infra*, pp. 13–14].
4. Demosthenes 19.273; see my remarks in "Herodotus and the Inscriptions," *University of London, Institute of Classical Studies, Bulletin* 8 (1961) 61 [*infra*, pp. 275–278].

πρὸς τὴν παροῦσαν δύναμιν τῶν δικαίων ἀξιουμένους (15.28) which, I think, alludes to the view expressed by Thucydides (5.89) δίκαια...ἀπὸ τῆς ἴσης ἀνάγκης κρίνεται, δυνατὰ δὲ οἱ προύχοντες πράσσουσι καὶ οἱ ἀσθενεῖς ξυνχωροῦσιν, and again (5.105.2) οὗ ἂν κρατῇ ἄρχειν.

It is, however, not with these two famous treaties that I am concerned at the moment but with two others which have not fared so well in our literary tradition, ancient and modern; they are the treaty with Dareios concluded in the last decade of the sixth century and that with the second Dareios concluded shortly before 415 B.C.

Herodotus reports (5.73) that the Athenians sent an embassy to Sardis to conclude an alliance with the Persians, after they had driven out Isagoras and Kleomenes, and because they feared to be subdued by the Spartans under Kleomenes. This passage, as well as that other about the famous shield signal (6.121–131), has given rise to a protracted debate about the existence in Athens of a pro-Persian party.[5] What is important is the fact that the Athenians wished to conclude an alliance with Persia which should protect them against an impending Spartan attack, and that the ambassadors actually concluded an alliance by offering on their own account submission to Dareios. The concluding sentence of Herodotus's story ("these [namely the Athenian envoys] went back home and were then held greatly responsible [for what they had done]") has been taken to mean that the Athenians at once repudiated the agreement of the envoys with Artaphernes. B. D. Meritt pointed out, however, that "the Athenians were on good terms with Persia and that they were interested in maintaining friendly relations" when they once more sent envoys to Sardis complaining about the propaganda of Hippias and asked the Persians not to believe people who had been exiled from Athens (Herodotus 5.96).[6] Such a complaint was justified if Athens considered herself still an ally of Persia.[7] There is, however, pace Meritt and Robinson, not the slightest indication of any friendly attitude on the part of Athens toward Hippias. The refusal by the Athenians to restore Hippias and especially their participation in the Ionian Revolt (Herodotus 5.97) constituted a violation of their treaty with Persia, for which Dareios wanted to punish them at Marathon (Herodotus 5.105).

Closely connected with the question whether or not there was a pro-Persian party at Athens is another—whether Kleisthenes and his family (the Alcmeonids) were responsible for the first embassy to Persia, and thus for the conclusion of the treaty. Herodotus certainly does not say so (5.73), nor does he identify the envoys, either the first ones or the second ones (5.96), but in talking about the shield signal after the battle of Marathon (6.115) he emphasizes that the responsibility for the plot with the Persians was given to the Alcmeonids,

---

5. Most of the references have now been conveniently assembled by P. Lévêque and P. Vidal-Naquet, *Clisthène l' Athénien* [*Annales... de Besançon*] (Paris 1964) 113, n.2.

6. "An Early Archon List," *Hesperia* 8 (1939) 63–64.

7. While M. F. McGregor questioned this interpretation in "The Pro-Persian Party at Athens," *HSCP* Suppl. 1 (1940) 79, C. A. Robinson, Jr., in "Athenian Politics, 510–486 B.C.," *AJP* 66 (1945) 247, seems to accept it.

and in his lengthy apology (6.121 and 123) he repeatedly states that the Alcmeonids were antityrannical; it is significant that he has nothing to say about their connection with the Persians. The question may well be asked whether Herodotus did not deliberately suppress some information on the part played by the Alcmeonids in making and in keeping the treaty with the Persians, because he did not want to taint Pericles with the charge of a pro-Persian attitude (6.131).

There can be little doubt that the punishment to which the Alcmeonids were subjected because of the suspicion that they had plotted with the Persians was the ostracism to which Megakles and Xanthippos were condemned, the one a nephew of Kleisthenes, the other the husband of a niece.[8] Kleisthenes may have meant the law of ostracism to be strictly antityrannical, but it was used at once not only against the leader of the tyrant's party, Hipparchos, but also against those Alcmeonids who, by being pro-Persian, seemed to have contributed to the danger of the restoration of tyranny.[9] It may well be, therefore, that Kleisthenes, Megakles, and Xanthippos had something to do with the two embassies to Persia, and that the punishment of the envoys to which Herodotus alludes (5.73) consisted in the ostracism of Megakles and Xanthippos.

Seen in this way, the first Athenian embassy to Sardis seems less surprising because Alkmeon, after whom the family was named, was a friend of Kroisos and had visited him in Sardis.[10] Considering that the Persians were the successors of the Lydians and that Kroisos re|tained good relations with the Persians (Herodotus 3.36), the Alcmeonids may have suggested sending envoys to Sardis. The emotionally charged contrast between Greeks and barbarians belongs to a later time, that of the Persian Wars.

[154]

If Pericles' father, uncle, and grandfather had good relations with Persia before the Persian Wars, it is understandable that Pericles himself should have had little enthusiasm to continue the war against Persia when he rose to prominence. This was at a time when the Persian War had virtually come to a standstill, that is after Kimon's great victories in Pamphylia, toward the end of the sixties of the fifth century. Having had no share in the Egyptian expedition, and pursuing an active policy within Greece proper, Pericles was glad to see the Persian War come to an end and to make peace with Persia. The effect of this policy upon the Athenian alliance, which had been founded for the purpose of continuing the war against Persia, is a problem which must be discussed separately; for our purposes it suffices to place the Peace of Kallias into the context of the treaty relations between Athens and Persia which began soon after the return to power of the Alcmeonid Kleisthenes.

It is unfortunate, but perhaps not surprising, that the Athenian treaty with

---

8. See my remarks on "The Ostracism of Xanthippos," *AJA* 51 (1947) 259 [*infra*, pp. 110–111], and on the chronology "Die Rueckkehr des Aristeides," *Historia* 8 (1959) 127–128 [*infra*, pp. 26–28].

9. See my remarks in "The Origin of Ostracism," *AJA* 55 (1951) 221 and 225–226 [*infra*, pp. 65–67 and 73–76].

10. Herodotus 6.125; for the family see F. W. Mitchel in "Megakles," *TAPA* 88 (1957) 127–130 and T. L. Shear, Jr., in "Koisyra," *Phoenix* 17 (1963) 99–112.

Dareios has left so few traces in our literary tradition that it was not even included in Bengtson's *Staatsvertraege*. After all, it was first broken by Persia when she supported Hippias, who had been expelled from Athens, and later by Athens when she supported the Ionian Revolt, and it became quickly forgotten during the long Persian Wars which lasted almost a whole generation, that of Kimon. When Kimon died and a new treaty was drawn up, the power situation had radically changed. No longer did Athens seek protection against Sparta by attaching herself to Persia, but both powers were willing and anxious to recognize each other's spheres of influence and power in order to be able to devote themselves to other more pressing problems, the Persians to other parts of their far-flung empire, the Athenians to the organization of their allies and to the impending conflict with Sparta.

Considering the Persian successes in Egypt, where they were able to crush the native revolt and the Athenian expeditionary force, and considering the traditionally aggressive policy of Athens, one should recognize that the peace between Persia and Athens remained undisturbed for many years. The claim

[155] has been made that Pericles' | Pontic Expedition (Plutarch, *Pericles* 20.1–2) violated the peace with Persia and must have been undertaken before this peace was signed.[11] Since Sinope was a Greek city, the Athenians had a right to concern themselves with her, and the treaty did not limit the range of the Athenian fleet; in fact, in this very passage Plutarch emphasizes Pericles' restraint and his care not to provoke Persia (20.3). More serious, however, was the support given to the Samian oligarchs by Pissouthnes, the Persian satrap in Sardis;[12] the entry of the Phoenician fleet into the Aegean was feared,[13] but it may never have taken place. Even if the satrap had acted on his own, this was an unfriendly move, the danger of which was not lost on Pericles. The ruthlessness of his conduct in crushing the Samian revolt was caused to a large extent by Persia's support of the rebels. Years later, on the occasion of the revolt of Mytilene, it was suggested to ask for help from Pissouthnes, but nothing came of it (Thucydides 3.31.1). Pissouthnes himself revolted from the king after the accession of Dareios, and his son Amorges did the same, perhaps at the same time.[14]

It is a fair assumption that Athens was at peace with Persia ever since Kimon's death. This assumption is supported by several pieces of evidence, all coming from the mid-twenties of the fifth century. First of all, there is a scene in Aristophanes' *Acharnians* (61–127), which if nothing else shows clearly that the Athenians were on friendly terms with Persia, that the sending of ambassadors to Persia and the receiving of envoys from Persia was considered nothing unusual, and that the Athenians were not surprised at the notion that they were

---

11. See e.g. J. H. Oliver in "The Peace of Callias and the Pontic Expedition of Pericles," *Historia* 6 (1957) 254–255.
12. Thucydides 1.115.4–5; Plutarch, *Pericles* 25.3–4.
13. Thucydides 1.116.1–3; Plutarch, *Pericles* 26.1.
14. Ctesias, *Persica* 52, ed. Henry; Thucydides 8.5.5; 19.2; 28.2–5; 54.3.

asking for financial assistance from the Great King.[15] In full agreement with this passage in a comedy which was produced early in 425 B.C. is a story told by Thucydides, which speaks of another Athenian embassy that was sent to Susa but actually went only as far as Ephesos when word was received of the death of Artaxerxes; at this news the Athenian envoys returned home (4.50). Both Wade-Gery and Andrewes[16] have called attention to this episode, but they have not | emphasized that the Athenian envoys went under an existing [156] treaty and not in order to conclude a new one. Obviously, the Athenians were shocked to hear that the Spartans were trying to negotiate for aid from Persia,[17] and they were prepared to complain about it, just as eighty years earlier they had complained about the anti-Athenian propaganda of Hippias. When news of the king's death reached them, they returned home because they realized that this was no time to discuss this matter at the Persian court. Together with the scene in the *Acharnians*, the story in Thucydides testifies to the continuation of friendly relations between Athens and Persia during the Archidamian War.

We learn from Andocides (3.29) that his uncle Epilykos, who later died in Sicily (1.117), negotiated a treaty of friendship for all time between Persia and Athens, that the Athenians abandoned this treaty soon after it was concluded by supporting Amorges, and that the Great King accordingly agreed to give financial aid to the Spartans, as recorded by Thucydides (8.14.4–5; 17.4–18). It has been assumed by Wade-Gery and Andrewes[18] that this treaty was concluded "in the first half of 423" B.C. (Andrewes) because during this Attic year Epilykos was first secretary and Neokleides was secretary of the tribe Aigeis, and we have a decree in which a treaty with the Great King is mentioned and which was passed on the day on which Neokleides was *epistatēs*;[19] Thucydides proposed this decree, and a Thucydides was treasurer during the year 424/3. All this is good but circumstantial evidence, and Andrewes pointed out its chief weakness when he observed that the "negotiations must have been rapid, since they necessarily begin after Dareios's accession (December 424 B.C.) and yet were complete in time for Herakleides' decree to be passed before the end of the Attic year 424/3 B.C."[20] I would prefer assuming that both the treaty with Dareios and the honorary decree for Herakleides (who had aided the Athenian envoys) belong to a somewhat later time, soon before the Sicilian Expedition in 415 B.C.[21] I also find it difficult to believe that this treaty was merely a renewal of the Peace of Kallias.[22] Not only was there no cause for a renewal

---

15. See also Thucydides 2.7.1, a passage in which the historian says that "both sides" (ἑκάτεροι) sought aid from Persia.
16. "The Peace of Kallias," *HSCP* Suppl. 1 (1940) 131; "Thucydides and the Persians," *Historia* 10 (1961) 1–5.
17. See also Thucydides 1.82.1; 2.7.1; Diodorus 12.41.1.
18. *Op. cit.* 127–132; *loc. cit.*
19. *IG*^{3} 2.8 = *SEG* 19.16.
20. See Andrewes *loc. cit.* 2–3.
21. See the doubts expressed by Gomme in "The Treaty of Callias," *AJP* 65 (1945) 332, and in his *Commentary* 1.333–334.
22. See Gomme *loc. cit.* 333 n.36 and Andrewes *loc. cit.* 5–6.

[157] (unless it had been broken) | but the new treaty was according to Andocides a treaty of friendship, while the Peace of Kallias was no such thing. If the report of Andocides can be trusted at all, we should assume that the Athenians sought to strengthen their hand by getting financial aid from Persia, and to protect their back while they were engaged in Sicily. True enough, Thucydides fails to mention that the Athenians had a new treaty with Persia when they embarked on their venture to Sicily, or that the Athenians had abandoned an alliance with Dareios and supported the rebellious Amorges when Alcibiades promised to reconcile them with Tissaphernes (8.45–47). This would, however, explain and clarify the policy of Alcibiades. One could assume that he was behind the treaty with Dareios, that this treaty was abandoned when Alcibiades fell into disgrace, and that he made his way from Sparta to Persia because he had previously established contact with Tissaphernes. There may be some support for these assumptions in several writings of Demosthenes (10 and 11), in which he emphasizes the fact that the Persian king supported at one time or another the Athenians or the Spartans in order to let neither of them get too strong (10.51 and 11.6), a policy the advocacy of which Thucydides attributed to Alcibiades (8.46).

The treaty between Persia and Athens may have been broken by the Athenian support of Amorges, and it may have been superseded by Dareios's treaty with Sparta (Thucydides 8.18). Thucydides may have failed to mention it for one reason or another, but the Athenians thought highly enough of it to honor the Clazomenian Herakleides for having aided in its conclusion; and when the stele recording these honors was destroyed, the Athenians renewed it early in the fourth century (*SEG* 19.16). At about the same time (392 B.C.), the orator Andocides made a special reference to the treaty which had been negotiated by his uncle.

In this connection belongs the hostile remark of Theopompus, who rejected as spurious not only the Oath of Plataea[23] but also the treaty with Dareios (*FGrHist* 115 F 153); the passage is quoted by Bengston (*Staatsvertraege*) in connection with the Hellenic Oath *before* the battle of Salamis (no. 130) with which it has nothing to do, but it is neglected in connection with both the Peace of Kallias and that of Epilykos (nos. 152 and 183). Ὁ Ἑλληνικὸς ὅρκος
[158] καταψεύδεται, ὃν Ἀθηναῖοί φασιν | ὀμόσαι τοὺς Ἕλληνας πρὸ τῆς μάχης τῆς ἐν Πλαταιαῖς πρὸς τοὺς βαρβάρους, καὶ αἱ πρὸς βασιλέα Δαρεῖον Ἀθηναίων πρὸς Ἕλληνας συνθῆκαι. This text has been unnecessarily emended and not fully interpreted, although Wade-Gery emphasized[24] that Theopompus refers here to the treaty between the Athenians and Dareios II which was concluded soon after the accession of this monarch; whether Theopompus refers in another fragment (154) to the Peace of Kallias (as is generally assumed) or also to the later treaty with Dareios is not possible to say with confidence. The important

---

23. See my remarks in "Herodotus and the Inscriptions," pp. 60–61 (*supra* n. 4) [*infra*, pp. 276–278].

24. *Op. cit.* 125–127.

thing, however, is the recognition of the contrast which Theopompus makes between the oath which was sworn before the *battle against the barbarians* (πρὸ τῆς μάχης... πρὸς τοὺς βαρβάρους) and the treaty which was made with Dareios (πρὸς βασιλέα Δαρεῖον) against the Greeks (πρὸς Ἕλληνας). The falsehood and duplicity of the Athenians lie not in the fabrication of documents (as has been generally thought) but in the changed attitude toward the barbarians. Oath and treaty are contrasted in one and the same sentence which has but one verb; Theopompus challenges here the Hellenic patriotism and the sincerity of the Athenians. In fact, he was not the only historian who pointed to the contrast between the bold antibarbarian language of the Greeks (and especially of the Athenians) during the Persian Wars and their willingness to establish peace and friendly relations with Persia not too long afterward. Diodorus reports (9.10.5), following a fourth-century source: ὡς ποιῆσαι τοὺς Ἕλληνας ὅτε κατηγωνίσαντο τὸν Ξέρξην. ὤμοσαν γὰρ ἐν Πλαταιαῖς παραδώσειν παίδων παισὶ τὴν πρὸς τοὺς Πέρσας ἔχθραν, ἕως ἂν οἱ ποταμοὶ ῥέωσιν εἰς τὴν θάλατταν καὶ γένος ἀνθρώπων ᾖ καὶ γῆ καρποὺς φέρῃ. τὸ δὲ τῆς τύχης εὐμετάπτωτον βεβαίως ἐγγυησάμενοι μετά τινα χρόνον ἐπρεσβεύοντο πρὸς Ἀρταξέρξην τὸν υἱὸν Ξέρξου περὶ φιλίας καὶ συμμαχίας.

The oath and the treaty mentioned by Diodorus and those referred to by Theopompus are different. The one oath was sworn before the battle of Plataea and its text is well known (*SEG* 19.167), the other was sworn after the battle under circumstances reported by Plutarch (*Aristides* 21).[25] Similarly, Diodorus mentions the treaty with Artaxerxes, thus the famous Peace of Kallias, the historicity of which is once more placed beyond all doubt, while Theopompus, as shown above, refers to the peace with Dareios II.

Looking back at the treaty relations between Persia and Athens, it appears that the first treaty was concluded with Dareios I at the request of the Athenians, who sought protection against the Spartans. This treaty was broken within ten years by the participation of the Athenians in the Ionian Revolt. Then came the great Persian Wars in which Persia sought to reduce Athens and to punish her. Instead, the power of Athens increased to such an extent that it was the Persian king who asked for a treaty, the famous Peace of Kallias, which ended a period of more than forty years during which Athens and Persia were at war with each other. From before the middle of the fifth century on, there existed once more treaty relations, but while at first the power of Athens dictated the terms, the situation changed during the Peloponnesian War, when Athens sought financial aid from Persia.[26] Accordingly, a new treaty was signed but hardly implemented, since the Athenians once more supported a revolt against Persia, and Persia herself, in turn, felt constrained to give aid to the Spartans. Subsequently, Persia's power increased until the famous Peace of Antalkidas

[159]

---

25. See my remarks in "The Covenant of Plataea," *TAPA* 91 (1960) 178–183 [*infra*, pp. 11–15].

26. It is difficult to fit into this story the notice of Satyros that Anaxagoras was accused of medism (Diogenes Laertius 2.12); see my remarks in "Theopompus on Thucydides," *Phoenix* 14 (1960) 84 [*infra*, pp. 308–309].

could be signed, which the Athenians sadly compared with the Peace of Kallias. As the century wore on, the threat of Macedonian power grew, and some Athenians considered once more the possibility of asking Persia for financial assistance. It is in this connection that Demosthenes' plea is to be understood, and his suggestion to send an embassy to Persia; he urged his audience to forget the ancient prejudices against the "barbarian" (10.33), and to remember that Persia had been of help in the past (10.34 and 51) and might aid Athens in the future (10.52 and especially 11.6).[27] In the more than one hundred and eighty years from the end of the sixth century to the twenties of the fourth, Athens and Persia had been at war for little more than forty years, from the Ionian Revolt to the Peace of Kallias.

27. For Demosthenes 10 (the 'Fourth' Philippic) see S. D. Daitz, *CP* 52 (1957) 145–162, and especially 159 on 10.31–34.

2

The Covenant of Plataea

The discovery and the prompt and competent publication of the Themistocles [178] Decree (M. H. Jameson, *Hesperia* 29 [1960] 198–223) have opened our eyes to the proper appreciation of the Oath of Plataea.[1] There can be little doubt that both inscriptions, though inscribed after the middle of the fourth century, record documents of the Persian Wars. Herodotus, who does not mention the documents in question, is not discredited by these discoveries; but our knowledge of the Persian Wars is greatly increased, especially by the realization (1) that the Attic orators and the writers of Athenian history possessed an account of the Persian Wars which was, if not at variance with, at least independent of that of Herodotus, and (2) that their account was often based on genuine documents. Since this applies also to other periods of Athenian history, such as the ages of Solon and Pisistratus, and even to the Peloponnesian War as described by Thucydides, it is now possible to write a history of Athens which closely corresponds to the popular tradition with which the Athenians of the fifth and fourth centuries were familiar.

As a small contribution to this larger task, a discussion of the "Covenant of Plataea" may be offered because there seems to be some evidence available which has not been considered so far in the controversy which has arisen between the distinguished ancient historian, J. A. O. Larsen, and one of the brilliant editors and authors of the *Athenian Tribute Lists*, H. T. Wade-Gery. Moreover, the topic itself is of considerable interest and importance since it illustrates the nature of the Athenian tradition which has so long been obscured by an otherwise justified admiration for Herodotus and an unjustified belief in the exclusiveness of his account.[2]

Transactions of the American Philological Association 91 (1960) 178–183

---

1. M. N. Tod, *Greek Historical Inscriptions* 2 (Oxford 1948) no. 204, with the necessary corrections made by G. Daux, *Studies Presented to D. M. Robinson* 2 (St. Louis 1953) 775–82.
2. For another example of a well-attested episode of the Persian Wars which is omitted by Herodotus, see A. E. Raubitschek, "Das Datislied," *Charites, Studien zur Altertumswissenschaft*, ed. by Konrad Schauenburg (Bonn 1957) 234–42 [*infra*, pp. 146–155].

[179] The "Covenant" is known only from Plutarch's *Life of Aristides* (21.1). The context in which it stands is independent of the account of Herodotus, although Plutarch followed him very closely, even to accepting his casualty figures on the Greek side (Herod. 9.70.5 = Plut. *Aristides* 19.5). After a lengthy discussion (9.71–85) of the most distinguished fighters and of the distribution of the spoils (interspersed with some anecdotes), Herodotus simply says (9.86–88) that the Greeks buried their dead and held a council at which it was decided to march against Thebes and to demand the extradition of the chief pro-Persians. If Thebes should refuse, they would take the city. So on the eleventh day after the battle they went to Thebes and besieged the city for twenty days, at which point one of the guilty escaped, while the others were surrendered and executed.

Before turning to Plutarch it may be well to compare the account of Diodorus, whose source Ephorus follows Herodotus but makes some additions (11.33): after the battle the Greeks buried their dead (more than 10,000 while Plutarch has only 1,360), divided the booty, and assigned awards. They dedicated a tripod at Delphi,[3] and the Athenians first held at that time their public funerals, the Epitaphia. Then Pausanias moved to Thebes and demanded the extradition, which was granted. The guilty men were executed. There is no mention of any joint resolution taken by the Greeks after the battle and before their attack upon Thebes, but Diodorus does report (11.29.1) that *before* the battle the Greeks vowed, in case of victory, to celebrate at Plataea on the anniversary of that day a festival of *Eleutheria* accompanied by games.

Plutarch's account is detailed and comprehensive. He first establishes the precise day of the victory (19.8), which he probably knew from the celebrations of the anniversary which were still held in his own day (21.6). He then tells the anecdote (20.1–3) explaining how and why the Plataeans received the prize of excellence and were given eighty talents to build a sanctuary of Athena; he adds significantly that the paintings in this temple were still in good condition in his day. Finally, he reports the story (20.4–6) of Euchidas, who in one day ran to Delphi and back and brought the sacred fire to light the new altar of [180] Zeus and the | other altars which had been defiled by the Persians. When Euchidas died of exhaustion, he was honored by a statue the inscription of which Plutarch saw and quotes.

All these details illustrate the actions of the Greeks assembled *after* the victory at Plataea. Plutarch refers to this assembly repeatedly, first (20.2) speaking of their council and then (20.5) of their leaders. Finally, the most important part of the meeting is reported, namely the establishment of the "Covenant" (21.1–2): when then a general assembly of the Greeks took place, Aristides proposed a decree that there should convene every year in Plataea representatives and delegates from Greece and that every four years the Games of the *Eleutheria* should be conducted. There should be a joint Greek assessment of ten thousand shields, a thousand horse, and a hundred ships for the war against the barbarian;

---

3. Diodorus gives the inscription and repeats with minor variations the two epigrams on the dead of Thermopylae, which Herodotus had quoted in 7.228.

and the Plataeans should be inviolate and sacred so long as they sacrificed to the god on behalf of Greece.

Nobody can seriously question either the establishment in 479 of the *Eleutheria* or the exceptional position of the Plataeans. The former is attested by Diodorus, by an inscription (*IG* VII.2509), by a fragment from one of the comedies by Poseidippus (frag. 29, Kock), and especially by Plutarch's own account, although we do not have any contemporary supporting evidence. The latter is attested by Thucydides, who refers several times to the guarantees given to the Plataeans (2.71.2 and 4); the doubts expressed by the authors of the *Athenian Tribute Lists* (3.102) do not seem to me to be justified.

It is the "Covenant" itself, with its representatives and the joint assessment of troops and ships, which has been subjected to severe and continued criticism.[4] George Grote said that "The defensive league against the Persians was again sworn to by all of them, and rendered permanent," but his editors J. M. Mitchell and M. O. B. Caspari (London 1907) 239 and note 3, declared that "Plutarch's statements... must be regarded with a good deal of suspicion." Busolt's footnote (*Gr. Gesch.* II² [Gotha 1895] 741, note 2) is, as usual, comprehensive but not decisive; he said | that the veracity has been denied without good cause, but he adds that Plutarch's source was probably the unreliable Idomeneus. In the *Athenian Tribute Lists* 3 (Princeton 1950) 101, we read: "we suspect that this covenant is not authentic." Finally, I. Calabi (*Ricerche sui rapporti fra le poleis* (Florence 1953) 63–69) completes her remarks with the assertion: "la lega ellenica è solo un'ipotesi di autori moderni, la festa panellenica una celebrazione nostalgica dei Greci dell'epoca Romana." P. A. Brunt closes the case (*Historia* 2 [1953] 157) with the declaration: "the evidence of Plutarch... is worthless."

Against this formidable array of adverse criticism, J. A. O. Larsen alone has upheld the reliability of Plutarch's account and the historicity of the "Covenant." Beginning in 1933 (*CP* 38 [1933] 262–64), continuing in 1940 (*HSCP* 51 [1940] 177–79), and completing his argument in 1955 (*Representative Government in Greek and Roman History* [Berkeley 1955] 48–50 and 208–210, notes 4 and 7), Larsen has patiently insisted that "if we did not have an account of the Congress... it would almost be necessary to postulate something of the kind" (*HSCP* 51 [1940] 179, note 2), a truly Cartesian syllogism.

Independent confirmation of Plutarch's account of the "Covenant" is provided by a curious Byzantine excerpt of the ninth book of Diodorus dealing with the seven wise men, in particular with Chilon (9.10.5): "he advises against assurances and strong pledges and decisions in human affairs, such as the Greeks made when they had defeated Xerxes. For they swore in Plataea to transmit to their children's children the enmity against the Persians, so long as

---

4. The figures and the proportions of the armament (10,000 men, 1,000 horses, 100 ships) have seemed schematic, but we find the same proportions in the force to be assembled against the Spartans in 377 B.C. (Diod. 15.29.7: 20,000 men, 500 horses, 200 ships) and later by Demosthenes (*Vitae X or.*, Decree 1, 851 B: 10,000 men and 1,000 horses). Thus the figures themselves are not schematic but conventional.

the rivers flow into the sea and the human race exists and the earth bears fruit. Yet having made firm pledges in face of the uncertainty of fortune, after some time they sent ambassadors to Artaxerxes the son of Xerxes concerning a treaty of friendship and alliance."

This is not the place to explore the wisdom of Chilon and its reflections in later Greek literature.[5] Nor do I propose to examine the significance of this passage for our knowledge of the famous Peace of Callias.[6] It may be

[182] appropriate, however, to | discuss briefly the nature of Chilon's "prophecy" and of the oath itself.

Diodorus himself credits Chilon also with the famous advice "nothing in excess" and says (9.10.3) that it implies one should not make final decision in any human affair, as did the Epidamnians, whose two parties swore eternal enmity only to be reconciled later. Although we cannot date the *stasis* in Epidamnus, we read of it in both Thucydides (1.24.3–4) and Diodorus (12.30.2). Similarly Chilon's most famous *dictum*, that the island Cythera should not have been created or once created submerged below the surface of the sea, is reported, following Herodotus (7.235), by Diogenes Laertius (1.71–72) in a context very similar to that of the other Chilon episodes of Diodorus. Here again later history bears out the wisdom of Chilon's pronouncement. Finally, we may recall the remark about Munychia made by Epimenides (Plut. *Solon* 12.5–6 = Diog. Laert. 1.114) which belongs in a similar context. Unfortunately we do not know the source of all these anecdotes about Chilon and about the other "wise" men, but the tradition goes undoubtedly back to the fifth and fourth centuries before Christ. It is quite possible that Ephorus himself illustrated the two statements of Chilon with historical events which seemed to him pertinent, the oath of the Epidamnians in the one case, that of the Greeks in the other. (For similar forward references in Ephorus, see *FGrH* 70 F 21 and 39.) At any rate there seems to be no good reason to doubt the historicity of either story.

The oath itself is curious in the vivid images which it presents to indicate the eternity of the hostility toward the barbarians; no doubt this is the formulaic

[183] language of archaic poetry.[7] This | everlasting enmity against the Persians was

---

5. Attention may be called, however, to the curious statement in Sophocles' *Ajax* (678–83) which seems to recall Chilon's *gnômê*.

6. Without anticipating further discussion, it may be suggested that the "Peace of Callias" is a genuine document of the middle of the fifth century, that its historical significance at the time was greatly reduced by the condemnation of its negotiator Callias (Demosthenes 19.273 *pace* M. Cary, *CQ* 39 [1945] 88–89, note 5), that it was inscribed or rediscovered at the time of the King's Peace and then given a historical importance which Theopompus (*FGrH* 115 F 153, 154) rightly denied. Thus the peace of Callias is one of the documents mentioned at the beginning of this paper.

7. To achieve the same end, the Phocaeans sank heavy iron weights into the sea before they left their city, saying that they would not return until this *mydros* reappeared on the surface (Herod. 1.165.3). The original members of the Athenian alliance did the same thing at the request of Aristides (Aristotle, *Const. of Athens* 23.5). Similarly the Athenians swore to pursue the war against the Persians "as long as the sun takes the same course" (Herod. 8.143.2, with R. W. Macan's note [London 1908] *ad loc.*; Plut. *Aristides* 10.6). The inscription from the tomb of Midas (Plato, *Phaedrus* 264D; Diog. Laert. 1.89–90; Preger, *Inscriptiones Graecae metricae* [Leipzig 1891] (188–92, no. 233) mentions a whole series of similar provisions beginning with "as long as the water runs" (presumably downhill, because the "uphill rivers" became proverbial; see Euripides, *Medea* 410; *Suppliants* 520;

well known in Athens: Isocrates, *Panegyricus* (4) 157, refers to it, and so does Demosthenes (10.33). The curse mentioned by Isocrates was first pronounced by Aristides before Salamis (Plut. *Aristides* 10.6), and it may have been used by him in the Hellenic oath sworn after Plataea. At any rate the "Covenant of Plataea" appears to have been a genuine document of 479 B.C. which was incorporated into the historical tradition, once in order to illustrate a saying of Chilon, and another time to give added splendor to the statesmanship of Aristides. Although Busolt (*Gr. Gesch.* II² [Gotha 1895] 740, note 5 [on 741] and 741, note 2) and Jacoby (on *FGrH* 338 F 7) strongly suggest that Plutarch's source was Idomeneus, we may confidently assert that the tradition surely goes back to the fourth century B.C., perhaps to the *Plataicus* of Hyperides which we know only from Plut. *De gloria Ath.* 350 B. It may be significant that Isocrates in his *Plataicus* (14) of *ca.* 371 B.C. does not mention the "Covenant" (see especially 57–63); did the document become known after this time?

To sum up, a future historian of Athens will remember that the Greeks after Plataea made a covenant which was very short lived if it was operative at all; it was, to be sure, the direct predecessor of the league of 478 B.C., which was also organized by Aristides (Aristotle, *Const. of Athens* 23.5). The festival *Eleutheria*, however, continued to be celebrated, and the special position of the Plataeans was remembered, though not always respected. Herodotus does not mention the "Covenant," but in the fourth century the contrast between the bold language of the "Covenant" and the "Peace of Callias" was noted, and the oath of the everlasting hostility against the barbarian was recalled in connection with the traditional anti-Persian policy of certain circles in Athens. While the "Covenant" had little historical significance, it is a genuine document of the Persian Wars.

Demosthenes 19.287; Hesychius, *s.v.* "*Anô potamôn*" [ed. K. Latte]). The most striking parallels come, however, from Latin poetry: Vergil, *Aeneid* 1.607; Ovid, *Amores* 1.15.10.

3

The Peace Policy of Pericles

[37] I have often wondered how the Athenians turned their wartime alliance which was directed against Persia into an instrument for their own protection which could be used in a war against their former ally, Sparta. Thucydides does not provide any help and it is therefore generally assumed that no essential change took place in the alliance. The silence of Thucydides includes, however, also the Peace of Callias, the Congress Decree of Pericles, and the Transfer of the League Treasury from Delos to Athens, events which are all closely connected with the reorganization of the league. Considering the authority of Thucydides, it is not surprising that none of these measures could be accurately dated or inserted into a meaningful sequence of events, because such a sequence must be essentially based on the account of Thucydides. Moreover, this historian repeatedly emphasizes the continuity of Athenian policy between the two wars and the gradual growth of Athenian power; thus the Athenians are made to say (1.75.3) that their *arche* was based at first on fear, then on ambition, and finally on advantage. They never mention that the war-time alliance directed against Persia had been changed into an organization for peace and security.

The silence of Thucydides would never have been challenged were it not for a series of inscriptions which support and illustrate the later tradition of the transfer of the treasury from Delos to Athens. These are the famous *Tribute Lists* (edited and discussed in four volumes by Meritt, Wade-Gery, and McGregor), records of annual first fruits (*aparchai*) made in cash by the members of the Athenian alliance to the goddess Athena. The lists were numbered and their beginning can be determined with certainty, since one of them has the number (34) as well as the archon's name (Ariston) preserved and can thus be dated in 421/0; the first contribution was accordingly made in 454/3. It is generally assumed that the series of donations began with the transfer of the treasury which is dated in 455/4.

American Journal of Archaeology 70 (1966) 37–41

The basis of these annual offerings was the bond existing between mother city and colony; and indeed most of the Ionian islands and cities of Asia Minor were considered as colonies (*apoikiai*) of Athens. Meritt and Wade-Gery put it very clearly when they say (*JHS* 82 [1962] 71) "when the treasure was moved from Delos to Athens an effort was made to give Athena, and the Panathenaia, the same kind of significance as the Delian Apollo had possessed... the idea was born that the League be assimilated to a system of colonies, with the four-yearly Great Panathenaia as their common feast." Actually, the cities participated not only in the Panathenaia and in the Dionysia (see Meritt and Wade-Gery *op. cit.* 70) but also in the Eleusinia (*IG* I^2, 76). Their most significant contribution was, however, the *aparche*, one-sixtieth of the *phoros*, annually paid and annually recorded from the very first year after the transfer of the treasury. This shows clearly that the members of the Athenian alliance, the "cities," were especially attached to Athens and to Athena from the very moment the seat of the alliance was changed from Delos to Athens; accordingly, Barron recently stated (*JHS* 84 [1964] 48) "As newly appointed patron of the League, Athena Polias received an *aparche* of the annual tribute immediately from the outset of the new arrangements." Nowhere in the long series of *Tribute Lists* and *Documents* is any mention made of the original purpose of the alliance, the "War against the Barbarian," and Wade-Gery correctly observed (*Hesperia* 14 [1945] 219) "since we know that Athens made peace with Persia after Kimon's death... there is good likelihood that it was this Peace which provided the occasion for revising the doctrine... that tribute was intended for the war against Persia."

This leads to the surprising conclusion that Kimon's death and the Peace of Callias belong shortly before the transfer of the treasury, thus to the middle of the 'fifties and not to the beginning of the 'forties, as is generally assumed. The chronology of these years as given by Thucydides and Theopompus (Nepos and Plutarch) posits a close sequence of the battle of Tanagra, Kimon's recall, the peace with Sparta, Kimon's expedition to Cyprus, and his death; see my note in *Historia* 3 (1955) 379–380. The Peace of Callias is absolutely dated only in Diodorus who presumably followed | Ephorus (who did not give absolute dates); Ephorus in turn followed mainly Thucydides (who did not mention the Peace of Callias). A comparison of Diodorus's account and that of Thucydides will show that Diodorus is spreading his account evenly over the years in order to have some story for each year; if he told the same story twice or if he was inaccurate, he is less to be blamed than is his modern reader who takes his absolute chronology too seriously.

[38]

	Diodorus	*Thucydides*	*date*
Tanagra	11.80 (458/7)	1.108.1	458/7
Oinophyta	11.83 (457/6)	1.108.2–3	
Tolmides	11.84 (456/5)	1.108.5	
Pericles	11.85 & 88 (455/4 & 453/2)	1.111.2–3	
Kimon's Peace	11.86 (454/3)	1.112.1	458/7

	Diodorus	*Thucydides*	date
Cyprus	12.3 (450/49)	1.112.2	
Kimon's death	12.4 (449/8)	1.112.4	
Peace of Callias	12.4 (449/8)	—	

It is quite evident that Kimon's Peace, his expedition to Cyprus, his death, and the Peace of Callias belong closely together. This is also indicated by three passages which connect the end of the Egyptian expedition and the death of Kimon.

    a. Thucydides reports (1.110.2) that Amyrtaios held out in the swamps and that Kimon sent sixty ships to him at his request (1.112.3); the implication is that this happened soon after the defeat of the Athenian forces.

    b. This is in fact asserted by Diodorus (12.3.1) who says that the Athenians sent Kimon to Cyprus shortly after the disaster in Egypt.

    c. A still more cogent synchronism is provided by the anecdote told by Plutarch (*Cimon* 18.7–8) according to which Kimon's messengers to the oracle of Ammon, having left Cyprus while Kimon was still alive, were told by the priests that Kimon was with the God, and learned on their arrival at the Greek camp which was then in the neighbourhood of Egypt that Kimon had died. This camp must have been the one in Cyrene where the last remnants of the Egyptian expedition assembled (Thucydides 1.110.1). Since Thucydides informs us that the duration of the Egyptian expedition was six years (1.110.1), the exact date of Kimon's death depends on that of the beginning of the expedition. According to Thucydides, the Athenians were about to sail for Cyprus when they decided to go to Egypt (1.104.2), and I believe that Kimon was in charge of this campaign and that Plutarch refers to it (*Cimon* 15.2–3); it took place during the summer of 462 B.C. at the same time as the Athenian second auxiliary expedition to Messene (Plutarch, *Cimon* 17.3) and before Kimon's ostracism which is dated in the spring of 461 B.C. In this case, the Egyptian expedition would have ended in 457/6, Kimon would have died in 456 B.C., and the Peace of Callias would have been negotiated in 456/5.

While Athens was engaged in making peace with Persia, she was also, as Wade-Gery aptly put it (*Hesperia* 14 [1945] 222), "forced to seek for a new basis for tribute... we have positive evidence about this search: Perikles' famous Congress Decree reported by Plutarch, *Pericles*, 17." Meritt and Wade-Gery have recently reaffirmed this interpretation by saying (*JHS* 83 [1963] 106–107) that the Peace of Callias "called for a reappraisal of Athens' relations with her allies, since the tribute contract must now be revised: we recognize two attempts at that reappraisal in two Periklean decrees, D 12 (the Congress decree) and D 13 (the Papyrus decree)." And yet, the Congress Decree has been considered a complete failure and only K. Dienelt recently suggested (*Die Friedenspolitik des Perikles* [Wien 1958] 22) that Pericles decided to apply the principles of his panhellenic plan to the Athenian alliance and to transform this organization into an instrument for peace. Pericles did this, according to Dienelt, by the foundation of a number of military colonies which were supposed to help maintain an organization which combined the features of an alliance and those of an amphictyony and the bond of which was the allegiance of colonies to their mother city, Athens.

The amphictyonic character of the Congress Decree has been stressed by the authors of *The Athenian Tribute Lists* (Princeton 1950) III 105 and 279–280, but they merely pointed to the fact that the cities and tribes to which envoys were sent represent a combination of the memberships of the Delphic Amphictyony and of the Delian League. The clauses of the decree, however, also reveal its amphictyonic character: rebuilding the sanctuaries destroyed by the Persians; conducting the sacrifices promised during the Persian Wars; securing free|dom of the sea and general peace. The first two of these clauses refer back directly to the oaths which were sworn by the Greeks on the occasion of the battle of Plataea. The sacrifices are especially mentioned in Plutarch's account of the Covenant of Plataea (*Aristides* 21), the historicity of which can now be considered assured; see my argument in *TAPA* 91 (1960) 178–183 [*supra*, pp. 11–15]. And the close connection between the Covenant and Delphi is shown by the anecdote of Euchidas which Plutarch reports in the preceding chapter (20): it was through the intervention of the Delphic Apollo that the cult of Zeus Eleutherios was established in Plataea. With the failure of the Congress Decree, this particular clause of the covenant was neglected, and this neglect has puzzled the authors of *The Athenian Tribute Lists* (III 101) and has led them to question the genuineness of the document.

[39]

More important because more specific is the first clause of the Congress Decree asking for a discussion of the sanctuaries burnt by the barbarian. The proposal as contemplated surely called for the restoration of these sanctuaries, and this would have been in clear violation of the oath in which the Greeks committed themselves not to restore the sanctuaries (Diodorus 11.29.2); for an inscribed version which lacks this clause, see *SEG* 19.167. It is not clear whether this prohibition was part of the oath sworn before the battle of Plataea or part of the Covenant of Plataea; in either case, it was part of an oath panhellenic in character and amphictyonic in spirit and form. After the failure of the Congress Decree, the rebuilding of the sanctuaries was undertaken as part of the building program of Pericles, and this building program was shared by the cities which participated in the great Attic festivals, the Panathenaia, the Dionysia, and the Eleusinia. This means not only that Pericles was able to adopt for Athens and her allies what he had planned for Greece as a whole, but that Athens and her allies must have revoked the oaths sworn at Plataea and reaffirmed at Delos; de Wet expressed this succinctly (*Acta Classica* 6 [1963] 107–108): "The wider ideal of Panhellenic co-operation and unity was then narrowed down to involve the Sea-league only." Thus, it is quite correct to see in the Congress Decree a "reappraisal" of the wartime alliance against Persia.

The third clause is the most important one because it is positive and directed to the future purpose of the alliance, to ensure safety and freedom of the sea. The "keeping the peace" has been taken by Wade-Gery (*HSCP* Suppl. 1 [1940] 150 n.2) to be a direct reference to the Peace of Callias, but it must be remembered that there had also been a good deal of recent fighting between Greeks and Greeks, and that it was highly desirable to keep the peace between Athens and Sparta which Kimon had negotiated shortly before his death. It is also natural that the Spartans had good reason to fear anew the power of Athens since the peace with Persia would free her for the extension of her power in Greece; and this is precisely the situation as Plutarch describes it when Pericles proposed the Congress Decree (*Pericles* 17.1).

As long as the Peace of Callias and the Congress Decree were dated after the middle of the century, it was necessary to find evidence for these momentous events in the record of the donations of the *aparche*, the Tribute Lists. Consequently, it was thought that after the lapse of a year of interruption and uncertainty tribute was once more imposed; on the "missing list" see Wade-Gery's positive statement in *Hesperia* 14 (1945) 212–229 and W. K. Pritchett's recent hypercritical comments in *Historia* 13 (1964) 129–134. It should be stressed that there is no evidence available to connect the "missing list" with the Peace of Callias and with the reorganization of the Athenian alliance, or to show that this alliance was different before and after the "missing list." It has become clear, moreover, that all our records, Lists and Documents, belong to the period after the reorganization of the League in 455 B.C.

The Tribute Lists recorded the donations of the faithful colonists who also brought a cow and a panoply to the Great Panathenaia, and a phallos to the Dionysia, and grain to the Eleusinia. In these donations lies the key to our understanding of the reorganization of the wartime alliance into a league of cities which accepted Athens as their leader and the gods of Athens as their gods, because they agreed to recognize Athens as the mother city from which they all had drawn their origin. This fact (or fiction) is well illustrated by a passage in the programmatic speech of the goddess Athena at the end of the *Ion* of Euripides: "and their sons [i.e., the grandsons of Ion] will in good time settle the island cities of the Cyclades and the plains of the | mainland [of Asia], which will give strength to my land [i.e., Athens]; and they shall settle the plains on both sides of the Straits, of Asia and of Europe" (lines 1581–1587). Evidently we have here three of the major divisions of the Athenian Empire: the Islands, Ionia, the Hellespont—Thrace and Caria being omitted. Dienelt was quite right, therefore, when he connected (*op. cit.* 22–23) the Periclean colonization (Plutarch, *Pericles* 11.5–6) with the reorganization of the alliance. The new colonies served so to speak as models of what the ancient "Ionic colonies" were supposed to have been.

In one respect did the new Athenian Amphictyony differ from similar older organizations, especially from the famous Delphic Amphictyony: Athens continued to collect the *phoros* but she promised to use it for peaceful purposes. In this, too, the new Athenian League was similar to the wartime alliances concluded during the Persian Wars which, to use Larsen's fine description (*CP* 39 [1944] 153–154), "largely appropriated and expanded the earlier program of the Amphictionic League" and "partook of the religious character of an amphictiony and the political character of a symmachy"; "it had taken over the old Amphictionic program for ameliorating war, transformed it into a program for peace within Greece, and had given the League a cohesive force in the shape of a national crusade against Persia." This description of the "Hellenic" League applies on the whole both to the Congress Decree and to the Athenian League as it was organized after 455 B.C. There was accordingly a close relationship between the contribution of *phoros* and of cow and panoply, the one being indicative of the continuation of the wartime alliance, the other of the ties between colony and mother city. The closeness of this relationship is illustrated by the famous decree of Kleinias (*The Athenian Tribute Lists* II 50–51: D 7 = *SEG* 19.8) in which the two contributions are discussed together. We have here the first enactment decree of the reorganized league, inscribed after the middle of the century but passed perhaps before that. Similarly, in the great assessment decree of 425/4 (*The Athenian Tribute Lists* II 40–41: A 9 = *SEG* 19.15), the newly "admitted" members of the league are ordered to pay not only the *phoros* but to bring also cow and panoply to the Panathenaia and to participate in the procession together with the colonists. Thus the concept is maintained that to be a member of the league means to be a colonist, just as originally the league was formed of cities and islands which were considered to have been colonies of Athens.

The close relationship between the Delphic Amphictyony and Athens during the fifties of the fifth century B.C. is not surprising if one considers the influence of Athens on the decisions of the council ever since the end of the Persian Wars; see G. Zeilhofer, *Sparta, Delphoi und die Amphiktyonen im 5. Jahrhundert vor Christus* (Dissertation Erlangen 1959) 41–50. It was not only Themistocles' defense of the medizing members (Plutarch, *Themistocles* 20.3–4) which raised the reputation of Athens in Central Greece where most of the members of the Amphictyony lived, but also the victory at Oinophyta in 458/7 which gave Athens control over Boeotia, Phocis, Locris, and Thessaly (except for Thebes and Pharsalos); see Thucydides 1.108.2–3; Diodorus 11.81.5–83. Gomme remarked (*Commentary* on 1.108.3) that "neither Phokis nor Boeotia was enrolled as a member of the Delian League," without realizing that the Congress Decree was especially designed to provide for those Athenian "allies" who were not members of the naval league.

To this period belongs the fragment of an Attic decree which has been interpreted as a treaty between Athens and the Amphictyony: *IG* I^2, 26 = *SEG* 16.2; see Zeilhofer, *op. cit.* 45–48; Bengtson, *Die Staatsvertraege des Altertums* (Munich 1962) II 142; for a photograph, see *AJA* 55 (1951) plate 37. Accepting the restorations suggested by Meritt and Wilhelm, there is in this decree no mention of Athens at all but only of the members of the Amphictyony. This means that a treaty was to be concluded between the members of the Delphic Amphictyony, including Athens, and that the inscription records the acceptance on the part of Athens of this treaty which had presumably been suggested to the Amphictyonic Council by the Athenian delegate; see Meritt, *AJP* 75 (1954) 372. This treaty may therefore be compared with the Hellenic League, with the Delian League, with the Peace of Antalcidas, and with the general peace of

346 B.C. (which was sponsored by the Amphictyony); see Bengtson, *op. cit.* nos. 130, 132, 242, 331. The question may well be asked whether this inscription did not record the Congress Decree itself | or was at least one of the steps taken by Pericles to establish his larger panhellenic organization.

[41]

The momentous events examined in this essay all took place between the death of Kimon and the transfer of the allied treasury from Delos to Athens; they consisted of the negotiations for a treaty with Persia and for a new international organization embracing the members of the Delphic Amphictyony and of the Delian League. The result was peace with Persia (*de facto* if not *de iure*) and the reorganization of the Delian League into what may be called an Attic Amphictyony. Thus the conditions were created for the rise of Periclean Athens.

4

Corinth and Athens before the Peloponnesian War

The relationship between the two cities is stated by the Corinthians, in their first speech at Athens, as being neither hostile nor friendly, but of common adherence to the Thirty Years' Treaty (Thucydides 1,40–41). In spite of this statement, the Corinthians claim that the Athenians owe them a debt of gratitude on account of two favors which they had done for the Athenians and which led to the subjection of Egina and to the punishment of Samos. In the one case, the Corinthians refused to join those Peloponnesians who voted to support the Samian revolt, and in the other, earlier case, they supplied the Athenians with twenty boats for the war against Egina.

The questions may be asked, what else do we know from other sources about the relations between Corinth and Athens before the Peloponnesian War, and how reliable the statement of the Corinthians really is.[1]

It is surprising that the planned Peloponnesian intervention in the revolt of Samos is not mentioned in any of our other sources; this is especially remarkable since Thucydides himself discusses the revolt thoroughly (1,115, 2–117) and emphasizes the threat of Persian intervention. Is it possible that the statement of the Corinthians concerning the planned intervention of the Peloponnesians in the Samian revolt and concerning their own opposition to it is completely false and that Thucydides by passing over it in silence in his account of the Samian revolt contradicted the Corinthian claim implicitly? Similarly, Thucydides flatly contradicts the Corinthians' assertion (1,41,1) that they were not hostile to the Athenians (which is also implied in their request to the older Athenians to persuade the younger to aid the Corinthians: 1,42,1) by stating

Greece and the Eastern Mediterranean in Ancient and Prehistory: Studies Presented to F. Schachermeyer (Berlin, 1977), 266–269

---

1. While D. Kagan (*The Outbreak of the Peloponnesian War*, 1969, 173–175), G. E. M. de Ste Croix (*The Origins of the Peloponnesian War*, 1972, 200–203 and 211–214), and R. Meiggs (*The Athenian Empire*, 1972, 190) accept the statement as reliable, R. Sealey (*CPh* 70, 1975, 106–107) admits that it is not clear whether the Corinthians spoke "the whole truth."

[267] that the Corinthians | hated the Athenians enormously ever since the Athenian acquisition of Megara (1,103,4), and that they participated during the revolt of Euboia in the defection of Megara (1,114,1). This shows clearly that at least one of the claims of Corinth that they supported Athens was false; Corinth was hostile to Athens at least since 459 B.C.

For the Persian Wars we have the detailed account of Herodotus which must be connected with the hostility between Athens and Corinth at the time when Herodotus wrote his book; see de Ste Croix's statement on this subject (*op. cit.*, 212). Herodotus himself makes this clear when he emphasizes on another occasion (6,89) that Corinth and Athens were friendly before the Persian Wars. This passage in fact refers to the second favor which the Corinthians later claimed to have done for the Athenians, the loan of twenty boats which led to the victory over the Eginetans (Thuc. 1,41,2). The actual events upon which this claim is based have been told both by Herodotus and by Thucydides, and the accounts of the historians do not quite bear out the claim of the Corinthians. True, the Eginetans were defeated in a sea battle (Hdt. 6,92,1), but they defeated the Athenians soon after and captured four of their ships with the men (6,93). There was no question of a conquest of the island since the Athenians had been warned by the oracle to wait thirty years for this (5,89). Thucydides refers to the same story (1,105) and describes accurately the ultimate conquest (1,108,4).

It appears that both examples offered by the Corinthians to show their past favors to the Athenians must have been unconvincing both to the historian Thucydides and to the Athenian audience. They knew that the aid of the Corinthian ships, however well meant, did not result in the conquest of Egina, and they had never heard of the proposed Peloponnesian intervention during the Samian revolt and they may not have believed it.

In contrast to these doubtful cases of Corinthian aid to Athens, there is a major contribution which the Corinthians made by which they virtually saved the Athenian democracy, namely their withdrawal from Eleusis during the vengeful attack by King Kleomenes which Herodotus reports in great detail (5,74–76). The decision of the Corinthians "that they were not doing the right thing" resulted in their own withdrawal and in that of King Demaretos and subsequently in the shameful collapse of the entire expedition.[2]

[268] This friendly disposition of the Corinthians showed itself on still another occasion which may belong into the same context, as G. Grote and after him G. Busolt suggested (*Gr. Gesch.* II^2, 399 n. 4). The story is told by Herodotus in connection with the battle of Marathon (6,108), and Thucydides refers to it in his account of the destruction of Plataea (3,55). When the Plataeans on the advice of Kleomenes appealed to the Athenians for help against Thebes and

---

2. Fritz Schachermeyer has recently suggested (*GB* 1, 1973, 211–220 = *Forschungen und Betrachtungen*, 1975, 75–84) that the Corinthians withdrew when the treaty between Dareios and the Athenians became known; for this treaty, see my remarks in *GRBS* 5, 1964, 151–159 [*supra*, pp. 3–10].

were subsequently attacked by the Thebans, the Corinthians appeared as arbitrators who favored so it seems the Plataeans and Athenians. Afterward the Boeotians attacked again, were defeated by the Athenians, and forced to accept the river Asopos as boundary between Plataea-Athens and Thebes. One would be inclined to identify this battle with the victory of the Athenians over the Boeotians of which Herodotus tells in connection with the rise of democratic Athens (6,77,2) and to assume that he told the same story twice without giving a cross reference, were it not for the fact that Thucydides specifically and precisely stated (3,68,5) that the destruction of Plataea took place in the 93rd year after the conclusion of the treaty between Plataea and Athens, i.e., in 426/25 B.C.[3]

How is one to explain the silence of the Corinthians concerning the aid they gave to the young Athenian democracy, an aid which was far more important than the loan of some boats for the war against Egina? Did the Corinthians not remember what they had done about eighty-five years earlier or did they not choose to mention it? Did Thucydides not know anything about it because he had not yet read Herodotus? The Corinthians saved Athenian democracy a second time, when they alone spoke against the Spartan proposal to return Hippias to Athens (Herodotus 5,92). The silence of the Corinthians and/or of Thucydides on this occasion is made up by Herodotus's awareness of the shift in the attitude of the Corinthians toward Athens. He puts into Hippias's mouth the statement (5,93) that the day would come on which the Corinthians will wish that the Peisistratids were back, when they themselves will suffer at the hands of the Athenians.

It appears that Corinth was on good terms with Athens from the expulsion of Hippias until the Persian Wars. During the Persian Wars, the Corinthians favored the defense behind the wall across the Isthmus, but there is no sign of any hostility on their part toward Athens; later, during Kimon's return from Ithome, there was no change in their attitude.[4]

[269]

The real break came over Megara, first its conflict with Corinth and then its alliance with Athens. This was according to Thucydides the origin of the strong hatred which the Corinthians held against the Athenians (1,103,4).

The speech of the Corinthians in Athens draws a different picture of their relationship to the Athenians, a picture which did not make a strong impression on the Athenian audience. How false it was is also shown by the speech which the Corinthians delivered a few months later in Sparta and which in Thucydides' account of it is replete with hatred (1,68–71).

---

3. A passage in Pausanias (9,6,1) implies that there was but *one* engagement between Athenians and Thebans over Plataea, and Thucydides himself seems to refer to an old treaty between Athens and Thebes according to which Panakton was to remain neutral (5,42,1). Finally, Herodotus emphasizes (5,78) that the military superiority of Athens over her neighbors showed itself only after the expulsion of the tyrants. Thucydides may have made a mistake but it is difficult to explain his text as it stands now.

4. For the anecdote told by Plutarch, *Kimon* 17,1–2, see de Ste Croix (*op. cit.*, 78 no. 34) and the idle speculations by J. Cole, *GRBS* 15, 1974, 374–378.

5

Die Rükkehr des Aristeides

[127] Herodot erzählt (8.79) daß Aristeides, der ostrakisierte Athener, ins Lager der Griechen vor Salamis kam und dem Themistokles persönlich mitteilte, daß die Perser die griechische Flotte im Kreise umgeben hätten und daß eine Flucht daher nicht mehr möglich wäre.[1] Plutarch hat diese Geschichte ausgeschmückt und dramatisiert (*Aristides* 8) und ihr die Angabe vorausgeschickt, daß die Athener die Verbannten im dritten Jahre zurückberiefen, als Xerxes durch Thessalien und Boeotien nach Athen zog.[2] Die Bedeutung dieser Angabe liegt in ihrem Zusammenhang mit dem Ostrakismos des Aristeides, den Plutarch im vorhergehenden Abschnitt besprach, und der damit zeitlich ins Frühjahr 481 festgelegt erscheint (das erste Jahr Plutarchs ist 482/1, das zweite 481/0, und das dritte 480/79).

Plutarch folgt hier augenscheinlich derselben Vorlage wie Aristoteles, der (*Staat der Athener* 22.8) angibt, daß die Athener wegen des Xerxeszuges im Archontat des Hypsichides alle die Ostrakisierten zurückberufen. Aristoteles datiert aber dieses Ereignis nicht in das Jahr 480/79, aber schon in 481/0, denn das muß wohl das Jahr des Hypsichides gewesen sein,[3] und er sagt auch noch am Anfang, daß dies im 4. Jahre stattfand. Man darf nun nicht glauben, daß hier ein Schreibfehler vorliegt oder daß Plutarch die Verbannung des Aristeides zwei Jahre später ansetzt als Aristoteles. Es ist einfacher, wieder anzunehmen, daß Aristoteles von der Verbannung des Xanthippos an zu zählen beginnt und demnach die Zurückberufung der Ostrakisierten in das Jahr unmittelbar nach der Verbannung des Aristeides setzt; das stimmt dann auch mit der Angabe des Plutarch überein: beide Schriftsteller setzen die Verbannung des Aristeides

---

1. Für den genauen Zeitpunkt dieses Ereignisses ("27. oder 28. September 480"), siehe G. Busolt, *G.G.*, II², S. 703 mit Anm. 3 (auch auf S. 704).
2. Für die Zeit dieses Zuges, August–September 480, siehe Busolt, S. 673 und Anm. 9.
3. Damit stimmen T. J. Cadoux (*J.H.S.*, 68 (1949) S. 123) und J. Labarbe (*La loi navale*, S. 107) überein; vgl. jetzt auch D. Hereward, *Palaeologia*, 1958, S. 336–337.

Die Rükkehr des Aristeides

in 482/1 und die allgemeine Zurückberufung der Ostrakisierten in den Sommer von 480 (481/0 oder 480/79).

Nun sagt aber Aristoteles (22.7), daß Aristeides im Archontat des Nikomedes ostrakisiert wurde, und man hat allgemein diesen Archon mit dem von Dionysios Hal. (8.83) genannten Nikodemos, dem Archon von 483/2, gleichgesetzt,[4] und danach auch die Verbannung des Aristeides in 482 datiert. Nimmt man jedoch an, daß Nikomedes in 482/1 Archon war (das Jahr ist nach Cadoux unbesetzt), dann muß man natürlich Aristoteles (22.6–7) folgen und den Ostrakismos des Xanthippos zwei Jahre früher ansetzen, nämlich in 484/3. Geht man noch weiter zurück, so liest man in Aristoteles (22.6), daß während dreier Jahre die Freunde der Tyrannen verbannt wurden, aber im vierten Jahre Xanthippos. Diese drei Jahre wären dann 485/4 486/5, und 487/6. Diese Rechnung würde den Ostrakismos des Megakles, der im Archontat des Telesinos stattfand (Aristoteles, 22.5), ins Jahr 486/5 bringen, ein Jahr das auch noch unbesetzt ist (Cadoux), und würde den ersten Ostrakismos, den des Hipparchos, dem Jahre 487/6 zuweisen. Nun sagt aber Aristoteles (22.3), daß nach dem Siege bei Marathon die Athener zwei Jahre ausließen und dann zum ersten Male das Gesetz vom Ostrakismos gebrauchten.[5] Diese zwei aus|gelassenen [128] Jahre müßten nun 489/8 und 488/7 sein, so daß auch von dieser Richtung her der erste Ostrakismos, nämlich der des Hipparchos, ins Jahr 487/6 gehören würde.[6]

Es ist gelungen die Angaben des Aristoteles und des Plutarch über die Zeit des Ostrakismos des Aristeides und der Zurückberufung der Verbannten in Einklang zu bringen, ohne der schriftlichen Überlieferung Gewalt an zu tun. Es bleibt jedoch die klare Behauptung des Nepos (*Aristides* 1.5) "*hic* (d. h. Aristeides) *decem annorum legitimam poenam non pertulit. nam postquam Xerxes in Graeciam descendit, sexto fere anno quam erat expulsus, populi scito in patriam restitutus est.*" Da Ordnungszahlen nicht in Ziffern geschrieben wurden, darf man "*sexto*" natürlich nicht mit dem Verweis VI = III in "*tertio*" ändern

---

4. So Auch Labarbe (S. 83, Anm. 2), der Plutarch, *Themistocles*, 32.2, hätte berücksichtigen sollen, wo der Schwiegersohn des Themistokles, Nikomedes, erwähnt wird, dessen Name jedoch in einer guten Handschrift als Nikodemos erscheint.
5. Meinen früheren Bemerkungen zu dieser Stelle (*A.J.A.*, 55 (1951) S. 221 [*infra*, pp. 65–67]) möchte ich jetzt nur hinzufügen, daß Aristoteles mit den zwei ausgelassenen Jahren anscheinend auf den zeitlichen Unterschied zwischen der Annahme des Gesetzes (die dann im Marathonjahre, 490/89, erfolgt wäre) und seinem ersten Gebrauch hinweist, einen Unterschied, den Androtion (*F. Gr. Hist.*, 324 F 6) eben nicht machte. Siehe auch Anm. 6.
6. Daß Aristoteles hier wirklich von zwei ausgelassenen Jahren spricht machen die folgenden Stellen klar. In *hist. an.*, 3, 523a 7–8, sagt er, daß Kühe vor der Geburt der Kälber einige Tage auslassen (d. h. keine Milch geben) und (5, 546b 6) daß das arabische Kamel nach der Geburt eines Jungen ein Jahr ausläßt und dann wieder copuliert. An einer dritten Stelle (*Pol.*, 4, 1299a 7–8) sagt er, daß in einigen volkreichen Städten die Bürger eine lange Zeit auslassen müssen (zwischen Ämterbekleidungen), und schließlich (*Oec.*, 2, 1349a 28–29) liest man, daß der Tyrann Dionysios zwei oder drei Tage ausließ, bevor er das Geld zurückgab. In allen diesen Fällen handelt es sich um eine Unterbrechung für eine bestimmte Zeit, während der etwas anderes geschieht als bevor und nachher. Daher wird man auch annehmen, daß der erste Ostrakismos der Annahme des Gesetzes nach einer Pause von zwei Jahren folgt, wenn kein Ostrakismos stattfand. Diese Aristoteles Stellen bestätigen somit meine Erklärung von Thucydides, 1.112, 1–2 in *Historia*, 3 (1955) S. 380.

(Labarbe, S. 95–96; die anderen Versuche sind nicht besser), sondern man muß den überlieferten Text zu verstehen versuchen. Nun wissen wir von Aristoteles und Plutarch, daß nicht nur Aristeides sondern alle Ostrakisierten (aber nur diese; siehe *Rh.Mus.* 98 (1955) S. 259, Anm. 2) damals zurückberufen wurden, und wir haben eben gesehen, daß Aristoteles diese Zurückberufung ins 4. Jahr nach der Verbannung des Xanthippos setzt. Vielleicht bezog sich auch die Quelle des Nepos auf alle Ostrakisierten und Nepos hätte schreiben sollen "*hi* (statt "*hic*") *decem annorum legitimam poenam non pertulerunt* (statt "*pertulit*"). *nam postquam Xerxes in Graeciam descendit, sexto fere anno quam erant expulsi* (statt "*erat expulsus*"), *populi scito in patriam restituti sunt* (statt "*restitutus est*")." Leider ist die Schwierigkeit damit noch nicht ganz gelöst da das sechste Jahr vom Sommer 481/0 auf 486/5 führt, d. h. auf den Ostrakismos des Megakles und nicht des Xanthippos, und man daher annehmen müßte, daß die Quelle des Nepos wußte, daß Hipparchos, der im Jahre vor Megakles verbannt wurde, von der Amnestie ausgeschlossen war; siehe *Class. et Med.* 19 (1958) S. 34–36.

Zusammenfassende Tabelle:

Jahr	Archon	Ereignis
490/89	Phainippos	Marathon und Ostrakismos Gesetz
489/8	Aristeides	—
488/7	Anchises	—
487/6	—	Ostrakismos des Hipparchos
486/5	Telesinos	Ostrakismos des Megakles
485/4	Philokrates	
484/3	Leostratos	Ostrakismos des Xanthippos
483/2	Nikodemos	
482/1	Nikomedes	Ostrakismos des Aristeides
481/0	Hypsichides	Rückberufung der Ostrakisierten
480/79	Kalliades	Salamis und Rückkehr des Aristeides

6

War Melos tributpflichtig?

Diese Frage ist in neuerer Zeit wiederholt aber widersprechend beantwortet *[78]*
worden (siehe zuletzt ausführlich W. Eberhardt, *Historia* 8, 1959, 284–314;
denn dem anscheinend eindeutigen Zeugnis des Thukydides (5.84.2: οὐκ ἤθελον
ὑπακούειν) steht das ebenso eindeutige einer Inschrift entgegen (IG² 1.63.61 u.
65: Νεσιοτικὸς φόρος... ⌂⊞Μέλιο[ι]; siehe SEG 17.3) und der durch die
Autorität beider eingeschüchterte Forscher möchte da lieber aufgeben als noch
einen Beitrag über dieses Problem lesen geschweige denn schreiben.

Doch scheinen trotz allen ehrlichen Bemühungen vieler die Quellen selber
nicht vollkommen erschöpft worden zu sein, und wenn man bedenkt, was die
Zeitgenossen selber gesagt und gedacht haben, bekommt man ein Bild, das der
Wirklichkeit näher kommt.

Herodot spricht von den Meliern zusammen mit den Seriphiern und Siphniern
als den einzigen Inselgriechen, die dem Perserkönig Erde und Wasser
verweigerten und die sich mit zwei Fünfzigruderern an der Schlacht bei Salamis
beteiligten (8.46 u. 48); er fügt hinzu, daß sie von lakedämonischer Abkunft
waren, und er kennt daher die Geschichte, die wir noch in Konon (*F. Gr. Hist.*
26 F 1/XXXVI) und Plutarch (*Virt. mul.* 247 D) lesen. Von Pausanias lernen
wir, daß sich die Melier zusammen mit den Keiern, den Teniern, den Naxiern,
und den Kythniern an der Schlacht von Plataä beteiligten (5.23.2), und in der
Inschrift der Schlangensäule (*Syll.*³ 31; siehe *SEG* 16.337) liest man noch die
Namen der Keier, Melier, Tenier, und Naxier in derselben Ordnung, die man
auch in Pausanias findet. Nach dieser großen Leistung in den Perserkriegen
hört man von den Meliern nichts mehr, und das wird von Thukydides bestätigt,
der sagt, daß die Melier am Anfang zu keiner Partei gehörten und Ruhe
bewahrten (5.84.2).

Für die Stellung von Melos während des Peloponnesischen Krieges muß man
die verschiedenen Quellen, und besonders Thukydides gesondert behandeln.
Zu Anfang des Krieges gehören alle Kykladen (mit Ausnahme von Melos und

Thera) zum Attischen Bunde (2.9.4), aber im Sommer des 6. Jahres (3.88.4–89.1), 426 v. Chr., wurde eine Flotte von sechzig Schiffen unter dem Kommando des Nikias gegen Melos geschickt, um die Melier, die sich geweigert hatten, dem Bunde beizutreten, hineinzubringen (3.91.1–3); trotz Verwüstung des Landes schlossen sich die Melier nicht an, und die Athener fuhren nach Oropos weiter. Nichts wird hier über den unmittelbaren Grund des gewaltigen Zuges noch seines plötzlichen Abbruches gesagt, doch macht Thukydides die Erfolglosigkeit des Unternehmens ganz klar.

[79] Zehn Jahre später, im Frühjahr 416 v. Chr., haben die Athener wiederum eine Flotte von 38 Schiffen gegen Melos geschickt, und an diesem Zuge beteiligten sich auch Bundesgenossen und Inselgriechen (5.84.1). Thukydides erwähnt jetzt (5.84.2), daß die Melier seit dem Nikiaszuge von 426 v. Chr. mit den Athenern in offenem Krieg standen, aber er gibt keine Erkärung, warum dieser neue Zug gerade damals stattfand. Die Expedition war unter der Leitung des Kleomedes und des Teisias, deren Namen auch in einer der Urkunden dieses Jahres erscheinen (*IG*² 1.302.28–32; siehe *SEG* 17.6).

Der Melierdialog selber enthält kaum etwas tatsächlich Neues; daß Melier und Athener miteinander im Krieg standen (5.94), wußten wir schon (5.84.2). Es ist aber vielleicht bemerkenswert, daß die Athener es zurückweisen, darüber zu diskutieren, ob die Melier ihnen gegenüber ein Unrecht begangen hätten oder nicht (5.89); das muß demnach ein mögliches Argument gewesen sein, und man fragt sich, auf welche Tatsache es sich gründet.

Sobald die Verhandlungen abgebrochen wurden, belagerten die Athener die einzelnen melischen Städte, ließen eine Besatzung zurück und zogen mit der Mehrzahl der Truppen ab. Es gelang den Meliern jedoch, den Belagerungsring zu durchbrechen, einige der Feinde zu töten und sich mit Nahrungsmitteln und anderen Lebensnotwendigkeiten zu versorgen. Nach Ende des Sommers, wohl schon im Jahre 416/5, wiederholten die Melier den Anschlag, aber als eine neue athenische Flotte unter Philokrates eintraf, wurde die Belagerung mit solchem Eifer betrieben, daß Melos sich ergeben mußte; da mag auch Verrat eine Rolle gespielt haben. Die Athener beschlossen, die erwachsene männliche Bevölkerung zu töten, die Frauen und Kinder als Sklaven zu verkaufen, und das Land mit fünfhundert Kolonisten zu besiedeln (5.114–116).

Leser des Melierdialogs haben sich oft gefragt, ob Thukydides die Bedeutung der Unterwerfung der verhältnismäßig kleinen Insel nicht überschätzt und zu sehr vergrößert hat. Dieser Vermutung könnte man entgegenhalten, daß schon Xenophon die Bedeutung von Melos betont hat, als er erzählte, wie die Athener am Ende des Krieges sich an Melos erinnerten und sein Schicksal für sich selber fürchteten (*Hellenica* 2.2.3); und so hat denn auch Lysander gleich nach der Einnahme von Athen die noch lebenden Melier in ihre Heimat zurückgeführt (2.2.9; Plutarch, *Lysander* 14.3). Die Erinnerung an Melos ist von den Feinden Athens wachgehalten worden wie die Proteste des Isokrates noch nach der Mitte des vierten Jahrhunderts zeigen (4.100–101 u. 112; 12.63 u. 89). Isokrates könnte da aus eigener Erfahrung sprechen, erinnert er sich doch auch an die Kosten der melischen Expedition, zu deren Zeit er schon ein junger Mann war (15.113).

Doch lernen wir weder von ihm noch von Xenophon etwas, das wir nicht schon von Thukydides wissen, und die Möglichkeit muß offen bleiben, daß der Fall von Melos durch das Werk des Thukydides berühmt wurde und daß sich Isokrates unmittelbar auf ihn bezieht.

Die ältesten und von Thukydides sicher unabhängigen Zeugnisse stehen in einer Inschrift, in der vierten Rede des Andokides, und in den "Vögeln" des Aristophanes; von diesen drei Stellen hat nur die Inschrift besondere Beachtung gefunden. Diese vielzitierte und -diskutierte Inschrift (IG^2 1.63) enthält den Text der Schatzung von 425/4 und eine Liste der Städte mit den Beträgen, die sie zu zahlen hatten. Im Inselkatalog findet man Melos mit fünfzehn Talenten eingeschätzt (Z. 65). Man hat darüber gestritten, ob Melos von dieser Schätzung etwas wußte und mit ihr einverstanden war, und man war außerstande, eine überzeugende Antwort auf diese Frage zu geben. Wenn man aber die Frage so stellt, ob Athen Melos 425/4 (und vielleicht früher und auch später) als Mitglied seines Bundes betrachtete, dann kann man sie einfach und klar bejahen und damit sofort sehen, daß die Inschrift und Thukydides einander widersprechen. Der einzige Ausweg wäre anzunehmen, daß die Inschrift eine momentane Forderung der Athener auf Stein verewigte, die weder vorher noch nachher eine geschichtliche Bedeutung besaß und demnach dem Thukydides unbekannt war oder als belanglos übergangen wurde. Hier muß man aber gleich warnen, daß die oben erwähnte Bemerkung der Athener über das ihnen von den Meliern zugefügte Unrecht (Thukydides 5.89) doch besser verständlich wäre, wenn die Melier einer von ihnen vielleicht gar nicht anerkannten Verpflichtung nicht nachgekommen wären. Daß diese Warnung, wenigstens historiographisch berechtigt ist, wird sich gleich zeigen.

[80]

Zunächst müssen wir anführen, daß Andokides behauptet, Alkibiades sei an dem Beschluß, die Melier in die Sklaverei zu verkaufen, schuld gewesen (4.22; siehe meine Bemerkungen in *Transactions of the Am. Philol. Ass.* 79, 1948, 200–201 [*infra*, pp. 123–124]); Plutarch, der hier nicht nur aus Andokides schöpft, gibt an (Alcibiades 16.5), daß Alkibiades den Beschluß unterstützte, die erwachsenen Melier zu töten. Diese Angabe steht natürlich mit dem thukydideischen Bericht nicht im Widerspruch, sondern ergänzt ihn einfach; doch darf shon hier auf den Ausdruck ἡβηδὸν ἀποσφαγῆναι hingewiesen werden, der verrät, daß Plutarch aus Ephoros schöpft und daß daher Ephoros die Einnahme von Melos mindestens teilweise unabhängig von Thukydides erzählt hat.

Eine weitere Ergänzung des thukydideischen Berichtes findet sich in den "Vögeln" des Aristophanes, in denen wenigstens zweimal auf den Fall von Melos angespielt wird. Schon in Z. 186 heißt es, daß man die Götter durch eine melische Hungersnot (λιμῷ Μηλίῳ) umbringen wird, und dieser Ausdruck ist dann später sprichwörtlich geworden, wie sein Vorkommen in Hesych, Photius, in der "Suda" und in den Sprichwörtersammlungen des Diogenian und Zenodot zeigt. Das heißt doch, daß die Melier 416 v. Chr., zwei Jahre vor der Aufführung der "Vögel," durch Hungersnot zur Übergabe gezwungen wurden. Das liest man nun nicht im Thukydides, es steht aber mit seinem Bericht nicht im

Widerspruch. Anders wird es jedoch, wenn man die Scholien zu dieser Stelle liest, die auch in der "Suda" etwas ausführlicher stehen: ἐν τοῖς [81] Πελοποννη|σιακοῖς κατὰ πάντων Μηλίων Νικίαν πέμψαντες 'Αθηναῖοι, ἐπὶ τοσοῦτον ἐπολιόρκησαν αὐτοὺς ὥστε λιμῷ διαφθεῖραι. Τῷ δὲ πρώτῳ ἔτει Νικίας Μῆλον παρεστήσατο, οὐ μόνον μηχανῶν προσαγωγῇ, ἀλλὰ καὶ λιμῷ, διὰ τὸ ἀποστῆναι αὐτῶν, πρώην ὑποτελῆ οὖσαν... καὶ οἱ Μήλιοι πολιορκούμενοι ὑπὸ 'Αθηναίων λιμῷ ἐπιέσθησαν καὶ παραδεδώκασιν ἑαυτοὺς ὡς Θουκυδίδης ἐν τῇ πέμπτῃ. Darf man dem Scholiasten wirklich glauben, daß der große Nikias sich an der endgültigen Unterwerfung beteiligte, und daß der Grund für den athenischen Angriff darin lag, daß Melos als Bundesmitglied abtrünnig wurde? Und woher hat der Scholiast denn diese ganz unglaubliche Information genommen?

Auf die Beteiligung von Nikias an der Belagerung von Melos scheint schon Aristophanes in derselben Komödie anzuspielen. Da läßt er einen Charakter lobend zu einem anderen sagen (Z. 363), daß er den Nikias in Belagerungsmaschinen übertreffe: ὑπερακοντίζεις σύ γ' ἤδη Νικίαν ταῖς μηχαναῖς, in Worten, die man mit dem eben zitierten Scholion zu Z. 186 vergleichen (Νικίας... μηχαμῶν προσαγωγῇ) und auf die Belagerung von Melos durch Nikias beziehen darf. In dieser Interpretation ist uns jedoch schon der gelehrte Symmachos zuvorgekommen, auf den die in der "Suda" besser erhaltenen Scholien zu dieser Stelle (s.v. ὑπερακοντίζεις) verweisen: ὑπερακοντίζεις σύ γ' ἤδη Νικίαν τὸν στρατηγοῦντα μηχαναῖς καὶ τοῖς στρατηγήμασι. Σύμμαχος πρὸς τὴν Μήλου πολιορκίαν. Φρύνιχος Μονοτρόπῳ· ἀλλ' ὑπερβέβληκε πολὺ τὸν Νικίαν στρατηγίας πλήθει τε κἀξευρήμασιν. ἢ ὅτι φρνιμώτατα Μηλίους ἀνεῖλεν. Es kann wohl kaum einem Zweifel unterliegen, daß beide Scholien auf den Aristophaneskommentar des Symmachos zurückgehen, der, um die beiden Komikerstellen zu erklären, sich auf einen Bericht der Belagerung von Melos bezog, der von dem des Thukydides völlig verschieden ist.

Wer hat nun erzählt, daß Nikias die abtrünnigen Melier durch Hunger zur Übergabe zwang? Schon Busolt hat erkannt, daß diese Angabe, die auch im Diodor steht, schon bei Ephoros gestanden hat (Gr. Gesch. 3.1271 Anm. 1; siehe auch Riencke, RE s.v. Nikias Nr. 5, 327.36–39), doch hat et sie als falsch bezeichnet und sich um die Aristophanesscholien gar nicht erst gekümmert. Nun sagt Diodor (12.80.5, ein Jahr zu früh datiert): 'Αθηναῖοι δὲ Νικίου στρατηγοῦντος εἶλον δύο πόλεις, Κύθηρα καὶ Νίσαιαν· τήν τε Μῆλον ἐκπολιορκήσαντες ἡβηδὸν ἀπέσφαξαν, παῖδας δὲ καὶ γυναῖκας ἐξηνδραποδίσαντο. Die Einnahme von Kythera ist eine wohlbekannte Leistung des Nikias, aber auch die von Nisaia ist nicht unbekannt. Berichtet doch Diodor schon in 12.67.1, daß Brasidas Nisaia den Athenern wegnahm (was Thukydides ausläßt), und Plutarch erzählt in einer Liste der Taten des Nikias, daß er kurz nach der Einnahme von Minoa Nisaia überwältigte (Nicias 6.4). Daß es sich hier um eine gemeinsame bedeutende Quelle handelt, in der auch der Beschluß über die besiegten Melier erwähnt und dem Alkibiades zugeschrieben wurde, zeigt ein Vergleich der Diodorstelle (12.80.5) mit der Erwähnung in Strabo (10.484:

Ἀθηναῖοι δέ ποτε πέμψαντες στρατείαν, ἡβηδὸν κατέσφαξαν τοὺς πλείους
und in Plutarch | (Alcibiades 16.5), die schon oben zitiert wurde. Hier liegt [82]
zweifellos eine Quelle vor, und es ist höchst wahrscheinlich, daß die Quelle, aus
der Diodor, Strabo, Plutarch, und Symmachos schöpfen, das berühmte
Geschichtswerk des Ephoros war, der hier einen von dem des Thukydides völlig
verschiedenen Bericht über den Grund des athenischen Angriffs auf Melos gab:
die Insel hatte zum attischen Seebund gehört, ist aber abgefallen und darum
überwältigt worden. Hier ist das Zeugnis, das der Beweisführung von Treu
gefehlt hat (*Historia* 2, 1954, 272), mit dem sie aber außerordentlich an
Bedeutung gewinnt. Es sei hier dem spürsinnigen Freunde gewidmet.

Wie hat denn Ephoros die Stellung von Melos gesehen? Wenn Diodor
12.65.1–3 auch auf ihn zurückgeht, hat er berichtet, und zwar nach der
Pylos-Episode und nach dem Tode des Artaxerxes—also 424/3, daß Nikias
gegen die Melier auszog, da sie die einzigen Kykladen waren, die sich als
Bundesgenossen der Spartaner behaupteten. Da diese Expedition kaum von
der bei Thukydides berichteten (3.91.1–3) des Jahres 426 verschieden sein kann,
könnte man annehmen, daß Diodor (und Ephoros) hier ausschließlich und
unkritisch den Thukydides wiederholt. Wir besitzen jedoch eine Inschrift (*IG*
5/1.1; siehe *SEG* 11.456), die zwei Geldzahlungen der Melier in die spartanische
Kriegskasse verzeichnet, und so mag wenigstens Ephoros nicht viel von der
Neutralität der Insel gehalten haben. Wann ist sie dann den Athenern untertan
geworden, und wann ist sie dann wieder abgefallen? Das erste kann leicht nach
dem Sieg bei Pylos, das zweite nach dem Frieden des Nikias geschehen sein,
aber im Augenblick muß man sich hier mit dem Zugeständnis des Unwissens
begnügen. Daß Melos jedoch kurz nach dem Siege des Kleon auf der
Schatzungsliste erscheint, spricht für unsere Vermutung.

In den Augen des Thukydides scheint jedoch weder die frühere Zugehörigkeit
der Insel zum spartanischen noch ihre spätere zum attischen Bunde eine
genügend große Bedeutung gehabt zu haben, um erwähnt oder gewürdigt zu
werden. Melos war und blieb für ihn die kleine Stadt, die allein gelassen werden
wollte, aber nicht in Ruhe gelassen werden konnte; der Einzelne im
innerpolitischen Streite wie die einzelne Stadt im Kriege der Städte mußte sich
der einen oder der anderen Seite anschließen. Zweifellos hat Thukydides gefühlt,
daß die Athener zu weit gegangen sind, und sein melischer Dialog mußte und
konnte wie eine Anklage gegen Athen gelesen werden. So hat ihn denn auch
Isokrates verstanden und zurückgewiesen, und sein Schüler Ephoros hat die
Gelegenheit wahrgenommen und den Bericht des Thukydides korrigiert. Im
Altertum hat Ephoros anscheinend Recht behalten, denn der Melierdialog wird
nicht mehr berücksichtigt. Ob Ephoros heutzutage dasselbe Glück haben wird,
muß man abwarten.

Anhangsweise soll auf einige Stellen hingewiesen werden, in denen der Fall von [83]
Melos mit dem berüchtigsten Melier in Zusammenhang gebracht wird, nämlich
mit Diagoras. Schon in den Wolken (Z. 830) des Aristophanes wird Sokrates
ein Melier genannt, um ihn als Gottlosen zu kennzeichnen, doch in den "Vögeln"

(Z. 1071–1078), die wenigstens zwei Hinweise auf Melos enthalten (siehe oben), zitiert Aristophanes das damals aktuelle Ächtungsdekret gegen Diagoras und will es auf Philokrates beziehen, den Kommandanten der zweiten melischen Expedition, der nach Aristophanes auch am Zug nach Orneai im gleichen Jahre beteiligt war. Auch Lysias verbindet die Asebie des Diagoras, die bestraft wurde, mit der des Andokides, die nicht unbestraft bleiben sollte (6.17–18); heißt das, daß der Asebiebeschluß ins selbe Jahr wie der Mysterienprozess und der Zug nach Orneai gehört, nämlich ins Jahr 415 v. Chr.? Das ist augenscheinlich die Ansicht eines der Scholiasten zu der ersten Aristophanesstelle, der erzählt, daß die Athener Melos wegen der Gottlosigkeit des Diagoras zerstörten, und eines der Scholiasten zur zweiten Stelle, der erklärt, daß Diagoras (auch?) nach dem Falle von Melos in Athen lebte. Sogar in der neulich entdeckten und von Jacoby ausführlich besprochenen (*Abh. Ak. Berlin* 1959/3) arabischen Biographie des Zenon findet man die Ächtung des Diagoras genau ins Archontenjahr des Charias (415/4) gesetzt. Trotz Jacobys Bedenken mag in dieser Überlieferung etwas Wahres stecken; sie scheint aber mit den hier behandelten Fragen kaum etwas zu tun zu haben.

7

The Heroes of Phyle

Upon these men the crown of valor
Was placed by Athens' ancient people;
They were the vanguard of those stalwarts
Who crushed the cruel tyrants' power
And lawless rule; nor shunned they peril.

Only modest public honors were granted to Thrasyboulos and the small group of patriots who were the first to join him in his fight for the return of freedom and democracy in Athens. He and his friends, while living as refugees in Thebes, understood the execution of Theramenes as an indication that the rule of the Thirty had turned into a desperate tyranny. These exiles trusted that Attika was full of truly democratic citizens who disapproved of the terror that had swept over their country in consequence of the military defeat of Athens and the subsequent occupation by the army of the enemy. But the desired downfall of these well-established forces required more than disapproval, and Thrasyboulos was determined by a courageous effort to turn this silent hostility into open revolt. He had not trusted in vain either in the democratic spirit of his fellow-citizens or in the hatred aroused by the Thirty against their own rule. And yet the importance of the heroic action of this small group of men must not be underestimated; they turned despair into hope and inertia into courage.

It was in the early winter of the year 404/3 B.C. that Thrasyboulos set out from Thebes accompanied by seventy men.[1] He crossed into Attika and arrived in the mountain deme Phyle, a place that could easily be defended from all

---

1. Xenophon, *Hellenica* 2.4, 2; compare R. Ziebarth, *Ath. Mitt.* 23 (1898) p. 33; P. Cloché, *La restauration démocratique à Athènes*, pp. 13ff.; A. G. Roos, *Klio* 17 (1921) p. 13; P. Foucart, *Mém. Acad. des Inscr. et Belles-Lettres* 42 (1922) pp. 325f.; G. de Sanctis, *Riv. di Fil.* 51 (1923) p. 292; J. Beloch, *Gr. Gesch.*, III^2, 1, p. 9; W. S. Ferguson, *C.A.H.* 5 p. 369; Busolt-Swoboda, *Staatskunde*, p. 915; G. Colin, *Xénophon historien*, p. 57; W. Schwahn, *R.E.*, s.v. Thrasybulos, col. 571, 20ff.; G. Glotz, *Hist. Grecque*, III, p. 56; Th. Lenschau, *R.E.*, s.v. οἱτριάκοντα, col. 2369, 9ff.

sides.² The news of Thrasyboulos's arrival spread, and, in spite of the military measures taken by the Thirty, his small group increased daily. Some thirty men joined him even before the oligarchs were able to launch their first attack, and for their share in the victory these men were given the same honors as the seventy who had come with him from Thebes.³ There is no reason to assume [285] that all of these men were Athenian citizens; | it is quite likely that a goodly number of them were metics.⁴ Finally, Thrasyboulos felt strong enough to leave the mountains and to occupy the fortress of Mounichia. The outcome is well known: Athens was first freed from her military occupation and later she regained her full constitutional freedom.

This general introduction provides the background for the following report concerning the rediscovery of the stele erected in honor of Thrasyboulos and his companions.

The only literary account of the public honors granted to the heroes of Phyle is given by the orator Aischines in his speech against Ktesiphon delivered in the year 330 B.C.⁵ After referring to the memorials of the Persian Wars which were erected in the Agora, Aischines says (3.187): Ἐν τοίνυν τῷ Μητρῴῳ παρὰ τὸ βουλευτήριον, ἣν ἔδοτε δωρεὰν τοῖς ἀπὸ Φυλῆς φεύγοντα τὸν δῆμον καταγαγοῦσιν, ἔστιν ἰδεῖν. ἣν μὲν γὰρ ὁ τὸ ψήφισμα νικήσας Ἀρχῖνος ὁ ἐκ Κοίλης, εἷς τῶν καταγαγόντων τὸν δῆμον, ἔγραψε δὲ πρῶτον μὲν αὐτοῖς εἰς θυσίαν καὶ ἀναθήματα δοῦναι χιλίας δραχμάς, καὶ τοῦτ' ἔστιν ἔλαττον ἢ δέκα δραχμαὶ κατ' ἄνδρα, ἔπειτα κελεύει στεφανῶσαι θαλλοῦ στεφάνῳ αὐτῶν ἕκαστον... καὶ οὐδὲ τοῦτο εἰκῇ πρᾶξαι κελεύει, ἀλλ' ἀκριβῶς τὴν βουλὴν σκεψαμένην, ὅσοι ἐπὶ Φυλῇ ἐπολιορκήθησαν, ὅτε Λακεδαιμόνιοι καὶ οἱ τριάκοντα προσέβαλλον τοῖς καταλαβοῦσι Φυλήν. After a short digression, Aischines continues (3.190): Ἵνα δὲ μὴ ἀποπλανῶ ὑμᾶς ἀπὸ τῆς ὑποθέσεως, ἀναγνώσεται ὑμῖν ὁ γραμματεὺς τὸ ἐπίγραμμα, ὃ ἐπιγέγραπται τοῖς ἀπὸ Φυλῆς τὸν δῆμον καταγαγοῦσιν.

Τούσδ' ἀρετῆς ἕνεκα στεφάνοις ἐγέραιρε παλαίχθων
δῆμος Ἀθηναίων, οἵ ποτε τοὺς ἀδίκοις
θεσμοῖς ἄρξαντας πόλεως πρῶτοι καταπαύειν
ἦρξαν, κίνδυνον σώμασιν ἀράμενοι.

The account of Aischines seems to be based on the contents of a single document that was inscribed on a stele set up in the Metroön, and which contained the honorary decree, the name list, and the epigram.⁶ Each of the

---

2. See W. Wrede, *Ath. Mitt.* 49 (1924) p. 222.
3. See P. Cloché, *op. cit.*, p. 15; P. Foucart, *loc. cit.*, p. 326.
4. See P. Cloché, *op. cit.*, p. 460; P. Foucart, *loc. cit.*, p. 326; G. de Sanctis, *loc. cit.*, p. 292; H. Friedel, *Der Tyrannenmord*, p. 61.
5. A short reference to the crown of olive given to Thrasyboulos is made by Cornelius Nepos, *Thrasybulus*, 4, 1: Huic pro tantis meritis honoris corona a populo data est, facta duabus virgulis oleaginis.
6. Compare H. von Prott, *Ath. Mitt.* 25 (1900) p. 39; A. Wilhelm, *Jahreshefte* 21/2 (1922/4) p. 168; P. Foucart, *loc. cit.*, p. 325; G. de Sanctis, *loc. cit.*, p. 292; H. T. Wade-Gery, *J.H.S.* 53 (1933) p. 73; H. Friedel, *op. cit.*, p. 61.

men honored received besides the crown of olive a gift of somewhat less than ten drachmas to be used for a sacrifice and for an offering.⁷ More than a hundred names were inscribed in the name list— | thirty more than the number given by Xenophon, and these thirty men may have been democrats who joined Thrasyboulos in Phyle, but did not come with him from Thebes. The discussion of both the date and the significance of the Archinos decree has been greatly complicated by the discovery and study of a document (*I.G.*, II², 10) that is unquestionably related to, but certainly not identical with, the Archinos decree.⁸ This document (*I.G.*, II², 10) is now commonly dated in the archonship of Xenainetos (401/0 B.C.), and it is assumed that it revived, in moderated form, the grant of citizenship to all metics who actively fought for the return of democracy, a grant that originally had been proposed by Thrasyboulos himself, but which was successfully opposed by Archinos. Without going into detail, it may be suggested for consideration that the preserved document is in fact part of the proposal made by Thrasyboulos and contains on the reverse a small fraction of the once long list of names of those who would have received Athenian citizenship. It is strange, indeed, that the cancellation of Thrasyboulos's proposal is recorded but that no literary evidence remains of the fact that virtually the same proposal became a decree only two years later.⁹ Whatever may be the verdict on this hypothesis, there is no reason to assume that Archinos waited for two years, until 401/0 B.C., with his proposal to honor the heroes of Phyle.¹⁰ It is much more likely that the honors for Thrasyboulos and his companions were proposed and granted immediately after the re-

[286]

---

7. A fourth-century honorary decree from Oropos (W. Dittenberger, *Sylloge*³, no. 298) provides that ten drachmas should be paid to each of the ten honored men εἰς θυσίαν καὶ ἀνάθημα, and the publication formula indicates that the preserved stele was not the ἀνάθημα. For a discussion of the same provision in the Archinos decree, see R. Ziebarth, *loc. cit.*, p. 31; H. von Prott, *loc. cit.*, p. 39; P. Cloché, *R.E.G.* 30 (1917) pp. 387f.; A. Wilhelm, *loc. cit.*, pp. 160 and 169; W. S. Ferguson, *op. cit.*, p. 375; G. Colin, *op. cit.*, p. 96; W. Schwahn, *loc. cit.*, col. 572, 10ff.; H. Friedel, *op. cit.*, pp. 60ff.
8. See *S.E.G.* 3, no. 70; M. N. Tod, *J.H.S.* 49 (1929) pp. 184f., note 180; E. Nachmanson, *Hist. Att. Inschr.*², no. 23; G. Colin, *op. cit.*, pp. 99f.; A. Diller, *Race Mixture*, p. 110, note 44; F. Ferckel, *Lysias und Athen*, pp. 54ff.; H. Friedel, *op. cit.*, pp. 62f.; J. Hatzfeld, *Rev. de Phil.* 13 (1939) p. 241, note 1. An Eleusinian inscription (Κ. Κουρουνιώτης, Ἑλληνικά 2 (1929) pp. 5ff.), once thought to belong to this period and to refer to the same events, is certainly much older; see B. D. Meritt, *Epigraphica Attica*, pp. 100ff.
9. The objection that Thrasyboulos proposed Athenian citizenship also to slaves (an inaccurate interpretation of Aristotle, 'Αθ. Πολ., 40, 2) may be countered by a reference to the fact that even his promises (Xenophon, *Hellenica* 2.4, 25) provided only ἰσοτέλεια for the ξένοι, much less, therefore, for the slaves; compare A. Wilhelm, *Sitzungsber. Ak. Wien*, 202, Abh. 5, 1925, p. 9. Another objection is based on the assumption that Pythodoros's name must not be restored as that of the eponymous archon (*I.G.*, II², 10, line 2), because he held office only during the rule of the Thirty; compare G. Mathieu, *R.E.G.* 40 (1927) p. 91, note 2. But the author of the *vita* of Lysias (Plutarch, *X Orat. Vit.*, p. 835 F) dates the proposal of Thrasyboulos in the time of the ἀναρχία, and Aristotle declares ('Αθ. Πολ., 41, 1) that the democracy was reestablished in the archonship of Pythodoros; compare F. Ferckel, *op. cit.*, pp. 27ff.; Th. Lenschau, *loc. cit.*, col. 2357, 41ff.
10. This date has been suggested by R. Ziebarth, *loc. cit.*, p. 32; A. Koerte, *Ath. Mitt.* 25 (1900) p. 396; A. Wilhelm, *Jahreshefte* 21/2 (1922/4) p. 166; W. S. Ferguson, *op. cit.*, p. 375; H. T. Wade-Gery, *loc. cit.*, p. 74; G. Glotz, *op. cit.*, p. 66.

establishment of democracy at Athens, during the archonship of Eukleides, in the year 403/2 B.C.[11] | In the same year and, as will be shown, in the same prytany belong the honorary decrees for the Samians (*I.G.*, II², 1, lines 41ff.) and for a man from Boeotia (*I.G.*, II², 2; see the addenda on p. 655), who probably was active in support of Thrasyboulos.

The evidence concerning the honorary stele set up for the heroes of Phyle has here been reviewed because a considerable part of this document has been discovered in the Agora of Athens.

78. Three of the fragments have been known for several years;[12] these are referred to in the illustrations as Fragments *a* (Plate 1), *b* (Plate 2), and *c* (Plate 3). It appears from the photographs and from the restored drawing that Fragments *a* and *b* join, but this assumption has still to be verified by examination of the stones in Athens. The following two small fragments were assigned to the same monument by the excavators and by Meritt, but they have not yet been published; they join as shown in the restored drawing (Figure 7.1).

Fragment *d* (Plate 4): Height, 0.098 m.; width, 0.07 m.; thickness, 0.026 m.; height of letters, 0.011 m. Inv. No. I 16 *b*. Broken on all sides; joins Frag. *e*. Found on May 29, 1935, in Section E, in loose filling inside the base to the east of the preserved column base of the Hellenistic Metroön; for this part of the building, see H. A. Thompson, *Hesperia* 6 (1937) p. 129, Figure 70.

Fragment *e*: Height, 0.04 m.; width, 0.022 m.; thickness, 0.034 m.; height of letter, 0.01 m. Inv. No. I 93. Broken on all sides; joins Frag. *d* (Plate 4). Found on July 15, 1931, in Section E, above the stelai laid over the drain.

It appears that all fragments were found immediately to the east of the complex of buildings one of which H. A. Thompson has identified with the Metroön; and it is known from Aischines that the stele was set up in the Metroön.[13]

The assumption that the fragments are part of the monument erected in honor of Thrasyboulos and his companions is based on the identification of the beginnings of the two elegiac couplets (lines 73–76) with the epigram quoted by Aischines. Additional proof is provided by the occurrence of the names of at least five men from the small deme of Phyle (lines 43–47), which indicates that the event that caused the erection of the monument took place at, or near,

---

11. This date has been suggested by W. Kolbe, *Klio* 17 (1921) pp. 246f.; P. Foucart, *loc. cit.*, p. 348; F. Hiller von Gärtringen, *I.G.*, I², p. 301, 76ff.; *Hist. Griech. Epigr.*, no. 61; G. Colin, *op. cit.*, p. 115; H. Friedel, *op. cit.*, p. 61.

12. They were published by B. D. Meritt, *Hesperia* 2 (1933) pp. 151ff., no. 3; compare P. Roussel, *R.É.G.* 47 (1934) p. 219, no. 3; M. N. Tod, *J.H.S.* 55 (1935) p. 185, no. 3. For the place of their discovery, see T. L. Shear, *Hesperia* 2 (1933) p. 107; the building called "Stoa of Zeus" by Shear has later been identified with the east porch of the Metroön.

In preparing the present report I have enjoyed the constant help and advice of Meritt, who has kindly allowed me to use his notebooks. I wish also to acknowledge help received from T. L. Shear and O. Broneer.

13. See W. Kroll, *R.E.*, s.v. Metroön, cols. 1488ff.; H. A. Thompson, *Hesperia* 6 (1937) pp. 203ff.; A. Rumpf, *Jahrbuch* 53 (1938) pp. 116f.; H. A. Thompson, *Hesperia*, Supplement 4, pp. 148ff.

Phyle. The name of one of the leading | men, Archinos from Koile, can now [291] readily be recognized as the first entry in the panel of his tribe Hippothontis (line 55), and the name of Thrasyboulos himself can easily be restored in the first place after the name of his tribe Pandionis (line 24).

Lines 1–2: The text of the heading is based on the assumption that the two preserved letters of the first line may be restored Φυλ]ήν; this accusative necessitates the addition of a verb rather than of a preposition (like ἐπί), and the phrase καταλαμβάνειν Φυλήν is so commonly used in the various historical accounts that the suggested restoration of the first line appears to be preferable to any other.[14] It may be significant, moreover, that Aristophanes uses the phrase εἰσὺ Φυλὴν κατέλαβες as a proverb.[15] The restoration is determined also by the available space of approximately sixteen letters in front of Φυλ]ήν.[16]

Line 4: It is possible that the name of Agoratos, son of Eumares, from Anagyrous, was listed in the panel of the tribe Erechtheis.[17]

Line 14: Here may be restored the name of Θρασύβολος Θράσωνος Κολλυτεύς.[18]

Line 22: The restoration Γαργή]ττιος has already been suggested by A. W. Gomme.[19]

Line 24: The restoration of Thrasyboulos's name fits the available space [292] which is determined by the tribal name in line 35.[20]

Line 25: Here may be restored the name of Kephisophon from Paiania. Kephisophon was a member of the council under the archon Eukleides, and he was probably the man whom Thrasyboulos sent as envoy to Sparta.[21]

Line 33: For the restoration of this line, see B. D. Meritt, *Hesperia* 2 (1933) pp. 154f.

Line 37: For the restoration of this line, compare J. Kirchner, *P.A.*, no. 13733.

Line 45: The father's name Εὐδήμο is written *in rasura*; the name that originally stood in its place cannot be read, but it contained one letter more than Εὐδήμο.

---

14. Compare Xenophon, *Hellenica* 2.4, 2; Aischines 3.187; Demosthenes 24.135; Diodoros 14.32, 1.
15. *Plutus*, 1146; see the scholion on this line.
16. For the restoration of the second line, suggested by Meritt, compare Aischines 3.190: τοῖς ἀπὸ Φυλῆς τὸν δῆμον καταγαγοῦσιν. The same phrase occurs also in III, 181 (οἱ ἀπὸ Φυλῆς τὸν δῆμον καταγαγόντες), but it may be noticed that Aischines declares here (3.182): ἐπιδειξάτω τοίνυν Δημοσθένης εἴ που γέγραπταί τινα τούτων τῶν ἀνδρῶν στεφανῶσαι; later on, he himself refers to the crowns granted to the heroes of Phyle.
17. See Lysias 13.77ff.; A. Schweitzer, *Die 13. Rede des Lysias*, pp. 79f.
18. See Demosthenes. 24.134; R. Ziebarth. *loc. cit.*, p. 33, note 1; P. Cloché, *La restauration démocratique à Athènes*, p. 17; W. Schwahn, *loc. cit.*, col. 575, 12ff.
19. In a letter to B. D. Meritt, dated July 2, 1934.
20. For Thrasyboulos, see R. Ziebarth, *loc. cit.*, p. 33, note 1; P. Cloché, *op. cit.*, pp. 15f.; W. Schwahn, *loc. cit.*, cols. 568ff.
21. Compare Xenophon, *Hellenica* 2.4, 36; J. Kirchner, *P.A.*, nos. 8400 and 8416; W. Kroll, *R.E.*, s.v. Kephisophon, col. 240, 34ff.; U. Kahrstedt, *R.E.*, s.v. Meletos, col. 503, 39ff. Since Meletos was the envoy sent by τοὺς ἀπὸ τῶν ἐν ἄστει ἰδιώτας, Kephisophon must have been a prominent member of Thrasyboulos's party.

[Ο ἱ δ ε κ α τ α λ α β ό ν τ ε ς Φ υ λ] ἠ ν
[τ ὸ ν δ ῆ μ ο ν κ α τ ή γ α γ ο ν].

['Ερεχθηίδος]
[-----]
5 [-----]
[-----]
[-----]
[-.---]
[-----]
10 [-----]
[-----]
[-----]
[Αἰγηίδος]
[-----]
15 [-----]
[-----]
[-----]
[-----]
[..........²⁶........]ς
20 [-----]
[.......¹⁸.......Γαρ]γή[ττι]ος
[.......¹⁵......Γαργή]ττιος
[Πανδιονίδος]
[Θρασύβολος Λύκο Στειρ]ιεύ[ς]
25 [.......¹⁶.......Π]αιαν[ιεύ]ς
[.......¹⁶.......]ωνίδο Πρ[οβα]λίσιος
[.......¹⁵.......]μένος Μ[υρρι]νόσιος
[.......¹⁵.......]δο 'Αγγ[ελῆθ]εν
[.......¹⁴.......]ο Κυθήρ[ριος]
30 [Λεωντίδος]
[......¹¹.....]ρτο Παιον[ίδης]
[......¹¹.....]φρονος Φ[ρεάρρι]ος
[Θεοκλῆς Λευ]κίο Σονι[εύς]
[....⁹....] Διονυσί[ο---]
35 ['Ακαμαντί]δος
[...⁷....]ος Διοδώρ[ο---]
[....κρ]άτης Τιμο[κλέος Εἰτεαῖ]ος

[Οἰνηίδος]
[.........²⁰.........] 'Αχαρνεύς
[.........¹⁷.........] 'Α]χαρνεύς   40
[.........¹⁷.........]ο 'Αχαρνεύς
[.........¹¹.....Γ]λαυκίππο Λακιάδης
[.........¹¹.....]ος Ἱερίο Φυλάσιος
[.........¹⁰..... 'Α]ντικράτος Φυλάσιος
[.........¹¹.....] Εὐδήμο Φυλάσιος   45
[.........¹².....]δήμο Φυλάσιος
[.........¹⁴.....]ωνος Φυλάσι[ος]
[-----]
[Κεκροπίδος]
[..]μοκλ[-----]   50
Σμικύω[ν-----]
Εὔφορ[-----]
Σωφιλ[-----]
Ἱπποθω[ντίδος]
'Αρχῖνο[ς Μυρωνίδο ἐκ Κοίλης]   55
Οἰνηί[δης-----]
Αἰαντ[ίδος]
'Ανδρο[-----]
Εὔκλε[ι-----]
[Τ]ιμο[-----]   60
['Αντιοχίδος]
[-----]
Δ[......¹⁶.......]ι[-----]
Λυ[.....¹²......β]όλο Σ[ημαχίδης]
'Αν[....¹⁰....Μεσ]σηνί[ο----]   65
Πόλ[ων.... Εἰτεαῖ]ος
Κν[ίφων.... 'Ατην]εύς
vacat
'Ε[λευθεράθεν]
Π|-----]   70
[-----]
[-----]

Figure 7.1.

Line 48: Here could be restored the name of Atrometos from the deme Kothokidai, the father of the orator Aischines; it may be doubted, however, whether Atrometos's name occurred in this list.[22]

Line 55: The name of Archinos from Koile can be restored with certainty; but the restoration of his father's name remains hypothetical.[23] The names of several other of Thrasyboulos's companions are known, but they cannot be

22. Compare J. Kirchner, *P.A.*, no. 2681; P. Cloché, *op. cit.*, p. 17; G. Mathieu, *R.É.G.* 40 (1927) p. 83, note 3.
23. Compare W. Judeich, *R.E.*, s.v. Archinos, cols. 540f.; R. Ziebarth, *loc. cit.*, p. 33, note 1; J. Kirchner, *P.A.*, no. 2526; P. Cloché, *op. cit.*, pp. 16 and 149ff.; V. Ehrenberg, *R.E.*, s.v. Myronides, col. 1131, 56ff.; *R.E.*, Suppl. 7, s.v. Myronides, col. 512, 34ff.

```
                Τού[σδ' ἀρετῆς ἕνεκα στεφάνοις ἐγέραιρε παλαίχθων]
                δῆμ[ος Ἀθηναίων οἵ ποτε τὸς ἀδίκοις]
75              θε[σμοῖς ἄρξαντας πόλεως πρῶτοι κατέπαυεν]
                ἦρ[ξαν κίνδυνον σώμασιν ἀράμενοι].
```
403/2 B.C.   ΣΤΟΙΧ. 81

```
[Ἔδοξ]εν τ[ῆι βολῆι καὶ τῶι δήμωι Πανδιονὶς ἐπρυτάνευε Ἀγύρριος Κολλυτεὺς ἐγραμ-
            μάτευε Εὐκλείδης]
[ἦρχε] Κηφ[ισοφῶν Παιανιεὺς ἐπεστάτε Ἀρχῖνος εἶπε· -----]
[....]τη[-----]
80 [...]ον[-----]
```

ΟΙ ΔΕ ΚΑΤΑΛΑΒΟΝΤΕΣ ΦΥΛΗΝ
ΤΟΝ ΔΗΜΟΝ ΚΑΤΗΓΑΓΟΝ

[inscription fragment with tribal lists: ΕΡΕΧΘΗΙΔΟΣ, ΟΙΝΗΙΔΟΣ, ΑΧΑΡΝΕΥΣ, ΑΧΑΡΝΕΥΣ, ΓΛΑΥΚΙΠΠΟΛΑΚΙΑΔΗΣ, ΣΙΕΡΙΟΦΥΛΑΣΙΟΣ, ΑΝΤΙΚΡΑΤΟΣ ΦΥΛΑΣΙΟΣ, ΕΥΔΙΜΟ ΦΥΛΑΣΙΟΣ, ΔΗΜΟΦΥΛΑΣΙΟΣ, ...ΝΟΣ ΦΥΛΑΣΙΟΣ]

ΑΙΓΗΙΔΟΣ   ΚΕΚΡΟΠΙΔΟΣ
                                   ΜΟΙ...ΙΚΥΑΝ
                                   ΕΥΦΟΡ
                                   ΣΑΦΙΛ
                                   ΙΠΠΟΟΛΝΤΙΔΟΣ
                                   ΑΡΧΙΝΣΜΥΡΛΝΙΔΟ ΕΚΚΟΙΛΗΣ
          ΓΑΡΓΗΤΤΙΟΣ   ΟΙΝΗΙΔΗΣ
          ΓΑΡΓΗΤΤΙΟ   ΑΙΑΝΤΙΔΟΣ
                                   ΑΝΔΡΟ
ΠΑΝΔΙΟΝΙΔΟΣ   ΕΥΚΛΕ
ΟΡΑΣΥΒΟΛΟΣ ΛΥΚΟΣ ΤΕΙΡΙΑΣΥΣ   ...ΜΟ
                    ΠΑΙΑΝΙΕΥΣ   ΑΝΤΙΟΧΙΔΟΣ
                    ΛΝΙΔΟΠΟΒΑΛΙΣΙΟΣ
                    ΞΕΝΟΣ ΜΥΡΡΙΝΟΣΙΟΣ
                    ΔΟΑΓΓΕΛΗΟΕΝ   ΒΟΛΟΣ ΗΜΑΧΙΔΗΣ
                    ΟΚΥΟΗΡΡΙΟΣ   ΜΕΣΣΗΝΙΟ
ΛΕΩΝΤΙΔΟΣ   ΑΧ   ΕΙΤΕΑΙΟΣ
                    ΤΟΠΑΙΟΝΙΔΗΣ   ΠΟΛΑΝ   ΑΘΗΝΕΥΣ
                    ΠΡΟΝΟΣ ΦΡΕΑΡΡΙΟΣ ΚΝΙΦΑΝ
ΟΕΟΚΛΗΣ ΛΕΥΚΙΟΣΟΝΙΕΥΣ   ΕΛΕΥΟΕΡΑΟΕΝ
                    ΔΙΟΝΥΣΙΟ
ΑΚΑΜΑΝΤΙΔΟΣ
                    ΟΣΔΙΟΑΛΡΟ
                    ΚΡΑΤΗΣ ΤΙΜΟΚΛΕΟΣ ΕΙΤΕΑΙΟΣ

ΤΟΥΣ ΔΑΡΕΤΗΣ ΕΝΕΚΑΣ ΤΕΦΑΝΟΙΣ ΕΓΕΡΑΙΡΕ ΠΑΛΑΙΧΟΛΝ
ΔΗΜΟΣ ΑΟΗΝΑΙΛΝΟΙ ΠΟΤΕ ΤΟΣ ΑΔΙΚΟΙΣ
ΟΕΣΜΟΙΣ ΑΡΞΑΝΤΑΣ ΠΟΛΕΛΣ ΠΡΛΤΟΙΚΑΤΑΠΑΥΕΝ
ΗΡΞΑΝ ΚΙΝΔΥΝΟΝΣΩΜΑΣΙΝ ΑΡΑΜΕΝΟΙ

ΕΔΟΞΕΝ ΤΗΙ ΒΟΛΗΙ ΚΑΙ ΤΛΙ ΔΗΜΟΙ ΠΑΝΔΙΟΝΙΣ ΕΠΡΥΤΑΝΕΥΕ ΑΓΥΡΡΙΟΣ ΚΟΛΛΥΤΕΥΣ ΕΓΡΑΜΜΑΤΕΥ ΕΕΥΚΛΕΙΔΗΣ
ΗΡΧΕΚΗΦΙΣΟΦΛΝ ΠΑΙΑΝΙΕΥΣ ΕΠΕΣΤΑΤΕ ΑΡΧΙΝΟΣ ΕΙΠΕ
ΤΗ
ΕΧ

Figure 7.1 (continued). The inscription honoring the heroes of Phyle.

placed with certainty in the preserved part of the name list: Aisimos, Anytos, Epikrates, Ergokles, Phormisios.[24]

Line 56: Meritt has here restored the name Οἰνηί[δης], although the spelling of this name is given by J. Kirchner (*P.A.*, nos. 11346 and 11347) as Οἰνείδης; but this spelling occurs once in a fifth-century inscription (*I.G.*, I², 324, line

---

24. Compare R. Ziebarth. *loc. cit.*, p. 33, note 1; P. Cloché, *op. cit.*, pp. 15–18 and 137–161.

82) which does not distinguish between epsilon and eta, and a second time in the third century of our era.[25]

Lines 63–67: The placing of the two Fragments *d* and *e* presents considerable difficulty, and the suggested solution must not be considered final. The wider spacing of the lines on these fragments agrees only with the lower part of the name list, and enough is preserved of the left column to exclude the possibility that the fragments belong there. The restoration is based on the assumption that the last two lines (66 and 67) contain the ends of demotics which belong to the tribe Antiochis, while the letters of the two preceding lines (64 and 65) belong to proper names. The first preserved letter of line 65 (on Frag. *d*, Plate 4), read as part of a sigma, may possibly have been a zeta.

Line 69: The one preserved letter of this line must not be restored as a proper name, but it was, as Meritt has observed, a caption. Meritt's restoration ἔ[γγραφοι] would imply that the names that followed were those of metics.[26] But it is preferable to assume that this list contained only the names of Athenians. It has been suggested that Thrasyboulos's original group of seventy men was joined, immediately upon its arrival at Phyle, by about thirty more men who afterward shared the honors with the first occupants of Phyle. The assumption that these thirty men came partly from Athens but mainly from the territory near Phyle is supported by the name list (lines 39–41 and 43–47) with its eight names of men who came from Phyle itself and from the nearby deme of Acharnai.[27] The suggested restoration of line 69 to 'E[λευθεράθεν] or 'E[λευθερεῖς] would imply that Thrasyboulos's group was joined also by several (probably three) men from Eleutherai which is near Phyle. Eleutherai was not an Attic deme, but its inhabitants must have been considered as Athenians.[28]

It is quite obvious that this name list cannot originally have contained as many as a hundred names; yet the latter must be concluded from Aischines' statement that the gift of a thousand drachmas meant that each of those honored received a little less than ten drachmas. It is true that the preserved fragments do not join, and that it is uncertain how many names were listed in the panels of the tribes Oineis and Kekropis. Yet it is extremely unlikely that members

---

25. The restoration Οἰνηΐ[δος] would certainly be wrong, although the name of Oineis occurs in this position in *I.G.*, II², 2369.

26. For the meaning of ἔγγραφοι, see S. Wenz, *Studien zu attischen Kriegergräbern*, p. 44; A. v. Domaszewski, *Sitzungsber. Ak. Heidelberg*, VIII, Abh. 7, 1917, pp. 16f.; G. Smith, *Cl. Phil.* 14 (1919) pp. 358f.; U. Kahrstedt, *Staatsgebiet und Staatsangehörige*, p. 84.

27. See the interesting observations made by P. Cloché, *R.É.G.* 30 (1917) p. 400, concerning the occupations of the metics in Thrasyboulos' army. It may be significant, incidentally, that all but one (line 42) of the preserved demotics belong to noncity demes, and that only Thrasyboulos from Kollytos and Archinos from Koile come from city demes. The most populous demes of the tribes Akamantis and Hippothontis belong to the harbor area, and these tribes are represented each with only two members in the present list; for the composition of these tribes, see A. W. Gomme, *The Population of Athens*, pp. 60 and 63.

28. Compare *I.G.*, I², 400, 537, and 943, line 96; U. von Wilamowitz-Möllendorff, *Hermes* 22 (1887) p. 242, note 2; L. Chandler, *J.H.S.* 46 (1926) pp. 9f.; G. Lippold, *R.E.*, s.v. Myron, col. 1124, 3ff.; U. Kahrstedt, *Staatsgebiet und Staatsangehörige*, pp. 351ff.; O. Walter, *Arch. Anz.*, 1940, cols. 171ff.

of the tribes Erechtheis, Aigeis, Oineis, and Kekropis accounted for almost eighty names, while the remaining six tribes furnished only twenty-three. The present restoration assumes that not more than fifty-eight names were inscribed. It so happens that Pausanias, who may have seen the monument, reports (I, 29, 3) that Thrasyboulos left Thebes with only sixty followers. The restoration of the name list with only fifty-eight names makes it necessary to assume that the monument originally contained another (a second) list with the names of more than forty noncitizens who received the same honors but were separately listed. This second list may have been inscribed below the decree.

Lines 73–76: The restoration of the epigram is based on the text given by Aischines. No explanation can be offered for the spelling τού[σδ'] which contrasts to the spelling of omikron for omikron-upsilon in the rest of the inscription. Wade-Gery preferred the spelling πόλιος in line 75 (*J.H.S.* 53 (1933) p. 74), and it may be doubted whether ἐγέραιρε in line 73 should not also be changed to ἐγέραρε, as F. Blass has suggested. Wade-Gery pointed out that the use of the word ποτέ in line 74 indicates that the monument was erected a considerable time after the event to which it refers took place. This view has not been generally accepted, and in this particular | case certainly not [295] more than a year elapsed between the occupation of Phyle and the erection of the monument in honor of the heroes of Phyle.[29]

Lines 77ff.: The few letters which are preserved of these lines are all that is left of the decree in honor of Thrasyboulos and his first companions. It is known from Aischines that Archinos proposed this decree, and it has been suggested above that it belongs to the year 403/2 rather than to the year 401/0 B.C. The restoration is naturally uncertain, but it so happens that the required space is exactly filled by a prescript that would date the proposal in the same year and in the same prytany as *I.G.*, II², 1, lines 41ff. Kephisophon from Paiania was a member of the council in that year, and he could have been the presiding officer only during the prytany of his tribe Pandionis. Not enough is preserved of the body of the decree to justify a restoration, but the main contents of the decree are known from the account of Aischines.[30]

A few words may be added concerning the reliability of Aischines' report of the monuments set up in the Agora of Athens. It is true that no doubt has been cast so far on the correctness of his account of the honors granted to the heroes of Phyle, and the recovered monument confirms it in every detail. But Aischines did not fare so well with his quotation of the famous epigrams inscribed on the Kimonian herms (3.183–185). A. von Domaszewski was probably the first to declare not only that two of these epigrams did not really exist on stone, but that they had to be substituted by two others; and he was

---

29. For comments on the meaning of ποτέ, see L. Weber, *Hermes* 52 (1917) pp. 551f.; P. Friedländer, *Studi ital. di Fil. Class.*, N.S. 15 (1938) p. 97; W. Peek, *Athenian Studies Presented to W. S. Ferguson*, p. 100, note 1.

30. Compare R. Ziebarth, *loc. cit.*, p. 31.

followed by H. T. Wade-Gery and E. Löwy.³¹ Only L. Weber has energetically protested against this mistreatment of a good literary tradition.³² I think that Aischines can be trusted in view of his obviously accurate account of the honorary monument for the heroes of Phyle. One historian has recently claimed that the excavations of the past hundred years have added very little to our knowledge of antiquity. It cannot be denied, however, that the veracity of a great many of the literary accounts the reliability of which was questioned by philologists has been confirmed by archaeological discoveries.

31. See A. von Domaszewski, *Sitzungsber. Ak. Heidelberg*, 5, Abh. 10, 1914, pp. 12ff.; W. Uxkull-Gyllenband, *Plutarch und die griech. Biographie*, pp. 35–39; H. T. Wade-Gery, *J.H.S.* 53 (1933) pp. 82f., 87f., and 93ff.; E. Löwy, *Sitzungsber. Ak. Wien*, 216, Abh. 4, 1937, pp. 25–30.
32. See L. Weber, *Philologus* 74 (1917) pp. 248ff., 253, and 257ff.; *Hermes* 52 (1917) pp. 551f.; *Hermes* 57 (1922) p. 377; *Rh. Mus.* 75 (1926) pp. 45ff., 295,, note 1, and 325; *Solon und die attische Grabrede*, pp. 51f.; compare also A. Wilhelm, *Anz. Ak. Wien*, 1934, p. 97; W. Dörpfeld, *Alt-Athen*, I, pp. 67f.; II, p. 153; W. Peek, *Athenian Studies Presented to W. S. Ferguson*, p. 105, note 1.

8

Review of A. Masaracchia, *Solone*

Solone. By Agostino Masaracchia. Florence: "La Nuova Italia," 1958. Pp. 5 + 394. L. 2600 (paper). [137]

This book consists of three parts, devoted to the literary tradition (pp. 1–78), to the biography (pp. 79–200), and to the poems (pp. 201–362) respectively. The following comments, however, are concerned only with the first part, since a better understanding of the original sources can advance our knowledge of Solon. Our aim must be to find out what an intelligent and informed Greek during the classical period knew and thought of Solon. The contemporary tradition of the sixth century is lost, except for fragments of the poems, and the later tradition represented mainly by Diogenes Laertius and Plutarch evidently goes back to sources of the fourth century. Thus we may never know more about Solon than what Aristotle knew, but we may find out that.

Herodotus. Herodotus is our first important witness (pp. 2–17), and his account of Solon (1.29–33 and esp. 88) offers two problems, one of which, the chronology, already engaged the curiosity of the ancients; see Plutarch *Solon* 27–28. The second problem concerns the question whether Herodotus used one or several literary sources or whether the Solon Logos is a creation of the historian.

M. thinks that Herodotus made up the whole story, and that the chronology is completely wrong. It is more likely, however, that Herodotus used a literary, probably a poetic, source which reported the famous episode in connection with the spectacular scene of Croesus on the pyre exclaiming three times "O Solon." It should be remembered that this scene loses none of its effect if Croesus is made to recall Solon's advice given in a poem and not to him personally; thus Herodotus represents a poetic elaboration of a much simpler story, and he in turn makes it historical by placing the encounter between Solon and Croesus in a chronological system which is consistent within the framework of his entire account of the sixth century. Solon's advice to Croesus should therefore be considered genuine Solonian poetry but not necessarily history.

The sentiments expressed in the Solon Logos are not unique, however, and especially the concept of the "jealousy of the gods" occurs very significantly also in the stories of Polycrates (3.40), of Battus (4.204), and of Artabanus (7.10 and 46). All these passages have a poetic flavor but only in the case of the account of the Tyche of Polycrates can we assert that it probably goes back to Anacreon's poem on this subject; see Strabo 14.1.16 (p. 638) and especially Himerius 28.2 (ed. A. Colonna); T. Lenschau, *RE*, s.v. "Polykrates" no. 1, 1730, 12–35; Schmid-Stählin *Gesch. d. griech. Lit.*, I, 431, n. 9. It is against this peculiar theology that Aeschylus takes a firm stand (*Agamemnon* 750–62), thus showing that it is old, perhaps as old as Solon (*Const. of Athens* 12.1). Equally well known is the concept that nobody is to be praised as happy before his death which occurs not only in the Solon Logos and is implied in the Polycrates Logos but which resounds also in the exodus of the Sophoclean *King Oedipus*; this is another piece of Solonian poetry.

M. emphasizes that Herodotus, in a passage evidently added by the historian himself in order to introduce the famous anecdote, speaks of Solon merely as lawgiver (that is, that he says nothing of *seisachtheia* and constitution) and that he fails to mention Solon's alleged relationship to Peisistratus, of which so much was made later that Aristotle had to deny it (p. 17). Actually, these two points are of different significance. The second indicates ignorance of a story such as Herodotus would have liked to include had he known it. The first, however, conveys in abbreviated form the outline of Solon's career: he was called upon to make laws, he fulfilled the assignment by a legislation, he went away for ten years during which the laws were not to be changed. This summary is independent of the Solon-Croesus episode, but there is also no link between it and the immediately following (1.59) story of the rise of Peisistratus. We cannot date the period of Solon's *apodemia*, except for insisting that according to Herodotus it fell within the reigns of Croesus, Amasis (1.30; 2.177), and Philocyprus (5.113).

Plato. The absence of all references to Solon in Thucydides and in the Old Oligarch, and the insignificant allusions to him in Comedy are noted by M. (pp. 18–25), but not used to emphasize how little known Solon must have been in fifth-century Athens. Even the publication of Solon's Laws at the end of the century, so important for the politics of the time and for the legal tradition of later times, was not immediately accompanied by the formation of a fuller biography of Solon. It was in the circle of Plato (pp. 58–65) that the recollection of Solon was kept alive and that his poems may have been kept. M. rightly points out that Plato's Solon is not unlike that of Herodotus, but he fails to recognize the genealogy and the chronology which lie at the base of Plato's account. The Critias of the *Timaeus* is not the tyrant (who was young when he died) but an old man whose great-grandfather Dropides was a friend of Solon's (20E) and whose grandfather Critias was admired by Solon (*Charm.* 157E); the conversation between this Critias and his grandson must have taken place, therefore, about 500 B.C., and at that time the poems of Solon were "new" but widely sung (21B). Solon is said to have found a revolutionary situation in Athens upon his return from Egypt (21C), and this must refer to the struggle out of which Peisistratus emerged as victor. When the accounts of Herodotus and Plato are combined, it appears that Solon left Athens for ten years soon after his legislative activity had come to an end, and that he returned soon before the first tyranny of Peisistratus.

The Orators. These provide reliable evidence on the knowledge of Solon during the fourth century, but M. (pp. 31–32 and 54–57) is unaware of the fact that the name of Solon does not appear prominently until after 360 B.C.; see E. Ruschenbusch, *Historia* 7 (1958), 399–408. This suggests that about that time Solon's poems became widely known and anecdotes about his life began to circulate in public.

The Atthidographers. If the works of these writers were completely preserved, they would provide us with all the necessary information on Solon; as it is, the absence of fragments dealing with Solon, except for one of Androtion, leaves us much in the dark. M.'s discussion (pp. 32–47) deals more with Isocrates than with the Atthis, but he does not follow Max Muehl (whom he calls "Von der Muehll," pp. 94 and 128), *Rh. Mus.* 96 (1953), 214–23, who preferred Androtion's interpretation of the *seisachtheia.* Evidently, Androtion could not have read Solon's poem which is quoted by Aristotle (12.4); this is another indication that Solon's poetry was not well known before the middle of the fourth century, and that it had not yet been subjected to a careful historical interpretation.

Aristotle. Aristotle provided this interpretation in his long account (5–14) which is studded with quotations, probably because the Solonian poems were used here for the first time for this purpose. M.'s account (pp. 48–54) is disappointingly brief, although this is the most important and most authoritative report on Solon we have. All M. is interested in is the democratic and antidemocratic tendencies in Aristotle's treatment of Solon, a distinction which is significant only in the discussion of the *seisachtheia* (6) in which Aristotle rightly rejects the anti-Solonian anecdote which probably comes through Theopompus from comedy, and accepts the democratic account which may or may not have been set down in writing before.

The great problem of the Aristotelian report is the chronology of the *apodemia* and of the events which followed Solon's departure (13). It is generally assumed (*a*) that Solon departed immediately after his archonship (because of 13.1), (*b*) that the Damasias who was archon twelve years after Solon's departure (13.2) held this office in 582/1 (according to the *Marmor Parium*), (*c*) that Peisistratus became tyrant thirty-one years after the *legislation* of Solon (14.1). These assumptions leave no time for all of Solon's political activities, which are described by Aristotle (10) as if they were spread over some years. They also leave a gap of about twenty years between the archonship of Damasias (13.2) and the rise of Peisistratus (13.4). Plato, however, implies (*Timaeus* 21C) that Solon's return was followed by the rise of Peisistratus, and Aristotle himself introduces Solon in 14.2 in such a way that it is hard to believe that nothing had been said about him for thirty-two years; see also Plutarch *Solon* 29. While it may not be possible to find out what really happened, it is possible to find out what Aristotle thought had happened. It must be remembered that he had before him a table, probably taken from Theopompus (see my remarks in *Phoenix* 14 [1960], 93–95 [*infra*, pp. 316–319]), which gave distance dates for Attic political history from Solon on to the politicians of the fourth century. It may be suggested that all three above mentioned assumptions be abandoned as certainly or probably false: (*a*) Aristotle does not say in 13.1 that no archon was elected in the fifth year after Solon's *archonship* but in the fifth year after his *period of power*, *arche* being used in this meaning in 17.1; (*b*) Damasias the archon of 582/1 may not be the same man who tried to become a tyrant; there was still another Damasias archon in 639/8 (see T. J. Cadoux, *JHS* 68 [1949], 120–22); (*c*) Aristotle refers in 14.1 not to the legislation of which he speaks in 10.1 but to the first laws (including the *seisachtheia*) which are mentioned as such in 6.1 and 7.1. We are now in a position to present a simple and clear chronology such as Aristotle had in mind when he wrote about Solon (see Cadoux, *op. cit.*, pp. 93–103).

561/0	Archon Komeas	14.1
562/1	Archon Damasias (two months)	13.2
	Ten Archons	13.2
563/2	Archon Damasias	13.2
564/3	Archon Damasias	13.2
567/6–565/4	Interval of three years	13.1
568/7	Anarchia	13.1
571/0–569/8	Interval of three years	13.1
572/1	No archon	13.1
575/4–573/2	Interval of three years	13.1
576/5	End of Solon's *arche*	11.1
	Beginning of *apodemia*	5.1
592/1 (594/3)	Archon Solon; *seisachtheia*	14.1

Heracleides Ponticus is mentioned by M. (pp. 71–72) who correctly emphasizes that we may owe to this author many (if not all) of the stories connecting Solon and Peisistratus, including the one that Solon was Peisistratus's lover, which is so violently rejected by Aristotle (17.2). M. did not notice, however, that we possess two long papyrus fragments (*P Oxy.*, 4, 664) which have been attributed to Herakleides; see *RE*, s.v. "Herakleides" no. 45, 475. 28–36; Beloch, *Griech. Gesch.* I².2.277; Schmid-Stählin, 2.1.72, no. 7. These fragments undoubtedly belong to a dialogue of the fourth century in which Solon and Peisistratus are linked. The speaker reports in the first fragment (which may belong to the introductory narrative) that Solon departed before Peisistratus took control because he could not persuade the Athenians that Peisistratus was aiming at tyranny; this version seems to underlie the account of Diogenes Laertius 1.50 who places Solon's visits to Egypt, Cyprus, and Lydia in this period, about 560 B.C. The speaker (Megacles?) then continues by saying that at the urging of Peisistratus (who was a friend or relative of his) and of Solon he returned to Athens to look after young Thrasybulus the son of

Philomelus and grandson of Hagnotheus, who had gotten into trouble owing to his love affair with the younger daughter of Peisistratus. | This story is also told, following the same tradition and completing the account of the papyrus which breaks off here, by Polyaenus *Strategemata* 5.14; Diodorus 9.37.1; Plutarch *Apophth. regum et imp.* 189C; Peisistratus 3 *De cohibenda ira* 457F–458A; Valerius Maximus 5.1. The second part of the papyrus consists of a dialogue in which the speaker, Ariphron, Adeimantus, and Peisistratus discuss the *arche* of Periander; hence the papyrus fragments were attributed by F. Blass to Heracleides' dialogue *peri arches* which was used by Diogenes Laertius (1.94). It is possible that the peculiar connection between Periander, Peisistratus, and Solon to which Dio Chrysostomus (37.4) refers belongs to the same tradition. At any rate, this dialogue, whether or not by Heracleides, is a valuable link in the formation of the Solon legend. To it may belong the closing paragraph of Plutarch's biography (p. 32) where we read that Heracleides said that Solon lived a long time after Peisistratus came to power.

Plutarch. His biography of Solon is appreciated by M. (pp. 75–77) who is aware of its great impact on our concept of Solon, but instead of penetrating deeper into the tradition which Plutarch represents, M. merely insists that the modern historian cannot be satisfied with accepting the Plutarchian account. Actually, a great deal more could be done in identifying and examining the sources used by Plutarch. For instance, the legislative measures described in chapters 20–24 with their references to Heracleides and Demetrius of Phalerum may well come from one of the great collections of laws such as the *Nomoi* of Theophrastus; this assumption is supported by the presence of parallel material which shows that we have here a selection of Laws on various subjects by various legislators.

To sum up, M.'s *Solone* is not the book on Solon which is most urgently needed, but its main body dealing with Solon's biography and with his poetic work is a fully adequate treatment designed to inform the Italian reader of most of what has come down to us about the great Athenian statesman and poet.

II

INSTITUTIONS

9

Ostracism: The Athenian Ostraka

As I was preparing the following report on the study of Athenian ostraka, I realized that the progress made in this small field of study follows a pattern set by the general course of Greek epigraphical studies. I may be permitted, therefore, to characterize briefly the progress in Greek epigraphy in order to convey to you more adequately the aim pursued in the work on the Athenian ostraka which has been undertaken by Eugene Vanderpool and by myself.

Greek epigraphy has been for a long time and to a large extent an ancillary discipline. After the great collections of Greek inscriptions were published during the nineteenth century and revised and augmented during the past fifty years, historical studies, large and small, general and specialized, have used epigraphical evidence to an ever-increasing extent. Inscriptions are mentioned not only in footnotes but even in the text of historical discussions. On the basis of inscriptions, the literary tradition has been not only illustrated and enriched but also corrected. In this way, epigraphy tried to answer many questions posed by the literary tradition. The problems, however, faced by the epigraphist were either strictly technical or generally historical but only rarely peculiarly epigraphical. The individual inscription was examined less in connection with other similar or related inscriptions but rather in relationship to certain passages of the literary tradition. In fact this isolation of the single inscription caused some epigraphists to praise the light thrown by one inscription upon an individual or upon an event otherwise unknown or dimly known from the literary tradition. In this phase of epigraphical studies, the epigraphist received his problems from the historian who himself formulated them on the basis of his study of the ancient historians and, in general, of the ancient literary tradition.

If epigraphy was to make any substantial progress, it had to free itself from the domination of literature; the epigraphist had to seek and treat his own problems derived from the epigraphical evidence, which has grown so enormously that it begins to overshadow the literary evidence. Inscriptions had to be considered no longer as isolated and unique documents but in groups of similar and related monuments. As a result it became possible not only to improve the text and the interpretation of the individual inscription but to derive from the study of the group as a whole new information concerning the political, administrative, social, economic, and religious institutions of the Greeks.

This new approach to epigraphical studies has been tried long ago, especially in the field of Attic epigraphy. German, French, and, more recently, American scholars have examined and, as new evidence appeared, reexamined the various lists and catalogues of the fourth century and of the Hellenistic period. The Athenian calendar has been reconstructed on the basis of epigraphical evidence, and the most spectacular accomplishment of modern epigraphical studies is the extensive collection and the complete edition of the documents pertaining to the Athenian Empire. Not only was it possible to reconstruct the tribute quota lists and some of the assessment decrees, but the capable editors of these inscriptions have had the courage to rewrite a chapter of Attic history which was already well known from the brilliant account of Thucydides. No doubt, epigraphy has come into its own, and although much work remains still to be done we can now see more clearly the path along which to proceed.

These introductory remarks on the general course of epigraphical studies are particularly relevant to the work on the Athenian ostraka on which I wish to report to you now. For the progress in the study of Athenian ostracism follows, at some distance, the general progress of epigraphy. This distance should enable us to recognize better the task which is ahead of us.

The first comprehensive account of the institution of ostracism was written exactly fifty years ago and published, after some delay, in 1909. The distinguished author, Jérôme Carcopino, had at his disposal a certain amount of literary evidence which had been greatly enriched, only a | short time before, by the discovery of Aristotle's *Constitution of Athens*. Moreover, Carcopino could rely on a number of special studies some of which were published as far back as the second half of the seventeenth century. Finally, Carcopino had available exactly 5 published ostraka of which only 4 had been identified as such at the time.[1] These four ostraka contained the names of Megakles, Xanthippos, and Themistokles, thus of men known to have been ostracized. No wonder, the epigraphical evidence played almost no part in Carcopino's discussion, and the then known ostraka were used merely as illustrations.

When Carcopino wrote the introduction to the second edition of his fine book, in July 1933, he could report that the number of ostraka found had increased from 4 to 62. Although he gave an accurate description of the available ostraka, he did not think that any of the conclusions reached in his first edition had to be revised. He continued to insist that neither Damon, the son of

---

1. For the fifth, see *DAA*, p. 339; *Hesperia Suppl.* 8, p. 412.

Damonides, nor Kallias, the son of Didymias, were actually victims of ostracism, and he was not disturbed by the occurrence on ostraka of names whose bearers were unknown from the literary tradition. The epigraphical evidence remained to be used as illustration, and, considering the small number of ostraka, this was justified.

Only four years after the publication of Carcopino's second edition, that is in 1939, Oscar Reinmuth composed his comprehensive article on ostracism for the *Realencyklopaedie* (issued in 1942). At that time, he was able to report that 527 ostraka had been found up to that date; all but 50 of these had been discovered in the American excavations of the Athenian Agora. Reinmuth listed faithfully the various names found on the ostraka, but he made hardly any use of the considerably increased archaeological and epigraphical evidence. To be sure, in the case of the ostracism of Damon, he upheld the literary evidence and insisted against Carcopino that Damon was actually ostracized, but the one ostrakon of Damon is not mentioned prominently in this connection. The 2 ostraka of Kallias the son of Didymias are mentioned by Reinmuth, but he did not dare to assert that Kallias, too, was ostracized | as is stated in the fourth oration attributed to Andocides. Valuable as Reinmuth's article is, he kept the epigraphical evidence still in the background. *[62]*

In contrast to Reinmuth, Robert Cohen in his magnificent and comprehensive account of Greek history, re-issued in 1939, emphasized the importance of the growing archaeological evidence for our understanding of the institution of ostracism. "It is not impossible," Cohen declared (p. 129), "that the result of the American excavations in the Agora will modify our views on ostracism." But Cohen's keen perception was unique. Even Arthur Gomme, in his article on ostracism in the *Oxford Classical Dictionary* published in 1949, makes but a passing reference to the discovery of ostraka.

It is necessary for me to mention also the book by Aristide Calderini which was published in 1945, although Calderini relies for the epigraphical evidence on Reinmuth's article. He was the first, however, to combine the literary tradition with what little was known to him of the archaeological finds. His book, it would seem to me, marks the end of the second stage of the study of ostracism. From now on, the epigraphical evidence will have to be considered not only in relation to its size and value, but especially in its own right and not merely as illustrative material.

The number of ostraka found up to July [1952] stands at 1,650 of which 1,426 were found in the American excavations in the Agora and on the North Slope of the Acropolis. The names of sixty different Athenians are now known from ostraka; half of these occur only once, while there are at least 570 ostraka of Themistokles. Next in number come the ostraka of Kallixenos the son of Aristonymos from the deme Xypete of whom there are more than 260 ostraka. We still do not think that he was ostracized, but, although nothing is known about him from the literary tradition, it seems clear that he was a member of the Alkmeonid family and a close relative of Kleisthenes himself.

[63] Before presenting some of the problems which the students of the ostraka are now facing, I should like to communicate to you some of the names which have not been included in the lists of either Reinmuth or Calderini; many of these can be found in the fine catalogue which Eugene Van|derpool wrote in 1946, and which was published in the *Shear memorial volume* (*Suppl.* 8) of *Hesperia* (408–412). Only one of the certain victims of ostracism is among the new names: Menon son of Menekleides of Gargettos. Eighty ostraka with his name have been found in the German excavations in the Kerameikos and were published in an exemplary fashion by Werner Peek in the third volume of *Kerameikos* (1941). The same volume contains 66 Kimon ostraka. In fact the names of all the Athenians who were in some way or another connected with ostracism by the literary tradition are now represented on ostraka—all, that is, with 3 noteworthy exceptions: Kleisthenes, Miltiades, and Nikias. This may mean nothing in the case of Nikias since we have only 4 Phaiax ostraka, 3 with the name of Alcibiades, and 2 with that of Hyperbolos, all belonging to the same occasion. The ostracisms of Kleisthenes and Miltiades, however, have been called imaginary by Carcopino; that of Kleisthenes is indeed suspect, but the reference in the third oration of Andocides to the ostracism of Miltiades the son of Kimon remains puzzling and tantalizing.

While the excavations have not helped us in this matter, they have added many new problems by presenting us with ostraka bearing the names of well-known Athenians whom we never suspected of having been involved in ostracism. One of these is Socrates of Anagyrous who was an Athenian general in the war against Samos and who may have been a partisan of Pericles. The most striking example, however, is an ostrakon bearing the name of Kleophon, son of Kleippides, no doubt the famous demagogue of the last decade of the fifth century. It now appears that Kleophon must have been prominent enough before the Sicilian expedition to be mentioned on an ostrakon which was undoubtedly cast during the famous ostracism of Hyperbolos. Moreover, the name of Kleophon's father, which was unknown before, permits us to identify Kleophon as the son of that Kleippides son of Deinias who was general in 428 B.C. and whose name appears on 25 ostraka found in the Kerameikos. Brückner may not have been correct in assuming that Kleippides *was* ostracized, but Körte's suggestion that the man was a partisan of Perikles has been vindicated by the discovery that he was the father of the popular leader Kleophon.

[64] Any additions to our knowledge of the Athenian prosopography of the fifth century are of course welcome, but the new names found on the ostraka are especially valuable because they are the names of politically active Athenians. In fact, the large number of different names offers the first significant problem posed by the archaeological evidence. According to the literary tradition, ostracism was a contest between the people and one man, or, occasionally, a contest between two powerful politicians who were able to command popular support. Only in the case of the Hyperbolos' ostracism are four people mentioned, and this case has always been treated as an exception, and as

symbolic of the breakdown of the whole institution of ostracism. Faced by the archaeological evidence, we shall have to reconsider this view. It appears that at all times there was a considerable scattering of votes against many persons in addition to the one main candidate. Not only were many individuals involved who may have received no more than a few votes, but there seem to have been very often alternate and opposition candidates who made the contest lively and its outcome unpredictable. In no other way can we explain the large number of ostraka inscribed with the names of people who were certainly not ostracized and who are often not even known otherwise. In this category belong Habron the son of Patrokles from Marathon (11); Boutalion, also from Marathon (7); Hippokrates the son of Alkmeonides (122); Hippokrates the son of Anaxileos (8); Kallixenos the son of Aristonymos of Xypete (262); Kydrokles the son of Timokrates (19). All these men were candidates in the eighties of the fifth century, and although one of them may have been actually ostracized, the others were never more than strong candidates.

It may be permitted to draw certain tentative conclusions from the character of the epigraphical evidence regarding the procedure of ostracism itself. It is highly improbable that the often repeated statement is correct, namely that public opinion was not influenced in favour of or against certain candidates. The opposite view is more apt to be the true one; this is indicated by certain passages in Plutarch, and it is confirmed by a careful interpretation of the fourth oration attributed to Andocides as well as by the epigraphical evidence itself.[2] The large number of ostraka cast against men who were evidently not candidates themselves, but rather the opponents of candidates, indicates that public opinion must have been influenced by many groups and by many individuals. This is confirmed by Broneer's discovery in 1938 of a hoard of close to 200 Themistokles ostraka which consisted of several groups, each of which was evidently inscribed by one person. Broneer rightly deduced that we have here evidence of an organized effort to manufacture Themistokles ostraka with the purpose of distributing them at the right moment. It is now possible to explain the ostracisms of Menon, Kallias, and Damon as the result of successful organization rather than as indications of strong popular feeling against any of these men. In fact, the ostracism of Hyperbolos belongs into the same class. Our evidence is clear on this point: Hyperbolos was ostracized not because the Athenians wanted to be rid of him, but because a well-organized political machine was able to deliver the number of votes sufficient to exile an otherwise popular leader.

The whole problem of the voting procedure will have to be reconsidered. The large scatter vote indicated by the epigraphical evidence can be interpreted in two ways. If a quorum of six thousand was necessary for a valid ostracism and if the losing candidate had to receive no more than a simple majority of the votes cast, the scatter vote can be understood only as an attempt at reaching the quorum. If, however, it was necessary for the losing candidate to receive

---

2. *Archaeology* 1 (1948) p. 79; *TAPA* 79 (1948) p. 196 [*infra*, p. 120].

at least six thousand votes, the scatter vote could be understood as an attempt at preventing the accumulation of so large a majority. We have not been able to reach a satisfactory decision between these two alternatives, but a decision will have to be reached before the final publication of the ostraka can be undertaken.

[66] Our own work so far has been twofold, historical and archaeological. On the historical side, we have tried to establish the chronological limits of the operation of ostracism, and I myself have suggested that the law of ostracism was enacted after the battle of Marathon and shortly before | the first ostracism, that of Hipparchos in 487 B.C., took place (*AJA* 55 (1951) [*infra*, pp. 65–76]). The archaeological evidence certainly confirms the statement of Aristotle that Hipparchos was the first man to be ostracized; for most of the Agora ostraka belong to one or the other of the ostracisms which are known to have taken place during the eighties of the fifth century. I have also argued that the last ostracism, that of Hyperbolos, was held in 415 B.C., immediately preceding the Sicilian expedition.[3] Here again, the discovery of 4 ostraka cast against Phaiax, the speaker of the fourth oration attributed to Andocides, confirms the stories told about Phaiax by Plutarch and the reliability, if not the genuineness, of the Phaiax speech. The next task will be a renewed study of the ostracisms which took place between the battle of Salamis and the outbreak of the Peloponnesian War. The three main questions to be answered concern the date of the various ostracisms and their general cause. It must be attempted to establish more securely the dates of the ostracisms not only of Themistokles, Kimon, and Thukydides, but especially of Menon, Damon, and Kallias. It must also be possible to answer once and for all the question of whether or not the length of exile was ever reduced from ten to five years; the answer is important not only for our knowledge of Kimon's last years but also for our understanding of the career of Thukydides and of the possible motive for the assassination of Hyperbolos. Finally, it will be necessary to examine and to answer the question of whether the law of ostracism was not only established but also continued to be operated as a purely internal measure. In that case, neither the alleged treason of Themistocles nor Cimon's pro-Spartan attitude were decisive causes of the exile each of these men suffered. Similarly, the ostracisms of Menon, Damon, Kallias, and Thukydides, not to mention those of the eighties, would each have had primarily an internal political cause. I personally believe that this can be shown, and that the story of Athenian ostracism will offer a distinct contribution to our knowledge of the working of Athenian democracy from its establishment before the Persian Wars to its temporary collapse under the impact of the Sicilian expedition.

[67] Another point of interest concerning which the ostraka have greatly increased our knowledge is the formation of the Attic name during the fifth century, consisting of the proper name with the addition of either the demotic or the

---

3. *TAPA* 79 (1948) [*infra*, pp. 116–131].

father's name or both. I have had occasion to comment at length on this problem in the publication of the Akropolis dedications (472–6), and on the basis of the material from the stone inscriptions I came to the conclusion that the full Attic name before Kleisthenes consisted of the proper name and the father's name. This practice continued for a whole generation after the reforms of Kleisthenes, because most people were loath to abandon their pre-Kleisthenian name with which they were born and brought up. There are, moreover, a great number of examples of the use of demotics instead of father's names, used especially by people who were born after the reforms of Kleisthenes; The most significant development, however, is the occurrence after 480 B.C. of names with father's names and demotics, long before the Periclean citizenship legislation of the middle of the century. The use of demotic and patronymic about a generation after Kleisthenes indicates that the sons of the newly made citizens were anxious to mention the names of their fathers, and that from now on there was no difference between old and new citizens. The official documents, however, were slow in adopting this practice, and when they did, by the end of the fifth century, it was under the influence of the Periclean citizenship legislation. It appears, therefore, that the Athenians reluctantly accepted the use of their demotics but that they never completely abandoned the use of the names of their fathers. In this respect the private dedications provide a better indication than the public inscriptions, and the only period during which dedicators add only the demotic to their proper name is the first quarter of the fifth century, and even then the number of examples is small, and much smaller than the instances in which the proper names occur joined to the names of the fathers. The evidence of the ostraka, of which more than twelve hundred are sufficiently well preserved to be considered, supports substantially the general conclusions already reached.

Before it is possible, however, to make use of the ostraka, certain misconceptions must be cleared away. Alfred Körte | and, following him, Werner Peek have insisted that the exclusive or predominant use of patronymics on ostraka indicates that the bearers of these names were aristocrats while the addition of demotics without or even with the patronymics reveals democratic tendencies or low origin on the part of the bearers. This distinction, if correct and generally applicable, would be of the greatest importance, because it would permit us to classify politically many Athenians whose names are known to us only from ostraka, even from a single sherd. Körte and Peek have based their assumption on the ostraka of Kimon and Thukydides which contain only patronymics and on those of Kleippides and Menon which contain, especially the latter, a great number of demotics. To these instances could be added the cases of the aristocrats Aristeides, Hippokrates, Kallixenos, and Xanthippos of whom we have a total of close to 500 ostraka of which only 20 exhibit the proper name followed by the demotic. And yet, there is a flaw in this statistical argument as in most conclusions based on statistical evidence. In the first place, it seems strange that the usage of the ostraka should reflect the status of the man against whom they were written rather than that by whom they were

[68]

written. If we look at the grand total of the better-preserved ostraka, we find that out of more than 1,200 examined nearly 950 contain patronymics while 250 have demotics and a bare 18 both patronymics and demotics. It would be misleading and wrong to see in these figures support for the well-known fact that the institution of ostracism was primarily directed against aristocrats. Such an erroneous conclusion could be encouraged, moreover, by the observation that of the 250 ostraka with demotics 160, that is all but 90, were cast against the democrat Themistokles, and that of the remaining 90 another 53 were voted against Menon whom Peek called a leader of the democratic party. It is precisely the Themistokles ostraka which show an entirely different picture: 321 of them have only patronymics as against 159 with demotics; and what is more, only 8 of the 191 Themistokles ostraka found by Broneer have demotics. This means

[69] that the political organization which made such | elaborate preparations for Themistokles' ostracism gave this hated upstart an aristocratic name. Even more significant is the fact that the 2 ostraka of Hyperbolos and the 1 of Kleophon, both democratic demagogues of low origin, have patronymics and not demotics.

The conclusion cannot be avoided that the use of demotics or patronymics on ostraka throws no light upon the political complexion or social origin of the candidates but that it rather indicates the preference of the voters. Patronymics were used by the overwhelming majority of the voters throughout the entire period of ostracism, while the demotics occur more frequently in the earlier period and especially on the ostraka cast against Themistokles and Menon. This does not mean that either Themistokles and Menon or the people who voted against them were either democrats or of lowly origin, but it may have an entirely different explanation. Among the private dedications from the Akropolis there is a preponderance of out-of-town demes, especially in the earlier period, while a number of dedicators who are known to have belonged to town demes did not have their demotics recorded. I have already suggested that the use of demotics was favored by the out-of-town people whose demes had been well-known communities even before Kleisthenes, while the town people did not take enthusiastically to their demes' names which were more artificially created. The same may be true for the ostraka. Among the demotics mentioned on ostraka there are four different town demes as against ten different out-of-town demes; but if one considers the number of ostraka, there are 21 with town demes as against 233 with out-of-town demes. The ostraka confirm, therefore, strongly our concept of the use and popularity of demotics during the fifth century B.C. They were primarily used by and for people who lived outside of the city of Athens, and they were never very popular until they were combined with the patronymic to form the full name πατρόθεν καὶ τοῦ δήμου.

The archaeological part of our investigation is carried on primarily by Eugene Vanderpool. He is examining the topographical evidence in order to arrive at more precise information concerning the section of the Agora in which the
[70] ostracisms took place. The enclosed area must have | been considerable if it was

to hold at one and the same time more than six thousand people. Vanderpool is also giving considerable attention to the closed groups of ostraka which were found at times in well-dated contexts. So far, none of these groups has been found in the Agora proper, and it appears that the mass of ostraka was removed and discarded after each ostracism; there is no reason why large deposits of ostraka should not be found at a considerable distance from the Agora. On the other hand, Vanderpool has tentatively suggested that most, if not all, the groups found thus far in the Agora excavations belong to a single ostracism, that of Aristeides. If this is true, we may still expect to find great quantities of ostraka in hitherto unexcavated areas, because the small inscribed potsherds were normally used to fill holes and they survived, though often broken, surprisingly well wherever they were dumped in antiquity.

This archaeological investigation permits us to assign to the year of the ostracism of Aristeides, that in 482 B.C., not only the ostraka bearing the name of Aristeides but also the vast majority of the Themistokles ostraka, many of which were found together with those of Aristeides. Since we know that Aristeides was not a "candidate" at the time when Themistokles was ostracized but that Themistokles was actively engaged in the ostracism of Aristeides, those groups of ostraka which contain votes cast against both Themistokles and Aristeides must be assigned to the ostracism of Aristeides. At that time, however, also Kallixenos and Hippokrates were involved. More than 200 ostraka with the name of Kallixenos have been discovered, and nearly 150 with that of Hippokrates. Both these men were members of the Alkmeonid family and relatives of the Megakles who had been ostracized in 486 B.C. It is impossible, however, to say in what relation they stood to each other and to either Aristeides or Themistokles. The most likely assumption seems to be that the friends of Aristeides who made an organized campaign against Themistokles also tried to divert votes from Aristeides by calling attention to the two prominent Alkmeonids Kallixenos and Hippokrates.

Vanderpool is also examining the types of pottery used for ostraka. The importance of this study is both historical and archaeological. It has been suggested that the ostraka | used against democratic candidates came from fine *[71]* pottery such as aristocrats would have, while the ostraka against aristocratic candidates came from coarse and cheap pottery. Unfortunately this very attractive distinction is not supported by the evidence of the preserved ostraka, and it should therefore be abandoned. It has also been suggested that the few ostraka which belong to black and red figured pottery may be used to establish more firmly the date of the introduction of the red-figured style. This notion, too, has to be abandoned since we have no reason to assume that the pottery used for ostraka was brand new when it was broken up; moreover, some ostraka were inscribed on fragments of pottery which was made more than a hundred years before the beginning of ostracism. There is, however, one way in which the ostraka may add to our knowledge of pottery. This is the study of the profiles of the small cups so often used for ostraka. And indeed the closed group of Themistokles ostraka found by Broneer on the North Slope has already yielded some results, and a careful examination of the entire material will be highly useful.

Turning now from the more archaeological to the peculiarly epigraphical part of the study of the ostraka, I should like to mention especially the writing itself, that is spelling, letter forms, and literacy. In these fields the ostraka provide ample evidence which is superior to that of the formal stone inscriptions and the carefully painted inscriptions on vases. While the former were engraved by professional stonecutters, and the latter were drawn by well-trained artisans, the inscriptions on the ostraka were scratched in quickly and casually by individual citizens. They therefore reveal accurately the current spelling, the customary letter forms, and the state of literacy as they existed during the fifth century in Athens. Once the great mass of the ostraka has been carefully examined, we shall be able to date more accurately the beginning of the use of double consonants, and of the Ionic letters gamma, eta, lambda, sigma, and omega. It has been already stated (*DAA*, pp. 444–6) that the change from single to double consonants took place during the generation between 515 and 485 B.C., and Werner Peek has dated the Menon ostraka earlier than those against

[72] Thukydides by observing that only two Ionic gammas and | one Ionic sigma occur among the eighty Menon ostraka, while these Ionic letters are frequent on the ostraka cast in 443 B.C. It may be observed in general that changes in spelling and in letter forms do not take place from one year to the next, but rather over a period of thirty or forty years during which a new generation grew up whose members learned the new spelling in school. Of special importance for the study of Attic letter forms will be the evidence provided by the ostraka for the use of four-stroke sigma and dotted theta, for these two letters were introduced into Attic script during the period covered by the ostraka.

Equally important is the information we gain from the ostraka on the state of literacy, because it is reasonable to assume that most ostraka were inscribed by the individual voters themselves. This is borne out by the famous anecdote told of the Aristeides ostracism. It is confirmed by the great variety in spelling and script found on the various ostraka, and it is suggested by the small number of painted ostraka which may have been prepared by professional scribes. Vanderpool calls our attention, however, to the fact that the painted ostraka all belong to the period after the middle of the fifth century from which we have comparatively few ostraka preserved. He also cautions us by pointing out that the paint used on these ostraka was apt to fade easily or to be washed off or to vanish in the course of time. This means that painted ostraka may today be just plain broken pieces of pottery which we are unable to recognize and to identify. Even so, it must be assumed that the average Athenian was expected to be able to write the name of a candidate, and that very many Athenians were capable of performing this task correctly and competently. Thus the high state of literacy of the Athenian population is confirmed by the great number of well-written ostraka.

As we look upon the ostraka as inscriptions, we cannot fail to observe the significant difference between Greek inscriptions in general and those found on ostraka. I wish to discuss this difference because it is of importance for the form in which the final publication of the ostraka is to be presented. The few ostraka known at the time the first volume of the editio minor of the *Inscriptiones Graecae* was prepared were included in this edition by Hiller, but in more | recent days both the editors of the tenth volume of the *Supplementum Epigraphicum Graecum* and of the *Sources of Greek History* have summarily referred to the existence of vast numbers of ostraka without reprinting the text of any one ostrakon. It has become apparent that the ostraka cannot be published or republished in the same lavish completeness as the decrees or the financial accounts, or even as the funeral lists, the tomb inscriptions, and the dikast tickets. Each ostrakon is the vote cast by one Athenian on the occasion of an "election" at which a limited number of "candidates" was available. While to the epigraphist every single ostrakon is a separate inscription, to the historian only the names of the candidates and the number of votes cast against each one of them is of importance. And yet even the epigraphist is not so much interested in the text of each ostrakon but rather in those aspects in which it is unique. These are the case of the name, the addition or omission of demotic and father's name, the spelling, and the letter forms. This means that an epigraphically satisfactory publication of the ostraka must contain not only the various forms of the names and the various spellings but also illustrations of the individual ostraka. These illustrations do not have to follow the current practice of photographs. Not only do ostraka not photograph well, but the thinly incised letters of their inscriptions can best be illustrated by line drawings. It may be suggested therefore that the final publication will devote a minimum space to the description, identification, and interpretation of the individual ostraka but that it will contain facsimile line drawings, at a uniform but reduced scale, of all the ostraka.

[73]

I am well aware of the fact that some of the epigraphists of strict observance will not think much of the inscriptions on the Athenian ostraka. If one considers the epigraphist's task to lie in the reading, restoration, dating, and historical interpretation of the individual inscription, one Themistokles ostrakon will appear to be not different from another nor to offer any new problems. The ostraka bearing the same name will constitute a group as homogeneous as the coins struck from the same die, and it is indeed true that the study of the ostraka, as I have described it, resembles in certain ways the study of coins.

The modern epigraphists, however, consider the epigraphical evidence neither as independent of the historical tradition nor as subservient and inferior to it, but as a significant part of the historical evidence. There is perhaps no better example of this attitude toward the epigraphical evidence than the masterly volumes which have appeared under the title of *Hellenica*, written by the most distinguished of the living epigraphists. There, the literary, archaeological, numismatic, and epigraphical evidence has been combined to solve problems

[74]

of social, political, economic, and religious history. As I have suggested at the beginning that epigraphy must seek and solve its own problems, so I am suggesting now that the work of the epigraphist be part of the work of the historian; it is historical knowledge which is our goal, not epigraphical, literary, archaeological, or numismatic knowledge. The study of Athenian ostracism offers a good example of the way in which this historical knowledge can be attained.

10

The Origin of Ostracism

ANDROTION AND ARISTOTLE

The date of the enactment of the law of ostracism is known from a fragment of Androtion who said that Hipparchus, a relative of Pisistratus, was the first to be ostracized, and that the law of ostracism was then at first enacted on account of the (public) suspicion toward the followers of Pisistratus who had established his tyranny through being a popular leader and military commander.[1] Aristotle (*Constitution of Athens* 22) repeated, almost word for word, this account, adding that the law was the work of Cleisthenes who had Hipparchus in mind when he introduced ostracism.[2] It is significant that Aristotle agrees with Androtion in associating the activity of Hipparchus with the enactment of the law, because he thereby seems to accept the date given by

American Journal of Archaeology 55 (1951) 221–229

---

1. Harpocration s.v. Ἵππαρχος. (Androtion Frag. 5 [Müller]–6 [Jacoby]) περὶ δὲ τούτου 'Ανδροτίων ἐν τῇ δευτέρᾳ φησίν, ὅτι συγγενὴς μὲν ἦν Πεισιστράτου τοῦ τυράννου, καὶ πρῶτος ἐξωστρακίσθη, τοῦ περὶ τὸν ὀστρακισμὸν νόμου τότε πρῶτον τεθέντος διὰ τὴν ὑποψίαν τῶν περὶ Πεισίστρατον ὅτι δημαγωγὸς ὢν καὶ στρατηγὸς ἐτυράννησεν.

G. Kaibel, *Stil und Text der ΠΟΛΙΤΕΙΑ ΑΘΝΑΙΩΝ des Aristoteles* (Berlin 1893) 174–175, considered the expression "enacted for the first time" as meaningless and indicative of the spurious character of the whole fragment. Androtion may have meant, however, that "the law was then enacted, at first because of the suspicion...," but later it was used also for other purposes; the missing part of Androtion's statement may be supplied from Aristotle's *Constitution of Athens* (henceforth referred to by the name of Aristotle) 22.6: μετὰ δὲ ταῦτα... καὶ τῶν ἄλλων εἴ τις δοκοίη μείζων εἶναι μεθίσταντο.

2. For recent treatments and summaries of earlier accounts, see G. Busolt and H. Swoboda, *Griechische Staatskunde* (Munich 1926) 884–887; G. Glotz, *The Greek City* (New York 1930) 169–173; V. Ehrenberg, *Gnomon* 5 (1929) 9–11; R. J. Bonner and G. Smith, *The Administration of Justice from Homer to Aristotle* 1 (Chicago 1930) 193–195 (hereafter referred to by the names of the authors); F. Schachermeyr, *Klio* 25 (1932) 346–347; J. Carcopino, *L'ostracisme Athénien* (Paris 1935) 15–28; O. W. Reinmuth, *RE* s.v. "Ostrakismos" 1675–1676; A. Calderini, *L'Ostracismo* (Como 1945) 27–30. I wish to thank F. E. Adcock, A. Andrewes, O. Broneer, N. Doenges, J. V. A. Fine, J. A. O. Larsen, H. W. Parke, H. Schaefer, and A. G. Woodhead for having read my manuscript and for having given me the benefit of their advice and criticism.

Androtion.³ The alternative interpretation, which is commonly accepted, has Hipparchus under suspicion twenty years before he was actually ostracized for the very same reason which caused the enactment of the law twenty years earlier. At that time Hipparchus could have been removed by Cleisthenes without resort to ostracism since only those of the tyrant's party were permitted to stay who had not been actively engaged in politics (Aristotle 22.4). Moreover, one should expect Aristotle to comment on this lag of twenty years between enactment and first use of ostracism and to register his disagreement with Androtion; in fact, Aristotle did not do either.

Were it not for his statement at the beginning of the chapter (22.1), nobody would have thought that the two accounts were at variance with each other. In this earlier passage, Aristotle gives a general introduction to the later legislation of Cleisthenes:⁴ "As a result of these events [described in 21], the constitution became more democratic than that of Solon. For it also happened that, on the one hand, the tyranny had obliterated Solon's laws by disuse, while, on the other hand, Cleisthenes, catering to the multitude, had instituted other new laws, among which was enacted also the law concerning ostracism." The law of ostracism evidently belonged to Cleisthenes' later legislative program, and there can be little doubt that the measures listed by Aristotle in chronological order (22.2–6) are, in his opinion, the new laws enacted by Cleisthenes, through which the constitution became more democratic.⁵

The real reason for separating the enactment and the first use of ostracism lies in the association of the law with the name of Cleisthenes. For if the law

---

3. See G. Mathieu, *Aristote, Constitution d'Athènes* (Paris 1915) 56–57; L. Pearson, *The Local Historians of Attica* (Philadelphia 1942) 85–86; cf. H. Bloch, *HSCP* Suppl. 1 (1940) 349, n. 3. F. Jacoby, *Atthis* (Oxford 1949) 383, n. 24, said that Androtion disputed the Cleisthenian origin of the law. The key difference lies in Androtion's assertion that the law was enacted immediately before it was used. Aristotle's statement however was thought to contradict Androtion, since it implied that the date of the law's first use was definitely not the occasion of its enactment. Yet, if this interpretation were to be accepted, one must also assume that the laws concerning the election of generals (22.2) and archons (22.5) were first used, but not enacted, at the respective dates given by Aristotle. Since this assumption is absurd, one must understand Aristotle's words as if he had said "they held the first ostracism"; he preferred the expression "they used the law for the first time" simply because he wanted to add a sentence giving the reason for the institution of ostracism at this time.

4. See Kaibel (*supra* n. 1) 22–23; Jacoby (*supra* n. 3) 206–208. In no other way can the beginning of the next sentence be understood. "First of all..." For these words have no meaning unless they introduce a summary of the new laws referred to in the preceding sentence; see T. J. Cadoux, *JHS* 68 (1948) 116. The κατάστασις mentioned by Aristotle (22.2) as a point of reference is of course the legislation of Cleisthenes described and dated by Aristotle in the preceding chapter (21). If any further proof were needed to show that 22.2–6 contains an account of the steps by which the constitution became more democratic, this is provided by 23.1 which sums up the preceding chapter; see also 41.2 where the constitutional reforms of the period 510–480 B.C. (described by Aristotle 21–22) are attributed to Cleisthenes.

5. One may, therefore, attribute to Cleisthenes all the measures mentioned by Aristotle in the first part of the chapter (22.2–5), including the famous electoral law of 487 B.C. I am here revising my earlier suggestion made in *AJA* 51 (1947) 258 [*infra*, pp. 109–110]; see also J. Gregor, *Pericles* (Munich 1938) 192. The second part of the chapter (22.7–8) deals with Themistocles' activities; the naval program is specifically connected with his name by Aristotle, while the amendment to the law of ostracism was passed also at the suggestion of Themistocles; see Plutarch, *Themistocles* 11.1; R. Goossens, *Chronique d'Égypte* 39/40 (1945) 125–133.

was passed in 488 B.C., Cleisthenes must have been alive and active after the battle of Marathon.[6] A reference to the career of men like Herbert Hoover, Winston Churchill, Carlo Sforza, and others, may give ample support to the assumption that Cleisthenes came out of retirement after Marathon, convinced by the activities of Hippias that Athens needed a legal safeguard against the reestablishment of tyranny.[7] The literary evidence certainly does not support the assumption that the law of ostracism was enacted twenty years before it was used for the first time.

ARCHONSHIP AND TYRANNY

In order to understand the original purpose of the law of ostracism, one must know the "legal" basis of Athenian tyranny, the restoration of which the law was to prevent. In addition to Aristotle's repeated insistence that the rule of Pisistratus was constitutional (14.3, 16.2, 16.8), we have Thucydides' statement that the laws of the land were kept in force during the time of tyranny (6.54.6). Thucydides added, however, a significant detail when he said that the tyrants saw to it that one of their group always held the archonship. This may mean that they controlled not only one or several of the highest annual offices but also the council of the Areopagus which was composed of ex-|archons and whose members served for life.

[222]

Any legislative action designed to prevent the restoration of tyranny must have been concerned with the way the archons were elected. Aristotle specifically stated (8.1) that Solon instituted the lot for the election of the nine archons. Each of the four tribes elected ten candidates, and the lot determined the nine who were to hold office. Since the candidates had to qualify mentally as well as physically, one may be tempted to assume that the individual candidates were designated for one particular job. Thus the lot determined which one of four (rather than forty) was to be eponymous archon, king archon, polemarch, one of the thesmothetae, or their secretary. This use of the lot was no innovation at all, but an application to the realm of politics of a principle of selection well known from the Homeric poems.[8] Although Solon evidently tried to ensure the election of qualified men, he made it comparatively easy to manage or to control the elections.

There is, perhaps, some evidence available to show the extent to which the tyrant was able to influence the lot. The names of six eponymous archons who

---

6. It has been generally assumed that he died not long after 508 B.C., or at any rate shortly after 500 B.C.; see K. Schefold, *Museum Helveticum* 3 (1946) 59–93, especially 68–70. Jacoby's repeated references (*supra* n. 3) to the "overthrow" of Cleisthenes (160–161, 339, notes 52–53) are not based on any evidence; it should be remembered that the cult of the tyrannicides is anti-Spartan rather then anti-Cleisthenian or anti-Alcmeonid (cf. *AJA* 44 [1940] 58, n. 2 [*infra*, p. 211, n. 24]).

7. From the accounts of Herodotus (5.65.2, 69.2) and Aristotle (20.1; 21.1; 22.1) one must infer that the Cleisthenian reform of 508 B.C. was primarily designed to prevent the aristocratic factions from regaining control of the state, but that it did not aim at forestalling the restoration of tyranny.

8. See H. T. Wade-Gery, *CQ* 25 (1931) 88–89. The sixth-century potsherds with names on them may well have been used in this process; see E. Vanderpool, *Hesperia* Suppl. 8 (1949) 407–408 on the "ostraka" of Pisistratus and Aristion.

held office in the years following the death of Pisistratus (528 B.C.) are known from an inscription.[9] The list contains not only the name of Miltiades (whose archonship in 524 B.C. had been known before) but also those of Cleisthenes and Hippias. It is difficult to imagine that Cleisthenes should ever have been a nominee of the tyrants, even if a reconciliation between his family and Hippias did take place after Pisistratus's death; the same applies presumably to Miltiades. The archonship of Hippias, however, is noteworthy for its date, two years after he became tyrant. Evidently he, the eldest son of Pisistratus, had not held any high office during his father's lifetime, nor did he succeed in making himself archon immediately after he assumed power. This list of archons shows that the use of the lot rather than direct election was in operation, but it also indicates that the lot introduced an element of uncertainty into the best planning.

The Solonian law governing the election of archons remained in force during the time of tyranny. This is implied by Herodotus (1.59.6), Thucydides, and Aristotle, who all affirmed that the tyrants did not alter or abrogate any existing laws, and also by Thucydides' reference to their manipulation of the elections. Moreover, Aristotle stated specifically (22.5) that the electoral law of 487 B.C. represented the second change since the end of tyranny, adding that between 510 and 487 B.C. the archons had been directly elected. The only possible interpretation of this passage is the assumption that during the tyranny the archons were not directly elected. Since the Solonian law combined tribal elections with the use of the lot, it was this law that was operative throughout the sixth century, until the expulsion of Hippias.[10]

The Athenians altered the law governing the election of archons after the expulsion of Hippias. This fact supports the assumption that the "legal" basis of Athenian tyranny consisted in the, perhaps illegally managed, control of the archonship. Yet it has become customary to insist, against Aristotle, that the Athenian archons were directly elected throughout the sixth century.[11] It has been pointed out that the archonship was so important an office that its occupants could not have been determined by the lot. The weakness of this

---

9. *Hesperia* 8 (1939) 59–65; no. 21; *SEG* 10.352 (with bibliography); Jacoby (*supra* n. 3) 371, note 99; Cadoux (*supra* n. 4) 71, no. 6, 77.

10. See V. Ehrenberg, *Klio* 19 (1925) 106–107, who claimed, however, that the change from lot to election was made by Hippias in 514 B.C. Larsen (*infra* n.11) 169, n. 13 (also on p. 170) insisted that "Solon transferred the election to the people," preferring the evidence of Aristotle's *Politics* (1273b36–1274a22) to that of the *Constitution* (8.1); see F. E. Adcock, *Klio* 12 (1912) 6–7, n. 4; *CAH* 4 (1939) 51. The passage in the *Politics* does not, however, distinguish between measures of the Solonian and later periods. Surely, Solon did not institute the dicasteries the members of which were elected by lot (1274a5–6). The important feature of Solon's electoral reform, emphasized both in the *Politics* and in the *Constitution*, was the transfer of the election from the hitherto self-perpetuating body of Areopagites to the people at large. It should be noticed, moreover, that there is no reference in the *Politics* to the peculiarly Athenian combination of election and lot, although this system was included, but not mentioned, in the discussion of the various modes of appointing magistrates (1300a10–1300b13).

11. See, for instance, R. J. Bonner, *Aspects of Athenian Democracy* (Berkeley 1933) 6–7 and 179, n. 14 (with some bibliography); J. A. Munro, *CQ* 33 (1939) 94–96; J. A. O. Larsen, *CP* 44 (1949) 169, n. 13.

argument should be apparent after what has been said above. It has also been pointed out that the use of the lot is a sign of advanced democracy such as could not have existed in sixth century Athens. Yet a reference to a very similar procedure used by the Homeric heroes (see especially *Iliad* 7.161–192) must remove any doubt about the conservative and traditional character of the lot.

By making the archons directly elected, it was thought that the "legal" basis of tyranny was removed. As long as this was considered a sufficiently strong safeguard against the return of Hippias, there was no need for the law of ostracism; for the same people who voted for the archons would also have been able to ostracize a potential tyrant or his chief supporter. Since tyranny had lost its popular support, the people were able to express their preference in the archon's elections, and ostracism would have been an unnecessary and negative confirmation of the election.

The archon elected for the year 496/5 (Dion. Hal. 6.1) was Hipparchus the son of Charmus, and his election may have foreshadowed the attempted restoration of Hippias. For he was | the man on whose account Cleisthenes [223] instituted the law of ostracism, and who was in fact its first victim (Aristotle 22.3–4). It must have been Hipparchus's activity during and after his archonship, and especially preceding and during the battle of Marathon, which showed both the danger of the man and the need for a new safeguard against tyranny. The second victim of ostracism was the man whom popular opinion, rightly or wrongly, connected with the famous signal given to Hippias after the battle at Marathon (Herodotus 6.115, 121.1), Megacles. For the first time since the expulsion of Hippias, the restoration of tyranny appeared as a distinct possibility. The law of ostracism was instituted by Cleisthenes in order to remove men like Hipparchus and Megacles (Aristotle, 22.6), and removed they were. It seems almost inconceivable that the law should have been enacted before the deeds were committed which it was designed to prevent and punish.

OSTRACISM AND THE ELECTIONS OF GENERALS

If one assumes that the power of the Athenian tyrants rested on their control of the archonship, he must consider the office of polemarch as well as that of eponymous archon. It was as polemarch that Pisistratus first gained prominence in Athens,[12] and Aristotle made it clear that his position of military leadership was later thought to have been the stepping stone to tyranny (22.3). Since the office of general did not exist in Pisistratus's day, Herodotus as well as Aristotle (who followed Androtion) must have referred to Pisistratus's position as polemarch.[13]

When, after the expulsion of Hippias, the archons were directly elected, the

---

12. See Herodotus 1.59.4–5, whose account was evidently used by Aristotle 14.1. G. V. Thompson, however, insisted that the office of general existed at that early time; see *PAPA* 1894 xviii–xx, *Hermes* 30 (1895) 478–480; cf. Jacoby (*supra* n. 3) 92–93.

13. It should not be surprising to find one of the chief magistrates going with the army into the field, considering the fact that the Spartan kings conducted virtually all military compaigns.

position of military command was brought under popular control. The creation of the board of ten generals was to have a far-reaching effect not only on the office of the polemarch but also on the development of Athenian democracy. Aristotle stated (22.2) that in 501 B.C. generals were elected by tribes, one from each of the ten tribes, but that the polemarch remained commander in chief. The first general whose name we know was Melanthius who commanded the Athenian auxiliary expedition during the Ionian revolt.[14] The sudden rise in importance of the generals is clearly indicated by Herodotus's account of the battle at Marathon (6.103.1, 104.1–2, 105.1, 106.1, 109–111.1, 114). After the victory, Miltiades was the man of the day (Herodotus 6.132), and he must have been reelected as general when he entered upon the fatal Parian expedition.

After the passage of the electoral law of 487 B.C., the archons were chosen by lot out of five hundred candidates. Thus the archonship was removed once and for all from the list of politically important offices. The polemarch, whose military qualification was now a matter of chance, ceased to be commander in chief, and the power of the generals increased enormously. They remained the only high officials who were directly elected, and the fact that they could be reelected must have strengthened their position. The decisive factor, however, was the mode of their election.

Aristotle's statement (22.2) is normally taken to mean that the assembly elected the generals by tribes, one from each tribe.[15] The candidates must have been chosen before the election by the tribes which accordingly had a considerable influence on the outcome of the election. Before 487 B.C., when the generals were merely the commanding officers of their tribal contingents, they may actually have been elected directly by their fellow tribesmen.[16] Thus a tenth part of the electorate was able to bring to power men who influenced the fate of the nation as a whole. A potential tyrant, or a leading supporter of tyranny, could have been placed in nomination or even maintained in office without the knowledge or even against the wish of the majority of the assembly. If such men had been in command at Marathon (and I suspect that Hipparchus led the contingent of his tribe, Aegeis), the outcome of the war might easily have been different.

---

14. Herodotus 5.97.3; see A. E. Raubitschek and L. H. Jeffrey, *Dedications from the Athenian Akropolis* (Cambridge, Mass., 1949) 191–194, no. 168; E. Vanderpool, (*supra* n. 8) 400–402, nos. 14–15.

15. This would have been done in ten separate elections. The text of Aristotle (44.4) may be emended to read ποιοῦσι δὲ καὶ δέκ⟨α⟩ ἀρχαιρεσίας στρατηγῶν; see also W. Schwahn, *RE* Suppl. 6.1074. The people showed their preference for one of two (or more) candidates presented to them by each of the tribes; see Xenophon, *Mem.* 3.4.1; S. Accame, *RivFil* 13 (1935) 350–351.

16. It is generally assumed that Aristotle (22.2) does not say this; see Accame (*supra* n. 15) 342; U. Kahrstedt, *Untersuchungen zur Magistratur in Athens* 2 (Stuttgart, 1936) 26–27; V. Ehrenberg, *AJPh* 66 (1945) 114–115. In that case, his account would be redundant; "one from each tribe" is a sufficiently accurate designation. Aristotle's description of the election of the sophronistae (42.2), which in many ways resembles that of the generals, may suggest that the words "by tribes" (22.2) are a short expression of "holding meetings by tribes" (42.2). In that case, the generals would have been elected (between 501 and 487 B.C.) by the tribes to which they belonged and the contingents of which they commanded; see Accame 352–353.

It was necessary to emphasize the importance the generals obtained at the very time when it became imperative to safeguard Athens against the restoration of tyranny. For it seems that the law of ostracism applied especially to those who might rise to power through the tribal elections or nominations of the generals.[17] In fact, Aristotle (22.3), who followed Androtion, declared specifically that the law of ostracism was enacted by Cleisthenes because he dis|trusted those who held military power, remembering that Pisistratus had used his military command as a stepping stone to tyranny. At a time when the office of general did not exist (or was insignificant, i.e., before 501 B.C.), and when the polemarch was directly elected (i.e., between 510 and 488 B.C.), the law of ostracism would have been ineffective. For if the number of people opposed to a man was sufficiently great to ostracize him, he could never have been elected polemarch. Yet the vote of ostracism was eminently suitable to prevent a general from being elected or placed in nomination by the votes of his tribe. And, indeed, ostracism was used during the fifth century almost exclusively against men who had served as generals. One may, therefore, conclude that the law of ostracism was out of place before the establishment of the board of generals (501 B.C.), or even before they had obtained great influence during the battle at Marathon (490 B.C.).

[224]

This close connection between the elections of the generals and the vote of ostracism is confirmed by the date within the year on which the two events took place. According to Aristotle (44.4), the generals were elected on the first good day after the sixth prytany.[18] This date was chosen in order to conduct the elections before the beginning of the spring campaign during which many of the candidates and some part of the electorate may have been absent. The preliminary vote on the question whether or not an ostracism was to be held took place at the full meeting of the assembly in the sixth prytany (Aristotle, 43.5), thus several days or weeks before the election of the generals. Concerning the date of ostracism itself one must rely on a fragment of Philochorus (79 b [Müller]–30 [Jacoby]) which has been emended and interpreted in several ways; see Carcopino (*supra* n. 2) 55–72. Philochorus said that the assembly cast the preliminary vote before the eighth prytany. Since the following lengthy description of the procedure does not contain any other reference to the date of ostracism itself, it has been generally assumed that the date "before the eighth prytany" actually referred to ostracism and not to the preliminary voting in the assembly which took place in the sixth prytany.

It appears, therefore, that both ostracism and the elections of the generals were held during the seventh prytany.[19] Considering this close association of the two events, one should assume that ostracism preceded the elections;

---

17. The system of primaries in this country [U.S.] shows how decisions of the electorate can be controlled by those in charge of the nomination of candidates.

18. See H. B. Mayor, *JHS* 59 (1939) 45, n. 1; W. K. Pritchett, *AJPh* 61 (1940) 469, n. 2.

19. It has been observed repeatedly that the two events took place at about the same time; see H. Swoboda, *H* 28 (1893) 548; R. W. Macan, *Herodotus, the Fourth, Fifth, and Sixth Books* 2 (London 1895) 142–145, §14; Carcopino (*supra* n. 2) 70; W. Peek, *Kerameikos* 3 (Berlin 1941) 104; A. E. Raubitschek, *Archaeology* 1 (1948) 79.

otherwise a successful candidate for the generalship could have become a victim of ostracism. In fact, it must have been the design of the legislator to place ostracism shortly before the elections in order to prevent a dangerous candidate from being elected general.[20]

The law of ostracism was instituted with particular reference to the generals as potential supporters of tyranny (or oligarchy). Moreover, the time within the year when ostracism took place was determined by that of the elections of the generals. Thus the law of ostracism cannot have been enacted until the office of general had been established. This is confirmed by the relative chronology given by Aristotle (22.2–3): generals were elected for the first time in 502/1 B.C., and the first ostracism took place in 487 B.C.

OSTRACISM AND THE LAW AGAINST TYRANTS

It may now be appropriate to investigate the law of ostracism as one of the legal measures by which the Athenians sought to protect themselves against attacks from within.

The most common form of political revolution during the seventh and sixth centuries was the establishment of tyranny. Bonner and Smith noted this fact (108), but they made only passing references (198 and 2.304) to the early Athenian law against tyrants (Aristotle 16.10): "Anyone who sets himself up as a tyrant or assists in the establishment of tyranny shall be deprived of all civil rights."[21] The continued operation of this law is assured by a phrase of the oath given by the Athenians in 410 B.C. (Andocides 2.97), which was based, according to Andocides (2.95), on a Solonian law. Aristotle makes reference to this law (8.4) which granted the Areopagus jurisdiction in cases involving to overthrow of the constitution.[22]

---

20. The elections had to be held at a full meeting of the assembly, and the vote of ostracism required at least as large a number of citizens as a full meeting. These two conventions may therefore have taken place on the same day, or on successive days. A detailed discussion of the procedure followed during ostracism must be postponed, and reference may be made, in the meantime, to the treatments by Carcopino (*supra* n. 2) 89–104, Reinmuth (*supra* n. 2) 1678–1679, and Calderini (*supra* n. 2) 37–39; see my forthcoming review of Calderini's book in *CP*.

21. Aristotle calls this a "mild" law, perhaps by mistake; see Usteri, *Ächtung und Verbannung im griechischen Recht* (Berlin 1903) 11–13, no. 4; G. Mathieu (*supra* n. 3) 39–40; Jacoby (*supra* n. 3) 364, n. 68. He used, however, a similar word to describe the "mild" treatment accorded to the friends of tyranny after the expulsion of Hippias (22.4). Since the old law was still in force, Aristotle's characterization of it was fully justified, considering its effectiveness. A good account of the operation of the Solonian law against tyrants has been given by T. Lenschau, *RE* s.v. τυραννίδος γραφή 1804–1808 (with the pertinent bibliography, mainly in German).

22. The phrase used by Aristotle ("overthrow of democracy") would have been out of place in the sixth century, but it was in common use after the overthrow of the oligarchic revolution of 411 B.C., and throughout the fourth century. One may assume that the older formula ("establishment of tyranny") was replaced by the new concept, and that this substitution took place not before 410 B.C.; see Usteri 16–17, no. 9. The Athenian decree for Erythrae, dated in 452 B.C., seems to contain a reference to the Solonian law against tyrants; see *SEG* 10.11; *Athenian Tribute Lists* 2 (Princeton 1949) 55, D 10, lines 32–34; cf. Lenschau (*supra* n. 21) 1807, lines 40–68. A passage in the *Wasps* of Aristophanes (463–507) shows clearly that in 422 B.C. "tyranny" was still a popular catch word; line 487, in particular, recalls the wording of the Solonian law. And even in 415 the accusation of tyranny had lost little of its force; see Raubitschek, *TAPhA* 79 (1948) 194 [*infra*, pp. 118–119].

The assumption that the Solonian law against tyranny which was administered by the Areopagus remained in force during the sixth and fifth centuries is also supported by the wording of the Solonian amnesty which was revived on the eve of the battle of Salamis and after the | defeat at Aegospotami (Andocides 2.77–79). From this amnesty were excluded those who had been condemned for tyranny but who had escaped capital punishment (Plutarch, *Solon* 19).[23] The identity of the court that issued the condemnation may be disputed (Bonner and Smith 106–107), but the court certainly was, directly or indirectly, connected with the Areopagus. At any rate, the Solonian amnesty law, which undoubtedly excluded the still surviving participants of the Cylonian conspiracy and their descendants, confirms the assumption that cases of sedition, especially tyranny, were judged by the Areopagus, or by some of its members; see the scholium on Aristophanes, *Knights* 445.

[225]

It is unknown whether the law against tyrants was invoked against Damasias (Aristotle 13.2), but the expulsion of Pisistratus was accompanied or followed by a formal verdict of condemnation (Herodotus 6.121). After the departure of Hippias, moreover, legal action was taken against Pisistratus and his descendants (Thucydides 6.55). The amnesty law of 481 B.C. (Andocides 2.78), which specifically excluded those condemned for tyranny, must have referred to the surviving members of Pisistratus's family. It is, therefore, evident that Hippias and his family were condemned by the Areopagus.[24] The law of ostracism has accordingly no place immediately after the overthrow of tyranny.

The only Athenian court trial for tyranny during the fifth century was that of Miltiades on his return to Athens in 493 B.C. Herodotus (6.104.2), who is our only witness for this incident,[25] reported that Miltiades was brought into court by his enemies and charged with tyranny in the Chersonesus.[26] If the law of ostracism had been in force at that time, Miltiades would have been subjected to its provisions.[27] If Miltiades' case came before a court, the conclusion is unavoidable that this was the court of the Areopagus.[28]

---

23. The text of the Solonian amnesty as reported by Plutarch should not be altered; see, however, Bonner and Smith 104–105. A. P. Dorjahn's assertion, in *Political Forgiveness in Old Athens* (Evanston 1946) 2, "also excluded were those who went into exile in the time of tyranny," is based on a faulty translation.

24. H. Swoboda, *Arch.-Ep. Mitt.* 16 (1893) 57–63 has tried to avoid this conclusion, and Usteri (*supra* n. 21) 40–41, no. 23, has followed his lead. It may, indeed, be surprising to find the Areopagus, which must have been packed with friends of the tyrants by 510 B.C. (Bonner and Smith 185–186), issue a verdict against Hippias and his family; but Aristotle comments specifically (22.4) on the mild treatment accorded to the friends of the tyrants (*supra* n. 22).

25. The story is repeated by Marcellinus (*Life of Thucydides* 13); Nepos (*Miltiades* 8), commenting on Miltiades' trial in 489 B.C., referred to his tyranny in the Chersonesus.

26. The doubts expressed by H. Berve, *Miltiades* (Berlin 1937) 23–24 and 66–67, were rightly rejected by A. V. Blumenthal, *Hermes* 72 (1937) 476–477; see also Lenschau (*supra* n. 21) 1808. Also erroneous in J. A. R. Munro's interpretation of the trial in *CAH* 4 (1939) 232.

27. See H. Schaefer in *Synopsis* (Festgabe für Alfred Weber, 1948) 491–492.

28. Notice the different terminology used by Herodotus in reporting Miltiades' trials before a court in 493 B.C. (6.104.2) and before the assembly in 489 B.C. (6.136.1). See, however, Bonner and Smith 197–198 and 299. The possibility exists, however, that it was the court of the prytaneum (consisting perhaps of the nine archons, also called prytanies; see Jacoby [*supra* n. 3] 369, n. 84), not that of the Areopagus, which judged cases of tyranny; see, in addition to Busolt and Swoboda,

It now appears that the law against tyrants as formulated by Solon and administered by the Areopagus was in force at least until the battle of Marathon. It was superseded by the law against the overthrow of democracy which was passed in 410 B.C.[29] There is no evidence to show that the Solonian law was used between 490 and 410 B.C., nor that it was abrogated or altered, or that its administration was transferred from the Areopagus to a popular court (as Bonner and Smith assumed, 197–198 and 299). The first ostracism took place in 487 B.C. and ostracism was used for the last time in 415 B.C. (*TAPhA* 79

(*supra* n. 2) 792–793, and Bonner and Smith 63–66 and 104–108, W. W. Hyde, *AJPh* 38 (1917) 167–175, and especially G. Smith, *CP* 16 (1921) 345–353. In this case, it would be easy to understand why the tyrants coveted the archonship, and why, after their expulsion, the electoral law concerning the archons was changed. This possibility also suggests a new interpretation of the Xanthippus ostrakon, published in *AJA* 51 (1947) 257–262 [*infra*, pp. 108–115]; see also *AJA* 52 (1948) 341–343 and 53 (1949) 266–268; J. Robert and L. Robert, *REG* 63 (1950) 25, no. 65. The prytaneum mentioned in the epigram may have been the court which used to deal with cases of tyranny.

29. See Hyperides, *For Euxenippus* 7–8, in the edition of V. de Falco (Naples 1947) 71–73 (with notes). Bonner and Smith have argued successfully (302–305) that this law must be dated in 410 B.C., but they have been less successful in their treatment of the preceding period, mainly because they did not separate the trials for tyranny from other legal actions, and because they did not consider the law of ostracism in this connection. The evidence presented by them may, therefore, be briefly reexamined.

> A. Isagoras is known to have tampered with the Athenian constitution (Herodotus 5.71.1; Aristotle 20.3), but his followers were tried and executed as traitors. The scholium on Aristophanes' *Lysistrata* 273 refers to the treason trial against the followers of Isagoras on the occasion of the next Spartan invasion which went only as far as Eleusis (Herodotus 5.74.1); see, however, P. Usteri (*supra* n. 21) 52–53 (d) and Bonner and Smith 199–200. Aristotle's remark (20.1) that Isagoras was a friend of the tyrants seems to be derived from the report of Herodotus (5.74.1), which was also repeated by the scholiast who observed that Cleomenes went against Athens in order to establish tyranny; see also Jacoby (*supra* n. 3) 337, n. 40.
> B. Hipparchus was convicted for treason with Persia after he had been ostracized for his dealings with Hippias; see Lycurgus, *Against Leocrates* 117; Bonner and Smith 199; Usteri 53 (e).
> C. Themistocles was living in exile because of his ostracism (the reason of which we do not know) when he was condemned for treason with Persia; see Thucydides 1.135.3, 138.6; Usteri 53–55 (f).
> D. Antiphon, who had played a leading part in the oligarchic revolution of 411 B.C. (Thucydides 8.90.1–2), was condemned and executed for treason with Sparta; see *Life of Antiphon* 11, 22–24; Bonner and Smith 2 (Chicago 1938) 27; Jacoby (*supra* n. 3) 208–209.

All of these condemnations were pronounced by a popular vote, but in none of them was there any question of tyranny or sedition. Certain other cases tried during the fifth century must also be left out of consideration, because they obscure the working of the Athenian law against sedition; see Bonner and Smith 299–304.

> E. Neither Phrynichus's presentation on the stage of the Fall of Miletus nor the failure of Miltiades' Parian expedition had anything to do with the overthrow of the constitution; see Bonner and Smith 197. It is reasonable to assume that the action against Miltiades (Herodotus [6.136]) followed the procedure described by Aristotle 61.2, by which commanding generals could be deposed and tried.
> F. Alcibiades was rumored to be engaged in seditious activities (Thucydides 6.28.2), but he was condemned on the charge of impiety; see Plutarch, *Alcibiades* 22.3; Bonner and Smith 301 and 304.

All these cases were tried by a popular court, but in none of them was there any question of subversive activities. They should, therefore, not be used as evidence that the law against tyrants was ever administered by a popular court.

[1948] 191–210 [*infra*, pp. 116–131]). Evidently, the history of ostracism fills precisely the gap between the earlier administration of the law against tyrants and the later use of the Athenian sedition act of 410 B.C. There was no place in Athenian legal procedure for the law of ostracism as long as the administration by the Areopagus of the law against tyrants was considered satisfactory. Nor is it possible to assume that the administration of this law was transferred to the popular assembly, because such a transfer would have made the law of ostracism entirely superfluous.

After the battle of Marathon, however, the old law against tyrants seemed to be singularly ineffective. In its long history it had never prevented the rise of tyranny, and recent events indicated that it was not even suitable to forestall the restoration of tyranny. Its main defect, as that of all sedition acts, lay in the simple fact that it was concerned with a crime which should be prevented rather than punished; it was a law which could not be justly administered as a preventive measure, that is, before the crime was actually committed. The law of ostracism avoided all the failures of the earlier law without necessitating its abrogation. There was no need of enforcing the old law as long as ostracism removed periodically those men who might have become violators of its provisions. It now becomes clear that the changing attitude toward subversive activities (symbolized by the terms "tyranny" and "overthrow of democracy") is reflected by the history of ostracism. At first it was used as a substitute for the law against tyrants,[30] but soon afterward men were ostracized because they opposed what was considered the policy supported by the majority of the electorate. It was felt, to an ever increasing degree, that opposition to this policy might (or must) ultimately lead to the overthrow of the particular form of democracy current at the time. The ostracism of Thucydides (443 B.C.) marked the end of this period. The circumstances surrounding the ostracism of Hyper|bolus showed clearly that the law of ostracism as preventive measure had outlived its usefulness. The experience of the oligarchic revolution of 411 B.C. demanded a new solution of the problem.

SUMMARY

The satisfactory operation of the Athenian law against tyrants depended, as does that of any law, on the composition of the court administering it and on the integrity of its members. The tyranny of Pisistratus could not have been avoided for the simple reason that it was supported by the people. Yet several attempts were made to get rid of the tyrant; two of them were crowned with success. After Pisistratus's death the composition of the court dealing with tyrants, rather than the integrity of its members, prevented any legal action from being

---

30. Notice that those convicted for tyranny under the old law were excluded from the amnesty of 481 B.C., while those who had been ostracized before that date were recalled on the motion of Themistocles. The phrase in the amnesty law of 405 B.C. (Andocides 2.78) referring to those condemned for tyranny must have been repeated from the amnesty of 481 B.C. where it referred specifically to the surviving members of Hippias's family.

taken. After the expulsion of Hippias, which had been the result of Spartan intervention, the already exiled members of the tyrant's family (including his father and brother who were dead) were condemned by court action. The old law against tyrants remained accordingly in force until at the time of the battle at Marathon a new situation was created by the activities of Hippias in the Persian camp and of his followers and supporters in the city of Athens. The old law did not seem to apply.

The urgency of the problem was increased by the attitude of some conservative Athenians who may have preferred, though with misgivings, the restoration of tyranny to the continuation of a form of government which relied to an ever increasing degree on popular support. Thus the law of ostracism, instituted at that time by the aged Cleisthenes, was designed as much to protect the young democratic institutions as to prevent the restoration of tyranny. In order to achieve the first mentioned aim, the legislator considered the danger necessarily inherent in the great power of the generals; their direct election did not and could not ensure the elimination of men who might become dangerous as supporters of tyranny or oligarchy. In fact, the very qualifications necessary for the generalship inevitably brought to power the leading members of the aristocracy. This condition remained in force until the Peloponnesian war when the rise of politicians to positions of military command contributed to the defeat of Athens without preventing oligarchic revolutions. Ostracism, however, became superfluous.

If looked upon in this way, ostracism worked surprisingly well. It eliminated during its time of operation the threat of tyranny, and it served as a check on the great power wielded by the political generals of the fifth century. Moreover, it ensured a vigorous, consistent, and progressive foreign and domestic policy by providing for the temporary removal of men who opposed the political trend of the time, and who otherwise would have been able to throw Athens into confusion and into continuous internal struggles. The accomplishments of the Periclean age are the best vindication of ostracism.

This study has resulted in the demonstration that the law of ostracism was enacted shortly before the first ostracism was held (487 B.C.). The literary evidence, including Aristotle's account in the *Constitution of Athens* (22), favors this date, while general considerations make a date immediately after the expulsion of Hippias very doubtful. Moreover, the close relationship between the vote of ostracism and the election of generals makes it virtually certain that the law of ostracism must be later than the establishment of the board of generals (501 B.C.). Finally, the law of ostracism was the second of three successive attempts to deal with sedition, and as such its introduction must be later than the last dated use (493 B.C.) of law which it replaced.

11

A Late Byzantine Account of Ostracism[1]

For Henry T. Rowell

I

Vaticanus Graecus 1144, a parchment codex of *saec.* XV,[2] contains on foll. 215ᵛ–225ᵛ a curious collection of gnomai, apophthegmata, and historical material. The collection has been published,[3] but publication was in an obscure Academy journal and the material has almost totally escaped notice. In particular, the item which I print here has played no part in the lengthy bibliography on ostracism.[4] The republication is intended to initiate discussion and my own remarks will be limited to a brief commentary.

fol. 222ʳᵛ

(1) Κλεισθένης τὸν ἐξοστρακισμοῦ νόμον ἐς ᾿Αθήνας εἰσήνεγκεν.

(2) ἦν δὲ τοιοῦτος·

(3) τὴν βουλὴν τινῶν ἡμερῶν σκεψαμένων (σκεψαμένην Sternbach)

American Journal of Philology 93 (1972) 87–91 [with J. J. Keaney]

---

1. The two parts of this article discuss the text from different approaches: the first part is by J. J. K., the second by A. E. R.
2. On the MS see now P. Canart and V. Peri, *Sussidi bibliografici per i manoscritti greci della Biblioteca Vaticana = Studi e Testi*, 261 (Città del Vaticano 1970), pp. 543–4.
3. By Leo Sternbach, "Gnomologium Parisinum ineditum, Appendix Vaticana," *Rozprawy Umiejetnosci Wydzial Filologiczny*, Serya II, Tom. V. (Krakowie, 1894), pp. 135–218 (App. Vat. on pp. 171–218). The edition is not very reliable. The portions of the historical material which deal with Alexander were republished by Sternbach in *Wiener Studien* 16 (1894), pp. 8–37.
4. Cf. A. E. Raubitschek, "Theophrastos on Ostracism," *Classica et Mediaevalia* 19 (1958), pp. 73–109 [*infra*, pp. 81–107]; W. R. Connor and J. J. Keaney, "Theophrastus on the End of Ostracism," *A. J. P.* 90 (1969), pp. 313–19; J. J. Keaney, "The Text of Androtion F 6 and the Origin of Ostracism," *Historia* 19 (1970), pp. 1–11; G. R. Stanton, "The Introduction of Ostracism and Alcmeonid Propaganda," *J.H.S.* 90 (1970), pp. 180–3; and R. Meiggs and D. Lewis, *A Selection of Greek Historical Inscriptions* (Oxford, 1969), pp. 40–7.

(4) ἐπιγράφειν ἔθος ⟨ἦν suppl. Sternbach⟩ εἰς ὄστρακα
(5) ὅντινα δέοι τῶν πολιτῶν φυγαδευθῆναι
[88] (6) καὶ ταῦτα ῥίπτειν εἰς τὸ τοῦ βουλευτηρίου περίφραγμα.
(7) ὅτῳ δὲ ἂν ὑπὲρ διακόσια γένηται τὰ ὄστρακα
(8) φεύγειν ἔτη δέκα,
(9) τὰ ἐκείνου καρπούμενον.
(10) ὕστερον δὲ τὸν δῆμον (τῷ δήμῳ Sternbach) ἔδοξε νομοθετῆσαι
(11) ὑπὲρ ἑξακισχίλια γίνεσθαι τὰ ὄστρακα τοῦ φυγαδευθῆναι μέλλοντος.[5]

(1) The specific attribution of the institution of ostracism to Cleisthenes is found only thrice elsewhere: in Aristotle, *Ath. Pol.*, 22, 1 and 3, in Aelian, *V. H.*, 13, 24, and in two lexical passages (cf. Philochorus, 328 F 30) which, I have argued,[6] are to be attributed to Philochorus, whose source was Androtion.
(2) Cf. Sch. Ar., *Eq.*, 855: τοιοῦτος ὁ τρόπος τοῦ ὀστρακισμοῦ and Diod., XI, 55, 1: ὁ δὲ νόμος ἐγένετο τοιοῦτος.
(3) Cf. Bekker, *Anecdota Graeca*, I, p. 248, 7: Ἐκφυλλοφορῆσαι· ... ἐσκόπει ἡ βουλὴ περὶ αὐτοῦ.
(6) Cf. Tzetzes, *Chil.*, 13, 449: ἐρρίπτουν εἰς Κυνόσαργες. The only other source to locate the procedure at the βουλευτήριον is Theodorus Metochites, *Miscellanea*, p. 609, Müller-Kiessling: ὡς ἤθροιστο μὲν ὁ δῆμος παντόθεν εἰς τὸ βουλευτήριον. Other sources say either that the agora was enclosed (e.g. Philochorus, F 30) or (correctly) that part of the agora was enclosed (e.g. Pollux, 8, 18). Close to the language of the MS is Plutarch, *Vit. Aristid.*, 7, 4: ἕνα τόπον τῆς ἀγορᾶς περιφραγμένον ἐν κύκλῳ δρυφάκτοις.
(7) W. R. Connor points out that σ' (two hundred) could readily be confused with ς (six thousand).
(9) Cf. Phil., F 30: καρπούμενον τὰ ἑαυτοῦ.
(11) Cf. Pollux, 8, 18: τοῦ μέλλοντος ἐξοστρακίζεσθαι. Unless the statement is overly compressed, the author holds the (mistaken) view that six thousand votes against a person were required for ostracism. Cf. the language of Timaeus, *Lex. Plat.*, s.v. Ἐξοστρακισμός· ... τούτων (sc. ὀστράκων) ὑπὲρ ἑξακισχίλια γενομένων. Timaeus supports the view that a quorum of six thousand was required.

[89] The account is drawn from no extant account of ostracism, but the parallels (a list of which could be extended) show affinities with several other accounts.[7] On the other hand, there are some details which are not found elsewhere. These are:

---

5. Sternbach's corrections are easy and—it would seem—necessary.
6. *Historia* 19 (1970), pp. 7–8. Cf. also Sch. Ar., *Eq.*, 855.
7. For these accounts, see Raubitschek and A. Calderini, *L'Ostracismo* (Como, 1945), pp. 99–130.

(3) that initially ostracism was conducted entirely by the βουλή;[8]
(6) that ostraka were thrown into an enclosure at the *bouleutêrion*;[9]
(7) that more than two hundred votes were initially required for ostracism;
(10–11) that later the *dêmos* increased the number of votes required.

The accuracy of these details depends, of course, upon the quality of their ultimate source: this source can be identified neither by content nor by context. The account (no. 213 Sternbach) does not occur within the chronologically organized historical material (nos. 153–194: from Homer to Alexander; for Athenian history, the major but not the sole source is Herodotus), but as one of a group of *kulturgeschichtliche* items which have no organization. Apart from the historical material just mentioned, the account is the only one which has to do with Athenian history and the only one in this whole section of the MS which has to do with constitutional history.

It is a ready assumption that the author had conflated and distorted two or more sources.[10] One could, *exempli gratia*, | hypothesize that the author took [90] (1) from Aelian, confused (3–6) ostracism with *ekphyllophoria* (the process by which the *boulê* voted out one of its own members),[11] and took (8, 9, 11) from one of various possible sources. But this kind of hypothesis appears desperate, especially when it comes to details. Although the author could have known of *ekphyllophoria* (from, e.g., the Suda, E 722 Adler) no extant description of that institution mentions a number of votes, and Tzetzes' account[12] does not even link *ekphyllophoria* with the *boulê*.

Whatever difficulties of text and interpretation subsist, the account has an internal consistency and has the appearance of being drawn from a single source.

II

This comprehensive though very compressed account of the early history of ostracism resolves at once the questions whether the law was instituted after the expulsion of Hippias or after the victory of Marathon, and whether or not Cleisthenes was its author. Evidently, the law was introduced (in the *boulê*?) by Cleisthenes in the short period between the end of tyranny and his own

---

8. In extant descriptions, the *boulê* and the archons preside over the procedure.
9. It is uncertain whether this detail refers to both stages in the development or only to the earlier stage.
10. That he used two sources for a single item is clear. E.g., no. 170 is a combination of Herod., 6, 96 and 120 and Diog. Laert., 3, 33. He is also capable of distorting his sources and perhaps even of invention. For curiosity's sake, I subjoin two items the sources of which cannot be identified: no. 212: Ξάνθος ὁ Λυδὸς τὰς Σάρδεις φησὶ καλεῖσθαι τὸ παλαιὸν Πάριον λόφον διὰ δὲ τὸ κάλλος ὀνομασθῆναι Σάρδεις· τὰ γὰρ διαφέροντα κατὰ τὸ εἶδος Σάρδεις καλοῦσιν, and no. 229: Ἐν τῷ Λυδίῳ Πακτωλῷ ψῆγμα χρυσοῦ καταφέρεται δαψιλῶς· γίνεται δὲ ἐν αὐτῷ λίθος θησαυροφύλαξ καλούμενος· δύναμιν δὲ ἔχειν φυλάττειν χρήματα, ὁσάκις ἂν φῶρες εἰσέλθωσι, σάλπιγγος ἦχον ἀποτελῶν.
11. The institutions of *ekphyllophoria* and ostracism are occasionally compared (see Raubitschek, p. 84, n. 5) but never confused.
12. *Chil.*, 13, 471–3. This work was available to the author.

exile, and it was administered by the *boulê* of Four Hundred. Its provisions were the same as those of the later law, except that a simple majority of the four hundred councillors determined the victim. After Cleisthenes returned and had his constitution enacted, nothing was said or done about the law of ostracism till the treason of Marathon raised the spectre of tyranny again, and all of our more detailed accounts of ostracism describe the working of the law as it was renewed immediately after Marathon.[13] Thus, Aristotle's two statements (*Ath. Pol.*, 22, 1 and 3) refer not only to two different periods in the history of ostracism but also to two different versions of the law.

It is also clear that the *boulê* of the Four Hundred existed until it was replaced by that of the Five Hundred, and that Herodotus (5, 72, 2) speaks of its activities at that very time.[14]

[91] Now that the law of ostracism is more firmly attributed to Cleisthenes and more securely dated in or shortly after 510 B.C., two other passages demand our attention.

(1) Ptolemaios Chennos, 6, 10 (ed. Chatzis [Paderborn, 1914)] from Photius, *Bibl.*, no. 190, p. 152a, lines 39–40 Henry: ὁ τὸν ὀστρακισμὸν ἐπινοήσας Ἀθήνησιν Ἀχιλλεὺς ἐκαλεῖτο, υἱὸς Λύσωνος. One may remember that Themistocles had such an advisor in Mnesiphilos of Phrearrioi (Herod., 8, 57–8; Plutarch, *Vit. Them.*, 2, 6, from a different source) and Pericles in Damonides of Oa (*Ath. Pol.*, 17, 4).

(2) Aelian, *V. H.*, 13, 24: Κλεισθένης δὲ ὁ Ἀθηναῖος τὸ δεῖν ἐξοστρακισθῆναι πρῶτος εἰσηγησάμενος, αὐτὸς ἔτυχε τῆς καταδίκης πρῶτος. F. Jacoby (on *F. Gr. H.*, 324F6): "if anybody can determine the source of Aelian, he may be able to solve the riddle." Aelian himself gives a clue (13, 23): λέγεται δὲ ὁ λόγος πρὸς τοὺς ἄλλα μὲν θελήσαντας, ἄλλων δὲ τυχόντας and the Suda (A 4101 Adler) the key: ὀστρακισθῆναι δὲ πρῶτον Ἀθήνησι Θησέα ἱστορεῖ Θεόφραστος ἐν τοῖς πρώτοις καιροῖς, supplemented by Eusebius, *Chron.*, p. 50 Schoene: πρῶτος ἐξωστρακίσθη, αὐτὸς πρῶτος θεὶς τὸν νόμον.[15] It was Theophrastus who wrote in his Πολιτικὰ τὰ πρὸς τοὺς καιρούς of the "ostracisms" of Theseus and Cleisthenes, referring to the exiles of these two democratic legislators as "ostracisms."

Theophrastus may in fact be the source of the new account of ostracism. His statement on ostracism in the Νόμοι has been recently reconstructed (see note 4), and the new information about the first law of ostracism would belong at the beginning of the systematic account of ostracism which is given by Philochorus, F 30.

---

13. *Historia* 9 (1959), pp. 127–8.
14. Thus the "grave doubts" expressed by J. Day and M. Chambers, *Aristotle's History of Athenian Democracy* (Berkeley and Los Angeles, 1962), pp. 200–1, are unjustified.
15. Raubitschek, p. 78, n. 3 [*infra*, p. 84, n. 3].

12

Theophrastos on Ostracism

PREFACE [73]

Students of Athenian history of the fifth century use not only the available primary evidence, namely Herodotos, Thucydides, Xenophon, the poets, orators, and essay writers of the fifth and fourth centuries, and the contemporary inscriptions and monuments. They are obliged to consider also the secondary evidence, namely the historians and antiquarians of the fourth century. Unfortunately, their books are lost to us, and the scattered fragments (though admirably assembled and critically discussed by F. Jacoby, *F. Gr. Hist.*) provide neither consecutive accounts nor even a clear notion of the general composition. To fill this important gap in our knowledge, students have to examine what may be called the tertiary evidence, namely the works of later historians and biographers, and the various anonymous entries in the Scholia and the Lexika. While it is generally recognized that the historians and the biographers, especially Diodoros, Nepos, and Plutarch, made use both of the primary and secondary evidence, there is little agreement on the extent of this use, and hence on the reliability of their narrative. It is, however, the Scholia and the Lexika which confront the student of Athenian history of the fifth century with even greater problems, and it is to the better understanding of this evidence that the following study offers a contribution.

The collections of source material for the history of the fifth century include passages from the various Scholia and Lexika which are listed most conveniently in alphabetical order. The students who use such collections are in no position to judge the value of any one of these entries and they are prone to disregard information | which is available only from late and anonymous sources. Little [74] do they realize that the Lexika depend ultimately on the Scholia, especially for their historical content, and that in many cases the Lexika are simply collections of alphabetically arranged Scholia. Once this is understood, each historically interesting entry of a Lexikon may be examined in order to determine to which passage or passages of an ancient author it may have served as a commentary (see *Hesperia* 24, 1955, pp. 286–287; *Hermes* 84, 1957, pp. 500–501). The Scholia themselves, especially those of historical interest, go ultimately back to the great commentaries of the Hellenistic and early Roman periods; unfortunately, only fragments of these valuable commentaries are preserved. It is possible to assert, however, that the authors of these commentaries, above all the great Didymos, used for their historical interpretation the same works of the primary and secondary evidence as did the later historians and biographers. This means that the historical information contained in the Scholia and Lexika of the Byzantine period may guide us back to the historians and antiquarians of the fourth century. Their guidance will be especially reliable whenever the information they present agrees in general with the accounts of later historians and biographers; for there can be no doubt that such an agreement indicates a common source, and a good source at that. Final proof may be attained in those rare cases in which an account composed from Scholia, Lexika, and later writers can be associated with a specific, sometimes short statement attributed to one of the fourth-century authors themselves; see *Phoenix* 9, 1955, pp. 122–126 [*infra*, pp. 320–324]. This is not to say that information so reconstructed must be historically true or accurate, but it does mean that this information was considered to be true by one of the great writers of the fourth-century who had excellent evidence available and whose books were read by people who were well informed.

[75]   In the course of my study of Athenian ostracism[1] I soon realized | that the

---

1. Since this is the last of a series of preliminary studies which E. Vanderpool and I have devoted to various problems connected with ostracism, it may be helpful to present here a brief bibliography, beginning with Vanderpool's 'List' which was composed in 1946 and published in 1949 in a volume dedicated to the memory of Theodore Leslie Shear whose work on the Agora ostraka Vanderpool and I are privileged to continue.

   1. E. V., List of known ostraka, *Hesperia* Suppl. 8, 1949, pp. 408–412.
   2. A. E. R., The ostracism of Xanthippos, *A.J.A.* 51, 1947, pp. 257–262 [*infra*, pp. 108–115].
   3. A. E. R., Ostracism, *Archaeology* 1, 1948, pp. 79–82.
   4. A. E. R., The case against Alcibiades, *T.A.P.A.* 79, 1948, pp. 191–210 [*infra*, pp. 116–131].
   5. G. A. Stamires and E. V., Kallixenos the Alkmeonid, *Hesperia* 19, 1950, pp. 376–390.
   6. A. E. R., The origin of ostracism, *A.J.A.* 55, 1951, pp. 221–229 [*supra*, pp. 65–76].
   7. A. E. R., Ostracism, *Actes du Deuxième Congrès Intern. d'Épigraphie Grecque et Latines*, Paris 1952, pp. 59–74 (also *C.J.* 48, 1953, pp. 113–122) [*supra*, pp. 153–164].
   8. E. V., The ostracism of the Elder Alkibiades, *Hesperia* 21, 1952, pp. 1–8.
   9. E. V., Kleophon, *Hesperia* 21, 1952, pp. 114–115.
   10. A. E. R., Philinos, *Hesperia* 23, 1954, pp. 68–71.
   11. A. E. R., Kimons Zurückberufung, *Historia* 3, 1955, pp. 379–380.
   12. A. E. R., Philochoros Frag. 30 (Jacoby), *Hermes* 83, 1955, pp. 119–120 [*infra*, pp. 132–133].
   13. A. E. R., Damon, *Class. et Med.* XVI, 1955, pp. 78–83 [*infra*, pp. 332–336].
   14. A. E. R., Theopompos on Hyperbolos, *Phoenix* 9, 1955, pp. 122–126 [*infra*, pp. 320–324].
   15. A. E. R., Menon, *Hesperia* 24, 1955, pp. 286–289.
   16. A. E. R., Zur attischen Genealogie, *Rh. Mus.* 98, 1955, pp. 258–262.
   17. A. E. R., (H)ABRONICHOS, *Cl. Qu.* 2, 1956, pp. 199–200.
   18. A. E. R., Das Datislied, "*Charites*" *für E. Langlotz*, pp. 234–242 [*infra*, pp. 146–155].
   19. A. E. R., The Gates in the Agora, *A.J.A.* 60, 1956, pp. 279–282.

primary and secondary literary evidence was meager indeed, especially when compared with the later tradition. It also became clear that the meagerness of the primary evidence was not entirely the result of the accident of preservation, for the Athenian authors took the institution of ostracism for granted and mentioned its operation only in passing; the fourth oration of Andokides is a notable exception (see *T.A.P.A.* 79, 1948, pp. 191–210 [*infra*, pp. 116–131]). On the other hand, the meagerness of the secondary evidence was indeed caused by the fact that we no longer have the books of two authors who were read and used by the commentators and the biographers of later antiquity: Theopompos and Theophrastos.

It has become my conviction that Theopompos's account of the Athenian statesmen of the fifth century contained most of what we know of the political activities of Themistokles, Aristeides, Kimon, Thucydides, Perikles, Kleon, and Hyperbolos, and especially of the ostracisms which ended or interrupted the careers of most of these men (see *Historia* 3, 1955, pp. 379–380). It may seem idle to spe|culate about the sources of Theopompos, but there are at least two of which we can be sure: Stesimbrotos's book on Themistokles, Thucydides, and Perikles, which contained also a good deal about Kimon, and Aischines' various Socratic writings which certainly dealt at some length with Themistokles, Aristeides, and Alkibiades. In turn, Theopompos's treatment of the Athenian statesmen of the fifth century was used already by Aristotle (in his account of Kimon) and by many later authors and commentators. It is a fair assumption, therefore, that most pertinent information on the political activities of these men, and especially on their ostracism, goes ultimately back to Theopompos. Later authors may have enlarged upon details, moralized on character traits, or dramatized situations, but they did not invent the basic elements of their accounts.

[76]

While the biographical information connected with the ostracisms in which the famous Athenian statesmen of the fifth century were involved can be traced back to Theopompos, another source must be sought for the various systematic accounts of the institution of ostracism which can be found in the later writers. The intermediary source of part of this evidence is actually known. Didymos, in his commentary on Demosthenes (23, 205) identified the expulsion of Themistokles, mentioned by the orator, as ostracism and gave a general account of ostracism which he quoted from the Atthis of Philochoros; three versions of this quotation are preserved (*F. Gr. Hist.* 328 F 30), two from Lexika and one from a papyrus containing the alphabetized commentary of Didymos on Demosthenes, 23. It is possible to associate with Didymos's statement on ostracism virtually all the accounts of ostracism found in the various Scholia and Lexika, and to assume that they all ultimately derived from this source. This reconstruction provides not only a fairly clear picture of the scholiastic and lexicographic tradition but it enables us also to attribute to Philochoros a more comprehensive account of ostracism than the known quotation contains. A further contribution of this reconstruction is the discovery that the statements on ostracism made by Diodoros, Plutarch, and Pollux are in substantial

agreement with that attributed in Philochoros, so much so that it can be claimed
[77] that they are not independent of one another. | While it is possible that Pollux
may have used Philochoros, this is unlikely in the case of Plutarch (who refers
to him only in his *Theseus*) and virtually impossible in the case of Diodoros.
The suggestion can be made, therefore, that the agreement between these authors
and Philochoros is explained by the assumption that they relied, directly or
indirectly, on Philochoros's source. This source, at least in the treatment of
ostracism, was Theophrastos whose great work on *Laws* contained an account
of ostracism. Theophrastos himself was a pupil and collaborator of Aristotle,
and the statement on ostracism attributed to Philochoros contains not only a
phrase known to have been used by Theophrastos but many other phrases
which in form and content agree with passages in Aristotle's *Politics* and with
expressions used by another Aristotelian, Demetrios of Phaleron.

The claim that Philochoros used, in at least one instance, Theophrastos's
Laws is significant for our understanding of the entire literary evidence on
Athenian political institutions. If this claim should be justified, one may turn
with confidence to a new collection of all the statements on Athenian political
and legal institutions found in the various Scholia and Lexika. Such a collection
may be the basis for a reconstruction not only of Philochoros's *Atthis* but also
of Theophrastos's *Laws*. The following study of the tertiary evidence on
ostracism may, therefore, be useful not only in itself but also, if its method
and conclusions are accepted, as a model of similar studies on other Athenian
political institutions.

THEOPHRASTOS ON OSTRACISM

Aristotle—Theophrastos—Philochoros. Herbert Bloch has shown (*H.S.C.P.*
Suppl. 1, 1940, pp. 355–376: 'Theophrastus' Nomoi and Aristotle') "that the
correlations between Aristotle's *Politeiai*, the empirical section of the *Politics*,
and Theophrastus' *Nomoi*, are very close" (p. 376).[2] Topics treated more
[78] ge|nerally in the *Politics*, and in a specific but greatly abbreviated form in the
Athenaion Politeia, were fully presented in the *Nomoi*; see also F. Jacoby, *Atthis*,
pp. 210–211. One of these topics was ostracism (p. 358), and Bloch has been
able to present "A New Fragment of Theophrastus' Nomoi" dealing with
ostracism (pp. 357–361).[3] He observed that the statement in the Scholia on

---

2. To the passages from Isokrates and Aristotle discussed by Bloch (pp. 361–366) may be added
Isokrates, *Philippus*, 12, which seems to refer not to Plato but rather to Aristotle and his school.
3. Theophrastos mentioned ostracism in another of his books, Πολιτικὰ πρὸς τοὺς καιρούς,
for the meaning of this title, see L. Radermacher, *Artium Scriptores*, pp. 48–49, on no. 24. Bloch
himself (p. 358, note 1) has attributed to this work the account of the ostracism of Hyperbolos
which Plutarch read in Theophrastos (*Nicias* 11 = frag. 139 Wimmer; see also *Alcibiades* 13);
compare my comments in *Phoenix* 9, 1955, pp. 122–123 [*infra*, pp. 320–322]. Bloch considered
this attribution as certain because Theophrastos treated in the same work "the alleged ostracism
of Theseus" (p. 358, note, *in fine*). Since Theophrastos has been taken to task for this piece of
information (especially by K. Mittelhaus, "De Plutarchi Praeceptis Gerendae Reipublicae," Diss.
Berlin 1911, p. 32, and F. Jacoby, *Commentary on F. Gr. Hist.* 328 F 19, pp. 311–312), it may be
appropriate to clear him of the charge that he was "little concerned about history" (Jacoby).

Lucian, Timon, 30 (pp. 114-115 Rabe = frag. 26 H. Hager, Journ. of Philol. 6, 1876, pp. 6 and 23-24) ἐπὶ τούτου δὲ καὶ τὸ ἔθος τοῦ ὀστρακισμοῦ κατελύθη, | ὡς Θεόφραστος ἐν τῷ περὶ νόμων λέγει occurs, in a systematic treatment of ostracism, in the Scholia on Aristophanes, Equites, 855: ὁ δὲ τρόπος τοιοῦτος τοῦ ὀστρακισμοῦ. προεχειροτόνει ὁ δῆμος ὄστρακον εἰσφέρειν, καὶ ὅταν δόξῃ, ἐφράττετο σανίσιν ἡ ἀγορὰ καί κατελείποντο εἴσοδοι δέκα. δι' ὧν εἰσιόντες κατὰ φυλὰς ἐτίθεσαν ὄστρακον, ἐντιθέντες τὴν ἐπιγραφήν. ἐπεστάτουν οἵ τε θ' ἄρχοντες καὶ ἡ βουλή. ἀριθμηθέντων δὲ ἑξακισχιλίων, τοῦτον ἔδει ἐν δέκα ἡμέραις μεταστῆναι τῆς πόλεως. εἰ δὲ μὴ γένοιτο ἑξακισχίλια, οὐ μεθίστατο. οὐ μόνον δὲ 'Αθηναῖοι ὀστρακοφόρουν, ἀλλὰ καὶ 'Αργεῖοι καὶ Μιλήσιοι καὶ Μεγαρεῖς. σχεδὸν δὲ οἱ χαριέστατοι πάντες ὠστρακίσθησαν, 'Αριστείδης, Κίμων, Θεμιστοκλῆς, Θουκυδίδης, 'Αλκιβιάδης. μέχρι δὲ 'Υπερβόλου ὀστρακισμὸς προελθὼν ἐπ' αὐτοῦ κατελύθη, μὴ ὑπακούσαντος τῷ νόμῳ διὰ τὴν ἀσθένειαν τὴν γεγενημένην τοῖς τῶν 'Αθηναίων πράγμασιν ὕστερον.

[79]

Bloch rightly associated the statement about ostracism in Argos, Miletos, and Megara with a parallel passage in Aristotle's *Politics* (1302 b 18; for Megara reference may be made to 1304 b 35-39), and the statement about the "weakness" with a passage in Aristotle's *Rhetoric* (1360 a 25f.; the same sentiment was expressed also by Theopompos; see my comments in *Phoenix* 9, 1955, p. 125 b [*infra*, p. 323]); he could have referred the reader also to *Politics*, 1284 b 21-23, where Aristotle mentions the abuse of ostracism (ἀλλὰ στασιαστικῶς ἐχρῶντο τοῖς ὀστρακισμοῖς). Bloch reached the convincing conclusion that "Aristotle goes back in his remarks to the same material which was discussed afterwards by Theophrastus," and he suggested that "the second part of the scholion... represents a last extract from this chapter" (namely of Theophrastos's *Nomoi* on the subject of ostracism). It is possible to go beyond these conclusions and to claim that the entire Scholion (on *Equites*, 855, quoted

According to Suidas (s.v. 'Αρχὴ Σκυρία, no. 4101 Adler = frag. 131 Wimmer) ὀστρακισθῆναι δὲ πρῶτον 'Αθήνησι Θησέα ἱστορεῖ Θεόφραστος ἐν τοῖς πρώτοις καιροῖς; this agrees with Eusthatius, I, p. 782, lines 52ff., ὅτι ὀστρακισθῆναι πρῶτον 'Αθήνησι Θησέα ἱστορεῖ Θεόφραστος. To the same source, namely Theophrastos, may be attributed the statement by Suidas (s.v. Θησείοισιν, no. 368 Adler) μετὰ γὰρ τὸ χαρίσασθαι τὴν δημοκρατίαν τοῖς 'Αθηναίοις τὸν Θησέα, Λύκος τις συκοφαντήσας ἐποίησεν ἐξοστρακισθῆναι τὸν ἥρωα (which is repeated verbatim from the Scholia on Aristophanes, *Plutus*, 627), the Scholia on Aristeides, 46, 241, 9-11 (vol. 3, p. 688 Dindorf) Θησεὺς ὑπὸ Λύνου (sic) 'Αθήνησιν εἰς τυραννίδα συκοφαντηθεὶς ἐξωστρακίσθη τῆς πόλεως and a passage in Eusebius (*Chron.*, p. 50 Schoene) Θησεὺς, 'Αθηναίους κατὰ χώραν διεσπαρμένους εἰς ἓν συναγαγὼν ἤτοι εἰς μίαν πόλιν, πρῶτος ἐξωστρακίσθη, αὐτὸς θεὶς τὸν νόμον. We are dealing here with an event of Theseus's Life which is reported by various other authors, none of whom employs, however, the word "ostracize." There can be little doubt that Theophrastos did not mean to say that Theseus was ostracized; the word ὀστρακίζω is evidently used metaphorically. This is made clear by Theophrastos himself who says about Theseus in the *Characters* (26, 6): τοῦτον γὰρ ἐκ δώδεκα πόλεων εἰς μίαν συναγαγόντα ⟨τὸν δῆμον⟩ λῦσαι τὰς βασιλείας. καὶ δίκαια αὐτὸν παθεῖν, πρῶτον γὰρ ἀπολέσθαι ὑπ' αὐτῶν, see H. Herter, *Rh. Mus.* 88, 1939, p. 311. Evidently, Theophrastos considered Theseus the founder and the first victim of democracy; the same view is also expressed by Aristeides, 46, 241 (p. 315 Dindorf) who refers (ὡς σύ που φῇς) to Plato, *Menexenus*, 238 C. Later authors, and especially the lexicographers, were puzzled by Theophrastos's use of the word ὀστρακίζω; some authors evidently misunderstood the statement and assumed that Theseus was actually ostracized and that he enacted the law.

previously) goes back to Theophrastos, through the medium of Philochoros; see F. Jacoby's introduction to Philochoros, *F. Gr. Hist.* 328 (p. 231).

Bloch insisted (p. 359) that the sentence listing the various victims of ostracism among the χαριέστατοι "is an insertion of the scholiast because the order... is wrong" and "the mention of Alcibiades in the place of Hyperbolus" is "the best proof for the spurious character of the cancelled passage." The list is certainly not in chronological order, but there is no reason why it should be; [80] nor is it true | that Alkibiades is mentioned in place of Hyperbolos, since Alkibiades (the Elder) was ostracized (see E. Vanderpool, *Hesperia* 21, 1952, pp. 1–8; A. E. Raubitschek, *Rh. Mus.* 98, 1955, p. 260, note 4), while Hyperbolos certainly was not one of the χαριέστατοι. Moreover, the following sentence which emphasizes the unique character of Hyperbolos's ostracism (μὴ ὑπακούσαντος τῷ νόμῳ) requires an antecedent describing the kind of people who were normally ostracized. Finally, the occurrence of the word χαριέστατοι should have discouraged Bloch from canceling the sentence in which it stands. This word is used by Plato, Isokrates, and Aristotle with reference to the better sort of people, the upper class, or, generally speaking, to gentlemen. It is significant that the view expressed here, namely that the victims of ostracism belonged to distinguished families, can also be found in a passage specifically attributed to Demetrios of Phaleron (*F. Gr. Hist.* 228 F 43).[4] We may conclude that the second half of the Scholion (*Equites*, 855) in its entirety (beginning with οὐ μόνον...) can be attributed to Theophrastos's account of ostracism.

The question arises whether the first part of the Scholion (*Equites*, 855, quoted above) does not also go back to Theophrastos's *Nomoi*. Bloch noted [81] (p. 358) that this first part "agrees with a quotation from | Philochoros" (*F. Gr. Hist.* 328 F 30); but although he recognized (p. 359, note 1) that its last sentence (μόνος δὲ Ὑπέρβολος...·μετὰ τοῦτον δὲ κατελύθη τὸ ἔθος...) "reflects the quotation from Theophrastos," he insisted that it "obviously has nothing to do with the previous quotation from Philochoros." Jacoby (*F. Gr. Hist.* 328 F 30, p. 107, line 24; Commentary, p. 316, lines 8–13; Notes, p. 228, note 5), without referring to Bloch, excluded the last sentence from the Philochoros

---

4. Bloch himself (p. 357 and note 3) emphasized the relationship between the Laws of Demetrios and the *Nomoi* of Theophrastos, and there can be little doubt that both men held the same political views and hence would agree on the significance of ostracism. Demetrios is credited by Plutarch (*Aristides*, 1, 2–3 = *F. Gr. Hist.* 228 F 43 = frag. 95 Wehrli) with the statement that ostracism was used against men who belonged to families which were great and envied because of the influence of their members. To the same context belong two other passages in Plutarch which may also be attributed to Demetrios's book on Socrates. In the *Aristides* (1, 7), Plutarch says that all those fell victim to ostracism who were thought to be above the mass either on account of reputation, or family, or ability to speak; hence Damon, the teacher of Perikles, was ostracized because he seemed to be outstanding in his thinking. The same idea is expressed by Plutarch in the *Nicias* (VI 1) where he says that the demos was constantly suspicious toward those who were powerful and curtailed their thinking and reputation; this was made clear by the conviction of Perikles, by the ostracism of Damon. Both passages elaborate Aristotle's more general statement (*Politics*, 1284a 17–23), and their connection with Sokrates (and hence with Demetrios's book on Sokrates) is provided by the reference to Damon (see also Plutarch, *Pericles*, 4, 1–2) whose case was used in the accusation against Sokrates; see my comments in *Class. et Mediaevalia* 16, 1955, pp. 78–79 [*infra*, pp. 332–333], and, below, Appendix IV.

quotation, although he saw that this sentence must have been preceded by the list of ostracized preserved in the Scholion on Aristophanes, *Equites*, 855; he discredited, however, the value of this list by claiming (by sheer oversight) that the name of Thucydides is missing. Instead of assuming that the quotation of Philochoros contains a final sentence which does not belong to it (but may go back to Theophrastos) and that the Scholion (*Equites*, 855) combines material taken from Philochoros and Theophrastos, it may be suggested that Philochoros's account of ostracism is based on the treatment which Theophrastos had given this subject in his *Nomoi*. This would mean that the Scholion in its entirety goes back to Philochoros and, possibly, to Theophrastos.

The transmission of the quotation from Philochoros (*F. Gr. Hist.* 328 F. 30) and its text may be briefly reexamined. Two almost identical texts are preserved, one in the Lexicon Rhetoricum Cantabrigiense (ed. A. Nauck, *Lexicon Vindobonense*, 1867, pp. 354–355 = E. O. Houtsma, 1870, pp. 23–24), the other in the fragmentarily preserved work of the lexicographer Claudius Casilo (or Claudius Alexandros, the son of Kasilo; see *R.E.* Suppl. I, col. 318) found in a codex from Mount Athos (ed. M. E. Miller, *Mélanges de littérature Grecque*, 1868, p. 398); the text of this version may be repeated here: ὀστρακισμοῦ τρόπος· Φιλόχορος ἐκτίθεται τὸν ὀστρακισμὸν ἐν τῇ τρίτῃ γράφων οὕτω· προχειροτονεῖ μὲν ὁ δῆμος πρὸ τῆς ὀγδόης πρυτανείας, εἰ δοκεῖ τὸ ὄστρακον εἰσφέρειν· ὅτε δὲ δοκεῖ ἐφράσσετο σανίσιν ἡ ἀγορά, καὶ κατελείποντο εἴσοδοι δέκα, δι' ὧν εἰσιόντες κατὰ φυλὰς ἐτίθεσαν τὰ ὄστρακα στρέφοντες τὴν ἐπιγραφήν· ἐπεστάτουν δὲ οἵ τε ἐννέα ἄρχοντες καὶ ἡ βουλή. διαριθμηθέντων δὲ ὅτῳ πλεῖστα γένοιτο, καὶ μὴ ἐλάττω ἑξακισχιλίων, τοῦτον ἔδει τὰ δίκαια δόντα καὶ λαβόντα ὑπὲρ τῶν ἰδίων συναλλαγμάτων ἐν δέκα ἡμέραις μεταστῆναι τῆς πόλεως | ἔτη δέκα, ὕστερον δὲ ἐγένοντο πέντε, καρπούμενον τὰ ἑαυτοῦ, μὴ ἐπιβαίνοντα εἰς τὸ (ἐντός codd.; see Appendix II) πέρα[ς] τοῦ Εὐβοίας ἀκρωτηρίου. Μόνος δὲ Ὑπέρβολος ἐκ τῶν ἀδόξων ἐξωστρακίσθη (διὰ [- - -] ἐξοστρακισθῆναι Lex. Rhet. Cant.) διὰ μοχθηρίαν τρόπων, οὐ δι' ὑποψίαν τυραννίδος· μετὰ τοῦτον δὲ κατελύθη τὸ ἔθος, ἀρξάμενον νομοθετήσαντος Κλεισθένους, ὅτε τοὺς τυράννους κατέλυσεν, ὅπως συνεκβάλοι καὶ τοὺς φίλους αὐτῶν.

[82]

The second version, also attributed to Philochoros, is preserved on a papyrus which contains alphabetically arranged notes on Demosthenes' speech against Aristokractes (23) in which a reference to the exile (i.e., ostracism) of Themistokles occurs (23, 205); this papyrus was published by F. Blass (*Hermes* 17, 1882, p. 152, lines 28–39, with comments on pp. 159–160) and republished by H. Diels and W. Schubart (*Didymi de Demosthene comm. cum anonymi in Aristocrateam lexico*, 1904, p. 47 = *Berliner Klassikertexte*, 1, 1904, pp. 81–84). Jacoby, beyond recording its text in his apparatus and saying that it (as well as the first version) goes "almost certainly" back to Didymos, has not paid much attention to this older, longer, and better version, which is unfortunately very fragmentary. The text of the lemma itself may be repeated here, with certain small changes and one more important one which will be discussed below.

καὶ Φιλόχο-
[ρος ἐν τῆι γ' τῆς 'Ατθίδος οὕτως φη]σίν· ὁ δ'ὄστρα-
[κισμοῦ τρόπος· προχειροτον]εῖ [μὲν ὁ δῆμος]
[πρὸ τῆς η' πρυτανείας, εἰ δοκεῖ τὸ ὄστρακον]
[εἰσφέρειν· ὅτε δὲ δοκεῖ, ἐφράσσετο σανίσιν ἡ ἀγο]-
[ρά, καὶ κατελείποντο εἴσοδοι ι', δι] ὧν [εἰσ]ερχόμε-
[νοι οἱ πολῖται κατὰ φυλὰς] κα[τ]ετίθεσαν τὸ
[ὄστρακον καταστρέφοντες τ]ὴν ἐπιγραφήν· ἐπε-
[στάτουν δ'οἵ τε θ ἄρχοντες κ]αὶ ἡ βουλή· διαριθμη-
[θέντων δέ, ὅτωι πλεῖστα γέ]νοιτο, καὶ εἰ μὲν ἑ-
[ξακισχιλίων μὴ ἐλάττω, τοῦτ]ον ἔδει, τὰ δίκαια δόν-
[τα καὶ λαβόντα...

[83] The important new restoration is in lines 9–10, where Blass restored εἰ μὲν ἐ[λάττω, ζ, ἀτελὴς ἦν· εἰ δὲ μή], Diels and Schubart εἰ μὲν | ἔ[φερε πλεῖον τῶν, ζ], assuming that the losing candidate had to receive not less than six thousand votes. I have argued (*Hermes* 83, 1955, pp. 119–120 [*infra*, pp. 132–133]) that Philochoros speaks of a quorum of six thousand, and the restoration of the papyrus suggested here would confirm this interpretation while retaining the vocabulary employed by Philochoros. In this sentence, as in lines 6 and 7, the version of the papyrus is somewhat fuller, thus confirming Jacoby's view (Notes, p. 228, note 4) that "the quotation gives the impression of being verbatim, but it seems to be abbreviated."

If Philochoros's account of ostracism is (at least in part) based on Theophrastos's *Nomoi*, and if Aristotle's *Athenaion Politeia* relies upon the same collection of material as Theophrastos's *Nomoi* (see Bloch, pp. 367–376), it may be presumed that Aristotle's statement on ostracism (*Ath. Pol.*, 43, 5: ἐπὶ [δὲ] τῆς ἕκτης πρυτανείας πρὸς τοῖς εἰρημένοις καὶ περὶ τῆς ὀστρακοφορίας ἐπιχειροτονίαν διδόασιν, εἰ δοκεῖ ποιεῖν ἢ μή; this statement is repeated in the *Lexicon Rhetoricum Cantabrigiense*, p. 348 Nauck = p. 20 Houtsma = V. Rose, *Aristotelis...Fragmenta*, 436) belongs in the same context as the beginning of the quotation from Philochoros (*F. Gr. Hist.* 328 F 30: προχειροτονεῖ μεν ὁ δῆμος πρὸ τῆς ὀγδόης πρυτανείας, εἰ δοκεῖ τὸ ὄστρακον εἰσφέρειν). And, indeed, Jacoby (Commentary, p. 316) sought to combine the two statements, saying that Philochoros mentioned the "preliminary vote in the 6th, the final voting in the 8th prytany." The reference in both passages to the prytany and the agreement in tense and terminology allow the assumption that Aristotle included in his summary only the first sentence of a more extensive treatment of ostracism which was subsequently presented by Theophrastos in his *Nomoi* and repeated by Philochoros; in fact, the agreement between Aristotle and Philochoros confirms the assumption that the latter drew upon Theophrastos. It has been thought (e.g., by J. Carcopino, *L'ostracisme*[2], pp. 234–235; see also M. Ostwald, *T.A.P.A.* 86, 1955, p. 110) that the preliminary vote was still taken in the time of Aristotle, but we now know that the full account on ostracism as it was given by Theophrastos contained the statement that ostracism ceased to be used after the case of Hyperbolos.

Pollux. In addition to the evidence presented so far, there are other statements [84] on ostracism which can be traced back to Theophrastos's *Nomoi*. Bloch himself (pp. 358-359) referred to Pollux, VIII 19-20, and Jacoby reprinted part of his text (Commentary, p. 315, lines 15-22) without commenting on it. R. E. Wycherley has recently tried to amend it (*The Athenian Agora, III, Testimonia*, p. 164, no. 534; see *A.J.A.* 60, 1956, p. 279, note 2) without considering its source. Pollux gives in the preceding paragraph (8, 18-19) an account of ekphyllophoria which is significant because it occurs in a number of other passages and because ekphyllophoria and ostracism are compared with one another.[5] Then he says: κοινῇ μέντοι πᾶς ὁ δῆμος ὀστράκοις ἐψηφίζετο, καὶ τὸ ἔργον ἐκαλεῖτο ὀστρακοφορία, καὶ τὸ πάθος ὀστρακισμός, καὶ τὸ ῥῆμα ἐξοστρακίσαι καὶ ἐξοστρακισθῆναι. περισχοινίσαντας δέ τι τῆς ἀγορᾶς μέρος ἔδει φέρειν εἰς τὸν περιορισθέντα τόπον Ἀθηναίων τὸν βουλόμενον ὄστρακον ἐγγεγραμμένον τοὔνομα τοῦ | μέλλοντος ἐξοστρακίζεσθαι. ὅτῳ δὲ ἑξακισχίλια [85] γένοιτο τὰ ὄστρακα, τοῦτον φυγεῖν ἐχρῆν, οὐχ ὡς κατεγνωσμένον, ἀλλ᾽ ὡς τῇ πολιτείᾳ βαρύτερον, δι᾽ ἀρετῆς φθόνον μᾶλλον ἢ διὰ κακίας ψόγον.[6] The

---

5. The first part of Pollux's statement deals with a form of ekphyllophoria used by the rural judges but otherwise unknown; for the revision of the deme lists, see Jacoby on *F. Gr. Hist.* 324 F 52 and 328 F 119. The last sentence of Pollux's statement (καὶ ἡ βουλὴ δὲ τῶν πεντακοσίων φύλλοις ἀντὶ ψήφων ἐχρῶντο) is a summary of a much longer account. The fullest statement is found in 1. Bekker's *Anecdota Graeca* 1, 248; see also Suidas and the *Et. Mag.* s.v. ἐκφυλλοφορῆσαι; *Tzetzes Chiliades*, 13, 477-482. Ekphyllophoria is connected with a certain Xenotimos (on him, see Isokrates, 18, 11). Aischines refers to ekphyllophoria (1, 111-112), and the Scholia explain this reference, as does Harpokration (s.v. ἐκφυλλοφορῆσαι) who says that Deinarchos mentioned ekphyllophoria in his speech against Polyeuktos. According to Suidas and the *Et. Mag.* Demosthenes is said to have mentioned ekphyllophoria in his speech against Neaira (59), and I feel confident that this refers to 59, 14-15, where Apollodoros appears as synegoros in court, although he had been expelled from the council (59, 3 and 6). Demosthenes does not actually use the word ἐκφυλλοφορῆσαι, but this is the legal term for "expelling from the council." The entire evidence concerning ekphyllophoria resembles that concerning ostracism: passages in the orators (Aischines, Deinarchos, Demosthenes) are explained, probably by Didymos, on the basis of a handbook, probably the Atthis of Philochoros who himself used Theophrastos's *Nomoi*, and the commentary on the orators was then incorporated in the various dictionaries and encyclopedias. The fact that both Pollux and Tzetzes discuss ekphyllophoria and ostracism together indicates that they found the two similar institutions explained in the same or a similar source. This is confirmed by the fact that the lexicographers (see above) compare ostracism with ekphyllophoria: . . . ὥσπερ ἐπὶ ὀστρακισμοῦ ὀστράκοις (*An. Gr.*); ὥσπερ ὀστράκῳ ἐπὶ τοῦ ὀστρακισμοῦ (Suidas, *Et. Mag.*).

6. It does not seem to have been noticed that Pollux's account of ostracism is repeated in the Scholia on Olympiodoros's Commentary on Plato's *Gorgias* (p. 156,18 ed. Norvin): κοινῇ πᾶς ὁ δῆμος ὀστράκοις ἐψηφίζετο καὶ τὸ ἔργον ἐκαλεῖτο ὀστρακοφορία καὶ τὸ πάθος ὀστρακισμός· ἐγίνετο δὲ οὕτως· περισχοινίσαντές τι μέρος τῆς ἀγορᾶς ἔδει φέρειν εἰς τὸν περιορισθέντα τόπον Ἀθηναίων τὸν βουλόμενον ὄστρακον ἐγγεγραμμένον τοὔνομα τοῦ μέλλοντος ἐξοστρακίζεσθαι· ὅτῳ δ᾽ ἑξακισχίλια γένοιτο τὰ ὄστρακα, τοῦτον φεύγειν ἐχρῆν οὐκ ὡς κατεγνωσμένον ἀλλ᾽ ὡς τῆς πολιτείας βαρύτερον, δι᾽ ἀρετῆς φθόνον μᾶλλον ἢ διὰ κακίας ψόγον. It is possible to identify the scholiast (homo quidam haud indoctus, according to W. Norvin, *op. cit.*, p. vii) as Arethas whose Scholia on Plato contain two quotations from Pollux (see W. C. Greene, *Scholia Platonica*, p. 24, note 2) and who owned the archetype of our Pollux manuscripts (see E. Bethe, *Pollucis Onomasticon*, 1, p. vi); there is another quotation from Pollux in the *Scholia on Olympiodorus*

first sentence, whether it is a quotation or a rephrasing of Pollux's source, suggests that the original statement on ostracism did not begin with an account of the procedure (τρόπος). This is also indicated by the beginning of the quotation from Philochoros (ὁ δ'ὀστρα[κισμοῦ τρόπος]) which must have been preceded by a more general characterization of ostracism such as is found in Pollux. Pollux's account of the procedure of ostracism follows in general the statement attributed to Philochoros, as Bloch and Jacoby have noticed, but there are differences which deserve attention because they are not confined to this version. While the quotation from Philochoros and the Scholion (*Equites*, 855) say that the Agora was fenced in, Pollux speaks of some *part* of the Agora, and this restriction is not only correct but also repeated by Timaios and Plutarch (see below, pp. 91–92 and 98–100). Pollux also adds that on the ostrakon was written the name of the person whom the citizens wished to exile; this statement, which is missing from the other two more detailed versions, is not only correct
[86] and necessary but also repeated | in various other sources (see below, pp. 91–93). Finally, Pollux concludes his account by saying that ostracism was not a punishment for a crime but directed against men who were unbearable; it was an expression of envy for virtue rather than of censure for wickedness. This general observation not only agrees with the characterization of ostracism given by Aristotle, Theophrastos, and Demetrios, but also with a number of passages to be discussed separately (see below, pp. 96–97). Thus our knowledge of ostracism has been enriched by the examination of Pollux, 8, 19–20, who evidently drew upon the full text of Theophrastos (?) which was available to him, but abbreviated it in a different way from the other excerptors, and presented it in his own terminology.

Tzetzes. At this point, mention may be made of Tzetzes' account of ostracism (*Chiliades*, 13, 441–455) which, as far as I can see, has not been taken into consideration in any modern treatment of the institution.

> Ὁπόθεν τὸ ὠστράκισεν ἐλέχθη, μάνθανέ μοι,
> Καὶ τὸ ἐφυλλοβόλησεν ὡσαύτως δὲ σὺν τούτῳ.
> Οἱ Ἀθηναῖοι μέλλοντές τινα ὑπερορίζειν,
> Οὐ πρότερον ἐξώριζον ἐκεῖνον τῆς πατρίδος,
> 445 Μέχρις ἂν συνετάξαντο ἡμέραν ὡρισμένην.
> Καὶ μέχρι ταύτης ἤκουσαν χιλίων κατηγόρων,
> Συνποσουμένων ἀριθμῷ τῶν κατ' αὐτοῦ λεγόντων.
> Εἰς ὄστρακον γὰρ γράφοντες τὸ ὄνομα ἐκείνου,

Commentary on Plato's *First Alcibiades*, p. 29 (ed. L. G. Westerink). This means that the best Olympiodoros manuscript which contains his commentaries on the *Gorgias*, the *Alcibiades*, and the *Phaedo* was owned by Arethas who added to it Scholia of his own. Norvin observed that most of these Scholia (he says, *loc. cit.*, "omnia") are taken from neo-Platonic commentaries on Plato; attention may be called, therefore, to the following significant exceptions: *In Plat. Gorg.*, p. 47, 15 = Hellanikos, *F. Gr. Hist.* 323 a F 1; p. 201,8 = Aristotle, *Ath. Pol.*, 43, 2–3; p. 235, 18 = Herodotos, 1, 47; *In Plat. Alc.*, p. 43 (65,5) = Herodotus, 7, 141, and 8, 96; p. 85 (129,13) = Aristophanes, *Ach.*, 530–534.

450   Οὗ καὶ τοὺς νόθους ἔρριπτον τοῖς χρόνοις τοῖς προτέροις.
      Ἐκεῖσε καὶ τὰ ὄστρακα ἐξοριῶν ἐρρίπτουν.
      "Ἂν οὖν ἡμέρᾳ τῇ τακτῇ χίλια ἐφευρέθη,
      Ἀσυμπαθῶς ἐστέλλετο πρὸς τὴν ὑπερορίαν.
      Εἰ δ'ἦσαν ἀποδέοντα τὰ ὄστρακα χιλίων,
455   Ἐν τῇ πατρίδι μένων ἦν, ἐπιτυχὼν συγγνώμης.

Tzetzes continues with an account of the ostracism of Aristeides (456–476) which follows in the main the story told by Plutarch (*Aristides*, 7, 5–6), and he concludes his account with a brief state|ment on ekphyllophoria which, however misunderstood, seems to go back to the same or a similar source as does Pollux's account (see note 5). Christian Harder assumed (*De Ioannis Tzetzae Historiarum fontibus...*, 1886, p. 60) that Tzetzes used Plutarch (*Aristides*, 7) for his account of ostracism, that he erroneously wrote χιλίων (in lines 446 and 454) for Plutarch's ἑξακισχιλίων (p. 14), and (pp. 42 and 49) that he added on his own the false information (lines 449–450, 465, 481) that ostracism took place in the Kynosarges, and the fanciful ending of the story of Aristeides' ostracism (lines 468–476). Apart from the mention of ekphyllophoria, which does not occur in Plutarch but is found in several lexicographers (see note 5) some of whom (Harpokration, Et. Mag., Suidas, Anecdota Bekker) compare this institution with ostracism, there is a significant statement in Tzetzes' account of ostracism which is not found in Plutarch or in any other source save Aristotle and Philochoros. This is the reference to the choice of a special day on which ostracism is to be held (lines 445–446, 452); for the statement of Aristotle-Philochoros on this point, see above, pp. 88–89. Since it seems unlikely that Tzetzes should have invented this detail, which happens to be correct, one must assume that he used a source other than Plutarch, or in addition to Plutarch, which contained the information in question. The same may also apply to the repeated references to the Kynosarges as the place of ostracism. What Tzetzes says about the Kynosarges can also be found in the Scholia to Plato's *Axiochus* (p. 409 Greene): τόπος ἐν τῇ Ἀττικῇ Κυνόσαργες καλούμενος, ἐν ᾧ τοὺς νόθους τῶν παίδων ἔταττον) and in Suidas (s.v. εἰς Κυνόσαργες). Perhaps Tzetzes knew that ekphyllophoria was practiced to separate the citizens from the nothoi, and that the nothoi were ordered to the Kynosarges, and he then assigned erroneously ostracism to the Kynosarges. At any rate, Tzetzes' treatment of ostracism is not derived solely from Plutarch, although it need not be considered as a source of independent evidence.

Didymos. The next passage to be discussed has been neglected by Bloch and Jacoby but mentioned by Wycherley (*loc. cit.*, p. 117, note 2, | *in fine*). Timaios (Lexicon vocum Platonicarum, s.v. ἐξοστρακισμός G. A. Koch) commented on ostracism because Plato had mentioned the ostracism of Kimon (*Gorgias*, 516 d): Ἐξοστρακισμὸς φυγὴ δεκαετής. πῆγμα δὲ γίνεται ἐν τῇ ἀγορᾷ εἰσόδους ἔχον, δι' ὧν εἰσιὼν πολίτης ἕκαστος ὄστρακον τίθησιν ἐπιγεγραμμένον· τούτων δὲ ὑπὲρ ἑξακισχίλια γενομένων, φυγὴ δεκαετὴς καταψηφίζεται τοῦ κρινομένου.

This is evidently a greatly abbreviated version of the account of Philochoros. Timaios (about him, see *R.E.*, s.v. Timaios, no. 8) agrees with Pollux (see above) in placing the enclosure in the Agora; moreover, he supports the assumption that Philochoros said that ostracism required a quorum of six thousand voters. Timaios's version is repeated by Zonaras (s.v. ἐξοστρακισμός) and, with a significant addition, in the *Etymologicum Magnum*, s.v. ἐξοστρακισμός (recently reprinted by A. Calderini, *L'ostracismo*, p. 117): πῆγμα δὲ γίνεται ἐν τῇ ἀγορᾷ εἰσόδους ἔχον, δι' ὧν εἰσιὼν πολίτης ἕκαστος ὄστρακον ἐτίθει ἐπιγεγραμμένον. ἑξακισχιλίων δὲ γινομένων, φυγῇ δεκαετὴς ψηφίζεται τοῦ κρινομέμου. ὠνόμασται δὲ ἀπὸ τοῦ ὀστράκου, εἰς ὃ ἐνέγραφεν ἕκαστος Ἀθηναίων εἰ δέον μεθίστασθαι τῆς πόλεως. ἐνομίσθη δὲ τοῦτο ἐπὶ τῷ τοὺς ὑπερέχοντας τοῖς φρονήμασι κωλύειν. ζήτει εἰς τὰς διαφοράς. There is no reason to assume that the last two sentences concerning the meaning of the word exostrakismos and concerning the significance of the institution go back to a different source, and it may be suggested that we have here another part of the original statement which has been omitted in the other versions; this is confirmed by a reference to Aristotle, *Politics*, 3, 1284 a 36–38: ὁ γὰρ ὀστρακισμὸς τὴν αὐτὴν ἔχει δύναμιν τρόπον τινὰ τῷ κολούειν τοὺς ὑπερέχοντας καὶ φυγαδεύειν. In fact, there is a further addition to which the *Et. Mag.*, refers by closing its account with the words ζήτει εἰς τὰς διαφοράς. The explanation of the "difference" between ostracism and exile is given in I. Bekker's *Anecdota Graeca*, 1, 285, a passage which conveniently repeats part of the text given in the *Et. Mag.*: ἔστιν οὖν ὁ ὀστρακισμὸς φυγῆς εἶδος, ὠνόμασται δὲ ἀπὸ τοῦ ὀστράκου εἰς ὃ ἐνέγραφεν ἕκαστος Ἀθηναίων εἰς ὁ δέων (sic) μεθίστασθαι τῆς πόλεως. ἐνομίσθη δὲ τοῦτο ἐπὶ τῷ τοὺς ὑπερέχοντας τοῖς φρονήμασι κωλύειν. διαφέρει δὲ φυγῆς, ὅτι τῶν ὀστρακιζομένων αἱ οὐσίαι οὐ δημεύονται, ἀλλὰ καὶ τόπον καὶ χρόνον ὡρισμένον ἔχουσι, τῶν δὲ φευγόντων οὐδέτερον τούτων πρόσεστιν. | A slightly different version of the first part of this entry is preserved by Photios (s.v. ὀστρακισμός Naber) which is worth repeating here because it is preceded by a sentence which links it to a statement attributed to Theophrastos, and because it offers a more complete and more intelligible text: οἱ κακονούστατοι τῷ δήμῳ ἐξωστρακίζοντο καὶ κατεδικάζοντο ὀστράκοις ἐγγραφόντων τὸ ὄνομα τοῦ φευξομένου. φυγῆς ἐστιν εἶδος, ὀνομασθὲν ἐκ τοῦ εἰς ὄστρακον ἐγγράφειν ἕκαστον τῶν Ἀθηναίων, εἴ τις αὐτοῖς ἐνομίζετο πρέπειν μεθίστασθαι τῆς πόλεως; see also Hesychios, s.v. ὀστρακισμός Schmidt. A rephrasing of this statement is given by Thomas Magister, *Ecloga Vocum Atticarum*, s.v. ἐξοστρακίζομαι (p. 24 Ritschel) ... ἐγένετο δὲ ὁ ὀστρακισμός οὐκ ἐπὶ τῶν τυχόντων, ἀλλ' ἐπὶ τῶν ἐνδόξων καὶ γένει λαμπρῶν. παρήχθη μέντοι ἡ λέξις ἀπὸ τῶν ὀστράκων ἐν οἷς γράφοντες ἐδίδουν τὰς καταδίκας τοῖς ὑπευθύνοις. A slightly different version of the second part of the *An. Gr.* entry is given by Suidas (s.v. ὀστρακισμός, no. 717 Adler) which may be repeated here because it is preceded by the same statement as the notice in Photios (thus linking the two together) and because it gives a somewhat better text than the *Anecdota Graeca*: οἱ κακονούστατοι τῷ δήμῳ ἐξωστρακίζοντο καὶ κατεδικάζοντο· ὀστράκοις γὰρ ἐγράφετο τὸ ὄνομα τοῦ φευξομένου.

ὀστρακισμὸς φυγῆς διαφέρει, ὅτι τῶν μὲν φυγῆς ἁλόντων αἱ οὐσίαι δημεύονται, τῶν δὲ ὀστρακισμῷ ἀποστάντων οὐκ ἀφαιρεῖται τὰ χρήματα ὁ δῆμος· καὶ τοῖς μὲν χρόνος ἐνδείκνυται καὶ τόπος, οἱ δὲ φεύγοντες οὐδέτερον τούτων ἔχουσιν. Finally, the Scholion on Aristophanes, *Vespae*, 947, explains the difference between exile and ostracism with nearly the same words: εἶδος γάρ τι φυγῆς ἐστιν ὁ ὀστρακισμός. ἐν δὲ τοῖς εἴδεσι περιέχεται τὰ γένη. καὶ τὸ μὲν ἐξωστρακίσθαι φεύγειν ἄν τις εἰκότως εἴποι, τὸ δὲ φεύγειν οὐκέτι ἐξωστρακίσθαι. διαφέρει γὰρ φυγὴ ὀστρακισμοῦ, καθὸ τῶν μὲν φευγόντων αἱ οὐσίαι δημεύονται, τῶν δὲ ὀστρακισμῷ μεταστάντων οὐκέτι κύριος ὁ δῆμος. καὶ τοῖς μὲν καὶ τόπος ἀπεδίδοτο καὶ χρόνος, τοῖς δὲ οὐδέτερον τούτων.

The passages assembled here agree so much in content and in terminology that they may be fitted together to form a coherent statement on ostracism which may be given in English:

> An enclosure is made in the Agora, with entrances through which each citizen passes and deposits an inscribed ostrakon. In case more | than six thousand votes are cast, the victim receives exile of ten years. The people most hostile to the demos were ostracized and condemned, the ostraka containing the name of the person to be exiled. Ostracism is a form of exile, called from the ostraka on which each Athenian wrote the name of the person who in his opinion should leave the city. It was designed to curb those who were outstanding in their arrogance. Ostracism is different from exile inasmuch as the property of the exiled is confiscated but the property of the ostracized is not taken by the demos. Moreover, the ostracized receive a period of exile and a place of residence, but those exiled get neither.

[90]

It is possible to show that this statement goes back, if not to Theophrastos and Philochoros, so at least to Didymos's commentary on Demosthenes, 23 205, to which we owe Philochoros's account of ostracism (see above, pp. 87–88). This can be demonstrated by an examination of the complete entries in Suidas and in the Aristophanes Scholia, parts of which have been quoted so far.

After describing the difference between ostracism and exile (see above, p. 92f.), Suidas, *loc. cit.*, continues ὅτι Κίμων τῇ ἀδελφῇ Ἐλπινίκῃ συγκοιμηθεὶς καὶ διαβληθεὶς πρὸς τοὺς πολίτας ἐξωστρακίσθη. This statement is repeated by Suidas, s.v. ἀποστρακισθῆναι· ... ὅτι ἀποστρακισθῆναί φασι τὸν Κίμωνα τῇ ἀδελφῇ Ἐλπινίκῃ συγκοιμηθέντα ὑπὸ Ἀθηναίων, and s.v. Κίμων (no. 1621 Adler), Ἀθηναῖος. οὗτος τῇ ἀδελφῇ Ἐλπινίκῃ συγκοιμηθεὶς διεβλήθη πρὸς τοὺς πολίτας καὶ διὰ τοῦτο ὠστρακίσθη πρὸς τῶν Ἀθηναίων. The relation between these passages and the account of the difference between exile and ostracism is made clear by a fragment of Didymos (p. 324, frag. 5, M. Schmidt) which is evidently the source of all the entries in Suidas quoted above: Δίδυμος δέ φησιν οὐχ ὅτι ἐλακώνιζεν, ἀλλ᾽ ὅτι Ἐλπινίκῃ τῇ ἀδελφῇ συνῆν. αἴτιοι δὲ τῆς διαβολῆς οἱ κωμικοὶ καὶ μάλιστα Εὔπολις ἐν Πόλεσι. Καλλίας δέ, ὁ υἱὸς (*sic*; repeated by Tzetzes, *Chiliades*, 1, lines 582–593) αὐτοῦ κατέβαλλεν ὑπὲρ αὐτοῦ πεντήκοντα τάλαντα· κατέβαλε δὲ ἐπὶ τὸ πρὸς γάμον λαβεῖν τὴν Ἐλπινίκην. Ἐξοστρακισθεὶς δὲ κατῆλθε, καὶ στρατηγήσας ἐνίκησεν ἐπ Εὐρυμέδοντι, καὶ κατὰ γῆν καὶ κατὰ θάλατταν. τέθνηκε δὲ τὸ Κήτιον τῆς

Κύπρου πολιορκῶν. Ἔφορος δέ...(F. Gr. Hist. 70 F 64); see E. Meyer, *Forschungen* 2, pp. | 36–43. It is still possible to ascertain the precise place where Didymos made this comment on Kimon, for the Scholion on Aristides (3 p. 515 Dindorf) attaches the quotation from Didymos to one from Demosthenes (23, 205). It so happens that Demosthenes speaks at the beginning of this very sentence of the ostracism of Themistokles, and Didymos's comments on this passage contain Philochoros's famous account of ostracism (see above, p. 00). Didymos himself gave the complete text of Philochoros including the distinction between exile and ostracism, which we know only from the *Anecdota Graeca*, Suidas, and the Scholia on Aristophanes (see above). Suidas (s.v. ὀστρακισμός) combined the end of Didymos's comments on ostracism with Didymos's immediately following comments on Kimon which rely on a different source, namely Theopompos; see H. T. Wade-Gery, *A.J.P.* 59, 1938, pp. 133–134; A. E. Raubitschek, *Historia* 3, 1955, p. 379. This combination is meaningful only if it is understood that the entry goes back to a commentary on Demosthenes, 23, 205, where the ostracism of Themistokles and the condemnation of Kimon are mentioned in one and the same sentence.[7]

It is also clear that the Scholion on *Equites*, 855 (which has been quoted and discussed above, pp. 84–87) goes back to Didymos's commentary on Demosthenes, 23, 205, since one of the attested versions of Philochoros's account of ostracism is actually preserved as a commentary on this passage of Demosthenes (see above, pp. 84–87). Jacoby observed, moreover, in the commentary on *F. Gr. Hist.* 328 F 30, p. 315, that "the three excerpts... go back to the same source (almost certainly Didymos)." In the case of the Scholion on *Vespae*, 947, the same conclusion can be reached, for this Scholion continues, after describing the difference between exile and ostracism (see above, pp. 92–93): ὅτι δὲ ὁ Ἀθηναίων δῆμος ἀειφυγίαν αὐτοῦ καταγνοὺς ἐδήμευσε τὴν οὐσίαν, καὶ πρὸς Ἀρταξέρξην ἧκε φεύγων, σαφὲς ποιεῖ Ἰδομενεὺς διὰ τοῦ β' τὸν τρόπον τοῦτον, οἱ μέντοι Ἀθηναῖοι αὐτοῦ καὶ γένους ἀειφυγίαν κατέγνωσαν, προδιδόντος τὴν Ἑλλάδα, καὶ αὐτοῦ ἡ οὐσία ἐδημεύθη. Jacoby, who commented on this passage (on *F. Gr. Hist.* 328 F 120; 338 F 1), insisted that it refers to Themistokles, but he did not notice that we have here the end of Didymos's commentary on Demosthenes, 23, 205: ἐκεῖνοι Θεμιστοκλέα

---

7. A few words may be added here concerning the reliability of Demosthenes' statement (23, 205) that Kimon escaped the death penalty but was fined fifty talents; see G. Busolt, *Gr. Gesch.* 3, p. 255, note 1; E. Meyer, *Forschungen* 2, p. 25, note 1; A. Fuks *The Ancestral Constitution*, p. 28, note 37. Actually, not only Kimon's father Miltiades, but also Kallias (Demosthenes, 19, 273) and Perikles suffered the same fate; in all these cases (see Busolt, *op. cit.*, p. 245, note 3) commanding generals (or an envoy) were called to account, pronounced guilty, and fined (but not executed). In the case of Kimon, we know (Aristotle, *Ath. Pol.*, 27, 1) that the trial was connected with the *euthyna*, and this may apply also to the other cases; see J. H. Lipsius, *Das Attische Recht* II/1, pp. 294–298. This means that Kimon was convicted and not acquitted as has generally been deduced from Plutarch, *Cimon*, 15, 1. Plutarch himself says (*Pericles*, 10, 5) that Kimon was tried on a capital charge, and it was therefore the death penalty which he escaped. Aristotle (*Ath. Pol.*, 27, 1) confirms Demosthenes' statement when he says that Perikles distinguished himself at the trial; a complete acquittal of Kimon would not have added to Perikles' reputation (compare now R. Sealey, *Hermes*, 84, 1956, p. 238, note 1).

λαβόντες μεῖζον αὐτῶν ἀξιοῦντα φρονεῖν ἐξήλασαν ἐκ τῆς πόλεως καὶ μηδισμὸν κατέγνωσαν. At the beginning of his commentary, Didymos interpreted the word ἐξήλασαν as referring to Themistokles' ostracism, he added Philochoros's statement on ostracism, and appended (perhaps also from an earlier source) some remarks on the difference between exile and ostracism; at the end, he explained the word κατέγνωσαν as referring to Themistokles' condemnation for treason, and he added the quotation from Idomeneus. The scholiast on Aristophanes (perhaps Symmachus; see I. H. White, *Scholia on the Aves*, pp. xlix–liii) found in an earlier commentary on *Vespae*, 947 the statement τοῦτον δὲ ἐξωστράκισαν 'Αθηναῖοι τὰ ί ἔτη κατὰ τὸν νόμον (referring to Thucydides the son of Melesias), and added from Didymos's commentary on Demosthenes (or from a Lexikon based on it) the information on the difference between exile and ostracism, appending the remarks on the treason of Themistokles. Similarly, Suidas, in composing his article on ostracism (see above) drew ultimately on Didymos's commentary on Demosthenes, 23, 205, but he appended the remarks on Kimon which followed, in Didymos's commentary, those on Themistokles. One may assume, therefore, that the two Scholia on Aristophanes (*Equites*, 855, and *Vespae*, 947) belong together and contain a somewhat abbreviated account of ostracism which is based on Didymos's commentary on Demosthenes, and the first part of which (at least) goes back to Philochoros.

Diodoros. The preceding section contains a collection of most of the late and anonymous statements on ostracism taken from the Scholia and the Lexika. It has been argued that all this evidence belongs together and depends ultimately on Didymos's commentary on Demosthenes, 23, 205. The first part of this commentary (as represented by the Scholia on *Equites*, 855) contained a quotation of Philochoros who himself goes back to Theophrastos (see above, pp. 84–89). Whether the second part (discussed in the preceding chapter) can also be traced back to Philochoros has still to be shown; it is obvious that it contains some Aristotelian material (see above, pp. 91–92). In order to prove that this second part goes also back to Philochoros (and perhaps Theophrastos), it will be shown that the information contained in it was familiar to authors who were not likely (or not able) to use either Didymos or any of the later Scholia or Lexika.

The testimony of Diodoros is especially important since he was a historian (and not an antiquarian, commentator, or lexicographer) and used historical sources, and because Jacoby (Commentary on *F. Gr. Hist.* 328 F 30, p. 317) confidently asserted (see also p. 315: "surely from Ephoros") that Diodoros's source on ostracism was Ephoros. If it can be shown that Diodoros's account of ostracism (11, 55, 1–3; 87, 1–4) belongs to the same tradition as that of Didymos (and Theophrastos-Philochoros), Ephoros cannot have been his source.

The first point of interest are the historical incidents which occasioned Diodoros's statements on ostracism: the exile of Themistokles (11, 54,5–55,3)

[93]

and the history of petalism in Syracuse (11, 87, 1-4). The former indicates that Diodoros (and his primary source for Athenian history: Ephoros) did not mention any ostracisms before that of Themistokles; nor did he or his source mention any of the later ostracisms (especially those of Kimon, Thucydides, Hyperbolos). In this respect, Diodoros agrees with Ephoros's main source of fifth-century Athenian history, namely Thucydides the historian; see my remarks in *Historia* 3, 1955, p. 379. If Ephoros had been sufficiently interested in ostracism to give a general account of its working, he surely would have mentioned some of the more im|portant ostracisms in their historical context. Since he did not do this, one may suspect that the passage on ostracism in Diodoros is not taken from Ephoros; moreover there is no other evidence which shows that Ephoros was interested in constitutional antiquities. Finally, Diodoros's account of ostracism in 11, 55 depends on the same authority as does his account of ostracism and petalism in 11, 87, and this authority was, according to R. Laqueur (*R.E.* s.v. Timaios, col. 1093, lines 42–63), Timaios who was much interested in constitutional antiquities and who is known to have used the political works of Aristotle; on petalismos, see H. Wentker, *Sizilien und Athen*, 1956, pp. 56–58; T. S. Brown, *Timaeus*, 1958, and *A.J.P.* 73, 1952, pp. 340–355. This assumption explains the agreement of Diodoros with the statement on ostracism that is here attributed to Theophrastos-Philochoros. The best way to show the common source of Diodoros, 54,5–55,3 and 87,1–3, may be to place the two passages side by side.

11, 54,5-55,3:

μετὰ δὲ ταῦτα οἱ μὲν φοβηθέντες
αὐτοῦ τὴν ὑπεροχήν, οἱ δὲ φθονήσαντες
τῇ δόξῃ, τῶν μὲν εὐεργεσιῶν ἐπελάθοντο,
τὴν δὲ ἰσχὺν αὐτοῦ καὶ τὸ φρόνημα
ταπεινοῦν ἔσπευδον.

Πρῶτον μὲν οὖν αὐτὸν ἐκ τῆς πόλεως
μετέστησαν, τοῦτον τὸν ὀνομαζόμενον
ὀστρακισμὸν ἐπαγαγόντες αὐτῷ ὃς
ἐνομοθετήθη μὲν ἐν ταῖς Ἀθήναις μετὰ
τὴν κατάλυσιν τῶν τυράννων τῶν περὶ
Πεισίστρατον. ὁ δὲ νόμος ἐγένετο
τοιοῦτος· ἕκαστος τῶν πολιτῶν εἰς
ὄστρακον ἔγραφε τοὔνομα τοῦ δοκοῦντος
μάλιστα δύνασθαι καταλύσαι τὴν
δημοκρατίαν· ᾧ δ' ἂν ὀστρακαπλείω
γένηται, φεύγειν ἐκ τῆς πατρίδος
ἐτέτακτο πενταετῆ | χρόνον. νομοθετῆσαι
δὲ ταῦτα δοκοῦσιν οἱ Ἀθηναῖοι, οὐχ
ἵνα τὴν κακίαν κολάζωσιν, ἀλλ' ἵνα
τὰ φρονήματα τῶν ὑπερεχόντων
ταπεινότερα γένηται διὰ τὴν φυγήν.

11, 87,1-2:

παρὰ γὰρ Ἀθηναίοις ἕκαστον τῶν
πολιτῶν ἔδει γράφειν εἰς ὄστρακον
τοὔνομα τοῦ δοκοῦντος μάλιστα
δύνασθαι τυραννεῖν τῶν πολιτῶν,
παρὰ δὲ Συρακοσίοις εἰς πέταλον
ἐλαίας γράφεσθαι τὸν δυνατώτατον
τῶν πολιτῶν· διαριθμηθέντων δὲ τῶν
πετάλων τὸν πλεῖστα πέταλα λαβόντα
φεύγειν πενταετῆ χρόνον. τούτῳ γὰρ
τῷ τρόπῳ διελάμβανον ταπεινώσειν
τὰ φρονήματα τῶν πλεῖστον ἰσχυόντων
ἐν ταῖς πατρίσι· καθόλου γὰρ οὐ
πονηρίας κολάσεις ἐλάμβανον παρὰ
τῶν παρανομούντων, ἀλλὰ δυνάμεως
καὶ αὐξήσεως τῶν ἀνδρῶν ἐποίουν
ταπείνωσιν.

The two parallel passages of Diodoros are linked to Theophrastos's account of ostracism through the statements that ostracism was introduced after the expulsion of the tyrants and that after a count of the votes (διαριμηθέντων) the man receiving a majority had to leave the city for a certain number of years; Diodoros is mistaken in saying that the period of ostracism was five years (see below, Appendix I, p. 102). The link between the passages in Diodoros and the account attributed to Didymos consists in the statement that each of the citizens had to write on an ostrakon the name of the person who in his opinion had the greatest power to destroy the democracy or to become a tyrant; Didymos merely says that it was the person who should leave the city. Another link between Diodoros and Didymos can be found in the statement that the Athenians adopted this law not to punish wickedness but to lower the arrogance of those who were outstanding. It should be noticed that Pollux closes his statement (above, p. 90) with a sentence saying the same thing with different words. Since it is likely that Diodoros used one and the same source for his statements on ostracism, and since this statement contains sentences which can be associated with statements attributed to Theophrastos and to Didymos, it may be presumed that Diodoros used an authority (presumably Timaios) who had before him a comprehensive statement on ostracism such as that of Didymos (the first part of whose account we have been able to attribute to Theophrastos). This means that even the second part of the statement on ostracism attributed to Didymos (with the possible exception of the section dealing with the difference between exile and ostracism) can be traced back to the late fourth century B.C.

R. Laqueur recognized (*Hermes* 46, 1911, p. 205) that Diodoros used the same source (as in 11) also in the introduction to the | story of Agathokles (19, 1,1–3), but he identified its source, I think mistakenly, as Ephoros. This passage merits consideration since it employs the same terminology as the other passages so far considered and because it contains other evidence which links it to Aristotle and his school; see R. M. Geer, *Loeb Classical Library, Diodoros*, IX p. 224, note 1. The entire passage may be compared with Aristotle's discussion of "the causes of revolutions" (*Politics*, 5, 1302 a 16ff.), but special attention may be called to the introductory sentence: παλαιός τις παραδέδοται λόγος ὅτι τὰς δημοκρατίας οὐχ οἱ τυχόντες τῶν ἀνθρώπων ἀλλ' οἱ ταῖς ὑπεροχαῖς προέχοντες καταλύουσιν. Reflections of this sentence occur in Andokides (4, 24: ἔστι δὲ σωφρόνων ἀνδρῶν φυλάττεσθαι τῶν πολιτῶν τοὺς ὑπεραυξανομένους, ἐνθυμουμένους ὑπὸ τῶν τοιούτων τὰς τυραννίδας καθισταμένας), in Aristotle (*Politics*, 5, 1304 a 34–37: οἱ δυνάμεως αἴτιοι... στάσιν κινοῦσιν), and in a passage attributed by Stobaios to Aristotle (4, 4,21: αἱ πλεῖσται στάσεις διὰ φιλοτιμίαν ἐν ταῖς πόλεσι γίγνονται· περὶ τιμῆς γὰρ οὐχ οἱ τυχόντες, ἀλλ' οἱ δυνατώτατοι διαμφισβητοῦσι); see also Thomas Magister, *Ecloga*, s.v. ὀστρακίζομαι... ἐγένετο γὰρ ὁ ὀστρακισμὸς οὐχ ἐπὶ τῶν τυχόντων ἀλλ' ἐπὶ τῶν ἐνδόξων καὶ γένει λαμπρῶν. Equally significant is Diodoros's statement on ostracism itself (19, 1,3): τοιγαροῦν Ἀθηναῖοι μὲν διὰ ταύτας τὰς αἰτίας τοὺς πρωτεύοντας τῶν πολιτῶν ἐφυγάδευσαν, τὸν λεγόμενον παρ' αὐτοῖς ὀστρακισμὸν νομοθετήσαντες. καὶ τοῦτ' ἔπραττον οὐχ ἵνα τῶν προγεγενημένων ἀδικημάτων λάβωσιν τιμωρίαν, ἀλλ' ὅπως τοῖς δυναμένοις παρανομεῖν ἐξουσία μὴ γένηται κατὰ τῆς πατρίδος ἐξαμαρτάνειν. This statement is in full agreement not only with Diodoros's earlier comments (quoted above) and with Aristotle's characterization of ostracism in the *Politics*, but also with the statement attributed to Didymos; notice the use of οἱ χαριέστατοι in Diodorus, 11, 87, 4 (see above, pp. 85-86). There can be no doubt that we have here different expressions of one and the same opinion: ostracism was not a punitive but a preventive measure; nor can there be any doubt that this opinion can be traced back to Aristotle and Theophrastos. It may even be that the wording of Diodoros is closer to the original than the text of the lexicographers, and this possibility is strongly supported by the fact that Plutarch employs the same terminology as does Diodoros.

[97] **Plutarch.** The final proof that the various statements on ostracism which have been assembled so far have a common origin is provided by Plutarch whose several comments on ostracism go evidently back to one source; see my note in *Phoenix* 9, 1955, p. 122, note 2 [*infra*, p. 320, n. 2]. Jacoby, in reprinting Plutarch, *Aristides*, 7, in his Commentary on *F. Gr. Hist.* 328 F 30, and in remarking (Notes, p. 228, note 3) that "the last three words" (καρπούμενον τὰ ἑαυτοῦ) "belong to the text of the law," has already indicated the close relationship between Plutarch and Philochoros; I was, however, unable to persuade him (*Hermes* 83, 1955, p. 119) that Plutarch agrees substantially with Philochoros (Notes on the Commentary on *F. Gr. Hist.* 297-607, p. 405, on 328 F 30, note 14). It is now possible to suggest that Plutarch used not Philochoros but Theophrastos (or Demetrios; see above, note 4) with whose works he is known to have been familiar. The main passage in the *Aristides* (7) contains several statements which can be associated with the account of ostracism attributed here to Theophrastos. *Aristides*, 7: μοχθηρίας γὰρ οὐκ ἦν κόλασις ὁ ὀστρακισμός, ἀλλ' ἐκαλεῖτο μὲν δι' εὐπρέπειαν ὄγκου καὶ δυνάμεως βαρυτέρας ταπείνωσις καὶ κόλουσις, ἦν δὲ φθόνου παραμυθία φιλάνθρωπος, εἰς ἀνήκεστον οὐδέν, ἀλλ' εἰς μετάστασιν ἐτῶν δέκα τὴν πρὸς τὸ λυποῦν ἀπερειδομένου δυσμένειαν. ἐπεὶ δ' ἤρξαντό τινες ἀνθρώπους ἀγεννεῖς καὶ πονηροὺς ὑποβάλλειν τῷ πράγματι, τελευταῖον ἀπάντων Ὑπέρβολον ἐξοστρακίσαντες ἐπαύσαντο. The following account of the ostracism of Hyperbolos may be attributed to Theopompos; see my suggestions in *Phoenix* 9, 1955, pp. 123 and 126 [*infra*, pp. 321-322, 324]. Plutarch then continues his general account of ostracism: Ἦν δὲ τοιοῦτον, ὡς τύπῳ φράσαι, τὸ γινόμενον. ὄστρακον λαβὼν ἕκαστος καὶ γράψας ὃν ἐβούλετο μεταστῆσαι τῶν πολιτῶν, ἔφερεν εἰς ἕνα τόπον τῆς ἀγορᾶς περιπεφραγμένον ἐν κύκλῳ δρυφάκτοις. οἱ δ' ἄρχοντες πρῶτον μὲν διηρίθμουν τὸ σύμπαν ἐν ταὐτῷ τῶν ὀστράκων πλῆθος· εἰ γὰρ ἑξακισχιλίων ἐλάττονες οἱ φέροντες εἶεν, ἀτελὴς ἦν ὁ ὀστρακισμός· ἔπειτα τῶν ὀνομάτων ἕκαστον ἰδίᾳ θέντες τὸν ὑπὸ τῶν πλείστων γεγραμμένον ἐξεκήρυττον εἰς ἔτη δέκα, καρπούμενον τὰ αὑτοῦ. The following account of the ostracism of Aristeides

[98] may go back through Demetrios (see above, note 4) to Aischines the Socratic (see my notes in the *"Charites" für E. Langlotz*, p. 241 [*infra*, pp. 153-154]). The second part of the foregoing quotation from Plutarch follows so closely the first two sections of the statement attributed to Theophrastos-Philochoros that

they must go back either to this very statement or to a parallel account, perhaps by Demetrios of Phaleron; notice that the statement, that each citizen wrote on an ostrakon the name of the person whom he wanted to remove, agrees not only with the account of Diodoros (see above, pp. 96–97) but also with the text attributed to Didymos (see above, pp. 92–93): εἰς ὃ ἐνέγραφεν ἕκαστος Ἀθηναίων εἴ ιτς αὐτοῖς ἐνομίζετο πρέπειν μεθίστασθαι τῆς πόλεως. The first part, however, requires some further comments, since it occurs also in other biographies of Plutarch; the pertinent passages may be presented side by side.

Themistocles 22, 3:	Nicias, 11, 1,5,6:	Alcibiades, 13, 4:
Τὸν μὲν οὖν ἐξοστρακισμὸν ἐποιήσαντο κατ᾽ αὐτοῦ κολούοντες τὸ ἀξίωμα καὶ τὴν ὑπεροχήν, ὥσπερ εἰώθεσαν ἐπὶ πάντων, οὓς ᾤοντο τῇ δυνάμει βαρεῖς καὶ πρὸς ἰσότητα δημοκρατικὴν ἀσυμμέτρους εἶναι. κόλασις γὰρ οὐκ ἦν ὁ ὀστρακιμσμός ἀλλὰ παραμυθία φθόνου καὶ κουφισμὸς ἡδομένου τῷ ταπεινοῦν τοὺς ὑπερέχοντας καὶ τὴν δυσμένειαν εἰς ταύτην τὴν ἀτιμίαν ἀποπνέοντος.	...καὶ γιγνομέης ὀστρακοφορίας, ἣν εἰώθει διὰ χρόνου τινὸς ὁ δῆμος ποιεῖσθαι, ἕνα τῶν ὑπόπτων ἢ διὰ δόξαν ἄλλως ἢ πλοῦτον ἐπιφόνων ἀνδρῶν τῷ ὀστράκῳ μεθιστὰς εἰς δέκα ἔτη,... μᾶλλον δὲ κόλασιν τὸν ἐξοστρακισμὸν ἡγούμενοι Θουκυδίδῃ καὶ Ἀριστείδῃ καὶ τοῖς ὁμοίοις, Ὑπερβόλῳ δὲ τιμὴν καὶ προσποίησιν ἀλαζονείας, εἰ διὰ μοχθηρίαν ἔπαθε ταὐτὰ τοῖς ἀρίστοις... καὶ τὸ πέρας οὐδεὶς ἔτι τὸ παράπαν ἐξωστρακίσ|θη μετὰ Ὑπέρβολον, ἀλλ᾽ ἔσχατος ἐκεῖνος, πρῶτος δ᾽ Ἵππαρχος ὁ Χολαργεὺς συγγενής τις ὢν τοῦ τυράννου.	...ᾧ κολούοντες ἀεὶ τὸν προὔχοντα δόξῃ καὶ δυνάμει τῶν πολιτῶν ἐλαύνουσι, παραμυθούμενοι τὸν φθόνον μᾶλλον ἢ τὸν φόβον.

It is evident that Plutarch is repeating the same information in all four biographies, and that the core of this information goes ultimately back to Aristotle, *Politics*, 3, 1284 a 36–38: ὁ γὰρ ὀστρακισμὸς τὴν αὐτὴν ἔχει δύναμιν τρόπον τινὰ τῷ κολούειν τοὺς ὑπερέχοντας καὶ φυγαδεύειν).[8] The same information is found in the *Anecdota Graeca* and in the *Etymologicum Magnum*,

---

8. Aristotle himself borrowed the words κολούειν and ὑπερέχοντες from Herodotus (5, 92 ξ 2) to whom he refers in the immediately preceding sentence (1284 a 26–33); see Herodotus 7, 10 ε: φιλέει γὰρ ὁ θεὸς τὰ ὑπερέχοντα πάντα κολούειν; compare H. Ryffel, Μεταβολὴ πολιτειῶν, p. 12, note 38.

and it has been taken from there and incorporated in the statement attributed to Didymos; the κωλύειν of the Lexika should be changed to κολούειν attested by Aristotle and Plutarch. The passages from the *Nicias* (11, 5 and 6) and from the *Aristides* (7, 3) contain the same information as the last section of the statement attributed to Theophrastos-Philochoros (see p. 87), showing that Plutarch was familiar also with this part of the statement. An analysis of the introductory paragraph of the passage on ostracism in the *Aristides* (7, 1–2) also shows that Plutarch relies here on material which goes back to Aristotle and his school: ἤδη δέ που καὶ ὁ δῆμος ἐπὶ τῇ νίκῃ μέγα φρονῶν καὶ τῶν μεγίστων ἀξιῶν αὐτὸν ἤχθετο τοῖς ὄνομα καὶ δόξαν ὑπὲρ τοὺς πολλοὺς ἔχουσι. καὶ συνελθόντες εἰς ἄστυ πανταχόθεν ἐξοστρακίζουσι τὸν Ἀριστείδην, ὄνομα τῷ φθόνῳ τῆς δόξης φόβον τυραννίδος θέμενοι. Plutarch emphasizes here the jealousy to which Aristeides was exposed, and the notion that φθόνος was the "cause" of ostracism can be attributed to Demetrios of Phaleron (*Aristides*, 1.2) and, through Pollux, to Theophrastos-Philochoros (see above, pp. 86–88). Another passage in the foregoing quotation which can be directly attributed to Aristotle is the observation that at the time of Aristeides' ostracism the demos was proud of its victory and was angry with those whose name and fame placed them above the mass of the people; this is exactly the way in which Aristotle describes the beginning of ostracism (*Ath. Pol.* 22, 3): θαρροῦντος ἤδη τοῦ δήμου. Considering the close relationship between Aristotle, Theophrastos, and Demetrios, it may be suggested that Plutarch's treatment of ostracism goes back to Demetrios of Phaleron and to Theophrastos. There still remains the question how to account for the peculiar verbal agreement between Diodoros and Plutarch, an agreement which is not shared by the statement attributed to Theophrastos-Philochoros. Both Diodoros and Plutarch employ the words κόλασις and ταπείνωσις to assert that "ostracism was not a punishment (κόλασις) for wickedness (κακίας, πονηρίας Diod.; μοχθηρίας Plu.) but a lowering (ταπείνωσις) of power (δυνάμεως)." It is possible that Diodoros and Plutarch have preserved more purely the terminology of their source; at any rate, the common source of Diodoros and Plutarch must be an author of the fourth-century B.C. who belonged to the circle of Aristotle.

Aristides and Theodoros Metochites. Plutarch himself may have served as a source for two later accounts of ostracism which are quoted here because they have not been considered so far in other treatments of ostracism. Aristides, 46, pp. 316–317 Dindorf: ἀλλὰ Θεμιστοκλῆς μὲν καὶ Κίμων ἐξωστρακίσθησαν. τοῦτο δ' ἦν οὐ μῖσος οὐδ' ἀλλοτρίωσις τοῦ δήμου πρὸς αὐτοὺς ἀλλ' ἦν νόμος αὐτοῖς περὶ ταῦτα, ἔχων μὲν ὁπωσδήποτε—ἐῶ γὰρ εἰ μὴ σφόδρ' ἄν τις ἐπαινέσαι τὸν νόμον—τὸ δ'οὖν ἁμάρτημα οὐκ ἀπαραίτητον αὐτῶν, ἀλλ' ἔχον ὡς ἐν τούτοις εὐπρέπειαν, νόμῳ γάρ, ὥσπερ εἶπον, ἐγίγνετο. ἦν δ' οὗτος ὁ νόμος· ἐκόλουον τοὺς ὑπερέχοντας μεθιστάντες ἔτη δέκα, ἄλλο δ' οὐδὲν ἔγκλημα προσῆν, οὐδ' ὡς ἐπ'ἐλέγχῳ πραγμάτων ὀργή.... ἀλλ' ἐκεῖσε ἐπάνειμι, ὅτι ὑπὲρ τοῦ τὰ φρονήματα ἐπισχεῖν τοῦτο τὸ εἶδος τῆς φυγῆς ἐνόμισαν.... δοκοῦσι γάρ μοι τὰς συμφορὰς ἐνθυμούμενοι τὰς ἐπὶ τῶν Πεισιστρατιδῶν γενομένας ἑαυτοῖς μηδένα βούλεσθαι μεῖζον ἐᾶν τῶν πολλῶν φρονεῖν, ἀλλ' ἐξ ἴσου εἰς δύναμιν εἶναι.... The last sentence which connects the institution of ostracism with the Peisistratids recalls vividly the statement of Aristotle (*Ath. Pol.* 22, 3): τότε πρῶτον ἐχρήσαντο τῷ | νόμῳ τῷ περὶ τὸν ὀστρακισμόν, ὃς ἐτέθη διὰ τὴν ὑποψίαν τῶν ἐν ταῖς δυνάμεσιν, ὅτι Πεισίστρατος δημαγωγὸς καὶ στρατηγὸς ὢν τύραννος κατέστη; for the dependence of Aristotle on Androtion (*F. Gr. Hist.* 324 F 6), see my comments in *A.J.A.* 55, 1951, p. 221 [*supra*, pp. 65–67]. There is only one indication that Aristides may have used a source other than Plutarch (or in addition to Plutarch); this is the occurrence of the words εἶδος τῆς φυγῆς which are not found in Plutarch but which have been attributed to Theophrastos-Philochoros (see previous discussion). The scholia on Aristides (vol. 3, p. 690 Dindorf) add some significant details: ἐκόλουον] ἁρμοδίως τῷ προκειμένῳ ἡ λέξις· οὐ γὰρ εἶπεν ἐκώλυον, ἀλλ' ἐκόλουον, τουτέστι τὸ φρόνημα ἐπεῖχον.... τοῦτο τὸ εἶδος τῆς φυγῆς] πάλιν οὐκ εἶπε φυγήν, ἀλλ'εἶδος φυγῆς; see also the Scholion on 46, 158,5 (vol. 3, p. 528 Dindorf) which is based on the text of Aristides quoted above (pp. 100f.) but which refers to two passages of Thucydides (1, 135; 8, 73) which are also mentioned by Thomas Magister (*Ecloga*, s.v. ἐξοστρακίζομαι, p. 264 Ritschel; see above, pp. 92–93 and 97–98). It seems that the scholiast, if not Aristides himself, knew the statement of Didymos, perhaps from one of the Lexika.

Here may be added a much later treatment of ostracism which has also escaped notice. Theodoros Metochites (*Miscellanea*, pp. 609–610, ed. Müller-Kiessling, 1821) tells the story of Aristeides' ostracism, following Plutarch's account but enlarging upon it, and introduces it with a brief account of ostracism which may also go back to Plutarch; see H.-G. Beck, *Theodoros Metochites*, pp. 72–75. The part dealing with ostracism in general is repeated here: λέγεται γέ τοι περὶ αὐτοῦ τόδε χαρίεν, ὡς ἤθροιστο μέν ὁ δῆμος παντόθεν εἰς τὸ βουλευτήριον, αὐτόθεν τε ἐκ τοῦ ἄστεος καὶ πολλοί γε ἥκοντες ἐκ τῶν ἀγρῶν, ὡς ἄρα νόμιμον ἦν ὀστρακισμοῦ γίνεσθαι μέλλοντος ἑνὸς δή τινος τῶν περιφανῶν ἐν τῇ πόλει· ἐπεὶ δ' ἦν τοῦτ' ἔθος, ἕκαστον τῶν ἠθροισμένων ὄστρακον λαμβάνοντα πολλῶν εἰς μέσον κειμένων, ἐγγράφειν αὐτῷ οὗ βούλοιτ' ἂν ὄνομα τῶν ἐκλογίμων ἀνδρῶν εἰς τὴν ἀπὸ τῆς πόλεως φυγήν· καὶ τῶν ὀστράκων μεθύστερον ἀριθμουμένων ἔδει φεύγειν τὸν ἐν πλείοσι καταγεγραμμένον, ὡς πλείοσιν ἄρα ψήφοις καὶ γνώμαις κατάκριτον τῇ φυγῇ· ἐντεῦθεν γὰρ καὶ τοὔνομα τὸν ὀστρακισμὸν τὴν ἀρχὴν κληθῆναι.... καὶ μέντοι καταριθμηθέντων ἔπειθ' ὕστερον τῶν ἐν τοῖς ὀστράκοις καταγεγραμμένων, ἐν πλείοσιν αὐτὸν Ἀριστείδην | εὑρηθέντα καταγεγραμμένον,...τῇ ὑστεραίᾳ τῆς πατρίδος ὀστρακίζεσθαι καὶ φεύγειν ἐπὶ ῥητοῖς ὡς νόμιμον εἰς ἔτη δέκα.

Conclusion. The presentation of all this material tends to show that there existed in antiquity a fairly comprehensive account of ostracism upon which all our extant testimonies are based. This account did not contain the story of any one individual ostracism but it did include an account of the procedure, a list of the better known victims, brief statements on the first and last ostracisms, and a general characterization of the institution. The reconstructions offered above should give an idea of this account; while they may contain a good number of words and phrases of the original version, they also contain later versions and even later additions. In order to identify the author of this account, one must remember that he must be older than Philochoros and Timaios, and younger than Aristotle, and that he evidently belonged to the circle around Aristotle. Since part of the account can be attributed to Theophrastos, and since it is known that Theophrastos dealt with ostracism in the *Nomoi*, it may be suggested that Theophrastos is the author of the entire account. His treatment of ostracism was used directly by Timaios and Philochoros, and perhaps by Pollux and Plutarch. It was used indirectly, through Timaios, by Diodoros, and, through Philochoros, by Didymos; all later scholiasts, and lexicographers depend on Didymos, all later authors on Plutarch. It is idle to speculate about the sources of Theophrastos beyond the fact that he used material which had been assembled by Aristotle (and included the Atthis of Androtion); for Aristotle and Plato were still able to draw upon the living traditions of the fifth century.

Appendix I. *The duration of ostracism*

The phrase ὕστερον δὲ ἐγένοντο πέντε (Philochoros) refers to the days within which the ostracized had to settle their private affairs (see above, pp. 86–87); this is shown not only by the fact that there is no evidence for ostracism lasting five years and also by the plural ἐγένοντο which must refer to ἡμέραι and not to the neuter ἔτη. For the phrase ὑπὲρ τῶν ἰδίων συναλλαγμάτων, see Demosthenes, 24, 213, and compare J. H. Lipsius, *Das Attische Recht* II/2, p. 683; *R.E.*, s.v. Symbolaion and συνάλλαγμα. For the meaning of τὰ δίκαια, see Thucydides, 3, 54, 1, and Xenophon, *Anabasis*, 7, 7, 14 and 17, and compare E. Weiss, *Griech. Privaterecht*, pp. 22, note 17 a, and 23. The whole phrase evidently refers to debts which could be quickly settled. There is probably no connection between the "five" (days) mentioned by Philochoros and the exile for five years of which Diodoros-Timaios speaks (see above, p. 97); petalism was for five years, and Diodoros evidently thought that this was also true for ostracism. Periods of ten days are known from fifth-century inscriptions (*I.G.*, I² 39, line 13; 71, line 12; 98, line 15), but so are periods of five days (*I.G.*, I² 55, line 8; 76, line 18; *S.E.G.*, 10, 20, lines 21–23; 12, 32, line 11). The change from the ten- to the five-day period may have been made when the residence requirement was changed just before the battle of Salamis.

Appendix II. *The residence requirement of ostracism*

The words used by Philochoros μὴ ἐπιβαίνοντα ἐντὸς πέρα τοῦ Εὐβοίας ἀκρωτηρίου (see above, p. 87) are difficult to understand, but ἐντός rather than πέρα τοῦ is corrupt, as Jacoby has seen (Commentary on *F. Gr. Hist.* 328 F 30, pp. 317–318), although he himself asserted (Notes, p. 228, note 18) that "the idea of the provision is...that one did not wish to have them in the immediate neighbourhood of Athens." I think the very opposite is the case.

According to Philochoros, the ostracized are not to set foot (embark) beyond the promontory of Euboia which would be in Greek μὴ ἐπιβαίνοντα εἰς τὸ πέραν τοῦ Εὐβοίας ἀκρωτηρίου, and this may be what Philochoros actually wrote; compare Xenophon, *Anabasis*, 3,5,2: διαβιβαζόμεναι εἰς τὸ πέραν τοῦ ποταμοῦ; *Hellenica*, 1, 3,17: διέβη ... εἰς τὸ πέραν. According to Aristotle (*Ath. Pol.* 22 8), the ostracized were ordered to reside on this side of Geraistos and Skyllaion or to be | without rights once and for all; the emphasis lies here clearly on the threat of ἀτιμία. The whole problem has been confused by Dobree's emendation of πέρα τοῦ into Γεραιστοῦ ⟨τοῦ⟩ which has been generally accepted, especially after the text of Aristotle was discovered. And yet, this emendation, in turn, caused the emendation of Aristotle's text (from ἐντός to ἐκτός, or the insertion of μή), so that the present view of the residence requirement is based on two emendations. P. Goossens has recently tried to defend the text of Aristotle (*Chron. d'Égypte* 39/40, 1945, pp. 125–133), and to maintain that according to Aristotle the victims of ostracism were to reside to the West (i.e., within) of a line determined by Geraistos and Skyllaion; this is in keeping with the general observation (above, pp. 92–93) that to the ostracized τόπος ἀπεδίδοτο. For the phrase ἐντὸς Γεραιστοῦ καὶ Σκυλλαίου, reference may be made to the clause of the Peace of Kallias according to which Persian ships were not to sail ἐντὸς Κυανέων καὶ Φασηλίδος (see H. T. Wade-Gery, *H.S.C.P.* Suppl. 1, 1940, pp. 134–136; R. Sealy, *Historia* 3, 1955, pp. 325–326; J. H. Oliver, *Historia* 6, 1957, pp. 254–255). Similarly, Herodotos, after giving a list of the Greeks who assembled at Troizen (8, 43–46), says (8, 47) that all these who were living on this side of the Thesprotians and the Acheron River (ἐντὸς ... Θεσπρωτῶν καὶ 'Αχέροντος ποταμοῦ) went to war; but of those living beyond them (τῶν δὲ ἐκτὸς τούτων οἰκημένων), only the Krotonians came to the aid of Greece. Herodotos evidently referred to the Thesprotians and to the Acheron River as to the same area which is once defined by the people who live in it, the other time by a river which passes through it. Aristotle's reference to Geraistos and Skyllaion may be understood in the same way: Geraistos was the easternmost point of Greece to the north of Attika, Skyllaion to the south of Attika. The ostracized were to stay within the bounds of Greece proper as she existed on the eve of Xerxes' invasion. According to Philochoros, they were not to go beyond the promotory of Euboia (Geraistos); for a similar negative provision, see Andokides, 1, 76: τοῖς δὲ μὴ ἀναπλεῦσαι εἰς Ἑλλήσποντον, ἄλλοις εἰς Ἰωνίαν. During the fifth century, the eastern frontier of Greece, especially of the Athenian Empire, was pushed to the shore of Asia Minor, and this explains why neither Kimon | nor Hyperbolos obeyed the residence requirement. It must also be remembered that after the ostracism of Themistokles the victims of ostracism (men such as Alkibiades, Kimon, Thucydides) were suspected of siding with Sparta rather than with Persia, and they may have been expected (and even ordered) to stay within the territory of the Athenian Empire (ἐντὸς τῶν πόλεων ὧν 'Αθηναῖοι κρατοῦσιν; see *I.G.*, I² 56, lines 14–15); this would explain the cases of Kimon (who went to the Cherronesos) and Hyperbolos (who went to Samos).

Appendix III. *The ostracism of Hipparchos*

It is generally assumed that the demotic of Hipparchos was Κολλυτεύς (Aristotle) and not Χολαργεύς (Plutarch, *Nicias* 11, 6, quoted above, p. 100), but this is by no means certain; Aristotle (*Ath. Pol.* 22, 4) may have made a mistake or a scribe may have mistaken ΧΟΛΑΡΓΕΥΣ for ΚΟΛΛΥΤΕΥΣ (the papyrus has ΚΟΛΥΤΤΕΥΣ). There can be no doubt that Theophrastos-Philochoros spoke not only of the "last" ostracism but also of its beginning; see above, pp. 87–88. Even Jacoby who claimed (Commentary on *F. Gr. Hist.* 328 F 30, p. 316) that the historical part of the quotation from Philochoros is "probably not (or at least not immediately) taken from Ph.," admitted that "it is, of course, not impossible that he gave a general summary, perhaps when the procedure was first, or when it was for the last time, put into practice; but he must have entered the passing of the law and each ostracism under its proper year." There is, in fact, no evidence that either he or anyone else did this. On the contrary, not a single ostracism is dated in our tradition by the archon under whom it occurred, and this is true even for Aristotle's *Ath. Pol.* 22 with its "dated" ostracisms. The first use (i.e., the introduction of the law; see my comments, *A.J.A.* 55, 1951, pp. 221–229 [*supra*, pp. 65–76], and Jacoby's "last word" in the Notes on the Commentary on *F. Gr. Hist.* 324 F 6, pp. 530–532; compare M. Ostwald, *T.A.P.A.* 86, 1955, p. 110, note 33) of ostracism is "dated" two years after Marathon, but no archon is mentioned. The second ostracism, that of Megakles, is "dated" in the archonship of Telesinos, but it is the electoral law which carried the archon's name, not the ostracism. The ostracism of Xanthippos is "dated" in the fourth year after that of Hipparchos, but the archon of this year is not mentioned. The ostracism of Aristeides, like that of Megakles, is "dated" in the archonship of Nikodemos, but it is actually the building of the fleet which carried the archon's name, not the ostracism. Finally, the recall of the ostracized is dated in the archonship of Hypsichides, but this was a decree and not an ostracism. While it is, therefore, unjustified to assume that each ostracism was entered under its proper year, either by Philochoros or by anybody else, it may be confidently asserted that Philochoros, and before him Theophrastos, gave a (perhaps incomplete) list of the victims of ostracism (see Jacoby's Notes on the Commentary of *F. Gr. Hist.* 328 F 30, p. 228, note 5) and mentioned especially the first and last men to be ostracized. Concerning the first ostracism, that of Hipparchos, Philochoros certainly said that the practice of ostracism began with the enactment of the law by Kleisthenes who wished to drive out the friends of the tyrants. This characterization of ostracism agrees so completely with the account of Aristotle (who followed Androtion) that it may be presumed that Philochoros (and before him Theophrastos) named the first victim Hipparchos. This assumption is in fact confirmed by Plutarch (see above, pp. 99–100) who mentions in this context that Hipparchos, a relative of the tyrant, was the first to be ostracized.

Before it can be argued whether or not Aristotle's account of ostracism as a whole (*Ath. Pol.* 22) was repeated by Theophrastos (and Philochoros), the identity of Hipparchos and the circumstances of his ostracism must be examined. We know from Kleidemos (*F. Gr. Hist*. 323 F 15 = 140 F 6) that Hippias married the daughter of Charmos; Charmos, the father of Hipparchos, was then an offspring of this marriage, and Hipparchos was a grandson of Hippias and a great grandson of Peisistratos. Hippias himself married a second time, Myrrhine the daughter of Kallias the son of Hyperochides, and one must not identify the two women, as Jacoby (following Wade-Gery) has recently reasserted (Commentary on *F. Gr. Hist*. 323 F 15, p. 71). One may even consider the possibility that Peisi|stratos the son of Hippias (Thucydides, 1, 54–55; *I.G.* I^2 761– *S.E.G.* X 318; *S.E.G.* X 352, line 7) was a brother of Charmos and not a son of Myrrhine. One can now understand why Hipparchos should have stayed in Athens when the sons and grandsons of Hippias (from Myrrhine) left; he was too young to share in the tyranny and was not even living in the tyrant's house. Later on, he became the ranking member of the family and the leader of the tyrant's friends in Athens. Hipparchos is mentioned only twice in our literary tradition (apart from Dion. Hal. 5, 77,6 = 6, 1,1; see T. J. Cadoux, *J.H.S.* 68, 1949, p. 116): by Androtion (*F. Gr. Hist*. 324 F 6; on him depend Aristotle and Theophrastos-Philochoros-Plutarch) and by Lykourgos (*In Leocratem*, 117–119). It has been suggested that the condemnation of Hipparchos (reported by Lykourgos) followed his ostracism either immediately (see my comments in *A.J.A.* 55, 1951, p. 229, note 29 B [*supra*, p. 74, note 29 B]) or after the battle of Salamis (J. Labarbe, *La loi navale de Thémistocle*, p. 100, note 2); in either case, Lykourgos would have made no mention of Hipparchos's ostracism or Androtion of his condemnation. The possibility must be considered that Lykourgos actually speaks of Hipparchos's ostracism, and that we can learn from him some important details concerning the early history of ostracism, namely the three years during which, according to Aristotle (*Ath. Pol.* 22 6), the friends of the tyrants were ostracized. The document used by Lykourgos (118) consisted of a bronze stele on which were engraved a) the decree ordering the removal of the bronze statue of the traitor Hipparchos and the erection of the bronze stele fashioned out of the melted down statue; b) the hypogramma consisting of the name of Hipparchos and of the names of others which had been added later (this list had the heading: ἀλιτήριοι καὶ προδόται). It so happens that the words ἀλειτηρὸς and προδότης occur on ostraka (*A.J.A.* 51, 1947, pp. 257–262 [*infra*, pp. 108–115]; *Hesperia* 19, 1950, p. 379, and note 13), and F. Walton in fact suggested (*Studies Presented to D. M. Robinson* 2, p. 604) that the writer of the Xanthippos ostrakon had the stele of Hipparchos in mind when he called Xanthippos ἀλειτηρόν. A further confirmation of the assumption that the first victims of ostracism were considered as traitors can be found in Herodotos who says (6, 124) that the | Alkmeonidai may have betrayed (προεδίδοσαν) the city. Aristotle makes it clear, moreover, that the first three victims of ostracism were considered as traitors since they were friends of the tyrants; their names could have stood on a stele such as described by Lykourgos. All this would agree with Lykourgos's account of the fate of Hipparchos, and his "trial" would be his ostracism. Only the death penalty pronounced against Hipparchos does not fit the picture, and one would have to assume either that Lykourgos made a mistake, or that the death penalty was pronounced after Hipparchos left Athens and went to Persia. According to the later law, Hipparchos would automatically have been declared ἄτιμος, but in the earlier period a special decree may have been necessary. The stele mentioned by Lykourgos contained accordingly only three names, and it may be significant that the first two names given by Aristotle are listed by him with father's name and demotic, while the names of Xanthippos and Aristeides have only patronymics.

If Theophrastos-Philochoros used Androtion-Aristotle for the account of Hipparchos's ostracism and of the enactment of the law by Kleisthenes, it may be presumed that they also added the information concerning the other ostracisms of the eighties which Aristotle offers. This assumption is made very unlikely by the following consideration. While the ostracism of Aristeides is well known to our literary tradition, it is nowhere, not even in Plutarch, related to the dispute over the use of the Laurion silver; Aristotle made this connection by saying that Aristeides was ostracized ἐν τούτοις τοῖς καιροῖς (which means "on that occasion" and is not a "vague phrase" as T. J. Cadoux, *J.H.S.* 68, 1949, p. 118, suggested). Moreover, the ostracism of Megakles is mentioned not only by Aristotle but also by Andokides (4, 34) and Lysias (14, 39) who used Andokides (see my comments in *T.A.P.A.* 79, 1948, pp. 203–204 [*infra*, pp. 125–127]), but there is no further mention of it in our literary tradition, although Megakles himself, his ancestors, and his descendants have received considerable attention, and his ostracism could have been mentioned in the Scholia on Pindar, *Pythia*, 7, lines 14–15. Finally, the ostracism of Xanthippos is entirely absent from our literary tradition (except for Aristotle) although both Xanthippos himself and his more famous son Perikles are well-known figures; it is curious that Plutarch tells the story of Xanthippos's dog (*Themistocles*, 10, 6 = Cato Maior, 5, 4) which is attributed by Aelian (*De Nat. Anim.*, 12, 35) to Aristotle and Philochoros (*F. Gr. Hist.* 328 F 116), and which presupposes (without mentioning) Xanthippos's return from ostracism before the battle of Salamis. For all these reasons, it seems highly unlikely that Theophrastos mentioned the ostracisms of Megakles and Xanthippos or, if he did, that Philochoros repeated this information.

Appendix IV. *The cause of ostracism*

Svend Ranulf devoted a chapter of his book (*The Jealousy of the Gods* I, pp. 132–142) to an account of ostracism in which he maintained that jealousy was its main cause. It has been pointed out before (pp. 86, note 4, and 100) that the ostracism of Aristeides was attributed to jealousy, and the same is true for that of Themistokles (pp. 96–97) and of Kimon (Nepos, *Cimon*, III); in fact, jealousy was given as the general cause of ostracism by Theophrastos (pp. 90–91). This political aspect of φθόνος appears already in the fifth-century, and jealousy was even then considered one of the causes of ostracism. Pindar speaks (*Pythia*, VII, line 19) of jealousy in connection with Megakles (who was ostracized), but the poem has now been dated in 490 B.C. (by H. C. Bennett,

H.S.C.P. 62, 1957, pp. 69–74), and if this date is correct the poem cannot refer to Megakles' ostracism which occurred later (see above, pp. 104–105). There is, however, a striking passage in Sophokles, *Ajax*, lines 154–163, which agrees with Pindar, *Nemea* 8, lines 21–23, and which illustrates the envy to which great men are exposed; see N. O. Brown, *T.A.P.A.* 82, 1952, pp. 16–17. It is hard to avoid the impression that Sophokles was thinking here of men like Themistokles, Kimon, and Thucydides, and especially of Themistokles who lent more than one feature to the picture of *Ajax*. It would seem, therefore, that we know one of the original causes of ostracism (φθόνος δόξης or ἀρετῆς) as well as its original purpose (κολούειν τοὺς ὑπερέχοντας; see above, p. 99, note 8).

13

The Ostracism of Xanthippos

[257] Within the last few years Athenian politics between 510 and 480 B.C. have been the object of many studies conducted by scholars on both sides of the Atlantic Ocean.[1] Without joining their rank, I am here submitting for their consideration a piece of evidence which may be of great value once its true meaning is understood.[2]

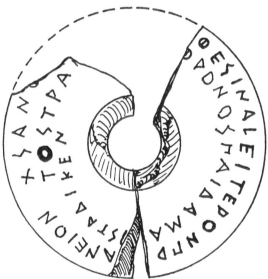

Figure 13.1.

American Journal of Archaeology 51 (1947) 257–262

1. The most recent note by C. A. Robinson, Jr. (*AJP*, 67, 1946, pp. 265–266) may be used as a guide to earlier bibliography.
2. This is the first in a series of preliminary studies on Athenian ostracism, designed to prepare the way for the final publication of the ostraka found at Athens. I wish to thank Professors W. S. Ferguson and C. A. Robinson for the help they have given me in the study of this ostrakon.

Among the ostraka found in the Agora Excavations there is one which is unusual both in the form and contents of its inscription[3] (see Figure 13.1). Instead of carrying the full name of the person to be ostracized, it contains a couplet consisting of an hexameter and a pentameter. Moreover, the Athenian who incised and probably also made the little poem apparently stated the reason why his candidate should be ostracized.

Χσάνθ[ιππον κατά]φεσιν ἀλειτερὸν πρ[υτ]ανεῖον
τὄστρακ[ον 'Αρρί]φρονος παῖδα μά[λ]ιστ' ἀδικέν.

The restoration of the pentameter is certain, since the name of Xanthippos's father is known, and since τὄστρακ[ον] is the only possible restoration of the first word. This restoration offers a slight metrical difficulty.[4] In the hexameter, the restoration of [κατά]φεσιν is uncertain; there may have been another four-letter word (like τόδε) beginning with a consonant.[5] The real difficulty lies in the last word of the hexameter, πρυτανειον. It is uncertain whether it should be transcribed as πρυτανεῖον (acc. sing. of πρυτανεῖον), πρυτανείων (gen. plur. of πρυτανεῖον), or πρυτανειῶν (gen. plur. of πρυτανεία). Moreover, it is equally uncertain whether it is a genitive (πρυτανείων or πρυτανειῶν) depending on ἀλειτηρόν or an accusative object (πρυτανεῖον) of ἀδικεῖν. All three possibilities will be considered in the following discussion which also contains a brief review of the life of Xanthippos.

In 489 B.C., Xanthippos served as the accuser of Miltiades (Herodotos, 6, 136; compare Nepos, *Miltiades*, 7), and at that time Miltiades would have been condemned to die had he not been saved by the presiding officer of the assembly (Plato, *Gorgias*, p. 516 E).[6] It is tempting to combine Plato's allusion to the

[258]

---

3. Inv. no. P. 16873. Diameter, 0.073 m. Found on May 2, 1940, on the lower part of the western slope of the Areopagus, near the road that skirts the west end of the hill, in Section NN, in early fifth-century fill, together with ostraka of Aristeides, Themistokles, and Hippokrates, son of Alkmeonides. This information was given to me by E. Vanderpool with whom I am studying the Athenian ostraka. The drawing illustrated in Figure 13.1 has been prepared by Mr. Travlos.

4. The second syllable of 'Αρρίφρονος may be long, but it must here be counted as short as in Simonides, no. 159 (Diehl), Epimenides, no. 19 (Diels-Kranz), *IG*, i², 700; in these instances, all dating from ca. 500 B.C., 'Αφροδίτη must be read with a short first syllable. The first syllable of 'Αρρίφρων, on the other hand, must be long, and this is confirmed by the spelling with double rho as it occurs on most of the ostraka found so far; see *IG*, i², 909; T. L. Shear, *Hesperia*, 5, 1936, pp. 39–40, Figure 39; *Hesperia*, 10, 1941, pp. 2–3, Figure 3. The literary sources, however, spell the name with only one rho, and it has been suggested to correct this spelling following the epigraphical evidence; see W. Dittenberger, *Sylloge*,³ no. 27, note 2: *distinguendum igitur hoc nomen* ['Αρρίφρων] *origine et usu ab illo* ['Αρίφρων; compare *IG*, ii², 3092, line 6], *et apud scriptores duplex liquida restituenda. Sine dubio hoc Attici sermonis legibus natum est ex* 'Αρσίφρων; see also U.v. Wilamowitz, *Arist. und Athen*, ii, p. 84, note 25. For the spelling Αρίφρων, see Herodotos, 6, 131, 2; 7, 33; 8, 131, 2; Aristotle, 'Αθ. Πολ., 22, 6; Aristodemos, *F. Gr. Hist.*, 2, no. 104, p. 496, lines 9–11; Diodoros, 11, 27, 3; Plutarch, *Alkibiades*, 1, 1 (referring to Perikles' brother); Pausanias, 3, 9, 7.

5. The only other occurrence of κατάφημι in the fifth-century is found in Sophokles' *King Oedipus*, line 505.

6. See W. W. Howe, *JHS*, 39, 1919, pp. 60–61; G. Busolt and H. Swoboda, *Griech. Staatskunde*, pp. 888–889, note 6; H. Berve, *Miltiades*, pp. 99–101; R. J. Bonner and G. Smith, *Administration of Justice*, 1, p. 207; A. W. Gomme, *AJP*, 65, 1944, pp. 324–325.

prytanis with the phrase of the ostrakon condemning Xanthippos as ἀλειτηρὸν πρυτανειῶν, but it is not quite clear what Xanthippos actually may have done to deserve this designation.[7]

Aristotle took notice of the hostility existing between Miltiades and Xanthippos when he stated ('Aθ. Πολ., 28, 2) that Xanthippos followed Kleisthenes as the leader of the people while Miltiades stood at the head of the aristocrats. We do not know the date of Xanthippos's leadership. It may have begun with the death of Kleisthenes (the date of which is unknown) and it certainly terminated with the ostracism of Xanthippos in 484 B.C. It is possible, therefore, that Xanthippos (not Themistokles; see C. A. Robinson, *AJP*, lxvii, 1946, pp. 265–266) was the leading democratic statesman when the reforms of 487 took place. These reforms undoubtedly increased the power of the boule and ekklesia, and consequently of the ruling committee of the boule, which presided over the meetings of both the boule and ekklesia. These committees were called πρυτανεῖαι, and it may be that the writer of the ostrakon had in mind Xanthippos's activity in 487 B.C. when he called him ἀλειτηρὸν πρυτανειῶν. It remains unclear, however, in what way Xanthippos may have done violence to the prytanies when he increased their power.[8]

[259] It may be remembered, in this connection, that Herodotos (6, 110) uses the term πρυτανηίη in order to designate the supreme command which rotated among the generals. Unfortunately, however, we do not know whether this rotation was based on the lot or (more likely) on the sequence of the tribes in their official order. It is barely conceivable that Xanthippos, who later served as military commander may have tried to violate this principle of rotation, just as it had been violated before the battle of Marathon in order to give the command to Miltiades.

The ostracism of Xanthippos, which had been mentioned in the summary of Aristotle made by Herakleides (*FHG*, ii, p. 209, no. 7), is fully discussed by Aristotle, 'Aθ. Πολ., 22, 6. Aristotle gives the date of the ostracism as 484 B.C., and he emphasizes that Xanthippos had nothing to do with the tyrant's party

---

7. It is possible that we should read the last word of the hexameter as πρυτανείων (gen. of πρυτανεῖα, always used in the plural) meaning "court fees." The ancient character of these fees is attested not only by their name (compare J. H. Lipsius, *Das attische Recht*, iii, pp. 824–825; R. J. Bonner and G. Smith, *Administration of Justice*, i, 63), but also by the occurrence of this term on several Attic decrees the earliest of which belongs to 485/4 B.C. (compare A. Wilhelm, *Sb. Ak. Wien*, 217, 5, 1939, pp. 19–21; *SEG*, 10, no. 4).

8. Unless one assumes that Xanthippos was responsible for introducing the lot in order to determine the sequence of the prytanies. This presupposes that in the Kleisthenian reform the ten prytanies followed each other in the official order of the tribes, an assumption once made by A. Mommsen but seriously questioned by W. S. Ferguson, *The Athenian Secretaries*, pp. 3–4. Ferguson now kindly informs me (in a letter dated September 8, 1946): "I see no reason for thinking that the allotted order of the tribes in the prytany did not originate with Kleisthenes; but I grant that the basis for this opinion—the lack of a report of subsequent change—is fragile, in view of the meagerness of such reports for the period prior to Pericles." On the other hand, he suggests (without approval): "if the prytaneis...of the sixth prytany had let it be known that they had Xanthippus in mind in authorizing resort to ostracism in 485/4 B.C., Xanthippus may have attacked the prytaneiai as partisan and hence improper custodians of the integrity of the voting."

in Athens, but he fails to offer a satisfactory reason for this ostracism.[9] One may consider, however, the fact that Xanthippos had married, perhaps some years before 490 B.C., Agariste the daughter of Hippokrates who was a brother of Kleisthenes; see Herodotos, 6, 131, 2. In this way, Xanthippos had associated himself with the family of the Alkmeonidai, and this association proved dangerous not only to his son Perikles but also to himself.

The Alkmeonidai had been under a curse since the Kylonian conspiracy, and this curse was responsible for the brief exile of Kleisthenes. Moreover, three other members of the family were either ostracized or came in danger of being ostracized during the years preceding the battle of Salamis. In addition to Megakles, who was presumably the brother-in-law of Xanthippos and who was ostracized in 486 B.C.,[10] we know of Hippokrates, son of Alkmeonides, and of Hippokrates, son of Anaxileos, who both belonged to the Alkmeonid family,[11] and whose names appear on more than eighty ostraka. The ostrakon of Xanthippos shows clearly that Xanthippos, too, was thought to be under the curse, for the adjective ἀλειτηρός was commonly used for those under the curse; see Thucydides, 1, 126, 11; Aristophanes, *Knights*, lines 445–446 (and the scholion on this passage).[12]

It may be assumed, therefore, that the occurrence of the word ἀλειτηρός on the ostrakon refers to Xanthippos's connection, through his marriage to Agariste, with the accursed clan | of the Alkmeonidai.[13] In accordance with this assumption, Ferguson and A. D. Nock suggested (in a letter dated September 30, 1946) that πρυτανεῖον should be taken as the inner object of ἀδικεῖν substituted for the usual πόλιν (or δῆμον) to satisfy poetic diction and the requirement of meter, and at the same time to give special point to ἀλειτηρόν by alleging a profanation of the sacred hearth. This accusation may have been

[260]

---

9. Modern commentators have not been more successful; see J. Carcopino, *L'Ostracism athénien*[2], pp. 148–149; O. W. Reinmuth, *RE*, s.v. ostrakismos, col. 1680, lines 48–52.

10. Aristotle links him, without giving any reason, with the party of the tyrants and indicates that this association accounted for his ostracism; Aristotle presumably believed in the rumor that the Alkmeonidai acted treacherously after the battle of Marathon.

11. See the stemma of the family given by E. Vanderpool, *Hesperia*, 15, 1946, p. 275. Anaxileos (see Vanderpool, *l.c.*, p. 272, no. 6) may have been a brother of Megakles (II) and Alkmeonides, and his name may be restored at the beginning of *IG*, i², 472, as will be discussed elsewhere.

12. See also Suidas, s.v. ἀλιτήριος. The form ἀλειτηρός, now attested by the ostrakon, should be retained not only in Alkman (no. 72 D.) but also in Sophokles' *King Oedipus*, line 371, where it is confirmed by the best manuscripts. F. Blass was right in correcting the texts of the orators, and W. H. P. Hatch's criticism (*HSCP*, 19, 1908, p. 162, note 3) is unjustified.

13. Kimon, the son of Miltiades, was technically in exactly the same position as Xanthippos, since he, too, was married to an Alkmeonid woman, Isodike, the granddaughter of Megakles; see A. W. Gomme, *AJP*, 65, 1944, p. 325. Kimon's sons Lakedaimonios, Oulios, and Thessalos were, therefore, just as accursed as Perikles. It may be suggested to amend the difficult passage in Plutarch's *Kimon* (16, 1) in which Stesimbrotos is credited with reporting that Perikles attacked the sons of Kimon by reproaching them for their μητρῷον γένος and by pointing out that they were born ἐκ γυναικὸς ... κλιτορίας (the emendation Κλειτορίας is based on Plutarch's *Perikles*, 29, 3, which is but an elaboration of the same story). Instead of supposing the existence of a "second" wife of Kimon (see *RE*, s.v. Oulios), one may read ΑΛΙΤΗΡΙΑΣ for ΚΛΙΤΟΡΙΑΣ and see in it a reference to Isodike who was indeed the mother of Kimon's three sons Lakedaimonios, Oulios, and Thessalos.

particularly justified if Xanthippos really belonged to the priestly family of the Bouzygai whose members had to perform certain sacred rites.[14] The fact that Xanthippos's wife, by being an Alkmeonid, was disqualified from participating in or performing certain sacred sacrifices would have been a source of great trouble to Xanthippos. Yet it should be understood that the term ἀλειτηρός was used in the fifth century also in a much wider meaning. The orator Lykourgos, referring to the fate accorded to Hipparchos son of Charmos, who was ostracized in 487 B.C., declares (*In Leocr.*, 117) that the Athenians removed Hipparchos's statue from the Akropolis and used the metal to make a stele on which were engraved the names of traitors and ἀλειτήριοι.[15] It is reasonable to assume that the word ἀλειτήριος was used in the ancient decree to which Lykourgos referred. Similarly, the philosopher Protagoras is called ἀλιτήριος by the comic poet Eupolis, possibly in connection with the ἀσέβεια trial he had to undergo.[16] We cannot be absolutely sure, therefore, that the term ἀλειτηρός was applied to Xanthippos only because he was married to Agariste and not because the writer of the ostrakon put him into the same group as those condemned before him, the ἀλειτήριοι καὶ προδόται. It is reasonable to assume, however, that the word ἀλειτηρός was employed because it had been used in contemporary documents referring to enemies of the state or of the Gods. If Xanthippos is attacked for violating the sanctity of the prytaneion, and this seems the most likely interpretation, his crime lay not only in his affiliation with the Alkmeonidai but especially in some official or religious position which he occupied and for which his marriage may have disqualified him.[17]

Xanthippos was recalled together with the other exiles just before the battle of Salamis, and a touching story is told by Plutarch (*Themistokles*, 10, 6) of the loyalty of Xanthippos's dog during the evacuation to Salamis. In the following year 479/8, Xanthippos was elected general, and in this capacity he defeated the enemy at Mykale and pursued him victoriously into the northern Aegean.[18] Before this, he had gone together with Kimon and Myronides | as Athenian representative to Sparta.[19] We do not know when Xanthippos died, nor do we know at what time his statue was erected on the Athenian Akropolis.[20]

---

14. See Schol. Aristeides, 3, p. 473 D.; compare F. Miltner, *RE*, s.v. Perikles, col. 748, lines 23–28.

15. For comments on this passage, see *Hesperia*, 8, 1939, p. 155–164 [*infra*, pp. 195–203]; W. B. Dinsmoor, *Studies in the History of Culture*, p. 196, note 14.

16. These and other less important occurrences of ἀλιτήριος and its more archaic form ἀλιτρός can be found in the good collection of the material made by W. H. P. Hatch, *HSCP*, 19, 1908, pp. 157–165 (but see above, note 12, and below, Appendix on the *Famine of Athens*).

17. For the part played by the prytanies in the Kylonian affair, see J. H. Wright, *HSCP*, 3, 1892, pp. 30–31, note 2; compare also R. J. Bonner and G. Smith, *Administration of Justice*, 1, pp. 104–111.

18. See Herodotos, 7, 33; 8, 131, 3; 9, 114, 2; 9, 120, 4; Aristodemos, *F. Gr. Hist.*, 2, no. 104, p. 496, lines 9–11; Diodoros, 11, 27, 3; 34, 2; 36, 5; 37, 5; Plutarch, *Perikles*, 3, 2; Pausanias, 3, 7, 9. It may have been Xanthippos's victory at Mykale which Phrynichos celebrated in the Phoenissae (see F. Marx, *Rh. Mus.*, 77, 1928, pp. 337–360) in 476 B.C., with Themistokles as choregos (Plutarch, Themistokles, 5, 4).

19. See Plutarch, *Aristeides*, 10, 8.

20. Pausanias still saw the statue (1, 25, 1).

There is no doubt that he belonged with Aristeides and Themistokles to the "democratic" party. Aristeides owed his ostracism to his opposition to Themistokles; we do not know whether the ostracism of Xanthippos was caused by the same split in the "party." The accusation hurled against Xanthippos by the writer of the ostrakon may have referred to the real reason of the ostracism, or it may have been inspired by personal animosity.

A NOTE ON THE "FAMINE OF ATHENS"

Hatch's study of the use of ἀλιτήριος (see above, notes 12 and 16) was confined to the "Greek authors from the earliest times down to about 300 B.C." He therefore put aside (*l.c.*, p. 160) a large group of references found in Plutarch, the scholiasts, and the lexicographers, references which apparently refer to an event in the early history of Athens. Since the usage of ἀλιτήριος is confined in the classical period to Attic writers, and since this word is considered Attic by the Atticists, it may be permitted to discuss in detail this group of quotations referring to a famine in Athens.

The story is told in various versions. Plutarch reports (*De curiositate*, p. 523 A–B): λέγεται δὲ καὶ τὸν ἀλιτήριον ἐκ φιλοπραγμοσύνης κατονομασθῆναι τὸ πρῶτον. λιμοῦ γὰρ ὡς ἔοικεν Ἀθηναίοις ἰσχυροῦ γενομένου, καὶ τῶν ἐχόντων πυρὸν εἰς μέσον οὐ φερόντων ἀλλὰ κρύφα καὶ νύκτωρ ἐν ταῖς οἰκίαις ἀλούντων, περιιόντες ἐτήρουν τῶν μύλων τὸν ψόφον εἶτ' ἀλιτήριοι προσηγορεύθησαν.[21] The same story is accredited to Ailios Dionysios by the scholiast of Plato's *Republic* (p. 470 d; compare also the scholion on Plato's *Laws*, p. 854 b): κατὰ Διονύσιον γὰρ τὸν Ἁλικαρνασέα, λιμοῦ γενομένου Ἀθήνησιν οἱ πένητες τὰ προφερόμενα ἄλευρα διήρπαζον... διέτεινε δὲ τὸ ὄνομα ὥστε καὶ ἐπὶ πάντων τῶν μετὰ βίας τι π⟨ο⟩ιούντων.[22] A somewhat more detailed account is given by Schol. Aristophanes, *Knights*, line 445 (see also Suidas, s.v. ἀλιτήριοι), where one reads λιμὸς κατέλαβέ ποτε τοὺς Ἀθηναίους, καὶ οἱ πένητες τὰ τῶν ἀλούντων ἄλευρα διήρπαζον· ἀπὸ γοῦν ἐκείνων καταχρηστικῶς τοὺς πονηροὺς ἀλιτηρίους ἐκάλουν. παρέτεινε δὲ τὸ ὄνομα καὶ ἐπὶ τῶν μετὰ βίας τι ποιούντων, ἀπὸ τῆς σιτοδείας τῆς κατὰ τὸν Αἰτωλικὸν πόλεμον γενομένης.[23]

21. Yet Plutarch himself rejects this etymology (*Quaestiones Graecae*, p. 297 A, no. 25): οὐ γὰρ πειστέον τοῖς λέγουσι ἀλιτηρίους κεκλῆσθαι τοὺς ἐπιτηροῦντας ἐν τῷ λιμῷ τὸν ἀλοῦντα καὶ διαρπάζοντας·... ἀλιτήριος δ' ὃν ἀλεύασθαι καὶ φυλάξασθαι διὰ μοχθηρίαν καλῶς εἶχε. This etymology closely resembles that given for πρυτανεῖον and discussed by E. Vanderpool, *Hesperia*, 4, 1935, p. 471, note 8: πρυτανεῖον δὲ ὠνόμασται ἐπεὶ πυρῶν ἦν ταμιεῖον (found in Timaios, Photios, Suidas).

22. For a discussion of the original authorship of this note, see L. Cohn, *Jahrb. f. Class. Phil.*, Suppl. 13, 1884, pp. 821–822; E. Schwabe, *Aelii Dionysii et Pausaniae Atticistarum fragmenta*, pp. 22 and 100, note 36, 1.

23. The same story is repeated in the *Anecdota Graeca*, ed. I Bekker, i, p. 377, lines 11–21, and in the *Etymologicum Magnum*, s.v. ἀλιτήριοι there the following notice is added (see also *Et. Gudianum*, s.v. Ἀλιτηρία: Ἀλιτηρία Δημήτηρ καὶ Ἀλιτήριος Ζεύς· ὅτι λιμοῦ συντόνου γενομένου, τοὺς ἀλοῦντας πάντας ἐφύλαττον, διὰ τὸ μὴ κλέπτειν τι τῶν ἀλουμένων. Ὡς οὖν ἐπόπτας καὶ τηρητὰς τῶν ἀλουμένων, τοὺς θεοὺς οὕτως ὠνόμασαν. This is the only reference to Ζεὺς Ἀλιτήριος) (and also to Δημήτηρ Ἀλιτηρία), but it is tempting to combine it with the similarly unique reference to Ζεὺς Ἀλεξιτήριος or Ἀλεξητήριος found in Aischylos's *Seven*

Whatever one may think of the etymology of ἀλιτήριος given in these accounts, it seems safe to assume that it was derived from the use of the term ἀλιτήριοι in connection with the story of the "famine of Athens" which took place during the "Aetolian" war. On the other hand, the additional information that certain people were called ἀλιτήριοι because they kept watch τηρεῖν of flour (ἄλευρα) or of those who ground it (ἀλεῖν) may have been derived | from the faulty etymology given for ἀλιτήριοι. Only one of the three essential elements of the story (λιμός, ἀλιτήριοι, Αἰτωλικὸς πόλεμος) can be dated. If the story is at all historical, it should belong to the sixth or fifth century B.C., since the term ἀλιτήριοι was an Attic term of this period, and was used later only with reference to its earlier usage. On the other hand, we do not seem to know of any occurrence of a famine in Athens during that period, nor do we know of any Aetolian war in which Athens was involved either then or at any other time.[24] The suggestion may be made, therefore, that the words λιμός and λοιμός were confused,[25] and that the Αἰτωλικὸς πόλεμος was in fact the Δωριακὸς πόλεμος to which Thucydides refers (2, 54, 1-2): ἐν δὲ τῷ κακῷ [the plague] οἷα εἰκὸς ἀνεμνήσθησαν καὶ τοῦδε τοῦ ἔπους, φάσκοντες οἱ πρεσβύτεροι πάλαι ᾄδεσθαι "ἥξει Δωριακὸς πόλεμος καὶ λοιμὸς ἅμ' αὐτῷ." ἐγένετο μὲν οὖν ἔρις τοῖς ἀνθρώποις μὴ λοιμὸν ὠνομάσασθαι ἐν τῷ ἔπει ὑπὸ τῶν παλαιῶν, ἀλλὰ λιμόν, ἐνίκησε δὲ ἐπὶ τοῦ παρόντος εἰκότως λοιμὸν εἰρῆσθαι· οἱ γὰρ ἄνθρωποι πρὸς ἃ ἔπασχον τὴν μνήμην ἐποιοῦντο. ἢν δέ γε οἶμαί ποτε ἄλλος πόλεμος καταλάβῃ Δωρικὸς τοῦδε ὕστερον καὶ ξυμβῇ γενέσθαι λιμόν, κατὰ τὸ εἰκὸς οὕτως ᾄσονται. It seems evident that there was a variant reading in the prophecy which combined a Dorian War and a famine. The possible link now established between ἀλιτήριοι and λοιμός immediately brings to one's mind not only the fact that Perikles was blamed for the plague (λοιμός) because he was related to the Alkmeonids (and thus an ἀλιτήριος), but also the story of Epimenides who came to Athens at the time of a pestilence (λοιμός) which was caused by those responsible for the Κυλώνειον ἄγος (that is: the ἀλιτήριοι).[26] In fact, it

against Thebes (line 8); compare J. W. Hewitt, *HSCP*, 19, 1908, p. 112, and the various forms ἀλεξητήρ ἀλεξητήριος, and ἀλεξήτωρ.

24. Mention may here be made of the locality called λιμοῦ πεδίον situated behind the prytaneion (Zenobios, iv, 93, in *Paroemiographi Graeci*, i, ed. E. L. Leutsch and F. G. Schneidewin): Λιμοῦ πεδίον· αὕτη τάττεται ἐπὶ τῶν ὑπὸ λιμοῦ πιεζομένων πόλεων· τόπος γάρ ἐστιν οὕτω καλούμενος. καὶ λέγουσιν ὅτι λιμοῦ ποτὲ κατασχόντος, ἔχρησεν ὁ θεὸς ἱκετηρίαν θέσθαι καὶ τὸν Λιμὸν ἐξιλεώσασθαι. οἱ δὲ Ἀθηναῖοι ἀνῆκαν αὐτῷ τὸ ὄπισθεν τοῦ πρυτανείου πεδίον. A similarly incomprehensible account is given in Anecdota Bekkeri (p. 278): Λιμοῦ πεδίον· ὄνομα τόπου. ἡ δὲ ἱστορία αὕτη· ἀφορίας γενομένης πεδίον κατὰ χρησμὸν ἀνέθεσαν τῇ εἰρεσιώνῃ, ὃν καταφεύγοντα ταῖς πλατάνοις ἀπὸ τοῦ συμβεβηκότος Λιμοῦ πεδίον ἐκάλεσαν. Suidas (s.v. εἰρεσιώνη, p. 532, no. 184 Adler) gives a similar account in which λιμός and λοιμός are confused and which omits all mention of λιμοῦ πεδίον; see also L. Deubner, *Att. Feste*, pp. 198–201; J. M. Edmonds, *Lyra Graeca*, iii², pp. 520–523. It may be that the "Paved Court of the Python" (*Hesperia*, xii, 1943, pp. 229–239) is in fact the Λιμοῦ πεδίον since it is situated behind the prytaneion, and since the prytaneion lies on the Tripod way.
25. See note 23 and G. Daux, *REG*, 53, 1940, p. 100.
26. The date of Epimenides is a problem of little significance in this connection, but it may be emphasized that the date at the end of the sixth century, first suggested by Plato (*Laws*, 1, 642D) and now almost generally accepted, is in good agreement with the thesis which is here presented, and especially with the testimony of Thucydides; compare, however, N. G. L. Hammond, *JHS*, 60, 1940, p. 81. For a good discussion of the evidence, see J. H. Wright, *HSCP*, 3, 1892, pp. 69–70.

seems quite reasonable to suppose that the prophecy which Thucydides quotes was made by Epimenides himself. We would thus obtain a new and significant fragment of Epimenides, and with it the earliest reference to his activity preserved in our literary tradition. Since the story of the λιμός (or λοιμός) and of the Aetolian (or Dorian) War, which Plutarch, the scholiasts, and the lexicographers report, cannot have been derived from the account of Thucydides, it may be assumed that this other account (Hellanicus?) also included a mention of the ἀλιτήριοι, and that this term may even have been used by Epimenides.[27] This lengthy excursus which is but little related to the ostracism of Xanthippos may help us better to understand the curse on the Alkmeonid family into which Xanthippos had married.

27. See L. Pearson, *The Local Historians of Attica*, p. 39, who suggests that Hellanicus may have discussed the "revolution of Cylon."

14

The Case against Alcibiades
(Andocides IV)

[191] One of the most puzzling documents of the internal history of Athens during the Peloponnesian War is the speech against Alcibiades preserved among the works of Andocides. It is generally assumed that the speech is not the work of Andocides and that it was composed early in the fourth century, as part of the Alcibiades literature, much of which has come down to us. The historical material contained in the speech has been, in part, accepted by modern scholars. There remain, however, a number of questions which have either never been asked or have not been satisfactorily answered. It is the purpose of this paper to present some of these questions and to examine them without coming, perhaps, to any definite answers. Jean Hatzfeld's admirable biography of Alcibiades has given me added encouragement to reexamine the evidence so completely and so critically assembled in his book.[1]

I

The authorship seemed to be settled once students agreed that the speech could not possibly have been written and delivered by Andocides himself.[2] The fact that the speech appears to have been attributed also to Lysias (among whose works two speeches against Alcibiades the younger are preserved) or to the Socratic Aeschines (who wrote a famous dialogue dealing with Alcibiades) may

---

1. *Alcibiade* (Paris, 1940) 116–142; see also L. Gernet, *RPh* 57 (1931) 313–326; J. Carcopino, *L'ostracisme Athénien* (Paris, 1935) 191–251; Th. Lenschau in *RE* s.v. "Phaiax," no. 4, cols. 1534–1536; W. Peek, *Kerameikos*, III (Berlin, 1941) 78–80 (no. 149) and 101–104; O. W. Reinmuth in *RE* s.v. "ostrakismos," cols. 1683–1684.
2. See the parallel accounts in F. Blass, *Die Attische Beredsamkeit*, I² (Leipzig, 1887) 336–339 (pp. 329–331 of the first edition), and R. C. Jebb, *The Attic Orators*, I (London, 1876) 134–139. Compare, however, A. Schroff, *Zur Echtheitsfrage der vierten Rede des Andokides* (Erlangen, 1901). The ancient testimonies and the modern bibliography have been treated in these studies (see also notes 1 and 3) so adequately that it will be necessary to document my own essay but sparingly.

have encouraged the assumption that it is the work of an otherwise unknown pupil of a sophist. One may wonder, however, why the speech was ever attributed to Andocides. Since the speaker | was evidently Phaeax, Andocides must have been thought to have composed the speech for Phaeax; and, indeed, the speech is called "a defense addressed to (or, concerning) Phaeax" (ἀπολογία πρὸς Φαίακα) in the biography of Andocides preserved under the name of Plutarch (§ 14). Since the mutilation of the herms, in which Andocides was personally involved, was perpetrated in order to prevent the sailing of the Sicilian expedition, the leader of which was Alcibiades, one may deduce that Andocides could have shown his hostility toward Alcibiades and his plans also by writing a speech for Phaeax designed to promote the ostracism of Alcibiades. If the extant speech was not actually written by Andocides, one can understand why it should have been attributed to him.

II

The dramatic date of the speech has been used as the main argument against its authenticity. Since Alcibiades is blamed for the enslavement of the defeated Melians (§§ 22–23), and the Sicilian expedition had not begun (or is not mentioned), the speech, if not fictitious, must have been written early in 415 B.C. Yet the last ostracism, that of Hyperbolus, is said to have taken place in 417 B.C., and accordingly no ostracism could have been held two years later. One may wonder whether the author of the speech was so ignorant of the events with which he dealt that he may be credited with so conspicuous an error in chronology. In fact, the reference to the capture of Melos is not the only part of the speech which must have been written after 417 B.C. The passage dealing with Alcibiades' conduct at the Olympic Games (§§ 25–31) refers presumably to 416 B.C. (see, below, IV G). Are we to assume that the author of the speech who dealt with this episode at such length was ignorant of this fact which modern students have been able to establish? One must suppose, therefore, that the speech was written with the ostracism of 415 B.C. in mind, and that this must have been also the date of the ostracism of Hyperbolus.[3] If Theopompus is credited with the statement that Hyperbolus lived in exile for six years (and was assassinated, according to Thucydides, 8.73.3, in | 411 B.C.), he or his copyist must either have made a mistake or have used a peculiar way of counting; see Aristophanes, *Wasps*, scholium on line 1007: ἐξωστράκισαν τὸν Ὑπέρβολον ἒξ ἔτη. If, however, quite apart from the authenticity of the speech, Hyperbolus was ostracized in 415 B.C., the circumstances of his ostracism are thrown into a new light. The conflict between Nicias and Alcibiades of which the speech tells, and their agreement to join forces against Hyperbolus, would belong

---

3. A. G. Woodhead's article "*I.G.*, I², 95, and the Ostracism of Hyperbolus" (to be published in *Hesperia*) contains proof that Hyperbolus could not have been ostracized before 416 B.C. I am grateful to Woodhead as well as to N. Doenges, J. V. A. Fine, R. Meiggs, H. W. Parke, G. Stamires, and E. Vanderpool for having examined my manuscript and for having given me the benefit of their advice and criticism.

immediately before the famous debate between the two men recorded by Thucydides.

It is true that modern historians have been able to draw a detailed and not altogether unconvincing picture of earlier differences between Alcibiades and Nicias concerning Athenian policy in the Peloponnesus and especially the Argive alliance (see note 1). They spoke of successes and setbacks in Alcibiades' struggle for domination, and they placed the attempted ostracism of Alcibiades in connection with either of them. Yet on closer inspection, it appears that apart from one incident (Thuc. 5.45f.), nothing is known of internal differences in Athens. The rather imaginative descriptions of the circumstances surrounding the ostracism of Hyperbolus, whether it is dated in 418 B.C. or in 417 B.C., read like doublets of the known conflicts which immediately preceded the Sicilian expedition.

III

One may wonder whether it is accidental that several of the topics of the Nicias-Alcibiades debate in Thucydides also appear in the speech of Phaeax. This is the more remarkable since there is no reason to assume that Thucydides knew the speech or that the author of the speech had read Thucydides.

The speech characterizes Alcibiades (§ 16) as a man who "refuses to be considered equal, or but little superior to, his fellows," and Thucydides lets Alcibiades assert (6.16) that it is "only fair that a man proud of his position should refuse to be upon an equality with the rest." The public attitude toward Alcibiades is characterized both in the speech (§ 21) and by Thucydides (6.15) as being dominated by anger, fear, and subservience. Alcibiades calls attention to his splendid performances of choregies (Thuc. 6.16), and the speech contains a lengthy illustration of this boast (§§ 20–21). The political differences between the younger and the older generations are stressed both by Thucydides (6.13 and 18) and in the speech (§ 22). The Melian affair, and Alcibiades' share | in it (§§ 22–23), is mentioned by Thucydides just before the Nicias-Alcibiades debate, but the historian gave no names.

More significant, however, than these rather casual and obvious similarities is the fact that both the speech and Thucydides' account contain explicit expressions of the popular feeling that Alcibiades was aiming at tyranny. This is not surprising in the speech (§§ 24 and 27) which purports to be concerned with ostracism, but it is significant in Thucydides where there is no mention of the ostracism of Hyperbolus and of the part Alcibiades played in it. In his general characterization of Alcibiades, Thucydides reported (6.15) that the people "were hostile to him because they thought he aimed at tyranny," and again when telling of Alcibiades' recall the historian indicated that the people of Athens thought of Alcibiades as a second Pisistratus (6.53), and Thucydides took the opportunity of recalling in detail the end of the Pisistratids (6.54–59). Intentionally or not, Thucydides at this point described the public sentiment as it must have existed not only after Alcibiades' departure for Sicily but also shortly before, when he was in danger of being ostracized.[4]

The most significant link between Thucydides' account of the Nicias-Alcibiades debate and the speech is provided by the references in both documents to Alcibiades' victory in the Olympic Games of 416 B.C. Thucydides referred to it not only in general terms (6.15) and when he had Nicias speak (6.12) of a certain young man's desire "to be admired because of his horses," but also most specifically at the very beginning of Alcibiades' speech (6.16). In fact, Alcibiades makes precisely the statement, basing his claim to leadership on his Olympic victory, which the author of the speech assumed he would make (§ 25).

To sum up, several passages in Phaeax's speech and Thucydides' account of the situation in Athens just before the Sicilian expedition agree to a remarkable extent, and this agreement may indicate either that the speech is genuine or that its author took great pains to produce a speech which would appear authentic.

IV

The various inconsistencies and blatant errors in the speech have been used to challenge not only its authenticity but also the reliability of such information which otherwise cannot be disproved. Yet there are inconsistencies and even errors in the two certainly genuine speeches of Andocides, and it should be remembered that a document like the speech of Phaeax must be compared with pamphlets like the Old Oligarch's or with speeches like those in Thucydides rather than with factual historical accounts like the main portion of Thucydides' work. If considered in this way, the question of authenticity loses some of its significance since the speech may be a pamphlet issued at the time (ca. 415 B.C.) and not the text of a speech actually delivered; or it may be a speech supposed to have been delivered in 415 B.C. but actually written ten or fifteen years later, just as most of the speeches in Thucydides' work are not accurate records of words actually spoken but accounts designed to convey the deeper meaning of the situation which they try to elucidate. What they lack in historical truth they gain fully in psychological and political insight. In the following paragraphs a few of the errors and inconsistencies contained in the speech may be examined in detail.

A. The speaker is evidently not a partisan of the radical democracy. The use of the adjective ἀγαθός referring to himself and to his audience should make this amply clear (§§ 1 and 2). This impression is strengthened by the speaker's frank and repeated confession that his antidemocratic attitude had resulted in four separate court actions, in which he was, however, acquitted (§§ 8 and

---

4. For a different interpretation, see L. Pearson, *AJPh* 70 (1949) 186–189.

35–37). The dates and circumstances of these trials are not known, but attention may be called to the speaker's obscure assertion that two other men accused on the same charges had been condemned and executed (§ 37); if we knew who these men were, we could judge more accurately the policy of the speaker.[5]

The criticism of the institution of ostracism (§§ 3–6) has often been used as evidence against the authenticity of the speech. The speaker declared bluntly (§ 3) that the law of ostracism is unconstitutional since it violates the famous principle that no man should be exiled, imprisoned, or executed without due process of law. The law of ostracism did not provide for a formal accusation, or for a formal defense, nor was there a provision in it securing the secrecy | of the ballot. This passage has been greatly misunderstood by most commentators. First of all, the unity of the three characteristic provisions of the due process (accusation, defense, secret vote) has been destroyed by all those editors and commentators who followed Schleiermacher in deleting the third οὔτε and in assuming that the law of ostracism provided for a secret ballot. Yet the voting during ostracism should not be contrasted to the open show of hands in the assembly but to the secret vote cast in the courts. There was nothing secret about any part of the process of ostracism except that the marked ballots were not shown to the magistrates who watched over the voting urns. Second, the passage in the speech was misinterpreted to mean that the law of ostracism specifically forbade any public discussion of the issue, and the author of the speech was accordingly accused of a flagrant and intentional error. In fact, all he stated was that the law did not provide for formal accusation or defense. It cannot be doubted, however, that a certain amount of propaganda, probably oral propaganda, was necessary to produce the large vote which the law required. The speech is undoubtedly a document designed to influence public opinion against Alcibiades, or else the author of the speech pretended that his work was of this type. The real or alleged occasion of the speech was, of course, not a formal assembly meeting but rather one of those informal but highly important meetings which are common to all democracies.

The next point made by the speaker (§ 4) is also significant because it shows that he contrasted ostracism to the due process of law (and not to decisions made in the assembly). In the law courts, the jury was determined by lot and therefore could not be influenced by partisan groups. Ostracism, however, with its complete freedom of propaganda and ballot could easily come under the influence of those who could count on the loyal following of a substantial number of people. The existence of factions and political clubs is well attested for the second half of the fifth century, and their operation indeed made ostracism meaningless.

The speaker's attack upon the law of ostracism is on the whole justified, though not very convincing; he himself gave a better characterization of the institution in §§ 35–36. One must not deduce from this criticism that the speech

---

5. Reference may be made to Thucydides 4.65.3, but the two men mentioned there suffered exile, not death.

is a forgery, nor can one learn very much from it about the provisions of the law. The influence of the clubs and factions is known from other sources, and this influence was particularly significant on the occasion of this ostracism when the factions of Nicias and Alcibiades made common cause against Hyperbolus.

B. Closely connected with the question whether or not the speech could have been delivered is the identification of the audience to whom it may have been addressed. If the law of ostracism did not provide for a formal accusation or defense, no speech like the one under discussion could have been made during a formal meeting of the assembly. And yet, it is generally assumed that the speech was delivered (or was alleged to have been delivered) at an assembly meeting. It seems clear, however, that the speaker addressed his audience (§ 7) not as he would have spoken in a formal assembly, but that he was rather merely pretending that his meeting had an official character. It is for this reason that he calls upon his audience to act as if they were all epistatai and archons. It is improper to think here of the role played by the archons and the members of the council during the ostracism itself, for the only duties of these functionaries were to guard the ballot boxes and to see to it that only qualified citizens cast their vote. The speaker refers to the fact that there may be another speaker following him on the platform (§§ 7 and 25), and it seems clear from one of these references (§ 25) that he is thinking of Alcibiades. Is one to assume, therefore, that the speaker and Alcibiades were both addressing this meeting, or is the reference to the second speaker purely rhetorical? We have no evidence to show that such informal meetings were held in connection with an ostracism, but meetings of this type were obviously necessary to concentrate public opinion on those few men who were likely "candidates" for ostracism.

C. While the speaker's accusation of Alcibiades on the count of adultery (§ 10), stated in general terms, was admittedly unsupported by facts, his attack upon Alcibiades because of his part in the assessment of the tribute is very precise (§§ 11–12). And yet it is the only noteworthy detail of Alcibiades' life contained in the speech which the biographer Plutarch failed to mention. If Plutarch used an account of Alcibiades the author of which was familiar with the speech, one must assume that this earlier biographer or historian discarded this particular story because he knew from | other sources that Alcibiades was not the person mainly responsible for increasing the tribute assessment. Modern scholars relying on the references in the plays of Aristophanes are firmly convinced that it was Cleon and not Alcibiades who should be attacked for this particular policy. Are we to assume that the author of the speech (whether he wrote at that time or some fifteen years later) was so ignorant that he unwittingly charged Alcibiades for an act which Cleon had actually committed? Is it not more likely that he knew that Alcibiades had played a minor role in the preparation of the assessment which became so hateful to the allies, and that he held Alcibiades alone responsible simply because Cleon was dead and Alcibiades' personal political position had to be attacked? This view seems to be supported by the fact that the speaker remains silent concerning all the other actions of Alcibiades during the period of the Archidamian War and the Peace of Nicias; he does this undoubtedly for the simple reason that these actions had been on the whole very creditable.

There are, moreover, two small details which indicate that the author of the speech was well acquainted with the famous assessment of 425 B.C. In the first place, he asserted that Alcibiades was elected chairman of a board of ten assessors. The very existence of such a board of ten men and its election are nowhere else attested in ancient literature, but the facts are now known also from the assessment decree which is still preserved.[6] Since there probably was no assessment after that of 410 B.C., the author, if he wrote ten or more years later, needed excellent information in order to cite this specific position held by Alcibiades.

The second point concerns the charge made against Alcibiades in the speech: he is said to have doubled the tribute assessment of each member of the alliance. This statement has often been challenged as incorrect because it was misinterpreted to mean that Alcibiades was responsible for a tribute assessment the total of which was double that of Aristides. Since the assessment of Aristides is known to have been 460 talents and since the literary tradition and the epigraphical evidence plainly indicate that the assessment of 425 B.C. was about three times that much, the author of the speech is accused of flagrant ignorance.

[199] It is evident, however, | that the doubling of the original individual assessment figures would have resulted in a trebling of the total tribute once it is realized that the membership of the alliance had greatly increased since the time of Aristides. An examination of the assessment decree itself and of the pertinent tribute lists reveals that the average individual assessments were doubled rather than trebled. It may be asked, moreover, why the author of the speech who desired to attack Alcibiades would name a lower figure if the higher figure were not only correct but also more damaging to the reputation of Alcibiades? Finally, it should be pointed out that the author of the speech showed great independence and originality when he described the character of the famous assessment as the doubling of the individual assessments rather than the trebling of the total amount which became the standard description in our literary tradition.

Thus, the passage referring to the tribute assessment shows the author of the speech well informed in matters of Athenian administration but ill disposed toward Alcibiades.

D. The story of Alcibiades' marriage (§§ 13–15) has also been questioned in one particular detail. The speaker claims that Alcibiades' father-in-law

---

6. See (also for the evidence mentioned in the following paragraph) B. D. Meritt, H. T. Wade-Gery, and M. F. McGregor, *The Athenian Tribute Lists*, 1 (Cambridge, Mass., 1939) and 2 (Princeton, 1949).

Hipponicus fell in battle while in command at Delium in 424 B.C.; it is known, however, from Thucydides (4.101.2) that it was the general Hippocrates who was killed in action during the battle at Delium. The author of the speech is accused, therefore, of having confused the two similar sounding names. Yet the only error he may have committed is the statement that Hipponicus was general when he was killed. For there is no reason for assuming that Hipponicus did not die at Delium. Since the question whether or not Hipponicus was general when he fell is of no importance to the argument, and since it is unlikely, moreover, that Hipponicus should have served as a common soldier or cavalry man, one could easily translate the words στρατηγοῦντος ἐπὶ Δηλίῳ with "while commanding his troops (in whatever capacity) at Delium."

E. The stories of Alcibiades' relationship to the painter Agatharchus, the choregus Taureas, and the woman from Melos, which occupy eight paragraphs of the speech (§§ 16–23), are summarized by Plutarch, in the same order, in one paragraph of his *Life of Alcibiades* (16.4–5). In fact, Plutarch devoted only a sentence each to Alcibiades' dealings with Agatharchus and Taureas, and two sentences to his affair with the Melian prisoner of war. In spite of this, Plutarch, who evidently either directly or indirectly relied on the speech, changed the significance of two of the episodes. While the speaker implied that Alcibiades and Agatharchus parted as foes, Plutarch stated that Alcibiades dismissed the painter with a handsome present after having forced him to decorate his house. Similarly, in the case of the Melian woman, Plutarch reported that Alcibiades' behavior was called by some people charitable, while the speaker (with an obscure reference to Aegisthus, derived perhaps from a tragedy well known at the time) asserted that the action of Alcibiades was more heinous than that of any villain in a tragic play. If it should become apparent that Plutarch's account (at least of the Melian episode) is more credible, one will have to conclude that the speaker was anxious to falsify the evidence, not out of ignorance but because of malice.

The story of the Melian woman is, in fact, one of the crucial passages of the speech, for if the story is true to the letter, the speech must be spurious (and with it, of course, the story itself; a good example of a Cretan syllogism). The known date of the conquest of Melos and the dramatic date of the speech (before or early in the eighth prytany of the year 416/5 B.C.) hardly allow for the required period between the conception and birth of a child. In fact, the child would have to be born such a short time before the speech was delivered that one should expect in the speech a reference to this fact. Some have assumed, therefore, that the woman was captured and bought by Alcibiades some time before the final conquest of the island, while others (with them, apparently, Plutarch and his source) maintained that Alcibiades was not the father of the child who may have been born a short time before or even after the conquest of Melos. Only this interpretation permits one to call Alcibiades' attitude charitable. Yet this interpretation is clearly contradicted by the speaker. Is it not possible that the speaker, in order to malign Alcibiades, asserted that the baby living in Alcibiades' house, the son of a Melian woman whom Alcibiades favored with his affection, was in fact Alcibiades' own son? Similarly, he claimed that Alcibiades was solely responsible for the cruel fate of the Melians, while Plutarch reported that he merely seconded the motion of somebody else; Thucydides (5.116.4), characteristically, mentions no names at all (see also L. Rademacher, *WSt* 67 [1939] 165; P. Treves, *JHS* 63 [1943] 133, note 1).

It appears that the obvious inconsistencies of the speaker in reporting the Melian episode may again be charged to his hostility against Alcibiades rather than to his ignorance of the facts involved.

F. If the speech was written many years after the event to which it purports to refer, it might reveal the knowledge of certain happenings which took place after this event but were so closely connected with it that mention of them must have been tempting to the author. Even Thucydides, intentionally or unconsciously, did not avoid such "prophetic" passages, and careful students have accordingly "dated" many of his speeches after the fall of Athens in 403 B.C. Considering this circumstance, it is surprising that the author of the speech almost entirely refrained from referring to the future or from showing that he was aware of what happened to Athens and to Alcibiades after 415 B.C.

The first passage to be mentioned in this connection is to be found in the discussion of the increase in the tribute assessment (§ 12). "The hostility of the allies," said the speaker, "will show itself clearly as soon as we and the Lacedaemonians become engaged in naval warfare." The presumption is, of course, that the allies would then desert Athens and join the enemy who up to that time had not dared to enter the Athenian home waters of the Aegean. It is true that many of the allies revolted after the Sicilian expedition, and that the Lacedaemonians manned a fleet and pursued the naval warfare so vigorously that they ultimately won the war by virtually destroying the Athenian navy. Yet the increase in the tribute assessment had little to do with the revolt of the allies, and at best one may assume that the author's knowledge of the Ionian War could have influenced his statement quoted above. There is, however, a distinct possibility that Isocrates referred to this passage of the speech when he stated (XVI, 10) that Alcibiades was accused of having caused the revolt of the islands (see, below, G).

Another passage in the speech is of a similar nature. In § 24, the speaker asserted that "the city will experience the greatest calamities from this man (Alcibiades), and he will be held responsible in the future for such deeds that nobody will remember his former villainies." This statement may be taken to refer to Alcibiades' treason or perhaps to his implication in the profanation of the Mysteries. Yet one is surprised that the author of the speech, if he really knew what happened after the spring of 415 B.C., did not refer more specifically to some of the misdeeds with which Alcibiades was charged in later years.

The conclusion cannot be avoided that the author of the speech either did not have any knowledge of the events which followed the dramatic date of his speech (simply, because the speech was written early in 415 B.C.), or that he was careful not to reveal such knowledge too clearly, more careful, in fact, than Thucydides. In spite of his oratory and dramatic technique, the author of the speech did not employ dramatic irony at all.

G. The long passage devoted to Alcibiades' behavior during the Olympic Games (§§ 25–32) contains several problems which are not without bearing upon the understanding of the speech as a whole. The Diomedes episode (§§ 26–28) is also known from other sources. Alcibiades' son was sued early in the fourth century by a certain Teisias who claimed five talents for having been cheated out of a team of horses by Alcibiades, almost twenty years earlier. The speech of Isocrates (XVI) in defense of the younger Alcibiades shows that the controversy had not died down, but that in fact it encouraged renewed attacks against the character and life of the great Alcibiades. This may be used as evidence that the speech of Phaeax was actually composed at that later time. Yet the man who sued the younger Alcibiades was Teisias and not Diomedes. Scholars have tried in vain to reconcile this difference which Plutarch had already found perplexing (*Alcibiades* 12.3). Whatever the answer may be, so much is clear, that the author of the speech knew nothing of Teisias, a very peculiar fact if he wrote at the time when Teisias filed his suit, and if his pamphlet was designed to discredit the memory of the older Alcibiades, that is, to obtain the same result as the accusation of Teisias. Another interesting deduction which may be drawn from the suit of Teisias concerns the date of the Olympic Games to which reference was made in the suit. If Alcibiades had entered the Olympic Games in 424 B.C. or even in 420 B.C. (both dates are made unlikely by other considerations), one would expect Diomedes to bring in his suit shortly after the incriminating action. Yet there was hardly time after the games of 416 B.C. to introduce a legal action against Alcibiades, who was general at the time, was reelected early in 415 B.C., and left Athens that very summer. Obviously, no suit could be initiated in this matter until Alcibiades' son had come of age. This consideration could be used as supporting evidence for the assumption that the Olympic Games in question were those of 416 B.C.

[203]

H. One of the most remarkable passages of the speech deals with several former victims of ostracism, among them two ancestors of Alcibiades himself (§§ 32–34). It has been pointed out by others that our knowledge concerning the ostracisms of Megacles and of Alcibiades the elder is based almost exclusively on this speech. In fact, our witnesses for these ostracisms are the ostraka of Megacles and Alcibiades found in excavations during the past sixty-five years, the text of Aristotle's treatise on the Constitution of Athens (22.5, where only the ostracism of Megacles is mentioned), and a passage in Lysias (XIV, 39) which seems to be derived from the speech of Phaeax (§ 34). The text of Lysias is corrupt at this point: Ἀλκιβιάδην μὲν τὸν πρόπαππον αὐτοῦ καὶ τὸν πρὸς μητρὸς Μεγακλέα οἱ ὑμέτεροι πρόγονοι δὶς ἀμφοτέρους ἐξωστράκισαν. The speech simply stated: ὁ τῆς μητρὸς πατὴρ Μεγακλῆς καὶ ὁ πάππος Ἀλκιβιάδης ἐξωστρακίσθησαν ἀμφότεροι. Since Lysias referred to the younger Alcibiades he had to add one generation to the relationship as defined in the speech. Referring to the elder Alcibiades he changed πάππος to πρόπαππος, but no such easy solution was possible in the case of Megacles who was the father of the mother of (the younger) Alcibiades' father. The text of Gernet and Bizos (Paris, 1943) τὸν πρὸς μητρὸς ⟨πάππον⟩ Μεγακλέα is impossible because Megacles was not the maternal grandfather of the younger Alcibiades. The text of Lamb (London, 1943) τὸν πατρὸς πρὸς μητρὸς ⟨πάππον⟩ Μεγακλέα is historically correct but it requires the addition of ⟨πατρός⟩ (not indicated as such by the editor) and ⟨πάππον⟩, and it conceals (as does the text of the French editors) the obvious fact that both men were great-grandfathers of the younger Alcibiades. What Lysias may have meant to say was τὸν πρὸς μητρὸς πατρὸς (πρόπαππον) Μεγακλέα, meaning "Megacles, the great-grandfather on the side of his father's mother." And, indeed, the repetition of πρόπαππον may have been unnecessary, and Lysias's text may be emended to read 'Ἀλκιβιάδην | μὲν τὸν πρόπαππον καὶ τὸν πρὸς μητρὸς ⟨πατρὸς⟩ Μεγακλέα. The repeated sequence of ΠΡΟΣ, ΤΡΟΣ, ΤΡΟΣ may easily have caused the omission of πατρός. A comparison of the versions of Lysias and of the speech of Phaeax seems to indicate that Lysias adapted the words of the speech to suit the status of Alcibiades the younger. Thus, were it not for comparatively recent and accidental discoveries, the passage in the speech would be our only evidence for these events. The text of Lysias, moreover, has misled editors and commentators into assuming that the orator said that the Athenians "ostracized both twice." This is, indeed, a literal translation of what Lysias said, but it does not mean that the Athenians "ostracized each of them twice." Peculiar though the expression may be, it can only be an attempt to say that ostracism was employed two times against members of Alcibiades' family, and that both grandfathers were thus affected. The words may be translated: "your ancestors (the Athenians) applied the law of ostracism on two (separate) occasions and (thus) ostracized both (these men)." It is, therefore, not only unnecessary but even faulty to insert the word δίς in the text of the Phaeax speech, where it could only mean that both men were simultaneously ostracized two times.

Although today nobody would doubt the reliability of the speaker's statement on the ostracisms of Megacles and Alcibiades the elder, the ostracism of Callias the son of Didymias (mentioned in § 32) was questioned until several ostraka bearing his name were found in the Agora excavations; see Reinmuth's article mentioned in note 1. Callias was another Athenian who brought home an Olympic victory, and it is fitting that the speaker should compare the fate of this famous athlete who was ostracized with that of Alcibiades who won a similar victory and for whose ostracism the speaker was pleading. Unfortunately, we do not know either the date or the occasion of Callias's ostracism. It is hard to believe that it took place after 443 B.C., the year of Thucydides' ostracism and of the beginning of Pericles' "principate."

The most peculiar part of the passage concerns the ostracism of Cimon (§ 33). While all other sources emphasize the political character of Cimon's ostracism, the speaker asserts that Cimon was exiled as a protest against his lawlessness consisting in his illicit relationship with his sister Elpinice.[7] In order to understand this extraordinary statement we must consider the context in which [205] it occurs. The speaker links the ostracism of Cimon, as he did in the case of Callias, with the fact that Cimon (as well as his father Miltiades) was victorious at the Olympic Games, thus avoiding any mention of Cimon's position as general and politician. It is known, moreover, that in Athens, as in modern democracies, political opponents often commented upon the private life of their adversaries. Pericles as well as Alcibiades are excellent examples of this rule, and Cimon was no exception to it. In fact, Plutarch mentioned Cimon's relationship with his sister not only in a lengthy passage (*Cimon* 4.5–7), but emphasized it again (15.3) when he discussed public sentiment at the time of the ostracism. The author of the speech therefore repeated a piece of gossip which was current in Athens. He did this in order to show, not very convincingly, to be sure, that other distinguished men had been ostracized because of their moral depravity.

I. From the points examined so far it appears that most of the critics were right in maintaining that the speech contains both errors of fact and historical inconsistencies and misinterpretations. And yet these objections do not allow us to state that Andocides could not have been the author of the speech, that it was not composed in 415 B.C. in the name of (or by) Phaeax, that it was not actually delivered or issued as a pamphlet, or, finally, that it was not written years later but with the full knowledge of the situation which it described. It may be compared with pamphlets like the Old Oligarch's, with speeches like those in Thucydides, and with a highly trustworthy but idealized document like Plato's *Apology*.

V

The most puzzling aspect of the speech is not the information which it contains but the facts which the author did not mention; this comment applies whether the speech was composed in 415 B.C. or early in the fourth century.

Aside from the more glaring omissions which must be discussed separately, there is a great deal of incidental information preserved in Plutarch's biography of Alcibiades but absent from the speech. Pericles and Socrates, who were so closely connected with Alcibiades, are not even mentioned in the speech. Are we to assume that the author of the speech was ignorant of all the episodes in [206] which Socrates or Pericles was linked with Alcibiades, episodes which must have formed the major portion of the so-called Socratic Alcibiades legend? Does this omission indicate that the speech was composed before any of the Socratic Alcibiades literature was published or had become known? Or were these episodes so favorable to Alcibiades that the author of the speech ignored

---

7. See also Suidas, s.v. Κίμων no. 1621.

them deliberately? Yet, in other instances, he turned a favorable story into an indictment.

It is also evident that the author of the speech made no use of Antiphon's attack upon Alcibiades. At least, there is no mention in the speech of the lurid details from this pamphlet reported in Plutarch's biography of Alcibiades (3) and in Athenaeus (12.525 B). The date of Antiphon's pamphlet is unknown, but if it is genuine it must belong to 415 B.C. or to one of the preceding years.

Of greater significance, however, is the omission in the speech of all mention of Alcibiades' activity as a politician and as a general. It is true that much of it was most creditable to Alcibiades, but the story of his duplicity towards the Lacedaemonian ambassadors (told by Thucydides, 5.45–46, and Plutarch, *Alcibiades* 14.6–9) could have been used to good advantage by the author of the speech. If Alcibiades' public record seemed so unassailable to the speaker that he preferred to ignore it altogether, this can only mean that he was either unaware of what happened after the dramatic date of the speech or that he was so clever as to put himself completely into the situation which he tried to recreate; even Thucydides failed often in this respect.

If the author of the speech knew what happened only a few weeks after the supposed date of the speech, how could he have avoided making reference to such incriminating evidence as the profanation of the Mysteries for which Alcibiades was held responsible? Even if this act of impiety did not become public knowledge until weeks later, the actions themselves must have taken place over a longer period of time, and it would have been tempting to the author of the speech to let the speaker pretend knowledge of them or at least hint at such knowledge. Yet there is not a word in the speech suggesting such a design on the part of its author.

The most astonishing omission, however, concerns the Sicilian expedition, the main topic of conversation during the early months of 415 B.C. The other

[207] omissions noticed so far could be explained | by the assumption that the speech was actually authentic and that its author had no knowledge of any of the events which were to follow. This explanation does not hold for the lack of any mention of the Sicilian expedition in the speech, because we know from Thucydides and Plutarch that conditions in Sicily engaged Athenian interests throughout the winter of 416/5 B.C., and it is certain that the Athenians dispatched a delegation to Segesta late in 416 B.C., thus a considerable time before the dramatic date of the speech. On the other hand, the delegation probably returned after that date, and thus the final decision had not yet been made. It must be remembered, moreover, that the election of generals followed (perhaps only by hours) the final vote of ostracism, and that the ostracism of Alcibiades would have eliminated him as a candidate for the generalship. It would also have ruined the whole plan of the Sicilian expedition which was championed by Alcibiades. Thus it would have accomplished more simply the very same thing which had to be carried out later by such devious means as the mutilation of the herms and the denunciation of Alcibiades for profanation of the Mysteries. Considering the general popularity of the aggressive policy

of Alcibiades, it would have been but wise for any enemy of his to avoid altogether the subject of foreign policy, although in truth foreign policy was the decisive issue. This holds true for the speech of Phaeax as well as for the various attacks against Alcibiades just before and just after the departure of the fleet for Sicily.

Whatever one may think of the various mistakes and omissions to be found in the speech, the complete lack of any reference to foreign policy in general and to the Sicilian expedition in particular can only have been deliberate. This consideration confirms our convictions that the speech must be either authentic or else must have been written by an extremely careful and well-informed man.

VI

There remains but one topic which has been purposely avoided in the preceding discussion, the ostracism of Hyperbolus. It provides the necessary link between the speech of Phaeax and the historical narrative of the events which took place early in the year 415 B.C. First of all, it must be pointed out that the speech contains no reference to Hyperbolus; in fact, the speaker stated very clearly at the beginning (§ 2) that the electorate had to choose between Alcibiades, Nicias, and himself. This may be another indication of the authenticity of the speech or of the care with which the author tried to recreate the situation. Quite obviously, the speech belongs to that period between the sixth and eighth prytanies of the year 416/5 B.C. when Alcibiades was opposed by Nicias, and before he and Nicias joined forces against Hyperbolus (see, above, II). We are not primarily concerned here with examining this agreement between Alcibiades and Nicias which resulted not only in the ostracism of Hyperbolus but also in the election as generals of both Nicias and Alcibiades, and in their joint preparation of the Sicilian expedition. It is clear, however, that the juxtaposition of the ostracism of Hyperbolus and the Sicilian expedition adds considerably to our understanding of Athenian policy at that time. It also relieves Thucydides from the charge of having completely ignored the ostracism of Hyperbolus. For what really mattered was not the disappearance of this popular demagogue but the peculiar association and disagreement between the two leading military men, Nicias and Alcibiades. This complex situation, however, was very cleverly described by Thucydides in the prologue to the Sicilian expedition itself (6.8–26).

The literary tradition concerning the ostracism of Hyperbolus is based on three fragments, one each of Androtion, Theopompus, and Philochorus, and on Plutarch's account in the biographies of Nicias and Alcibiades. Androtion merely stated that Hyperbolus was ostracized because of his bad character, and this inadequate information is repeated by Philochorus; it may be derived from passages like that of the comic poet Plato quoted by Plutarch (*Alcibiades* 13.5; see also W. Schmid, *Philologus* 93 [1939] 415–416). Theopompus is credited with the information that Hyperbolus lived in exile for six years; he also added some sordid details about the death of Hyperbolus. It seems that he was more interested in what happened to Hyperbolus after his ostracism than in the ostracism itself. It is evident that little can be learned from these fragments about the ostracism of Hyperbolus.

Plutarch's account needs a more careful analysis in order to determine to what extent either he or his source knew or used the speech of Phaeax. In the biography of Nicias (11), which was composed before that of Alcibiades, there is a lengthy account of the hostility between Nicias and Alcibiades which culminated in the fight over the ostracism. The story is told immediately before the Sicilian expedition, thus lending some support to the assumption that the two events followed each other in short order. The general characterization of Alcibiades (11.2) agrees not only with that given in the speech but also with the account of Thucydides (see, above, III). The name of Phaeax is not mentioned except at the very end (11.7), perhaps as an afterthought possibly written after the more detailed account in the biography of Alcibiades. In this supplementary note, Plutarch called his reader's attention to the fact that Theophrastus asserted that it was Phaeax and not Nicias who opposed Alcibiades. Could it be that Theophrastus knew and commented on the speech of Phaeax, and that all he said was that Phaeax opposed Alcibiades openly as shown by the speech which he delivered against Alcibiades? It appears, therefore, that Plutarch's account in the biography of Nicias was written without reference to the speech of Phaeax, except for the final paragraph which should be considered together with the author's account of the same affair in his biography of Alcibiades.

[209]

It has been often emphasized that Plutarch or the source which he used for his biography of Alcibiades relied upon the biographical information contained in the speech of Phaeax. The only noticeable exception is the story of Alcibiades' part in the tribute assessment; see, above, IV C. At the same time, it appears that Plutarch could not have used the speech exclusively since there are certain differences between the two accounts which can be most easily explained by the assumption that Plutarch had before him a source which was based on the speech but did not follow it in every detail. This source may have been less unfavorable to Alcibiades than the speech, and it may have elaborated certain episodes, perhaps by drawing upon other information. This impression is confirmed by an examination of that chapter of Plutarch's biography of Alcibiades (13) which deals specifically with the ostracism of Hyperbolus.

The accounts of the agreement between Nicias and Alcibiades and of the subsequent ostracism of Hyperbolus are told in much the same way as in the biography of Nicias, except that Plutarch stated that the ostracism involved three men (13.4). It is evident that these three men were Nicias, Alcibiades, and Phaeax, since the first two paragraphs (13.1-2) mention Phaeax in addition to Nicias and Alcibiades, and at the end of the chapter (13.4) Plutarch reported that some authorities claim that the agreement was made not between Alcibiades and Nicias but between Alcibiades and Phaeax. Is it too much to deduce that these authorities included Theophrastus whom Plutarch in the biography of Nicias had credited with similar information? If so, could it be that the basis for Theophrastus's modified version of the events was his knowledge of the

[210]

speech of Phaeax which had hitherto not been introduced into the evidence? This assumption would reveal the source of Plutarch's account of the controversy between Phaeax and Alcibiades (13.1–2), for it is evident that this passage merely explains in detail the statement attributed to Theophrastus in Plutarch's biography of Nicias (11.7). Actually, Plutarch or even his source did not offer much information about Phaeax. His father's name was Erasistratus, he was a young man of about Alcibiades' age, ridiculed for his babbling oratory by the comic poet Eupolis. It is at this point (13.2) that Plutarch referred to a certain speech of Phaeax in which a story about Alcibiades was told which actually occurs in a somewhat different form in the preserved speech (§ 29). The introductory sentence has been considered corrupt and accordingly has been emended: φέρεται δὲ καὶ λόγος τις κατ' 'Αλκιβιάδου καὶ Φαίακος γεγραμμένος. ... Editors have challenged the second καί and changed it to ὑπό or ὑπέρ or left it out altogether changing the following name Φαίακος to Φαίακι. Would it not be simpler to assume that the original text read φέρεται δὲ καὶ Φαίακος λόγος τις κατ' 'Αλκιβιάδου γεγραμμένος and that the words καὶ Φαίακος were omitted, reinserted, and ultimately misplaced? Whatever may be the original form of the sentence, it seems clear that Plutarch repeated (perhaps not literally) from his source without knowing that he also owed other material to the same document.

The conclusion which one is tempted to draw is this. All three passages in which Plutarch mentioned the name of Phaeax drew upon the same authority, probably Theophrastus; see H. Bloch, *HSCPh*, Suppl. vol. 1 (1940) 355, note 1. This author modified the commonly known version of the ostracism of Hyperbolus by introducing the person of Phaeax. He made this innovation because he believed in the authenticity of the speech of Phaeax, and because he associated it, perhaps for the first time, with the ostracism of Hyperbolus.

15

Philochoros *Frag.* 30 (Jacoby)

[119] Die Frage, wie die Stimmen beim Ostrakismos gezählt wurden, konnte bisher nicht eindeutig beantwortet werden, da die zwei Hauptzeugnisse einander zu widersprechen schienen.[1] Plutarch berichtet (*Aristeides*, 7.4–5), daß zwei Zählungen vorgenommen wurden. Zuerst haben die Archonten die Gesamtzahl der abgegebenen Scherben festgestellt, um die Gültigkeit des Verfahrens zu bestimmen; denn es mußten wenigstens sechstausend Bürger ihre Meinung ausgesprochen haben. Wenn die vorgeschriebene Anzahl erreicht war, wurden die Scherben nach den Namen gesondert und wiederum gezählt; derjenige, der die höchste Zahl erhielt, wurde für zehn Jahre ausgewiesen.

Gegenüber diesem klaren Bericht steht die Angabe des Philochoros (*F. Gr. Hist.* 328 F 30): διαριθμηθέντων δὲ ὅτῳ πλεῖστα γένοιτο, καὶ μὴ ἐλάττω ἑξακισχιλίων, τοῦτον ἔδει... μεταστῆσαι τῆς πόλεως ἔτη δέκα.... Man hat allgemein angenommen, daß Philochoros nur von einer Zählung spricht; der Ostrakisierte mußte die meisten Stimmen *und* nicht weniger als sechstausend erhalten. Die Zweifel an der Richtigkeit dieser Angabe waren bisher nur sachlich begründet; ich möchte aber vorschlagen, daß der Satz auch sprachlich anders erklärt werden kann. Man hat πλεῖστα und μὴ ἐλάττω ἑξακισχιλίων als Gegenstand der Aussage γένοιτο betrachtet und dabei übersehen, daß μὴ ἐλάττω ἑ. unmittelbar auf διαριθμηθέντων bezogen werden kann. Diese Verbindung ist möglich, da Formen wie ἐλάττω weitgehend als Beiwörter verwendet wurden; ἐλάττω würde hier dieselbe Bedeutung wie ἔλαττον haben.[2] Der Hauptgedanke des Nebensatzes (ὅτῳ... γένοιτο) stand voran, und die hinzugefügte Bemerkung (καὶ μὴ ἐλάττω ἑ.) bekam eine notwendige

Hermes 83 (1955) 119–120

---

1. Siehe O. W. Reinmuth, *R.E.*, s.v. ostrakismos, Sp. 1678–9; A. Calderini, *L'Ostracismo*, S. 37–9; A. E. Raubitschek, *A.J.A.*, 55 (1951), S. 228, Anm. 20 [*supra*, p. 72, n. 20]; id., *Cl. Ph.*, 47 (1952), S. 203–4; id., *Hesperia*, 23 (1954) S. 71, Anm. 18.

2. Siehe E. Schwyzer, *Gr. Gramm.*, I, S. 536, Anm. 3 (auf S. 537), mit reichlichen Verweisen, besonders auf W. Crönert, *Philologus*, 61 (1902), S. 161–92 (Die adverbialen Comparativformen auf —ω).

Einschränkung. Die bisherige Erklärung würde eine andere Wortfolge verlangen: ὅτῳ πλεῖστα καὶ μὴ ἐλάττω ἑξακισχιλίων γένοιτο. Eine gewisse Bestätigung der hier vorgeschlagenen Erklärung kann man in der wiederholten Verwendung des Ausdruckes μὴ ἔλαττον ἑξακισχιλίων bei Demosthenes (24.45, 46,59) sehen, in Stellen, die mit dem Ostrakismos oft in Zusammenhang gebracht wurden.[3]

Um die doppelte Bedeutung des fraglichen Satzes klar zu machen, sollen zwei (leider nicht sehr glückliche) Übersetzungen zusammengestellt werden.

"Nach einer Zählung, wem die meisten (Stimmen) zukamen und nicht weniger als sechstausend..."

"Nach einer Zählung, wem die meisten (Stimmen) zukamen und von nicht weniger als sechstausend..."

Da beide Übersetzungen sprachlich möglich sind, muß man die zweite vorziehen, die mit dem Bericht von Plutarch sachlich übereinstimmt.

3. Zum Beispiel von J. Carcopino, *L'ostracisme Athénien* (Paris, 1935), S. 98–103. So auch F. Jacoby (in den Erläuterungen zu Philochoros, III b I, S. 316–7), der jedoch annimmt, daß wir nur eine mißverständliche Abkürzung des Philochoros-Zitates besitzen; Jacobys Anmerkungen (III b II) sind mir leider noch nicht zugänglich.

16

A New Attic Club
(Eranos)

[93] We know a great deal about Greek Clubs and Associations, private and semipublic, professional, social, or beneficial, mainly from inscriptions. The material has been carefully collected and critically analyzed by Franz Poland in his *Geschichte des griechischen Vereinswesens* (1909); it has been significantly augmented by Mariano San Nicoló in his *Ägyptisches Vereinswesen zur Zeit der Ptolemäer und Römer, Münchner Beiträge zur Papyrusforschung*, 1 (1913) and 2 (1915). One of the clubs was called ERANOS, and a new inscription from Attica, now in the J. Paul Getty Museum, adds considerably to our knowledge of its organization. The *Eranos* in general has been briefly discussed by Poland (pp. 28–33) and by San Nicoló (1, pp. 212–225, and 2, pp. 188–191), and J. Vondeling has treated it fully in his *Eranos* (1961).

The main function of an *Eranos* has been stated by Aristotle (*Ethics* 1160a): some of the associations seem to exist for the enjoyment (δι' ἡδονήν) [of their members], for the sake of sacrifices and companionship (θυσίας ἕνεκα καὶ συνουσίας). Centuries later, the emperor Trajan offered another equally valid characterization (Pliny, *Ep.* 93): they serve not to promote disturbances and illegal associations (*ad turbas et inlicitos coetus*) but to alleviate the indigence of the poor (*ad sustinendam tenuiorum inopiam*). The new inscription illustrates these points very well.

The almost completely preserved text is engraved on two joining fragments constituting a completely preserved stele of Pentelic marble (h. max. 75 cm., w. max. 44.7 cm., th. max. 6 cm.—78.AA.377), crowned by a pediment with an akroterion (Plates 5, 6); within the pediment is a shield (*hoplon*) in relief on which a portrait (*eikon*) may have been painted. At the bottom, there is a broad and roughly picked tenon with which the stele was inserted into a stone base or into the living rock. During the past two decades the stone passed through several hands before it was acquired by a New York collector, who generously offered it to the Getty Museum. It is said to come from Liopesi, a village in central Attica, the site of the ancient deme Paiania; this location is of great significance, as will be seen.

In the reading, restoration, and explanation of the text I have been aided by a number of friends to whom I should like to express my gratitude: W. Burkert, A. Dihle, J. Frel, D. Geagan, Ch. Habicht, E. Kapetanopoulos, D. M. Lewis, R. Merkelbach, †J. H. Oliver, F. Sokolowski, R. E. Wycherley; also to E. Handley, L. Pearson, and S. Stephens for their suggestions.

The lettering is uneven, irregular, and crowded, and belongs to the period after A.D. 100. The text was inscribed by two hands. The first engraved lines 1–36, the second begins with the line 37 and corrected passages in lines 3, 13, 21, 24, 32, and 35; it is characterized by using $Σ$ instead of [, by using the punctuations $\rangle$ and = instead of)(, by writing I instead of EI, and by observing the line divisions of words more carefully. The two hands are, of course, contemporary.

     Ἀγαθῇ Τύχῃ. Ἐπὶ Τίτου Φλαβίου Κόνωνος ἄρχοντο-   [94]
     ς καὶ ἱερέως Δρούσου ὑπάτου, Μουνιχῶνος ὀκτὼ
     καὶ δεκάτῃ· ἔδοξεν τῷ ἀρχερανιστῇ Μάρκῳ Αἰμιλίῳ
     Εὐχαρίστῳ Παιανιεῖ συνόδου τῆς τῶν Ἡρακλιαστῶν τῶν
5   ἐν Λίμναις)(τάδε δοκματίσαι· ἐάν τις ἐν τῇ συνόδῳ
     μάχην ποιήσῃ, τῇ ἐχομένῃ ἡμέρᾳ ἀποτινέτω προστείμ-
     ου ὁ μὲν ἀρξάμενος δραχμὰς δέκα)(ὁ δὲ ἐξακολουθ-
     ήσας δραχμὰς πέντε)(καὶ ἐξάνανκα πραττέσθω τῶν σ-
     [υ]νερανιστῶν ψῆφον λαβόντων ἐκβιβάσαι)(τῆς δὲ ἐνθήκ-
10 ης τῆς τεθείσης ὑπὸ τοῦ ἀρχερανιστοῦ καὶ ὅση ἂν ἄλλη ἐν-
     θήκη ἐπισυναχθῇ, ταύτης μηθεὶς κατὰ μηδένα τρόπον ἁπτ-
     [έ]σθω πλείω τοῦ τόκου τοῦ πεσομένου)(μὴ πλέω δὲ δαπανάτ-
     [ω] ὁ ταμίας δραχμῶν Τ· ἔδοξε ἐκ τοῦ τόκου)(ἐὰν δέ τι πλείων-
     [ο]ς ἄψηται ἢ ἐκ τῆς ἐνθήκης)(ἢ ἐκ τοῦ τόκου ἀποτεινέτω προσ-
15 [τ]είμου τὸ τριπλοῦν· ὁμοίως δὲ καὶ ἂν ταμιεύσας τις ἐπιδειχθῇ
     [ν]ενοσφισμένος)(ἀποτινέτω τὸ τριπλοῦν. περὶ δὲ ἱερεωσυν-
     [ῶ]ν ὧν ἄν τις ἀγοράσῃ παραχρῆμα κατατιθέστω)(ἐν τῷ ἐχ-
     [ο]μένῳ ἐνιαυτῷ)(αὐτῷ τῷ ἀρχερανιστῇ καὶ λανβανέτω πρόσ-
     [γ]ραφον παρὰ τοῦ ἀρχερανιστοῦ, λαμβάνων δὲ ἐξ ἔθους τὰ διπλᾶ
20 [μ]έρη ἐκτὸς τοῦ οἴνου· οἱ δὲ ἐργολαβήσαντες ὑϊκὸν ἢ οἰνικὸν μ-
     [ή] ἀποκαταστήσαντες ἐν ᾧ)(δειπνοῦσιν ἐνιαυτῷ ἀποτινέτω-
     σαν τὸ διπλοῦν. οἱ δὲ ἐργολαβοῦντες ἐνγυητὰς εὐαρέστους
     παρατιθέτωσαν τῷ ταμίᾳ καὶ τῷ ἀρχερανιστῇ. καταστάνεσθαι δὲ Γ̄
     παννυχιστὰς τοὺς δυναμένους· ἐὰν δὲ μὴ θέλωσιν τότε ἐκ πάντ-
25 ων κληρούσθωσαν καὶ ὁ λαχὼν ὑπομενέτω· ἐὰν δὲ μὴ ὑπομένῃ ἢ
     μὴ θέλῃ παννυχιστὴς εἶναι λαχὼν ἀποτινέτω προστείμου δραχμὰς ἑκ-
     ατόν)(καταστάνεσθ{ωσαν}αι δὲ ἐπάνανκες ἐκ τῆς συνόδου πράκ-
     τορες δέκα)(ἐὰν δέ τινες μὴ θέλωσιν πράκτορες ὑπομένειν κληρούσθω-
     σαν ἐκ τοῦ πλήθους δέκα)(ὁμοίως δὲ καὶ ἐὰν ὁ ταμίας ἀποδιδοῖ λόγον ἀγ-
30 ορᾶς γενομένης καταστάνεσθαι ἐγλογιστὰς τρεῖς καὶ τοὺς ἐγλογιστὰς

ὀμνύειν αὐτόν τε τὸν Ἡρακλῆν καὶ Δήμητρα κα[ὶ] Κόρην)(κληροῦσθαι δὲ τῆς ἡμέρ-
ας ἑκάστης ἐπὶ τὰ κρέα ἀνθρώπους δύω)(ὁμοίως καὶ ἐπὶ τοὺς σ|τρε|πτού-
ς ἀνθρώπους δύω)(ἐὰν δέ τις τῶν πεπιστευμένων εὑρεθῇ ῥυπαρόν τ-
[ι] πεποιηκὼς ἀποτινέτω δραχμὰς εἴκοσι)(αἱρείσθω δὲ ὁ ἀρχερανιστὴς
35 οὓς ἂν βούληται ἐκ τῆς συνόδου |εἰς τὸ συνεγ|δανίσαι τὴν ἐνθήκην μετ᾽
αὐτοῦ
ἀνθρώπους Γ̅. διδότωσαν δὲ τὴν σιμίδαλιν πάντες τῇ δημοσίᾳ χοίνικι[.].
ἐγδιδόσθαι δὲ καθ᾽ ἕκαστον ἐνιαυτὸν ὑπὸ τοῦ ταμ[ίο]υ θῦμα τῷ θεῷ
κάπρον Μ̇ κ< /ἐὰν δέ τις τῶν ἐκ τοῦ ἐράνου τέκνον [τ]ίσι θέλῃ ἰσάγιν
διδότω υἱκοῦ Μ̇ 15 <', ἐὰν δέ τις ἐμβῆναι θέλῃ διδότω υἱκοῦ Μ̇ΛΓ.
40 καταβάλλεσθαι δὲ τὸν λόγον ὅταν οἱ ἐγλογισταὶ ὀμόσαντε[ς]
ἀποδῶσι τῷ ἀρχερανιστῇ τὸν λόγον καὶ ἐπιδίξωσι εἴ τι ὀφίλι ὁ
ταμίας. ξύλα δὲ ἐγδιδόσθαι ὑπὸ τοῦ καθ᾽ ἔτος ταμίου< τὰς δὲ φορὰς
καταφέριν τῷ ταμίᾳ ἐπάναγκες ἰς τὰς ἐγδόσις· ὁ δὲ μὴ κατενένκας
ἀποτινέτω τὸ διπλοῦν< ὁ δὲ μὴ δοὺς τὸ κάθολον ἐξέρανος
45 ἔστω¹ μὴ ἐξέστω δὲ τῶν ἐν τῷ ἅλσι ξύλων ἅπτεσθαι< στέφα-
[νο]ν δὲ φέριν τῷ θεῷ ἕκαστον¹

Translation:

[95] Good luck. When Titus Flavius Konon was archon and
priest of consul Drusus, on the eighteenth of Mounichion,
the archeranistes of the association of the Herakliastai in
5 the Marshes, Marcus Aemilius Eucharistos of Paiania, decided to lay down (or to have laid down) the following
order: If anyone during a meeting enters a fight, let him
pay on the following day a fine of ten drachmai if he
started it and of five drachmai if he participated in it, and
without fail let him be (made to be) expelled after his
10 fellow eranistai have cast a vote. With respect to the endowment deposited by the archeranistes and any other
that may be added, no one in any way shall touch it
beyond the amount of the interest which will fall due. The
treasurer shall not spend more than 300 drachmai (this
was ordered) from the interest. If he draws upon more
either from the endowment or from
15 the interest, he shall pay a threefold fine. Similarly, if
someone who has been treasurer shall be found guilty of
touching the fund, let him pay threefold. Concerning
priesthoods, if someone buys one at once, let him make
the payment during the following year to the archeranistes, and accept a receipt from the archeranistes; let him
receive, as is customary, double portions except for the
20 wine. The contractors of the pork and wine supplies who
do not restore the funds during the year in which they

provide the dinner are to be fined the double amount.
The contractors are to provide acceptable sureties to the
treasurer and to the archeranistes. Let them establish three
powerful men as nightwatchmen; if they should refuse the
assignment, then let three be chosen by lot out of all
members, and he who is chosen must accept the task. If he
25 will not consent or if he does not want to be a night-
watchman although he has been chosen by lot, let him
pay a fine of one hundred drachmai. It shall be com-
pulsory to establish from the association ten paymasters,
but if some are unwilling to serve as paymasters, ten shall
be chosen by lot out of all. Similarly, when the treasurer
renders his account after a transaction has taken
30 place, three auditors are to be appointed, and they are to
give an oath by Herakles himself and Demeter and Kore.
They shall choose by lot every day two persons in charge
of the meat and similarly two persons in charge of the rolls.
If anyone of those entrusted should be found having done
something dishonest, let him pay a fine of twenty drachmai.
35 The archeranistes shall select three people of his choice
from the association for helping him to lend out the endowment. Let all contribute the wheat flour according to
the public measure of a choinix. Let there be made annually by the treasurer a sacrifice to the god of a boar
weighing twenty minai. If a member of the eranos wishes
to introduce a child by making a payment, let him contribute $16\frac{1}{2}$ minae of pork; and if somebody wishes to
enter himself, let him give 33 minae of pork. Let the ac-
40 count be closed when the auditors after having given an
oath render the account to the archeranistes and indicate
if the treasurer owes something. Firewood should be issued
by the annual treasurer. The dues are to be brought to the
treasurer without fail for the (expenditure or for the) making of loans. He who does not bring his dues is to pay as
fine double the amount. He who does not pay at all is to
be expelled. No one is permitted to
45 touch the firewood in the grove. Everybody is to wear a
wreath in honor of the god.

COMMENTARY

Lines 1–2. The document is dated by the archon, Titus Flavius Konon, whose family tree I have reconstructed in *Jahreshefte des Österr. Arch. Institutes*, XXXVI (Beiblatt, cols. 35–39). Accordingly, his demotic was Σουνιεύς, and he may have been the brother of Titus Flavius Sophokles of Sounion who was archon in A.D. 121/2, and his archonship may fall in the same period because "the archon eponymos was also the priest of the consul Drusus from the time of the death of Drusus until the reign of Hadrian"; see D. J. Geagan, *The Athenian Constitution after Sulla* (1967), p. 8; Geagan also supplied me with a list of the priests the last of whom (*IG*, II², 3589) is dated A.D. 122/2.

Lines 2-3. The date within the month, the eighteenth of Mounichion, is significant for two reasons. Another document which was also found in Liopesi (*IG*, II², 1369) and also contains a νόμος ἐρανιστῶν, is also dated on the eighteenth of Mounichion, probably the date of the annual meeting of the association. Unfortunately, this document is no longer preserved but both A. Wilhelm (*Serta Harteliana*, 1896, pp. 231-235) and L. Robert (*AJP*, 100, 1979, pp. 153-157) have contributed to our better understanding of it. It is puzzling to find so many meetings of thiasoi, orgeones, and eranoi to take place in the month of Mounichion; this is especially true for the documents of the worshipers of Μήτηρ (Magna Mater), *IG*, II², 1314/5, 1327/8/9. Unfortunately, I could not find any special significance in the choice of the eighteenth of Mounichion.

Lines 3-4. The name Μάρκῳ is entered *in rasura* by the second hand; see p. 93. The archeranistes Marcus Aemilus Eucharistos of Paiania is not otherwise known, but the fact that this inscription and *IG*, II², 1369 were both found in his home deme is significant. One would like to assume that the association, founded or at least controlled by Eucharistos, was located in Paiania, were it not for the fact that it is expressly stated that its location was ἐν Λίμναις (line 5). Under these circumstances it may be best to assume | that the two inscriptions were set up in Paiania because Eucharistos was at home there. This would mean that *IG*, II², 1369 should also be connected, if not with him, then at least with his son or grandson.

Lines 4-5. The official title of the Association was σύνοδος ἡ τῶν Ἡρακλιαστῶν τῶν ἐν Λίμναις. Eucharistos was the archeranistes of this synodos, and the members were called Herakliastai. This is not surprising, because we find the same kind of organization in other, similar, earlier associations: *IG*, II², 1292 (Serapiastai with proeranistria), 1322 (Amphierastai with archeranistes), 1335 (Sabaziastai with eranistai), 1339 (Heroistai with archeranistes), 1343 (Soteriastai with synodos and archeranistes); all these inscriptions are honorary decrees of the Hellenistic period. The crux is the localization ἐν Λίμναις because I have not been able to find any place by that name except the famous one with the sanctuary of Dionysos, the location of which "still remains an insoluble riddle"; see R. E. Wycherley, *The Stones of Athens* (1978), p. 172; G. Neumann, ΣΤΗΛΗ (in honor of N. Kontoleon), 1980, p. 617: "immer noch ein Rätsel." Wycherley suggests, "maybe these Herakliastai had something to do with Kynosarges," (which was a gymnasium connected with a shrine of Herakles; see Wycherley, *op. cit.*, pp. 229-231), but it may be remembered that W. Judeich (*Topographie von Athen*, 1931, pp. 291–293) placed it on the west slope of the Acropolis and the Iobakcheion in it; see lines 5–7.

Line 5. The combination of ἔδοξεν ... δοκματίσαι (in classical Greek it would have been δογματίσαι) shows clearly that the Archeranistes decided to make this proclamation or caused it to be made. He was probably entitled to do so because of his personal share in the founding and maintenance of the synodos.

Lines 5–9. It is surprising that the first point of the edict concerns the prohibition of physical disturbances during a meeting of the association, but it will be noticed that the two most similar documents, *IG*, II^2, 1368 (concerning the Association of the Jobacchai, the charter of which was discovered in a house that was thought to be located ἐν Λίμναις; see on lines 4–5) and *IG*, II^2, 1369 (concerning the same Association of the Herakliastai but being perhaps of a later date) also have long passages dealing with μάχαι, πληγαί, θόρυβοι. In *IG*, II^2, 1369, lines 40–44, expulsion and a fine of 25 drachmai is threatened to the person who μάχας ἢ θορύβους κεινῶν (κινῶν) φαίνοιτο, while in *IG*, II^2, 1368 there is a long passage (lines 72–102) dealing with these and similar offenses; here there has to be a sworn testimony on the part of the injured, and the punishment of a fine of 25 drachmai which is leveled also against the instigator and not only against the perpetrator, and exclusion from meetings until the fine is paid. If it comes actually to blows, the victim has to appeal to the priest who has to call an assembly which is to decide by vote for how long the culprit is to be excluded from meetings and how much he has to pay (up to 25 drachmai). The victim himself is to be fined if he does not communicate with the priest or the archibackchos but denounces the attack in public (i.e., appeals directly to a court of justice). The members of the association are held responsible, under the threat of a fine, to remove people who are fighting, to attend the assembly voting on the expulsion of perpetrators of violence, and to carry out the verdict of the assembly. All this throws a significant light not only on what may have been going on at the meetings but also on the high moral tone of the charters of the associations. It may also be noted that the fine mentioned in our inscription (10 drachmai for starting a fight, 5 drachmai for joining in it) is comparatively small, while the threat of expulsion is greater than on the later documents. The word σ[υ]νερανιστῶν (lines 8–9) is attested by Poland (*op. cit.*, p. 32) and *I.G.*, II^2, 2721.

Lines 9–16. This passage about the endowment and its use is not only most important but also most informative since it spells out details which are quite new. In general, one gets the impression, which Poland and others expressed before, that the association's main purpose was the lending of money to its members, presumably without the collateral that a banker may have required. We do not know whether the founder of the endowment, the archeranistes, made money out of his investment; if so, this is not stated. The principal (which consisted probably of real estate) was not to be spent under any circumstances, and of the interest not more than 300 drachmai. The interest accrued of course from the payments of members who borrowed money. It is not easy to understand the word ἔδοξε (written *in rasura*) in line 13; Merkelbach suggested that it was a later insertion into the proposal, originally prepared by the archeranistes, and referred to the fact that it was "decided" to put a limit of 300 drachmai (to come from the interest) on expenditures made by the tamias.

Lines 16–20. About the sale of priesthoods, about the quick payment of the price, about the receipt issued by the archeranistes, and about the double portions (except for wine) to be given to the priests.

Lines 20–23. The question is here whether the contractors (ἐργολαβοῦντες) receive money from the treasurer (ταμίας) in order to provide pork and wine for the dinners during a year, sell these dinners to the member guests and return the original funds to the treasurer, or whether they use the funds in order to provide free meals to the members. The provision under discussion makes sure that they do either one. Merkelbach opts for the second possibility, I for the first. The OY in δειπνοῦσιν is *in rasura* by the second hand.

Lines 23–27. The παννυχισταί have been watchmen whose duty must have been very important, considering the careful way they were selected and the fine they had to pay in case they neglected their duty. They were to protect the property and the safety of the association and of its members, and I presume that they served only when there were meetings; nothing is said about their receiving any compensation. The otherwise unattested word παννυχιστάς is written *in rasura* by the second hand.

Lines 27–29. It is not clear what the duty of the πράκτορες was. It could have been the collection of the membership fees, but it appears from lines 43–44 that the treasurer received the φοραί directly from the members. It is therefore better to assume that the πράκτορες collected the various and many fines, since in no case is any mention made how and by whom the fines are to be collected.

Lines 29–31. The ἐγλογισταί were controllers and auditors who assisted the treasurer when he prepared and presented his account to a formal assembly (ἀγορά) and they had to give an oath, presumably to conduct their business properly, to Herakles, to Demeter and to Kore. We wish we knew why Demeter and Kore were included because their presence may give us a clue to the location of the meeting house of the Herakliastai.

Lines 31–34. "Every day" may refer to every feast day, and it would be natural to have two men in charge of the meat and two in charge of the twisted bakery (στρεπτοί). Since these men are threatened with fines of 20 drachmai each if they conduct their work dishonestly, they were probably members of the association. The letters TPE of στρεπτούς are written *in rasura* by the second hand.

Lines 34–36. The three men chosen by the archeranistes to assist him in making loans from the endowment were surely members of the association. This shows how the endowment could earn interest (lines 12–13); this may have been the main purpose and attraction of the association. The letters εἰς τὸ συνεγ are written *in rasura* by the second hand.

Line 36. σιμίδαλις is the fine wheat flour used to make the pastry. Every member was to contribute one choinix according to the public measure. For σιμίδαλις, see Athenaeus, *Deipnosophistae* III 109a and b, 112b, 115c and d. The phrase τῆ δημοσίᾳ χοίνικι is not quite clear in its meaning; the missing letter at the end of the line may have been a numeral. I assume that each member was to contribute one choinix of fine wheat flour, according to the public standard. The use of πάντες is puzzling; one would expect ἕκαστος. The last letter of this line may have been erased; it may have been a $\bar{\Gamma}$ (3).

Lines 37–38. The boar of twenty pounds, which the treasurer was to provide for an annual sacrifice, was not a "wild" boar but just a male piglet.

Lines 38–39. The text of this sentence is not certain since I am not at all confident that [τ]ίσι (=τίσει) is the right restoration. It is only natural that the association should receive new members from the ranks of the sons (there is no mention made of women) of the members, and the distinction between σύνοδος (the name of the association) and ἔρανος may be significant. The phrase occurring at the beginning of the νόμος ἐρανιστῶν (*IG*, II², 1369, lines 30–34), referring to the same association, is significant: [μη]δενὶ ἐξέστω ἰσι[έν]αι ἰς τὴν σεμνοτάτην | σύνοδον τῶν ἐρανιστῶν. ... The synodos consists of eranistai who are called Herakliastai and are also called οἱ ἐκ τοῦ ἐράνου. It is puzzling that the initiation fee consisted of meat; we may presume that it could be presented in the form of an amount of money with which the meat could be purchased.

Lines 40–42. This phrase refers back to lines 29–31 and may have to be added to the original statement on the auditors.

Line 42. The ξύλα are obviously firewood which is to be supplied by the treasurer (tamias) who held office for one year, as presumably all officials did, with the possible exception of the archeranistes.

Lines 42–44. The membership dues, the contributions or φοραί, are to be brought to the treasurer; it is not clear what the necessary connection with the ἐκδόσεις is, since ἔκδοσις seems to mean lending money rather than paying it out. The emphasis is on the payment of the dues, and the phrase ἐπάναγκες εἰς τὰς ἐκδόσεις may indicate the time when the members had to pay their fees. Delayed payments had to be double, and nonpayment resulted in automatic expulsion.

Line 45. It is not clear whether the ξύλα is firewood which was kept in the ἄλσος, the grove, or whether the word refers to the | trees of the grove which must not be cut down. The location of the grove is of course unknown, but Judeich pointed out (see on lines 4–5) that there was a grove in the Kynosarges (*op. cit.*, pp. 423–424).

Lines 45–46. It is not clear whether every member has to wear a wreath in honor of Herakles, or whether he had to bring a wreath to the god.

This inscription does not record the foundation of the Synodos of the Herakliastai in the Marshes but a significant event in the history of this association. The Synodos may have begun in the Hellenistic period, and it did continue later, since we possess one of its documents of the second half of the second century. The first clause concerning the maintenance of peace and order may have been required by the Roman government, but the second item, concerning the endowment and its use must have been the main cause for publishing the edict. Eucharistos, the archeranistes, is otherwise unknown; he or his family may have received Roman citizenship through the intervention of one of the famous Aemilii (Lepidus or Juncus) who were active in Athens. This means that he was not only rich but also respectable. It was his endowment (the size of which is unknown) that put the Synodos into business and ensured Eucharistos's position in it as archeranistes. The important thing was to ensure that the endowment would be increased, and virtually all the measures point in this direction. The ways the endowment was to be increased were these: interest payments on loans, membership dues, fines, sales (of offices and perhaps of services and food). Expenditures were severely restricted. It is possible that the members got free meals, but nothing is said of salaries or remunerations for services rendered; in fact, they even had to contribute flour for the baked goods. All that the treasurer was expected to issue was an animal of small size for an annual sacrifice and firewood. There is not much said about religious and social activities, but these may not have required specific regulations. It may be assumed that the members were devoted to Herakles and that they enjoyed each other's company.

17

Zur Frühgeschichte der Olympischen Spiele

Althistoriker und Archäologen zögern, die Ursprungssagen der Olympischen [64]
Spiele ernst zu nehmen, die einen, weil sie nicht an die Geschichtlichkeit solcher Heroen wie Pelops und Herakles oder sogar des Spartaners Lykurgos glauben wollen, die anderen, weil es keine Bodenfunde athletischen Charakters aus Griechenland und besonders aus Olympia vor dem siebenten Jahrhundert gibt. Diese vorsichtige Einstellung wird trotz der Zeugnisse eines Pindar (*Ol.* 10), Strabon, Phlegon, des Pausanias und des Plutarch eingenommen, und sie läßt sich im Grunde sehr schwer widerlegen oder auch nur in Frage stellen. Uns geht es daher gar nicht darum, diese alten Geschichten als wahr zu erweisen, sondern wir wollen sie zeitlich festlegen und inhaltlich erklären.

Den ausführlichsten Bericht verdanken wir Phlegon von Tralles, einem Freigelassenen des Kaisers Hadrian, der eine Buch $\Pi\epsilon\rho\grave{\iota}\ \tau\hat{\omega}\nu\ \ '\!O\lambda\upsilon\mu\pi\acute{\iota}\omega\nu$ geschrieben hat, von dem wir ein Stück der Einleitung in dem *Codex Palatinus Graecus* 398, fol. 244^r erhalten haben, abgedruckt von F. Jacoby in *FGH* 257 F 1. Da schreibt er, daß er es für nötig hält, den Ursprung ($\tau\grave{\eta}\nu\ \alpha\grave{\iota}\tau\acute{\iota}\alpha\nu$) der Olympischen Spiele darzulegen, und er erzählt, daß, nachdem Peison (der Eponym von Pisa), Pelops und dann Herakles die Festversammlung ($\pi\alpha\nu\acute{\eta}\gamma\upsilon\rho\iota\nu$) und den Wettkampf ($\dot{\alpha}\gamma\hat{\omega}\nu\alpha$) in Olympia begründet hatten, der Kult ($\theta\rho\eta\sigma\kappa\epsilon\acute{\iota}\alpha\nu$) eine Zeitlang aussetzte ($\dot{\epsilon}\kappa\lambda\epsilon\iota\pi\acute{o}\nu\tau\omega\nu$), bis (von Iphitos) achtundzwanzig Olympiaden (also 112 Jahre) bis Koroibos gezählt wurden (s. dazu Jacobys Kommentar), und daß wegen der Vernachlässigung des Wettkampfes ($\dot{\alpha}\mu\epsilon\lambda\eta\sigma\acute{\alpha}\nu\tau\omega\nu\ \tau\upsilon\hat{\upsilon}\ \dot{\alpha}\gamma\hat{\omega}\nu o s$) Unruhe ($\sigma\tau\acute{\alpha}\sigma\iota s$) innerhalb der Peloponnes ausbrach; darauf bezieht sich auch Pausanias 5, 4, 5, der die politischen Unruhen ($\sigma\tau\acute{\alpha}\sigma\epsilon\omega\nu$) und die von Phlegon später besprochene Pest ($\dot{\upsilon}\pi\grave{o}\ \nu\acute{o}\sigma o\upsilon\ \lambda o\iota\mu\acute{\omega}\delta o\upsilon s$) in einem Satz zusammenfaßt.

Phlegon gibt hier den Eindruck, daß, solange die olympischen Feste stattfanden, Friede im Land herrschte, und dieser Eindruck wird durch Lysias

Lebendige Altertumswissenschaft: Festgabe zur Vollendung des 70. Lebensjahres von Hermann Vetters (A. Hakkert, 1985), 64–65.

und Isokrates bestätigt. Sagt doch der eine in seiner zweiten Olympischen Rede, daß Herakles glaubte, daß die von ihm in Olympia gegründete Versammlung (σύλλογον) der Anfang der Freundschaft zwischen den Griechen sein werde (ἀρχὴν γενήσεσθαι τοῖς Ἕλλησι τῆς πρὸς ἀλλήλους φιλίας). Und Isokrates lobt im *Panegyricus* 43-44 die Begründer der panhellenischen Feste (πανηγύρεις) aus demselben Grund; wir werden uns dieser Stelle später eingehend widmen.

Phlegon wendet sich jetzt der Erneuerung der Olympischen Spiele durch den Herakliden Lykurgos aus Sparta, den Herakliden Iphitos aus Elis und den Pisaten Kleosthenes zu; durch die Betonung der Abstammung von Herakles scheint er die Verbindung mit dem Begründer der Spiele und mit der Heroenzeit unterstreichen zu wollen. Es war der Wunsch dieser Männer, die Bevölkerung (τὸ πλῆθος, wohl erstlich der Peloponnes) zur Eintracht (ὁμόνοιαν) und zum Frieden (εἰρήνην) zurückzubringen, und sie glaubten, dies durch das Zurückführen (ἀνάγειν) der olympischen Festversammlung (τὴν Πανήγυριν τὴν Ὀλυμπικήν) zu den alten Bräuchen (εἰς τὰ ἀρχαῖα νόμιμα) und durch die Einrichtung (ἐπιτελέσαι) eines gymnischen Wettkampfes (ἀγῶνα γυμνικόν) tun zu können. Man gewinnt den Eindruck, daß die Festversammlung in gewissem Sinn wichtiger war als die Wettkämpfe selbst, besonders wenn es sich um Eintracht und Frieden handelt, und daß der gymnische Agon eine Neueinrichtung der lykurgischen Wiederherstellung war; vorher mag es sich hauptsächlich um Wagenrennen der reichen Adeligen gehandelt haben. Diese Erklärung würde durch Überlegungen Strabons unterstützt, der (8, 3, 30) betont, daß die Wettkämpfe der Frühzeit unbedeutend (οὐκ ἔνδοξα) waren, wenn sie überhaupt stattfanden.

Auf die Anfrage in Delphi, ob Apollo die Pläne gutheiße, erhielten die Veranstalter einen günstigen Bescheid und die zusätzliche Bestimmung, den teilnehmenden Städten einen Waffenstillstand (ἐκεχειρία) zu verkünden. All das wurde in ganz Griechenland bekanntgegeben, und die Hellanodikai beschrifteten den Diskos mit diesen Verordnungen; Aristoteles hat den Diskos noch selbst gesehen (Plutarch, *Lycurgus* 1, 1), und er enthielt nur die Bestimmungen für die Ekecheiria.

[65] Phlegon berichtet weiterhin, daß die Peloponnesier ohne Begeisterung (οὐκ ἄγαν προσιεμένων) an | den Wettspielen teilnahmen und sie gar übel aufnahmen (δυσχεραινόντων); daher fielen sie einer Pest und der Vernichtung ihrer Ernte anheim (λοιμὸς καὶ φθορὰ καρπῶν). Auf ihr Verlangen schickten Lykurgos und seine Freunde wieder nach Delphi, um das Aufhören der Pest und eine Heilung von ihr zu erbitten. Die Pythia antwortete mit einem langen hexametrischen Gedicht, in dem sie die Vernachlässigung der olympischen Feierlichkeiten (ἀτιμάζοντες) als Grund für den Zorn des Zeus angab und nach einer kurzen Geschichte des Heiligtums (Peison, Pelops, Herakles werden als Gründer des Festes, hier mit dem alten Namen ἔροτις = ἑορτή, und der Wettspiele, hier ἔπαθλα genannt, bezeichnet) die Wiederherstellung des Festes anordnete. Als dieses Orakel den Peloponnesiern mitgeteilt wurde, waren sie noch immer ungläubig (ἀπιστήσαντες) und schickten ihrerseits nach Delphi; die Antwort war kurz und bündig: gehorcht den Worten der Seher (πείθεσθε τά κεν μάντεις ἐνέπωσιν). Und das geschah nun auch. Elis erhielt die Erlaubnis, die

Wettkämpfe zu veranstalten und die Ekecheiria bekanntzugeben, und mußte sich verpflichten, sich nicht an Kriegen zu beteiligen (πολέμον δ' ἀπέχεσθε).

Es genügt nicht, zu behaupten, daß dieser Bericht des Phlegon kaum etwas mit den geschichtlichen Ereignissen zu tun habe, die im achten Jahrhundert die Gründung oder die Erneuerung der Olympischen Spiele zur Folge hatten. Es genügt auch nicht, Einzelheiten aus diesem Bericht herauszuheben und zu behaupten, daß sie wahrscheinlich schon in diese Frühzeit gehörten, wie z. B. die Ekecheiria selber, die Prominenz der gymnischen Agone und vielleicht auch die politischen Unruhen in der Peloponnes während des achten Jahrhunderts.

Wir müssen statt dessen versuchen, den Bericht als eine Einheit zu verstehen, und ihn geschichtlich einordnen und erklären. Dabei müssen wir erkennen, daß das Hauptthema dieser Frühgeschichte der Olympischen Spiele ihre Verbindung mit panhellenischer Eintracht und panhellenischem Frieden ist. Da zeigt sich nun vorerst und ganz überraschend, daß unsere drei Hauptzeugen von der olympischen Festversammlung und von den Olympischen Spielen als Zeichen panhellenischen Friedens überhaupt nichts wissen oder wenigstens nichts sagen. Weder Pindar noch Herodot noch Thukydides erwähnen Olympia als ein Symbol gemeingriechischer Verwandtschaft und Einheit, obwohl alle drei an der Idee der Einheit der Griechen sehr interessiert sind.

Der peloponnesische Krieg mit seiner Selbstzerstörung des Griechentums hat manchen Dichtern und Denkern die Augen geöffnet, aber Thukydides war anscheinend nicht darunter. Dafür hat Aristophanes, der schon immer ein Gegner des Krieges von Griechen gegen Griechen war, in der *Lysistrata* (V. 1128–1134) seiner Heldin die schönsten Worte in den Mund gelegt, in denen sie darauf hinweist, daß die Griechen wie Mitglieder einer Familie (ὥσπερ ξυγγενεῖς) aus einer Schale an den Altären in Olympia, an den Thermopylen und in Delphi Opfer darbringen, aber dann Griechen und griechische Städte vernichten. Schon vorher hat vielleicht Gorgias in seiner Olympischen Rede ähnliches, möglicherweise zum ersten Mal, ausgesprochen. Jedenfalls hat Lysias in seiner Rede die panhellenische Freundschaft als den Zweck der Gründung des olympischen Festes durch Herakles bezeichnet, und Isokrates hat diese Idee im Panegyricus zu einem politischen Programm erweitert. So sagt er, daß die panhellenischen Feste eingerichtet wurden, um die zwischen Griechen bestehenden Feindseligkeiten zu beseitigen, um den Griechen die Gelegenheit zu geben, ihrer gemeinsamen Herkunft zu gedenken, und um die Griechen zu überreden, in der Zukunft die gegenseitige Freundschaft zu pflegen.

Wir haben augenscheinlich den Zeitpunkt gefunden, an dem die Friedensidee mit den panhellenischen Festen verbunden wurde. Es läßt sich leicht verstehen, daß damals die Behauptung aufgestellt wurde, daß die Olympischen Spiele von allem Anfang an dem gemeingriechischen Frieden dienen sollten. Der Bericht des Phlegon gehört also in diese Zeit und kann sogar dem elischen Sophisten Hippias zugewiesen werden, der für die Zusammenstellung der Olympionikenliste bekannt war; s. Jacoby, *FGH* 6 F 2 mit Kommentar. Hippias hatte dann eben dieser Liste eine historische Einleitung gegeben, in der die Friedensidee seiner eigenen Zeit schon in die Gründungszeit der Spiele vorverlegt wurde.

18

Das Datislied

[234] Wenn Trygaios erfährt, daß Kleon und Brasidas tot sind und daß die Vernichtung Griechenlands daher aufgeschoben ist (Aristophanes, *Pax*, 259–287) spricht er, jetzt allein auf der Bühne, die folgenden Verse an die Zuschauer (289–295):

Νῦν τοῦτ᾽ ἐκεῖν᾽ ἥκει, τὸ Δάτιδος μέλος,
ὁ δεφόμενός ποτ᾽ ᾖδε τῆς μεσημβρίας
"ὡς ἥδομαι καὶ χαίρομαι κεὐφραίνομαι."
νῦν ἐστι ὑμῖν, ὦνδρες Ἕλληνες, καλὸν
ἀπαλλαγεῖσι πραγμάτων τε καὶ μαχῶν
ἐξελκύσαι τὴν πᾶσιν Εἰρήνην φίλην,
πρὶν ἕτερον αὖ δοίδυκα κωλῦσαί τινα.

Die Scholien zu dieser Stelle (289) geben die folgenden Erklärungen des sonderbaren Ausdrucks τὸ Δάτιδος μέλος:

1. Δάτις σατράπης Περσῶν, εὐδοκιμήσας κατά τινα χρόνον πολεμῶν, ἑλληνίζειν δὲ βουλόμενος εἶπεν ἥδομαι καὶ χαίρομαι, καὶ ἐβαρβάρισεν· ἔδει γὰρ εἰπεῖν χαίρω. ἅμα δὲ οὗτος εἶπε πρὸς τὸ ὁμοιοκατάληκτον. οὐκ εἴρηται δὲ τὸ χαίρομαι· αὐτὸ γὰρ τὸ χαίρω αὐτοπαθὲς ὂν προϋφήρπασεν αὐτοῦ τὸ σημαινόμενον. λέγεται δὲ τὸ τοιοῦτο δατισμός.

2. Στοχάζεται ὅτι εἴη ἂν διαμνημονευόμενος περὶ Δάτιδος τοῦ Πέρσου, ὅτι εἰπέ ποτε παρά τινα συνουσίαν ἥδομαι καὶ χαίρομαι, ἑλληνίζειν τι θέλων. καὶ γὰρ καὶ παρὰ παλαιοτέρου τινὸς ἐμφαίνεται. εἴη δὲ ὕπαρχος τοῦ βασιλέως τῶν ἐλθόντων ἐπὶ Μαραθῶνα. ὃν καὶ φυγεῖν φασι πολὺν χρυσὸν καταλιπόντα.

3. Ἄλλως. Δάτις τῶν ἐν Μαραθῶνι παραταξαμένων Ἀθηναίοις στρατηγὸς Δαρείου, ὅν φασι ἀντιποιήσασθαι τῆς Ἀττικῆς ὡς ἰδίας· ἀπὸ Μήδου γὰρ τοῦ Μηδείας καὶ Αἰγέως ἐδόκει εἶναι.

4. Ἄλλως. Δᾶτις σατράπης Περσῶν βασιλέως, ὅστις συνεχέστερον πρεσβευόμενος εἰς Ἀθήνας ἠράσθη τῆς αὐτῶν πολιτείας, καὶ μείνας ἐκεῖ ἐβούλετο ἑλληνίζειν τῇ ὁμιλίᾳ, καὶ ἐβαρβάριζεν καὶ τοὺς τρόπους καὶ τοὺς λόγους.

Charites E. Langlotz: Studien zur Altertumswissenschaft, ed. K. Schauenburg (Bonn, 1957), 234–242

5. Τινὲς δὲ Δᾶτιν λέγουσι τὸν τραγικόν, κακῶς ὑπονοοῦντες· ἐκεῖνος γὰρ υἱὸς ἦν Καρκίνου οὗτος δὲ ὕπαρχος Περσῶν.

Die erste Hälfte des ersten Scholiums hat Suidas fast wörtlich wiederholt, *[235]* s.v. Νῦν τοῦτ' ἐκεῖν' ἥκει τὸ Δάτιδος μέλος: (No. 611 Adler) ἐπὶ τῶν χαρμοσυνῶν. ὁ γὰρ Δᾶτις σατράπης Περσῶν, ὃς εὐδοκιμήσας κατὰ τὸν πόλεμον, ἑλληνίζειν βουλόμενος εἶπεν, ἥδομαι καὶ εὐφραίνομαι καὶ χαίρομαι· καὶ ἐβαρβάρισε. An einer anderen Stelle findet man die folgenden Angaben, s.v. Δᾶτις: (No. 89 Adler) Πέρσης· ὃς ἐπετήδευσεν ἑλληνίζειν. ὃν λέγουσι φάναι χαίρομαι, ἀντὶ τοῦ εἰπεῖν χαίρω. καί ἐστι Δατισμός.

Suidas erzählt dann die von Herodot (6, 118, 1–2) und Pausanias (10, 28, 6) bekannte Geschichte von dem Traum des Datis und fährt dann fort: ὃς ἑλληνίζειν βουλόμενος, εἶπεν ἥδομαι καὶ χαίρομαι καὶ εὐφραίνομαι. Schließlich hat der Verfasser des Philetairos (Herodianus?), S. 433 (ed. J. Pierson, 1759) die folgende Bemerkung: χαίρω ἐρεῖς οὐ χαίρομαι· εἰ δὲ μὴ ἁμάρτημα ἁμαρτήσεις, ὃ καλεῖται Δατιασμὸς (sic) ἀπὸ Δατίδος (sic) τοῦ Πέρσου, ὅστις πρῶτον εἶπεν, ὅτι χαίρομαι, πλανηθεὶς τῷ ἔθει τῶν 'Αττικῶν.

Der Ausdruck Δάτιδος μέλος ist auch in die Sprichwörtersammlungen aufgenommen worden, aber nur in zwei Handschriften wiedergefunden worden, wie L. Cohn, *Breslauer Phil. Abh.*, 2/2, 1887, S. 24, gezeigt hat: einmal (im Codex Laurentianus 58, 24, von Cohn L² genannt; siehe K. Rupprecht, R.E. s.v. Paroimiographoi, Sp. 1747–1748) lesen wir nur Δάτιδος μέλος ohne irgendwelche Erklärungen, aber in einer anderen Handschrift (Codex Parisinus Suppl. gr. 676, von Cohn, S. 57, S genannt und S. 57–83 ausführlich behandelt; siehe auch Rupprecht, Sp. 1753, Z. 16–20) liest man s.v. Δάτιδος μέλος· (siehe auch Cohn, S. 76, No. 38) τὸν Πέρσην Δᾶτιν ἐπιτηδεύοντα ἑλληνίζειν ⟨φάναι⟩ ποτὲ καὶ χαίρομαι. Cohn fügt richtig hinzu (S. 24) daß das Sprichwort aus Aristophanes, *Pax*, 289, stammt und daß die Erklärung mit der in Suidas (s.v. Δᾶτις) übereinstimmt.

Mit dem Datislied haben sich vor allem die Herausgeber und Erklärer des Aristophanes beschäftigt, und die neueren haben im allgemeinen die Angaben der älteren (nämlich der Scholien) wiederholt; siehe zum Beispiel die Ausgaben von Ch. D. Beck (Invernizius, 1819, Bd. 6, S. 413–414), J. Richter (1860), F. H. M. Blaydes (1881, Bd. 3, S. 166–167), B. B. Rogers (1913). U. von Wilamowitz-Möllendorff wies (*Timotheos*, 1903, S. 43, fn. 1) die Angaben der Scholien als "absurd" zurück und nahm an, daß Aristophanes auf das Lied eines Asiaten in einer Komödie anspielt; P. Mazon, wohl unabhängig, machte denselben Vorschlag in seiner Ausgabe des Friedens (1904).

Die erste und einzige kritische Behandlung des Datisliedes verdanken wir J. van Leeuwen (*Mnemosyne*, 16, 1888, S. 435–438); sie ist dann in den Ausgaben von H. van Herwerden (1897), W. W. Merry (1900), und J. van Leeuwen (1906; siehe auch seine Bemerkungen in der gleichzeitigen Ausgabe der *Ranae*, zum Vers 86) wiederholt worden. Van Leeuwen führt nur den dritten Teil des Scholiums an und bezeichnet seinen Inhalt als vollkommen unhistorisch und unglaublich; im Gegensatz dazu schenkt er dem fünften Scholium vollen

Glauben das von der Gleichsetzung des Datis mit dem so genannten Sohn des Karkinos berichtet (aber diese Gleichsetzung zurückweist). Er behauptet, daß Aristophanes hier auf den Karkinossohn Datis anspielt und vermutet, daß es sich um den Dichter Xenokles handelt, der seiner barbarischen Sprache wegen den Beinamen Datis erhielt. Diese Gleichsetzung der Karkinossöhne Datis und Xenokles ist vollkommen überzeugend, aber die Erklärung des Beinamen Datis

[236] kann nur befrie|digen, wenn man auch annimmt, daß der Name Datis in den Athenern die Erinnerung an einen Griechisch sprechenden Barbaren wachrief. Van Leeuwens Beitrag ist daher unvollständig, und man muß die gesamte Überlieferung nochmals überprüfen.

Der Ausdruck τὸ Δάτιδος μέλος muß im Zusammenhang mit den folgenden ähnlichen Ausdrücken betrachtet werden: τὸ 'Αδμήτου μέλος, τὸ 'Αρμοδίου μέλος die sich bei Aristophanes (fr. 430 Kock) und Kratinos (fr. 236 Kock) finden (Schol. Aristophanes, *Vespae*, 1239). Wir haben demnach hier nicht ein Lied des Datis (als Dichter oder als Person einer Komödie) sondern ein Lied auf Datis in dem jedoch, wie die nächste Zeile zeigt, Datis sprechend (oder singend) angeführt wurde. Mehr kann man mit Sicherheit aus den Worten des Aristophanes nicht schließen; vielleicht haben das erste Scholium und Suidas jedoch recht, wenn sie annehmen, daß die Freude des Datis durch einen Sieg begründet war. In diesem Falle muß es sich um die Unterwerfung von Eretria handeln, die Datis den Athenern mitteilen ließ, wie wir von Plato wissen (*Leges*, 698 C-D; vergleiche Menexenus, 240 B-C und *Sophistes*, 235 B; siehe C. Ritter, *Platos Gesetze*, S. 103; G. J. DeVries, *Spel bij Plato*, S. 75; G. C. Whittick, *L'Ant. Class.*, 22, 1953, S. 27-31; hierher gehören auch Aristides, 13, 123, 6-7 mit den Scholien; weitere Belege bei Geyer, *R.E.* Suppl. 4, Sp. 378, Z. 41-45). Die "freudige" Botschaft des Datis ist uns nun wirklich in dem Geschichtswerk des Diodorus erhalten (10, 27), der hier wohl Ephorus folgt:

῞Οτι Δᾶτις ὁ τῶν Περσῶν στρατηγός, Μῆδος ὢν τὸ γένος καὶ παρὰ τῶν προγόνων παρειληφὼς ὅτι Μήδου τοῦ συστησαμένου τὴν Μηδίαν 'Αθηναῖοι καθεστήκασιν ἀπόγονοι, ἀπέστειλε πρὸς τοὺς 'Αθηναίους δηλῶν ὡς πάρεστι μετὰ δυνάμεως ἀπαιτήσων τὴν ἀρχὴν τὴν προγονικήν. Μῆδον γὰρ τῶν ἑαυτοῦ προγόνων πρεσβύτατον γενόμενον ἀφαιρεθῆναι τὴν βασιλείαν ὑπὸ τῶν 'Αθηναίων καὶ παραγενόμενον εἰς τὴν 'Ασίαν κτίσαι τὴν Μηδίαν. ῍Αν μὲν οὖ αὐτῷ τὴν ἀρχὴν ἀποδῶσιν, ἀφεθήσεσθαι τῆς αἰτίας ταύτης καὶ τῆς ἐπὶ Σάρδεις στρατείας. ἂν δὲ ἐναντιωθῶσι, πολὺ δεινότερα πείσεσθαι τῶν 'Ερετριέων. Ὁ δὲ Μιλτιάδης ἀπεκρίθη ἀπὸ τῆς τῶν δέκα στρατηγῶν γνώμης, διότι κατὰ τὸν τῶν πρεσβευτῶν λόγον μᾶλλον προσήκει τῆς Μήδων ἀρχῆς κυριεύειν 'Αθηναίους ἢ Δᾶτιν τῆς 'Αθηναίων πόλεως. τὴν μὲν γὰρ τῶν Μήδων βασιλείαν 'Αθηναίων ἄνδρα συστήσασθαι, τὰς δὲ 'Αθήνας μηδέποτε Μῆδον τὸ γένος ἄνδρα κατεσχηκέναι. ὁ δὲ πρὸς μάχην ἀκούσας ταῦτα παρεσκευάζετο.

Es kann keinem Zweifel unterliegen, daß das dritte Scholium den Bericht des Diodorus (Ephorus) zusammenfassend wiedergibt und augenscheinlich mit dem Datislied in Verbindung bringt. Es ist jedoch das vierte Scholium, dessen richtiges Verständnis am meisten zu unserer Kenntnis der Sachlage beiträgt.

Hier lesen wir, daß Datis öfters nach Athen Boten sandte und daß er begierig war Athen zu besitzen; die Worte ἠράσθη τῆς αὐτῶν πολιτείας sagen dasselbe wie die Angabe des dritten Scholium ἀντιποιήσασθαι τῆς 'Αττικῆς ὡς ἰδίας, und weisen auf die in Diodors Werk erhaltene Erzählung hin. Die folgenden Worte sind jedoch zweideutig; sie können so verstanden werden, daß Datis in Athen war (μείνας ἐκεῖ) und sich wie ein Grieche benehmen wollte, oder daß Datis den Wunsch hatte (ἐβούλετο) in Athen zu leben und sich wie ein Grieche zu benehmen. Die erste Möglichkeit muß zurückgewiesen werden, da sie eine geschichtlich unmögliche Sachlage erfordert, während die zweite mit der übrigen hier wiedergegebenen Überlieferung vollkommen übereinstimmt. In diesem Falle stimmt das vierte Scholium mit dem dritten überein, geht aber über dieses hinaus, indem es die Worte des Datisliedes unmittelbar mit der Botschaft des Datis verbindet von der Diodorus ausführlich berichtet. *[237]*

Dieser außerordentliche Bericht war, wie schon erwähnt, dem Plato bekannt, der auf ihn in den folgenden Worten anspielt (*Leges*, 698 C–D): καὶ ὁ Δᾶτις τοὺς μὲν 'Ερετριέας ἔν τινι βραχεῖ χρόνῳ παντάπασι κατὰ κράτος τε εἷλε μυριάσι συχναῖς, καί τινα λόγον εἰς τὴν ἡμετέραν πόλιν ἀφῆκε φοβερόν, ὡς οὐδεὶς 'Ερετριέων αὐτὸν ἀποφευγὼς εἴη· συνάψαντες γὰρ ἄρα τὰς χεῖρας σαγηνεύσαιεν πᾶσαν τὴν 'Ερετρικὴν οἱ στρατιῶται τοῦ Δάτιδος. ὁ δὲ λόγος, εἴτε ἀληθὴς εἴτε καὶ ὅπῃ ἀφίκετο, τούς τε ἄλλους "Ελληνας ...

Die Verbindung dieser beiden Stellen besteht nicht nur in der Übereinstimmung der Zeit (nach dem Fall von Eretria und vor der Schlacht bei Marathon) und der Hauptperson (Datis) sondern auch der Botschaft selbst. Man wüßte gerne, was Plato mit seinen abschließenden Worten gemeint hat: Diese Mitteilung (λόγος), ob sie nun wahr war oder wie sie übermittelt wurde (ἀφίκετο), ... Die Frage nach der geschichtlichen Wahrheit (ἀληθής) mag darauf begründet sein, daß Herodot diese Geschichte mit Schweigen übergangen hat, aber die Frage nach der Weise (ὅπῃ) in der des Datis Botschaft nach Athen gebracht wurde, legt die Vermutung nahe, daß Plato sich hier auf eine literarische, vielleicht dichterische, Behandlung dieser Geschichte bezieht).[1]

---

1. Vielleicht sollte man sich an die Anekdote erinnern, die Plutarch in seinem Leben des Themistokles (6, 2) berichtet: ἐπαινεῖται δ' αὐτοῦ καὶ τὸ περὶ τὸν δίγλωττον ἔργον ἐν τοῖς πεμφθεῖσιν ὑπὸ βασιλέως ἐπὶ γῆς καὶ ὕδατος αἴτησιν. ἑρμενέα γὰρ ὄντα συλλαβὼν διὰ ψηφίσματος ἀπέκτεινεν ὅτι φωνὴν 'Ελληνίδα βαρβάροις προστάγμασιν ἐτόλμησε χρῆσαι. Dieselbe Geschichte, ohne Nennung des Themistokles, erzählt Aristides, 13, 122, 6–14, und sie gehört zweifellos in die Zeit vor der Schlacht bei Marathon, denn später sandte der König keine Boten mehr nach Athen (Herodot, 7, 32), und die Boten des Mardonios (Herodot, 9, 1 und 5) wurden nicht belästigt. Man hat daher angenommen, daß sich diese Geschichte auf die ersten Boten bezieht, die Dareios nach Griechenland schickte, und die in Athen ins Barathron geworfen wurden (Herodot, 7, 133). Es ist jedoch bemerkenswert, daß Herodot, der für die schlimmen Folgen, die sich aus dem Gesandtenmord für die Spartaner ergaben, besonders betont (7, 133–137), ausdrücklich versichert, daß er nicht wußte, wie die Athener für ihre Untat bestraft wurden. Pausanias hat nicht gezögert (3, 12, 7), die ganze Sache dem Miltiades in die Schuhe zu schieben und sein unrühmliches Ende damit zu erklären. Anderseits weiß Suidas (s.v. Δᾶτις, Nr. 89 Adler) von der Hinrichtung der Boten des Datis und Artaphernes in Sparta, er sagt aber von den nach Athen geschickten: 'Αθηναῖοι δὲ βαρέως ἐνεγκότες τοὺς πρεσβευτὰς ἐξεκήρυξαν. Wie dem auch sei, zögert man die Verbindung des grichisch sprechenden Boten und der griechischen Botschaft des Datis außer Acht zu lassen.

Sehr geehrter Jubilar, Sie haben so viele alte Bildwerke aus zeitgenössischen und späteren Abbildungen wiedergewonnen und uns geschenkt, daß ich es wage, Ihnen hier die folgende Wiederherstellung einer alten Geschichte vorzulegen, die in die Zeit der Perserkriege gehört und die Beziehungen zwischen Persien und Athen beleuchtet.

Nach der Einnahme von Eretria hat der Meder Datis, der Befehlshaber der Persischen Truppen, einen Boten nach Athen geschickt, der nicht nur den Sieg verkündete, sondern auch eine persönliche Aufforderung von Datis überbrachte, die die Athener zur Übergabe aufforderte. Datis verwendete die griechische Sprache und betonte, daß er als Nachkomme des Medos des Sohnes des Aigeus und der Medea, ein Recht auf die Herrschaft von Athen hatte.

Das Alter dieser Geschichte und ihre Glaubwürdigkeit kann jetzt noch weiter untersucht werden. Sie war dem Aristophanes und seinen Zuhörern in den zwanziger Jahren des fünften Jahrhunderts bekannt, wie seine Anspielung auf das Lied des Datis beweist; siehe oben. Da diese Geschichte jedoch auch dem Ephorus und dem Plato bekannt war, ist es wahrscheinlich, daß sie schriftlich festgelegt war; siehe R. W. Macans kluge Bemerkungen, *Herodotus, The Fourth, Fifth, and Sixth Books*, Bd. 2, S. 212. Medos als Sohn des Aigeus und Stammvater der Meder war jedenfalls eine den Athe|nern des fünften Jahrhunderts geläufige Figur; siehe A. Lesky, *R.E.*, s.v. Medeia, Sp. 47–48. E. J. Jonkers, *Studia...C. G. Vollgraff...oblata*, S. 81–82; M. Valsa, *Lettres d'Humanité*, 14, 1955, S. 116–121. Ebenso bekannt war die "Griechenfreundlichkeit" des Datis, die er vor der Schlacht bei Marathon den Einwohnern von Lindos (Chr. Blinkenberg, *Lindos*, 2/1, Sp. 175, 32, Sp. 183–184, Z. 42–47; Sp. 192–198 = *F. Gr. Hist.*, 532 F 1, 32 und D 1, Z. 31–34) und Delos (Herodot, 6, 97; I.G., 11/2, 161 B 95–96; 199 B 24; Th. Homolles Zweifel, *B.C.H.*, 15, 1891, pp. 140–141, sind von Blinkenberg, Sp. 195–196, als unberechtigt zurückgewiesen worden) und nach der Schlacht dem Böotischen Delion (Herodot, 6, 118; Pausanias, 10, 28, 6) bewies).[2] Die Möglichkeit muß erwogen werden, daß uns noch ein Bruchstück des Datisliedes in einer Inschrift erhalten ist, die W. Peek veröffentlicht hat (*Ath. Mitt.*, 67, 1942, erschienen 1951, S. 159–160, no. 333), die mir Herr Stamires freundlicherweise gezeigt hat, und deren Bedeutung J. Pouilloux (*La forteresse de Rhamnonte*, S. 160–161, no. 52) bestimmen konnte. Es handelt sich um die Reste eines in äolischer Sprache abgefaßten Gedichtes, das vielleicht im Zeitalter der ersten Kaiser (für das damalige Interesse an den Perserkriegen siehe meine Bemerkungen in *Hesperia*, 23, 1954, S. 319 und fn. 5 [*infra*, p. 222, n. 5]) in einer Inschrift verewigt wurde. Dieses Gedicht (nach Peek wohl ein Hymnus) bezieht sich auf die Schlacht von Marathon, da der Name des Datis vorkommt (Z. 9); der Stein wurde in den Ausgrabungen gefunden, die B. Staïs in Sunion und Rhamnus (in 1891) veranstaltete. Peek erwog die Möglichkeit, daß das Bruchstück von

---

2. Die Angabe des Ctesias (*Persica*, 18 und 21), daß Datis bei Marathon fiel und seine Leiche den Persern nicht herausgegeben wurde, und daß dies einer der Gründe des Xerxeszuges war, kann nicht mit dem Bericht des Herodot (6, 118) vereinigt werden; siehe R. W. Macan, *op. cit.*, S. 233, und R. Henrys *Ausgabe des Ctesias*, S. 90, fn. 51 und 55. Vergleiche auch fn. 1.

Rhamnus stammt, und Pouilloux hat sie durch einen Verweis auf Pausanias, 1, 33, 2 als richtig erwiesen. Der Perieget erzählt, daß der Zorn der Göttin Nemesis (deren Heiligtum in Rhamnus er beschreibt) die in Marathon landenden Barbaren traf. Diese dachten mit Verachtung, daß ihrer Einnahme von Athen nichts im Wege stünde und brachten daher ein Stück Parischen Marmors mit, um daraus ein Siegesdenkmal zu schaffen. Diesen Stein hat nun Phidias für seine Statue der Nemesis (in Rhamnus) verwendet; es folgt eine Beschreibung der Statue. Diese Geschichte war auch dem Aristeides bekannt (13, S. 203, Z. 2–3, Dindorf; siehe E. Beecke, *Die historischen Angaben in Aelius Aristides Panathenaikos auf ihre Quellen untersucht*, S. 16, mit weiteren Verweisen) und sie wurde in einigen Epigrammen verwendet (*Anth. Pal.*, 16, nos. 221, 222, 263; siehe F. Dübners wertvolle Anmerkungen, S. 628); eines (222) ist von Parmenion (siehe W. Peek, *R.E.*, s.v. Parmenion, no. 3, Sp. 1566, Z. 31 bis 34), ein anderes (263) wurde von Ausonius ins Lateinische übertragen (*Epigr.*, no. 42, White). Ausonius hat dieselbe Geschichte auch in seinen *Epistulae* erzählt (27, Z. 53–57) und auf die prahlenden Worte (*grandia verba*) des Arsakiden (Perser) hingewiesen. Diese Betonung der überheblichen Sprache der Barbaren (siehe Aischylos, *Persae*, Z. 533: μεγαλαύχων) findet sich nun nicht nur in dem Scholion zur oben erwähnten Stelle des Aristeides (S. 133, Z. 31–32, Dindorf), sondern auch in der Inschrift selbst (Z. 11: ’Αχαιμενιδᾶν μεγαλαύχων); siehe auch E. Miller, *Mélanges de littérature Grecque*, p. 382: ’Αδράστεια· ἡ νεμεσῶσα τοῖς μεγαλαυχουμένοις ἀπὸ τούτων μηδὲν ἀποδιδράσκειν Dieses Scholion (S. 133–134, Dindorf) berichtet zwar, daß der Stein von Susa gebracht und daß die Statue in Marathon (nicht Rhamnus) errichtet wurde, es enthält jedoch auch die bemerkenswerte Angabe, daß die Barbaren mit einem schlachtenlosen Siege rechneten. All das mag in dem Gedicht gestanden haben, dessen Bruchstück wir besitzen; warum aber nicht mehr? Das Lied verherrlichte die Bestrafung des persischen Übermutes durch Nemesis, und es muß natürlich das Ausmaß dieses Übermutes eingehend beschrieben haben. Es ist glaublich, daß dieser nicht nur in dem voreiligen Mitbringen des Siegesdenkmals bestand, sondern auch in der stolzen Botschaft des Datis, die hier wiederherzustellen versucht wurde. In diesem Falle kann man annehmen, daß das Datislied ein Hymnus für Nemesis war, der bei Gelegenheit der Einweihung ihres Tempels in Rhamnus (ca. 436–432 v. Chr., nach W. B. Dinsmoor, *Hesperia*, 9, 1940, S. 47; siehe auch E. Kjellberg, *Studien zu den Attischen Reliefs des V. Jahrhunderts v. Chr.*, S. 112 und 114–116; W. Zschietzmann, *A.A.*, 1929, Sp. 448; G. M. A. Richter, *The Sculptors and Sculpture of the Greeks*,[2] S. 240–242) vorgetragen wurde und der in späterer Zeit auf einer Stele aufgezeichnet wurde; siehe die vielleicht gleichzeitige Weihung des Tempels an Livia, *I.G.*, II,[2] 3242. Es ist bemerkenswert, daß, während W. Gurlitt (*Über Pausanias*, S. 178–179) und H. Posnansky (*Nemesis und Adrasteia*, S. 40–42) die ganze Geschichte als eine späte Erfindung zurückwiesen, U. v. Wilamowitz–Moellendorff (*Antigonos von Karystos*, S. 12–13) sie zögernd auf Polemon zurückführte und H. Herter (*R.E.*, s.v. Nemesis, Sp. 2349, Z. 34–43, mit allen nötigen Belegen) sie ‚eine sehr bezeichnende und nicht ganz junge Legende‘ nannte. Der Parische

Marmorblock, der den Anstoß zur ganzen Geschichte gab, mag von den Pariern mitgebracht und auf dem Strande von Marathon zurückgelassen worden sein; dort haben ihn dann die Athener gefunden und ihm die vorliegende Erklärung gegeben. Man darf mit ihm vielleicht den Paroszug des Miltiades verbinden, von dem Herodot (6, 133) und Ephoros (*F. Gr. Hist.*, 70 F 63, dem Nepos, *Miltiades*, 7, 1–4, folgt) berichten; in der Erzählung des letzteren ist auch auf die Perserfreundschaft der Parier mit besonderer Erwähnung des Datis angespielt. Man darf daher annehmen, daß im Denken der Athener Datis eine ähnliche Stellung einnahm wie ihr alter Tyrann Hippias).[3]

Beide machten den Anspruch auf die Herrschaft von Athen und beide versuchten in verschiedener Weise die Athener zur Übergabe zu bewegen. Von einem dieser Versuche lesen wir in Herodot (6, 115; 121; 123; 124), der erzählt, daß die Alkmeoniden verdächtigt wurden, mit den Persern und Hippias im Einverständnis gewesen zu sein und den Feinden das berühmte Schildsignal gegeben zu haben. Herodot bemerkt abschließend (6, 124) daß er diese Verdächtigung nicht teile; er gibt aber zu, daß Verrat am Werke war, ohne daß er den Verräter nennen konnte (oder nennen wollte?); siehe auch 6, 109,5, eine Stelle auf die der Scholiast zu Aristides, 26, 185,13 (p. 593 Dindorf) anspielt, wenn er die Zurückberufung des Aristeides erwähnt. Wir wissen nun von Aristoteles (*Verf. v. Athen*, 22), daß dieser Verdacht im Ostrakismos seinen Ausdruck fand, dem Hipparchos und Megakles zum Opfer fielen. Aristoteles betont jedoch (22, 6), daß nur die ersten drei Verbannungen gegen die Freunde der Tyrannen gerichtet waren. Wir haben aber guten Grund, anzunehmen, daß der alte Verdacht noch während der weiteren Ostrakisierungen am Leben blieb, wenn er auch vielleicht nicht von entscheidender Bedeutung war. Xanthippos, der als vierter in die Verbannung ging (22, 6), war mit der Schwester des Megakles verheiratet und stand den Alkmeoniden nahe (Herodot, 6, 131). In der Tat ist er auf einem Ostrakon als ἀλειτηρός bezeichnet (A. E. Raubitschek, *A.J.A.*, 51, 1947, S. 257–261 [*supra*, pp. 108–115]; Ad. Wilhelm, | *Anz. Ak. Wien*, 1949, S. 240 und 242), ein Wort, das von den Alkmeoniden im allgemeinen gebraucht wurde. Ein anderer Alkmeonide, Kallixenos, wurde, wohl im Jahre des Ostrakismos des Aristeides, auf einem Ostrakon als προδότης, auf einem anderen als ἄτιμος bezeichnet (G. A. Stamires und E. Vanderpool, *Hesperia*, 19, 1950, S. 376–381, besonders S. 378–379). Man sieht demnach, daß die Ostrakisierungen der achtziger Jahre im allgemeinen im Schatten des angeblichen Marathonverrates standen; die Verbannten wurden als Verräter und Freunde des Hippias und der Meder angesehen.

In diesem Zusammenhang mag auch ein zeitgenössisches Zeugnis vorgelegt werden, das die ganze Untersuchung angeregt hat. Ein Ostrakon, das vor einigen Jahren zusammen mit Themistokles, Kallixenos, Hippokrates, und Aristeides

---

3. Vergleiche die von Herodot (VI, 107) und Suidas (s.v. Ἱππίας, Nr. 544 Adler), vielleicht nach Aelian (frag. 4 Hercher), berichtete Anekdote, in der Hippias augenscheinlich verspottet wird, wie Macan, *op. cit.*, S. 154, feinsinnig betont hat.

Ostraka gefunden wurde (P 9945), trägt die folgende Inschrift (Plate 7):

ΑΡΙΣΤ[---]   Ἀριστ[είδεν]
ΤΟΝΔΑ[---]   τὸν Δά[τιδος]
ΑΔΕΛΦ[---]   ἀδελφ[όν].

Figure 18.1.

Der mit Da- beginnende Name kann nicht der eines wirklichen Bruders des Aristeides sein, sondern muß eine höhnische oder feindselige Anspielung auf ein Verhältnis sein, das zwischen Da- und Aristeides bestand oder vermutet wurde. Wem die hier vorgeschlagene Ergänzung zu gewagt erscheint, mag für ἀδελφ[όν] ἴσ]/ἀδελφ[ον] einsetzen, ein Wort, das Euripides (*Orestes*, 1015) verwendet und auf des Pylades Verhältnis zu Orestes bezieht.

Es mag zuerst erstaunlich erscheinen, hier den Namen des berühmten Aristeides des Gerechten mit dem des Landesfeindes verbunden zu sehen. Wir besitzen jedoch nicht wenige Zeugnisse, die diese Verbindung, wenn auch nicht rechtfertigen, so doch erklären und begründen. Man muß sich erinnern, daß schon Aristoteles die Verbannung des Aristeides mit dem Bau der Athenischen Flotte zum Krieg gegen Aigina zusammengestellt hat (22, 7), und daß Thuckydides wußte (1, 14, 3), daß damals die Gefahr eines neuen Perserkrieges schon bestand (siehe auch Herodot, 7, 144). Auch wenn | Aristoteles von [241] der Zurückberufung der Ostrakisierten spricht (22, 8), bringt er sie mit dem bevorstehenden Perserkrieg zusammen, und Plutarch (*Aristides*, 8, 1; *Themistocles*, 11, 1; siehe auch die Scholien zu Aristides, 26, 185, 13, S. 593 Dindorf) sagt ausdrücklich, daß die Athener Aristeides wieder haben wollten weil sie fürchteten, daß er im Zorne über seine Verbannung mit den Barbaren gemeinsame Sache machen und auch andere seiner Mitbürger zum Abfall bewegen konnte. Auf jeden Fall hielt sich Aristeides während seiner Verbannung in Aigina auf, das damals mit Athen im Kriege stand (Herodot, 7, 145) und sich während des ersten Perserkrieges den Persern unterworfen hatte (6, 49); dies ist uns nicht nur von Herodot (8, 79, 1), Demosthenes (26, 6) und Aristodemus (*F. Gr. Hist.*, 104 F 1, I, 4) bezeugt, aber auch von Suidas (s.v. Ἀριστείδης) in einem merkwürdigen Zusammenhang erwähnt; nach einem Bericht über den Ostrakismos des Aristeides fährt er fort: ὁ αὐτὸς διέτριψεν ἐν Αἰγίνῃ φυγών. Ξέρξης (besser: Ξέρξου) δὲ ὡς αὐτὸν ἐν τῇ φυγῇ πρεσβευσαμένου καὶ τρισχιλίους δαρεικοὺς, ὅτε ἐπῄει τὴν Ἑλλάδα διδόντος, οὐδὲν ἐπιστρέφεσθαι ἔφη τοῦ Περσικοῦ πλούτου, τοιαύτῃ. χρώμενου διαίτῃ. ἔτυχε· δὲ οὐκ ἐπιμελῆ τὸν ἄρτον προσφερόμενος.

Dieselbe Anekdote ist von Suidas auch s.v. Δαρεικούς überliefert, aber ihren Ursprung hat man anscheinend nicht erkannt. Und doch hat schon Wilamowitz betont (*Aristoteles und Athen*, I, S. 160, fn. 65), daß das Bild des armen Aristeides von Aeschines dem Sokratesschüler wenn nicht geschaffen, so doch verbreitet wurde (siehe auch H. Dittmar, *Phil. Unters.*, 21, 1912, S. 206–207). Man könnte daher dem *Callias* des Aeschines nicht nur Plutarch, *Aristides*, 25, 3–6 (Suidas, s.v. Ἀριστείδης hat eine andere ergänzende Fassung) und V, 5–6 (Kallias und Aristeides nach der Schlacht bei Marathon) sondern auch die nur von Suidas zitierte und hier ausgeschriebene Geschichte von der Bestechung des Aristeides durch Xerxes zuschreiben.

Aus allen diesen Stellen möchte man schließen, daß Aristeides von manchen Leuten der Perserfreundlichkeit verdächtigt wurde. Seine vermutliche Stellungnahme gegen den Bau der Athenischen Flotte und seine Wahl von Aigina als Verbannungssitz müssen diesen Verdacht bestärkt haben, und die Kunde von dem Bestechungsversuch des Xerxes mag die Zurückberufung des Aristeides schnell hervorgebracht haben. In diesem Zusammenhang muß man das Ostrakon verstehen, das Aristeides beschuldigt, daß er sich wie ein Bruder des Datis betrage, d. h., daß er ein Freund der Perser sei.

Es ist natürlich eitel nach dem wahren Grund der Ostrakisierung des Aristeides zu fragen. Die tausenden Athener, die ihre Stimmen gegen Aristeides abgaben, hatten verschiedene Gründe, und wir können nur einige noch jetzt feststellen. Aristoteles (22, 7) bringt die Verbannung mit dem Streit über die Verwendung des Maroneia Silbers für den Flottenbau in Verbindung. Plutarch (*Aristides* 7, 1) behauptet, daß Themistokles die Verbannung des Aristeides in einer Rede forderte, in der er ihn tyrannischer Gerichtsführung anklagte. Die Grundlage dieser Anklage wissen wir nicht, noch ist Plutarchs Quelle bekannt, aber ein anderes Ostrakon mag diese sonderliche Angabe etwas beleuchten. Eines der Aristeides Ostraka, das in einer Gruppe mit achtzehn Themistokles Ostraka und vierzehn anderen Aristeides Ostraka gefunden wurde (P 5978), erlaubt die folgende Ergänzung (Plate 8):

[Ἀριστείδες]
[ℎο Λυσιμ]άχο
[ℎὸς τὸ]ς ℎικέτας
[ἀπέοσ]εν.

Die Ergänzung des letzten Wortes muß wohl unsicher bleiben, doch kann es keinem Zweifel unterliegen, daß der Schreiber des Ostrakon seiner Meinung

Ausdruck geben wollte, daß Aristeides die "Schutzflehenden" schlecht behandelte. Das stimmt mit | der von Plutarch berichteten Anklage des Themistokles überein. Es ist bestechend, des Aristeides Verhalten gegenüber den Schutzflehenden irgendwie mit den Aigineten in Zusammenhang zu bringen. Wir wissen von Herodot (6, 74, 1), daß die Athener um diese Zeit Aiginetische Geiseln hielten, die sie nicht freilassen wollten (6, 85–86), daß die Aigineten eine Anzahl hervorragender Athener gefangen nahmen (6, 87), und daß einige Aigineten in Athen Zuflucht nahmen und in Sunion angesiedelt wurden (6, 90); Herodot berichtet auch (6, 93) von einer Anzahl anderer Athener, die von den Aigineten im Kriege gefangen wurden. Obgleich der Zeitpunkt dieser Ereignisse nicht genau feststeht (siehe A. Andrewes, *B.S.A.*, 37, 1940, S. 6–7; G. Welter, *Arch. Anz.*, 1954, Sp. 30–31) kann man sich leicht vorstellen, daß unter diesen Umständen eine Attische Behörde, die dem Ex-Archonten Aristeides unterstand, sich mit Schutzflehenden zu befassen hatte, und daß des Aristeides Gerechtigkeit ihm von einem Bürger übelgenommen wurde.

19

Die Gründungsorakel der Dionysien

[300] H. v. K. in alter und getreuer Freundschaft

Es ist merkwürdig, daß die Angabe des Demosthenes über die gesetzlichen und religiösen Grundlagen der Dionysien (21.51) in den Theaterbüchern vernachlässigt wird. Sagt doch der Redner in der Anklage gegen Meidias: "Ihr wißt ganz genau, daß all die Chöre und Hymnen dem Gott von euch veranstaltet werden, nicht nur entsprechend den sich auf die Dionysien beziehenden Gesetzen sondern auch im Sinn der Orakel, in denen allen der Stadt sowohl von Delphi als auch von Dodona aufgetragen wurde, Chöre aufzustellen, wie es schon immer üblich war, die Straßen mit Opferrauch zu füllen und sich zu bekränzen."

Wenn auch die folgenden Orakel (52–53) nicht wortgetreu wiedergegeben sind, so läßt sich kaum daran zweifeln, daß die Angaben des Demosthenes über die Einrichtung der Dionysien verläßlich sind und daß der Redner auf Orakel anspielt, die nicht seiner eigenen Zeit angehören, sondern auf die Gründungszeit der Dionysien zurückgehen.

W. H. Parke nahm in seinem Buch *The Oracles of Zeus* (1967) auf diese Orakel von Dodona Bezug[1] und zitierte Goodwin mit Zustimmung: These oracles may be genuine[2]. Leider läßt er sich aber nicht darauf ein, sie zu datieren; er sagt nur (S. 86), daß die beiden Orakelsprüche möglicherweise zu verschiedenen Gelegenheiten gegeben wurden. Ein einziges Wort der Orakel, DEMOTELES, das E. Capps in der Überschrift der großen didaskalischen Inschrift (*IG* II² 2318) ohne Rücksicht auf das Orakel ergänzte,[3] wurde von J. Gould und D. M. Lewis[4] in ihrer Besprechung dieser Überschrift kurz erwähnt.

---

1. S. 84ff.
2. a.a.O.S.92 Anm. 12
3. *Hesperia* 12, 1943 S.9ff.
4. Sir Arthur Pickard-Cambridge, *The Dramatic Festivals of Athens*² (1968) S. 102

Auch in seiner letzten Veröffentlichung bespricht Parke die Einrichtung des öffentlichen Festes der Dionysien ohne die Verbindung mit Delphi and Dodona zu erwähnen.[5] Auch T. B. L. Webster, der einige Worte den Chören der Dionysien widmet, sagt nichts über ihren öffentlichen Charakter.[6]

Andrerseits betonen Parke und D. E. W. Wormell[7] den großen Einfluß des Orakels auf die Einrichtung von Kulten, aber in ihrem Verweis auf die von Demosthenes zitierten Orakel bezüglich der Dionysien[8] behaupten sie, daß das Orakel wegen der "uninspired diction of the verses"[9] in die Zeit des Demosthenes gehört.

Als Beispiel der Einführung eines Kultes in Athen durch einen Orakelspruch aus Dodona führt Parke[10] den Bendiskult an, wobei er allerdings auch auf die zeitgenössische attische Inschrift (*SEG* 10, 64) hätte verweisen können.[11] Die frühe und fortgesetzte Verbindung Athens mit Dodona wird von Parke (131–137) ausführlich besprochen, so daß es keiner weiteren Erklärung bedarf, wenn die Athener sowohl Delphi wie auch Dodona bezüglich der Dionysien befragten.

Um auf den Text der Orakel selber zurückzukommen, so berichtet Demosthenes (51) daß beide Orakel den Athenern auftrugen χοροὺς ἱστάναι κατὰ τὰ πάτρια, Delphis Orakel bestimmt (52): μεμνῆσθαι Βάκχοιο καὶ εὐρυχόρους κατ' ἀγυιὰς / ἱστάναι ὡραίων Βρομίῳ χάριν ἄμμιγα πάντας, / καὶ κνισᾶν βωμοῖσι κάρη στεφάνοις πυκάσαντας. Dodonas Orakel ist viel genauer: Διονύσῳ δημοτελῆ ἱερὰ τελεῖν καὶ κρατῆρα κεράσαι καὶ χοροὺς ἱστάναι.

Eine unvoreingenommene Betrachtung dieser Stelle zeigt, daß wir es hier mit drei Bestimmungen zu tun haben, die auf öffentliche Kosten ausgeführt werden sollen: Opfer, Weinschenke, Chöre—alle dem Dionysos geweiht und gewidmet. Es ist bemerkenswert, daß hier nichts von dramtischen Aufführungen gesagt wird; diese wurden ja nicht auf öffentliche Kosten veranstaltet. Wir besitzen somit in den von Demosthenes beschriebenen und zitierten Orakeln (21.51–53) ein wertvolles Zeugnis für die Einrichtung des öffentlichen Kultes des Dionysos auf Grund von Orakeln, die den Athenern vor 500 v. Chr. aus Delphi und Dodona gegeben wurden.

[301]

5. *Festivals of the Athenians* (1977) S. 126
6. *The Greek Chorus* (1970) S.92f.
7. *The Delphic Oracle I* (1956) S.330 u. 333
8. a.a.O.S.337f. und II (1956) S. 114f. nr. 282
9. Vgl. auch A. Mommsen, *Feste der Stadt Athen* (1898) S. 391 Anm. 3
10. a.a.O.S.149f.
11. Vgl. auch W. S. Ferguson, *Hesperia* Suppl. 8 (1949) S.133f.

20

The Priestess of Pandrosos

[434] The evidence concerning the cult of Pandrosos in Athens has been conveniently assembled by A. B. Cook.[1] Additional information is provided by an Attic decree concerning the genos of the Salaminioi.[2] This inscription contains the provision (lines 11–12) that the priesthood of Aglauros, Pandrosos, and Kourotrophos should be filled by a member (or by the daughter of a member) of the genos, and that the priestess should be elected by lot. It states, moreover (line 45), that the priestess of Aglauros and Pandrosos should receive a loaf of bread from the loaves kept in the sanctuary of Athena Skiras. Ferguson, who commented at length on this important document, reexamined (*loc. cit.*, pp. 20–21) the evidence concerning the cult of Aglauros and Pandrosos. He called attention to the only known priestess of Aglauros (*IG.* ii², 3458–3459), Pheidostrate, daughter of Eteokles from Aithalidai, and he apparently assumed that her family belonged to the genos of the Salaminioi.[3] It so happens that we know also of a priestess of Pandrosos from a fragmentary inscription now published as *IG.* ii², 3481. The text of this document can be completed by the addition of an apparently unpublished fragment in the Epigraphical Museum of Athens (E.M. 12364) (Figure 20.1).

It may be assumed from the demotic of Demochares and from the fact that his family belonged to the genos of the Salaminioi that he was a relative of
[435] Pheidostrate, | the priestess of Aglauros. This assumption is significant because we know the names of several other members of this family. An inscription from Delos (*I. de Délos*, no. 2381) records a dedication made while Demochares

---

1. *Zeus* 3, pp. 243–245; compare O. Broneer, *Hesperia* 1, 1932, p. 53; G. P. Stevens, *Hesperia* 5, 1936, p. 489; O. Broneer, *Hesperia* 8, 1939, p. 428.
2. W. S. Ferguson, *Hesperia* 7, 1938, pp. 1–74, no. 1.
3. H. R. Immerwahr, *Hesperia* 11, 1942, pp. 344–347, no. 3, published an inscription honoring another member of this distinguished family. According to the letter forms, the inscription seems to belong to the second century B.C.

```
Ο ΔΗΜΟΣ
ΦΙΛΙΣΤΙΟΝ ΔΗΜΟΧΑΡΟΥ
ΑΙΘΑΛΙΔΟΥ ΘΥΓΑΤΕΡΑ
ΙΕΡΕΙΑΝ ΠΑΝΔΡΟΣΟΥ
```

Figure 20.1. E. M. 12364 + *IG*. ii², 3481.

from Aithalidai was agoranomos.[4] This Demochares may be a descendant of Philistion's father, perhaps his grandson. His date can be determined because his colleague Charias, son of Charias, from Aithalidai (probably a relative), is known.[5]

Philistion, as a woman's name, is sufficiently rare both in Athens and elsewhere to warrant the addition of a few remarks concerning two other Athenian women who had this name. First may be mentioned the tombstone of Philiston, daughter of Demochares from Azenia, wife of Timotheos from Melite (*IG*. ii², 5315). It is puzzling to notice that the fathers of this woman and of the priestess of Pandrosos had the same name Demochares but that they belonged to different demes. This may be a mere coincidence which should warn us to be careful in making identifications. On the other hand, the change of the deme, accompanied by the identity of the names, may indicate that an adoption took place somewhere along the line.

The third inscription containing the name Philistion is published as *IG*. ii², 3475, and it can be completed by the addition of *IG*. ii², 3476 which obviously belongs to the same round pedestal.[6]

---

4. The name of Demochares' father may be restored as $\Delta[\omega\sigma\iota\theta\acute{\epsilon}]ου$ or $\Delta[\iota o\delta\acute{\omega}\rho]ου$ with reference to *IG*. ii², 2336, line 58, and *I. de Délos*, no. 2006. Straton, the father of Dosithea, is also known as a demesman from Aithalidai (*IG*. ii², 2336, lines 58–59).

5. The stemma of the family, drawn up by P. Roussel, *BCH*. 32, 1908, p. 367, no. 577, and based upon *I. de Délos*, nos. 2609 and 2381, may be enlarged by the inclusion of the men mentioned in *I. de Délos*, no. 2595, lines 7, 12, 33; *IG*. vii, 416, lines 27–28; *IG*. ii², 1009, line 67, 2336, lines 220 and 231 (*HarvSt*. li, 1940, p. 123, lines 227 and 238). This augmented family tree includes a brother of the younger Charias, Dositheos, who is mentioned as son of the older Charias (though without demotic) in *I. de Délos*, no. 2595, line 12, and whose complete name may now be confidently restored in *IG*. ii², 1009, line 67. The old reading of the demotic $[X]ολλίδης$ is epigraphically possible since only the tops of the letters are preserved, but it is made unlikely by the different spelling of the demotic in lines 75 and 78 of the same inscription.

6. This combination will be published elsewhere. Attention may be called to the fact that Kirchner's date of *IG*. ii², 3475/6 is too early, and that of *IG*. ii², 4690 is too late. The activity of the priestess Glauke may be dated at the beginning of the first century B.C.

21

Review of F. Sartori, La eterie nelle vita politica Ateniese del VI e V secolo A.C.

[81] Franco Sartori. Le eterie nella vita politica Ateniese del VI e V secolo A. C. Rome, "L'Erma" di Bretschneider, 1957. Pp. 169. (*Università degli Studi di Padova, Pubblicazioni dell' Istituto di Storia Antica*, III.)

Mabel Lang, in her fine study of "The Revolution of the 400" (*A.J.P.*, 69 [1948], pp. 272—89) summarized (pp. 278–9) the "softening-up activities of the oligarchic clubs in the ... order in which Thucydides presents them (first, a calculated kind of violence; second, a deceptively mild program; and third, the use of the first two as levers of persuasion and intimidation)." This account agrees so well with Plato's statement in the *Republic* (2, 365 D) that Plato may well have had the situation of 411 B.C. in mind, as Sartori himself now suggests (*Historia*, 7 [1958], pp. 164-8); see also *Theaetetus*, 173 D, and, not independent of these passages, Isocrates, 3, 54.

Students of Athenian politics have always been interested in the composition, the organization, and the activities of these clubs. George M. Calhoun's monograph on *Athenian Clubs in Politics and Litigation* (*Bull. of the Univ. of Texas*, no. 262, issued Jan. 8, 1913) has been able to satisfy this interest, and his book is used and quoted whenever these clubs are mentioned. Calhoun admits quite frankly (p. 4) that his "investigation is but incidently concerned with the origin or the history of the clubs," and his section on "Origin and Development" (pp. 10–17) and on "Political Tendencies" (pp. 17–24) are rather sketchy. He concludes (p. 24) that "the clubs were not restricted to any one party, (but) the majority of them seem to have been oligarchic." He is led to this conclusion (p. 18) by "the existence of hetaeries which supported popular leaders of the fifth century, Themistocles, Pericles, Alcibiades." Actually, our knowledge of the "clubmembership" of Themistocles is based on a passage in Plutarch's *Aristides* (2, 4) which goes back to Aeschines the Socratic, probably via Theopompus; in this passage the justice of Aristeides is contrasted to the partiality of Themistocles. It is doubtful, moreover, whether in that early period there was as yet any alternative to the aristocratic organizations of the Athenian nobles. The case of Pericles is different (see also below, note 5), for we know even from Plutarch (*Pericles*, 7; 9, 2–3) that Pericles was an aristocrat until the sixties when Aristeides had died and Cimon was away on military campaigns.[1] Alcibiades was, of | course, never a democrat, although he may have been [82] popular. It is, therefore, clear that the clubs as such, whenever they were active in the political field, should be considered as antidemocratic, whether or not certain of their members courted popular favor to achieve and to maintain their political positions.

Since Calhoun's excellent book did not deal with the "historical development" of the clubs, there was a need for a study of "The *hetaireiai* in the political life of Athens during the sixth and fifth centuries B.C." This need has now been fulfilled by Sartori's book under review; its subject is of sufficient importance to justify the preceding introduction and the following lengthy discussion. Both Sartori and Attilio Degrassi (in the brief preface) claim that new discoveries and recent research have made Calhoun's book obsolete; it is surprising, however, how little new evidence on the Athenian clubs has been brought to light either by excavations or by the reexamination of old material. Nor has Sartori made use of all of the new information which has become available. I am thinking of ostracism: the few references (collected in the index), all conventional, are based either on Calhoun (pp. 136–40, a very good account) or on Carcopino's second edition of *L'ostracisme Athénien* (1935) which is little more than a reprint of his splendid book of 1909. Actually, a great many ostraca have been found in recent years, and a great deal of work has been done on ostracism; see O. W. Reinmuth, *R.-E.*, s.v. Ostrakismos, and the list of articles in footnote 1 of my forthcoming article "Theophrastos on Ostracism" in *Classica et Medievalia*, 19 (1958) [*supra*, p. 82, n. 1]. If Sartori really believed (p. 45) that the clubs were active in the ostracisms not only of Hyperbolus but also of Hipparchus, Aristeides, Themistocles (see Ehrenberg, *People*, p. 340, note 4), Cimon, and Thucydides the son of Melesias, he should not have been satisfied with a passing reference to Calhoun's treatment, and with the addition of a few references in a footnote (p. 80, note 6) on the date of the ostracism of

---

1. This is confirmed by a passage of the Scholia on Aristides (3, p. 446, lines 17–26, ed. Dindorf) which does not seem to have been considered so far, although it appears to go back to Theopompus: καὶ τούτων μὲν (sc. τῶν δημοτικῶν) προίστατο Κίμων, πολλὰ διανέμων καὶ συγχωρῶν ὀπωρίσασθαι τοῖς βουλομένοις, καὶ ἱμάτια διανέμων τοῖς πένησι (see F. Gr. Hist., 115 F 89, and J. E. Sandys' comments on Aristotle's *Const. of Athens* 27, 3). τῶν δὲ ὀλιγαρχικῶν προίστατο Περικλῆς· κατηγορηθεὶς δὲ ὁ Κίμων ὑπὸ Περικλέους ⟨ἐπὶ⟩ Ἐλπινίκῃ (ΕΠΙΛΑΝΙΚΗ codd.) τῇ ἀδελφῇ καὶ ἐπὶ Σκύρῳ τῇ νήσῳ, ὡς ὑπ' αὐτοῦ προδιδομένου ἐξεβλήθη. δεδιὼς δὲ ὁ Περικλῆς μὴ ζητηθῇ ὑπὸ τῶν δημοτικῶν, πρὸς αὐτοὺς ἐχώρησεν· οἱ δὲ ὀλίγοι γαμβρὸν ὄντα Θουκυδίδην τὸν Μελησίου τοῦ Κίμωνος ἀπεσπάσαντο, σκυλακώδη ὄντα καὶ ὀλιγαρχικόν. The expulsion of Cimon evidently refers to his ostracism which has always been connected with Pericles (as accuser; see Plutarch, *Pericles*, 9, 4) and Elpinice (as cause; see Andocides, IV, 33; Plutarch, *Cimon*, 15, 3; Scholia on Aristides, III, p. 515, line 15, quoting Didymus), but it is difficult to understand the reference to Skyros and to Cimon's betrayal of Pericles. R. Sealey, *Hermes*, 84 (1956), pp. 234–47, discussed Pericles' entry into politics without considering his earlier aristocratic associations.

Hyperbolus (to which should be added *Phoenix*, 9 [1955], pp. 122-6 [*infra*, pp. 320-324]).

The first chapter (pp. 17-33) is devoted to an examination of the terms *hetaireiai* and *synomosiai*, and to an attempt to show that these terms are not only not similar but actually contradictory (p. 17).[2] Actually, the two terms were used indiscriminately in references to the political activities of the oligarchic clubs between 424 B.C. and 403 B.C., by Aristophanes, Andocides, Thucydides, Plato, and Aris|totle; the former term (*synomosia*) had often a derogatory meaning, while the latter (*hetaireia* and *hetairikon*) was the more formal designation. On the whole, the matter of terminology has already been well treated by Calhoun (pp. 4-9). Nor does Sartori's second chapter, devoted to the character of the Attic *hetaireiai* (pp. 37-49), go beyond what Calhoun has already stated in his book.

In the third chapter (pp. 53-7: the Attic *hetaireiai* until the time of Cleisthenes), Sartori passes from the conspiracy of Cylon (whose associates are called an *hetaireia* by Herodotus, 5, 71, *synomotai* by Plutarch, *Solon*, 12, 1) directly to the Alcmeonid attack upon Leipsydrion (called *prodosetairon* by Aristotle, *Const.*, 19, 3) and to the conflict between Isagoras and Cleisthenes in which, according to Herodotus, 5, 66 (and Aristotle, 20, 1), the *hetaireiai* were actively involved. Sartori mentions (p. 55), without approval and without reexamination, Beloch's old theory that the political divisions of the first half of the sixth century B.C. reappeared in the struggle for power after the expulsion of Hippias. Whatever may be the truth of this theory, a study of the *staseis* of old Athens should certainly have been included in an account such as Sartori's.

The meaning of *stasis* in the sense of "a group of people taking a certain political position" (we commonly use the related term "opposition") cannot be attested before the fifth century (see note 4), although it is probable that it originated in the political struggles of the late Solonian age.[3] *Staseis* are first mentioned as existing immediately after Solon's reforms (Aristotle, 11, 2), and they are identified, anachronistically, as *demos* and *gnorimoi*; actually, these groups are said (2, 1) to have been at odds with each other even before Solon (see also 5, 1-2). We next hear of the *staseis* from Herodotus (1, 59-62, whom

---

2. The evidence offered is three Hellenistic inscriptions (*Syll.*[8], 360, 526, 527) condemning *synomosiai*, one of which (527) testifies also to the existence of *hetaireiai* as official divisions of the people of Dreros, just as they are known from the Gortynian Laws (*I.C.*, III, 72, X, line 38; see M. Guarducci's commentary on p. 168). This merely shows that *synomosia* can mean "conspiracy"; actually, it can also mean "confederation" as Sartori recognizes (p. 33). See now J. and L. Robert, *R.E.G.*, 71 (1958), p. 195, no. 75.

3. See Kock, *Com. Att. Frag.*, I, p. 584, no. 859 (= J. M. Edmonds, *Frag. Att. Com.*, I, pp. 778-9, no. 859): στάσις· οὐχ ἡ φιλονεικία, ἀλλ' αὐτοὶ οἱ στασιάζοντες; for the formation of the word, see G. R. Vowles, *C.P.*, 23 (1928), p. 42; E. Schwyzer, *Gr. Gramm.*, I, pp. 504-6. L.-S.'s *Lexicon*, s.v., 3, 1, lists, I think incorrectly, Theognis, 1, lines 51-2, a passage to which Herodotus's (poetic?) source in 3, 82, alludes. This political view must be associated with Solon, fr. 3 (Diehl), lines 18-22, a passage to which, in turn, Herodotus refers in 8, 3 (A. W. Verrall, *C.R.*, 17 [1903], pp. 98-9); see also the related *gnome* in 1, 87. In all these passages, and in many others (see L.-S.'s *Lexicon*, s.v., 3, 2), *stasis* is used in the sense of "the action taken by the group called *stasis*, faction (which also has a double meaning), sedition, discord"; for Aristotle's analysis of this term, see M. Wheeler, *A.J.P.*, 72 (1951), pp. 143-61, who calls Herodotus, 1, 59, 3, the *locus classicus* but fails to notice that Herodotus uses the word here in a different meaning.

Aristotle follows, 13, 4–15) who reports that Peisistratus raised a third *stasis* in opposition to the two led by Lycurgus and Megacles. It is generally assumed that these three *staseis* comprised large segments of the population, as did presumably the two *staseis* of the earlier Solonian period, and that it is to these *staseis* that the Solonian law against "neutralism" refers (Plutarch, *Solon*, 20, 1); Solon himself would then have obeyed his own law (see Aristotle, 14, 2, and the parallel passages assembled in Sandys' edition). It is far more likely, however, that the three *staseis* were comparatively small groups led by ambitious aristocrats, and that | they were not at all different from the later *hetaireiai* of [84] the time of Cleisthenes or of Alcibiades; see also Plutarch, *Solon*, 29, 2. This interpretation is confirmed by Herodotus who speaks (5, 69–72) of the *hetairoi* of Isagoras and of Cleisthenes as *antistasiotai, systasiotai, stasiotai*, and by Plutarch who consistently (following here *one* source, probably Theopompus) refers (*Aristides*, 7, 3; *Nicias*, 11, 4; *Alcibiades*, 13, 4) to the *hetaireiai* of Nicias and Alcibiades as *staseis*; only in one significant passage (based on Theophrastus) does he mention the *hetaireia* of Phaiax.[4] It is clear, therefore, that the *staseis* of the first half of the sixth century should have been included in a historical account of the Attic *hetaireiai*; in fact, the political struggles of the age of Cleisthenes read like a repetition of those in the time of Peisistratus.

The sixty years from the battle of Marathon to the beginning of the Peloponnesian War (a better break would have been the death of Pericles) are treated by Sartori on five pages (pp. 61–6). This is the period during which the law of ostracism was enacted and employed; only one ostracism (the freak one of Hyperbolus) took place after the death of Pericles. This is the period during which Athenian democracy developed against the opposition of the aristocratic elements united in the *hetaireiai*. This is the period during which Athenian politics was directed by the generals who were chosen and elected under circumstances which made possible the activity of the *hetaireiai* either on behalf of certain candidates or against them. Of all this, one reads next to nothing in Sartori's account; but this is not the place to present the story which he failed to tell.[5]

After the death of Pericles, as after the expulsion of Hippias, a new political situation arose; only this time it was Cleon, not Cleisthenes, who was the leader

---

4. This is, of course, not the place to examine all the significant occurrences of *stasis* (and related terms) in Herodotus, Thucydides (e.g., 4, 71), Isocrates (e.g., 4, 79), Plato, Antiphon (Harpocration, s.v. *stasiotes*), Aristotle (e.g., *Oec.*, 2, 1348a35-b4), and in other authors; attention may be called, however, to a few occurrences in Aeschylus because these have been combined and given a separate meaning in L.-S., *Lexicon*, s.v., II; see now G. Italie, *Index Aeschyleus*, s. vv. *stasis* and *stasiarchos*. Actually, they are the earliest testimonies to the use of *stasis* as "group of people who stand in opposition." There can be no doubt that to Aeschylus and to his audience the word *stasis* had a political meaning, and that its use evoked recollections of the political struggles of Athens during the sixth and early fifth centuries.

5. The reader may merely be warned that the quotation of the description of Pericles' activity by Plutarch (following Critolaus), *Pericles*, 7, 5: τἆλλα δὲ φίλους καὶ ἑταίρους καθιεὶς ἔπραττεν (pp. 65–6) is incorrect. Calhoun (p. 18, n. 5) defends the MS reading φίλους καὶ ῥήτορας ἑταίρους, claiming that ἑταίρους is used here as an adjective; Lindskog-Ziegler (following Holzapfel) transpose ἑταίρους ῥήτορας; while B. Perrin (following Xylander) prints ῥήτορας ἑτέρους, an easy emendation which I consider to be correct.

of the *demos*. Accordingly, the aristocratic groups were pushed still further into the background, gaining in strength only through Nicias's successes (especially his "peace") and through Alcibiades' bold adventures. The ostracism of Hyperbolus, as masterly a stroke as that of Themistocles more than | fifty years earlier, was quickly followed by the bloody purge of the *Hermokopidai* and of the Mystery-Mockers. After the panic of the Sicilian disaster and again after the defeat at Aigospotamoi, the way stood open for an oligarchic revolution, and it was taken by men who formed no longer a loyal opposition but rather a subversive conspiracy. The overthrow of the Thirty marks the end of the political activities of the aristocratic clubs; they were and remained to be discredited.

The first part of this period (431–421 B.C.) is well treated by Sartori (pp. 69–76);[6] the main evidence is the "Old Oligarch" and the *Knights* and *Wasps* of Aristophanes. Sartori points out that the oligarchic clubs became more and more associated in the mind of the *demos* with conspiracies to establish a tyranny. Whether this was merely the result of propaganda or whether there was some factual evidence for this suspicion, we cannot tell; the "Old Oligarch" certainly discourages all hope that the hated democracy may be overthrown.

The next lustrum (421–415 B.C) stands in the shadow of Alcibiades who, like Pericles before him, was able to attain great popular support especially since the democratic "machine" was in the hands of Hyperbolus upon whom contemporaries and later generations have heaped abuse, perhaps not unjustly. Sartori passes quickly over the years following the peace of Nicias and even over the ostracism of Hyperbolus in which as many as three *hetaireiai* seem to have been involved, and he devotes this chapter (pp. 79–98) to a detailed but somewhat awkwardly presented account of the accusations made against Alcibiades and his friends, both before and especially after the departure of

---

6. He calls attention to a passage in Plutarch's *Praecepta gerendae reipublicae* (806F–807A) which, in his opinion, suggests that Cleon may have been associated with a "nonaristocratic" (p. 72) *hetaireia* before entering politics. It so happens that a fragment of Theopompus's account of Cleon (*F. Gr. Hist.*, 115 F 92) is repeated without the author's name in Plutarch's *Praecepta* (799D), and I suggest that the story of Cleon's entry into politics also goes back to Theopompus; compare the similar account of Pericles' entry into politics (treated above, note 1) which can be attributed also to Theopompus. The newly identified passage should be associated with *F. Gr. Hist.*, 115 F 93 which speaks of Cleon's first political activities. It is, therefore, significant that before that time (i.e. 428/7 B.C.), Cleon is said by Plutarch to have attacked Pericles (*Pericles*, 33, 6–7; 35, 4). If the reference to Cleon's friends implies, therefore, his membership in a *hetaireia*, this does not mean that it was a "democratic club"; in fact, the passage in question indicates that Cleon abandoned his "friends" to take as *hetairoi* the worst elements of the people, following, it would seem (at least in the account of Theopompus, the example of Cleisthenes and of Pericles. In any case, the passage in Plutarch (*Praecepta*, 806F–807A) must not be used as evidence for the existence of a "democratic" *hetaireia* attached to Cleon; see now M. L. Paladini, *Historia*, 7 (1958), pp. 54–6.

There is evidence, however, that Cleon did surround himself with a "gang" of supporters, which may have fulfilled a function similar to that of an *hetaireia*. In Aristophanes' *Knights*, the Sausageseller addresses Demos, saying (lines 850–7) that his adversary (Cleon) has a device by which he can avoid being punished by Demos, namely his gang (*stiphos*) of brush, honey, and cheese sellers, i.e. a private army. The use of this "squad" is indicated in the following lines (855–7): they will prevent their master from being ostracized, thus fulfilling the same function as the *hetaireiai* according to Andocides, 4, 4; see my comments in *A.J.A.*, 60 (1956), p. 279.

the Sicilian expedition. He comes to the convincing conclusion that the mocking imitation of the mysteries (without any particular political aim) was done at many times, in many places, and by many groups of people, and that many friends of Alcibiades were involved in these actions and in the Multilation of the Herms (which Sartori considers of political significance). The close association of the literary and the epigraphical evidence (following Pritchett's brilliant example) is certainly welcome and provides a better understanding of the composition of the *hetaireiai*. We need, however, a closer reexamination of the relation between the ostracism of Hyperbolus and the "purge" of the aristocrats on the charge of "impiety" not of "subversion" (Sartori never mentions this fact; see *A.J.A.*, 55 [1951], pp. 229f. [*supra*, p. 74, n. 29]). We also need a more thorough examination of all the individuals connected with the *hetaireiai*; Sartori merely makes some significant remarks on this point. Finally, one must distinguish between the cause and the effect of the Mutilation of the Herms; it seems, from Sartori's own excellent account, that the two were distinct, and that our only testimony for the cause, the story told by Andocides, is in many ways untrustworthy.

With the period from 415 to 412 B.C. (pp. 101–12) we enter the homestretch of the history of the Athenian oligarchs. Sartori examines Aristophanes' *Birds*, Eupolis's *Demoi* (see now J. M. Edmonds, *Frag. Att. Com.*, I, pp. 978–94), and Euripides' *Helen*, in order to extract from these plays some information on the working of the *hetaireiai*; the harvest is unfortunately small and unsatisfactory.[7] The rest of this chapter is devoted to the activities of Peisander who does not seem to have had any contact with the *hetaireiai* until his return to Athens (see below).

The next chapter deals with the revolution of the Four Hundred and the part in it played by the *hetaireiai* (pp. 115–26). Sartori's careful account is based primarily on Thucydides, without ignoring, however, the various other traditions and pieces of evidence. At the outset, he mentions C. Diano's startling thesis (*Dioniso*, 15, 1952) that Sophocles' *King Oedipus* belongs to the beginning of 411 B.C., and he lists (p. 104, and in the index) the poet as one of the *probouloi* of 413–411 B.C.; see now H. Schaefer, *R.-E.*, 45, cols. 1225–8. Next, he treats in some detail Aristophanes' *Lysistrata* (Lysimache, according to D. M. Lewis, *B.S.A.*, 50 [1955], pp. 1–7), without, however, making it clear that Lysistrata

[87]

---

7. Sartori claims, perhaps rightly, that the name of Peisthetairos referred to the trust among the members of a *hetaireia*, without noticing that the passage from the text of the introduction to the *Birds* (which he quotes on p. 102) does not read ὡς εἰ πεποιθοίη ἕτερος τῷ ἑταίρῳ, but ὡς εἰ πεποιθοίη ἕτερος τῷ ἑτέρῳ καὶ ἐλπίζοι ἔσεσθαι ἐν βελτίοσιν; see W. G. Rutherford, *Scholia Aristophanea*, I, p. 423 (whose unnecessary emendations have been accepted by J. van Leeuwen, on line 644 of his edition of the *Aves*); J. W. White, *The Scholia on the Aves of Aristophanes*, p. 8. It may be noticed, moreover, that we have a tombstone from before the middle of the fourth century B.C. (*I.G.*, II², 5347) recording the death of Πιστοκλῆς Πισθεταίρο(υ) 'Αθμονεύς; the Pisthetairos here mentioned was probably alive when the *Birds* was performed, and one may argue that Aristophanes had his name in mind; see, however, B. B. Rogers's introduction (pp. viii–x) to his edition of the *Birds* (1906), and White's comments (on line 1 of the Scholia) who gives the name confidently as Peithetairos. The name Πεισθ[έταιρος] has been restored in another inscription (*I.G.*, II², 12440/1).

and her women form a *synomosia* (182, 237, 914, 1007), seize the Acropolis (176, 241–2), like Cleomenes (274), wish to make peace with Sparta but not with Persia (1133–4), are accused of tyranny (618, 630), think little of *psephismata* (697, 703–4), in other words act like an oligarchic *hetaireia*. Sartori was therefore mistaken when he interpreted the reference to Peisander (489–90) as indicating his leadership of the oligarchs; see also Sartori's comments in his earlier book, *La crisi del 411 A.C.* (1951), p. 12, n. 7. On the contrary, Peisander was at that time still considered a popular leader. Thucydides' account (8, 54, 4) of Peisander as "approaching" (*epelthon*) the clubs shows clearly that he did not belong to them; see A. G. Woodhead's fine study of Peisander in *A.J.P.*, 75 (1954), pp. 137–8. It is hard to escape the impression that Aristophanes gives in this play a not unsympathetic account of the political position occupied by the oligarchic clubs on the eve of the revolution.

For the period from the overthrow of the Four Hundred to the Restoration of Democracy (pp. 129–43), Sartori uses, in addition to the well-known evidence of Lysias, Xenophon, and Aristotle, also Euripides' *Phoenissae*, Sophocles' *Oedipus at Colonus*, and Aristophanes' *Frogs*. Unfortunately, these plays add nothing to our knowledge of the *hetaireiai*, although they testify to the political unrest in Athens. For the period after 403 B.C. (pp. 147–52), Sartori only repeats what Calhoun has presented more fully, namely that the *hetaireiai* of which we read (mainly in the orators) concerned themselves with court matters, i.e. they returned to the activities originally attributed to them by Thucydides (8, 54, 4). Sartori's attempts to discover political significance in some of these references are unsuccessful;[8] he promises (p. 148, n. 6), however, to devote a separate study to Plato and the *hetaireiai*, and in it (*Historia*, 7 [1958], pp. 157–71) he gives an admirable account of the three meanings in which Plato used the word *hetaireia*: friendship, philosophical association (of the circle of Socrates), political association (disapproved by Plato).

Sartori's conclusions (pp. 153–5) repeat some of the weak points of his arguments: the essential difference between *hetaireia* and *synomosia*; the existence of democratic *hetaireiai*; Theramenes' membership in an *hetaireia*. On the other hand, Sartori emphasizes correctly the aristocratic (as distinct from oligarchic) character of the *hetaireiai*, and he suggests persuasively that the use of *hetaireiai* for political ends may be attributed to Thucydides the son of Mele|sias. If so, this would be a revival of the factional conflicts of the late Solonian and of the Cleisthenian periods, with that difference that the earlier *staseis* were between *hetaireiai*, the one at the end of the fifth century B.C. between the oligarchic *hetaireiai* and the democracy. And this was exactly Peisander's aim (Thucydides, 8, 54, 4) when he approached the *hetaireiai*, παρακελευσάμενος ὅπως ξυστραφέντες καὶ κοινῇ βουλευσάμενοι καταλύσουσι τὸν δῆμον.

---

8. His discussion of Andocides, 4, fails to consider the recent work done on this speech, both by myself (*T.A.P.A.*, 79 [1948], pp. 191–210 [*supra*, pp. 116–131]; *Hesperia*, 23 [1954], p. 68, nn. 2 and 3; *Phoenix*, 9 [1955], p. 123, n. 3 [*infra*, p. 321, n. 3]) and by A. R. Burn (*C.Q.*, 4 [1954], pp. 138–42).

22

Review of M. P. Nilsson, *Die hellenistische Schule*

Die hellenistische Schule. By Martin P. Nilsson. Munich: C. H. Beck. 1955. Pp. vii, 101, 8 plates. (DM 9)   [40]

We know only too well that the extent and quality of our elementary and secondary education determine not only the success of all higher learning but also the educational level of our adult population. It is, therefore, of considerable interest to know how the Greeks themselves carried on the education of their children.[1]

Nilsson's slim book on the Hellenistic school is the result of a life-long interest and the by-product of the author's monumental work on Greek religion. Accordingly, it is both limited in scope and far-reaching in the depth and breadth of the author's knowledge and wisdom; the reviews I have seen are almost all enthusiastic.[2]

As van Groningen rightly points out the "school" treated in this monograph [41] would more properly be called *gymnasium*, especially because of its strong emphasis on athletics. Actually, the combination of studies with athletics, religious practices, and military training can be found in one way or another in the American private secondary schools and colleges, a fact which should make Greek education especially relevant to us.

Nilsson's book consists of three chapters unequal in length and in treatment. The first (pp. 1–29) is devoted to the classical period and to the ephebic institution of Athens. The main point of this chapter is the claim that the Athenian ephebes of the classical period did not receive any public education but merely military training, and that the Athenian ephebic school of the second

---

Phoenix 12 (1958) 40–42

1. See J. H. Turner, *CW* 49 (1955) 65. H. I. Marrou's comprehensive book on this subject has been translated into English, *A History of Education in Antiquity* (New York 1956). Cf. G. Downey, "Ancient Education," *CJ* 52 (1957) 337–345.
2. G. Dunst, *DLZ* 77 (1956) Heft 6; B. A. van Groningen, *Gnomon* 28 (1956) 501–504; P. DeLacy, *AJP* 77 (1956) 438–440; A. G. Woodhead, *CR* 6 (1956) 257–259; C. B. Welles, *CP* 52 (1957) 53–54.

and first centuries B.C. imitated the Hellenistic schools of the East rather than served as their model. It is a pity that Nilsson did not examine the more recently published Attic ephebic documents; see the summary by O. W. Reinmuth, *Hesperia* 24 (1955) 226–227.

I am not sure whether inscriptions or excavations can give an adequate picture of the academic side of a school. And yet, it is largely the physical remains of the various Hellenistic schools which provide Nilsson with the material evidence for his highly synthetic and masterly second and third chapters (pp. 30–98). After a short account of the buildings, the age groups of the "students" are discussed: paides, epheboi, and neoi; for paides, see now Aug. Hug in *RE* Suppl. 8, cols. 374–400. Nilsson argues persuasively against Marrou that the ephebeia outside of Athens was not confined to a single year but included youngsters between the ages of fifteen and eighteen. It would have been worthwhile to reexamine the Athenian evidence in order to see whether or not the same situation prevailed also at Athens; see the comments on *IG* II^2, 956–966. Next comes the curriculum in which athletics and military training occupy more space than music and literature. There again, the evidence comes largely from inscriptions listing the winners in competitions (the Greek substitute for exams), and the "humanities" may not have been suitable for competition. Our information concerning teachers and visiting lecturers is meagre but sufficient to show that they occupied a place in the Hellenistic school not inferior to that of the coaches and trainers; the existence of libraries in many places confirms Nilsson's conclusion that the Hellenistic school was academically a respectable institution. The direction of the school was in the hands of an annually appointed gymnasiarch (corresponding to the Athenian kosmetes) who had not only administrative duties but also, especially in later times, financial obligations. To judge from various honorary inscriptions, the gymnasiarch paid for the operations of the school out of his own pocket, although some of the gymnasia were, as we would call it, privately endowed. In many places, the gymnasiarch was in charge both of paides and epheboi, but | in some instances paidonomoi were appointed to look after the elementary schools.

Religious education had an important place in the Hellenistic school, but it consisted in cult practice rather than in the study of dogma and doctrine. The whole gymnasium stood under the protection of Hermes, Herakles, Apollo, and the Muses, and of a number of local heroes and heroized men, and the students participated in the cult of these deities and heroes, but they were also called upon to take part in the public cults of their cities. In this way, the school prepared directly for life. Boys' choirs sang at religious festivals, and were active also in the cult of the Hellenistic princes, and later of Rome and of the Roman emperors. It is a pity that Nilsson did not consider the Athenian evidence which is plentiful on this point; he was probably influenced by his general view that the Athenian ephebate has little to do with the Hellenistic school, and he may have thought that Athenian education was already well known through various special studies.

Nilsson gives a separate treatment of the Hellenistic school within the Seleucid Empire and in Egypt, not so much because its organization was different but because it fulfilled a different function in countries in which the Greek population was in a minority. There, the schools not only provided education but also maintained the national and cultural tradition and integrity; moreover, the schools were also instruments in the Hellenization of the Orient. In Alexandria, in Egypt, the gymnasium and its gymnasiarch occupied a very important position, especially in Roman times, when the imperial administration found in the Greeks a source of unrest and discontent; accordingly, many a gymnasiarch had to pay with his life for his "political" activities.

There is still much about the Hellenistic school which we do not know, but Nilsson's essay should encourage students to add details and expand individual sections.

III

ART, MONUMENTS, AND INSCRIPTIONS

23

Inschriften als Hilfsmittel der Geschichtsforschung

Der moderne Geschichtsforscher und Geschichtsschreiber bemüht sich nicht [177] nur die tatsächlichen Ereignisse objektiv zu erfassen, sondern auch die Beweggründe der verschiedenen Handelnden klar zu erkennen. Denn er weiß, daß der Nutzen der Geschichte für uns darin liegt, die Verbindung von Motiv, Handlung und Ergebnis in die menschliche und politische Erfahrung einbeziehen zu können, so daß wir vernünftiger und erfolgreicher zu handeln imstande sind. Dabei sind wir uns als Historiker bewußt, daß die Beweggründe menschlichen Handelns zwar im Wesen subjektiv sind, aber doch objektiv erfaßt werden können. Es ist gerade die Spannung zwischen den Beweggründen entgegengesetzter Parteien, die nicht nur für den Historiker sondern auch für den Politiker von ausschlaggebender Bedeutung ist. Wenn zum Beispiel die eine Seite alles tut, um den Gegner völlig zu zerstören, während die andere Seite nur auf die eigene Sicherheit und das eigene Wohl bedacht ist, dann kann man leicht sehen, daß ein Krieg, wie der im Nahen Osten entbrannte, unvermeidlich ist. Wenn dagegen beide Seiten sich um das Wohlergehen eines Landes zu bemühen vorgeben, dann darf man mit Recht eine friedliche Lösung wie die der deutschen Einheit erwarten.

Eis ist natürlich nicht immer leicht die subjektiven Beweg gründe historischer Ereignisse objektiv zu erfassen. Der Ge schichtsforscher der Neuzeit kann hier die Zeitungen und die politischen Reden beider Seiten lesen und so ein verbindliches Bild menschlicher Mißverständnisse geben. Zu den objektiv feststellbaren Einzelereignissen und den subjektiven und gegensätzlichen Beweggründen kommen aber dann noch die Dokumente, nicht nur der Einzelstaaten, sondern auch die zwischenstaatlichen Verträge und Vereinbarungen. Die subjektive Natur dieser | Dokumente wird von ihrer objektiven [178] Erscheinung oft verdunkelt, und man vergißt leicht, daß eine Verordnung oder ein Vertrag zwar selber ein objektiv feststellbares Ereignis darstellt, aber seinem Inhalt nach Wünsche oder Hoffnungen wiedergibt, deren Erfüllung das

Dokument selber weder angibt noch nachweisen kann. Das heißt, daß eine Verordnung nur dadurch ihre volle historische Bedeutung erhält, daß sie befolgt wird, daß ein Vertrag nur dadurch volle geschichtliche Bedeutung gewinnt, daß er eingehalten wird. Ansonsten stellen solche Dokumente nur subjektive Einstellungen objektiv dar.

Der Historiker der Gegenwart ist sich all dessen so gut bewußt, daß er es oft nicht der Mühe wert findet, es besonders auszusprechen. Er weiß ja, welche Verordnungen und Verfassungen nicht eingehalten werden und welche Vereinbarungen und Verträge nur "Fetzen" von Papier sind. Wenn er solche Dokumente überhaupt erwähnt, so nur um ihre Erfolglosigkeit zu betonen und damit ihre Subjektivität; sie sind im Grunde genommen von politischen Reden oder Leitartikeln nicht verschieden, die alle die subjektive Einstellung des Verfassers oder seiner Partei wiedergeben. All das ist. wie gesagt, dem zeitgenössischen Beobachter, ob er nun ein Historiker ist oder nicht, ganz wohl bekannt, und ich erwähne es nur, um darauf hinzuweisen, daß in dem Studium der Vergangenheit solche Betrachtungen selten angestellt werden, das heißt, die subjektive Natur der angegebenen Beweggründe ebenso unbeachtet bleibt wie die nur bedingte Wirklichkeit der Dokumente. Die folgenden Ausführungen sollen andeuten, wie das besser gemacht werden kann.

In der griechischen Geschichte, die mir als Beispiel dienen möge, sind wir ja nur für das fünfte und vierte Jahrhundert gut genug unterrichtet, um uns über die Beweggründe historischen Handelns und über die geschichtliche Wirklichkeit erhaltener Dokumente ernstlich Gedanken zu machen. Nur aus jener Zeit besitzen wir nicht nur die Werke der großen Geschichtsschreiber Herodot, Thukydides und Xenophon, die politischen Erklärungen und Abhandlungen der großen attischen Redner von Andokides bis Demades, sondern auch eine Reihe von Inschriften, die die schriftliche Überlieferung ergänzen und zum Teil berichtigen. Außerdem haben wir für diese Periode zwei reiche und verläßliche Sekundärquellen, Diodor und Plutarch, die beide auf heute verlorene Werke des vierten und zum Teil auch des fünften Jahrhunderts
[179] zurückgehen, auf die Geschichts|werke des Theopomp und Ephoros und auf die Abhandlungen verschiedener Pamphletisten und Antiquare, wie des Stesimbrotos von Thasos, Aischines des Sokratikers, und der Atthidographen Androtion und Philochorus.

Es ist daher leicht verständlich, warum die Geschichte des fünften und vierten Jahrhunderts und besonders die Geschichte Athens während dieser Zeit von Studenten und Forschern und vom allgemeinen Publikum so eine große Beachtung erhält. Nicht nur gehören die Ereignisse dieser Periode von den Perserkriegen bis zum Tode Alexanders zu den wichtigsten der Weltgeschichte sondern die schriftliche und inscriftliche Überlieferung ist sehr reichhaltig. Der Abwehrkampf gegen die Perser, die Gründung der attischen Hochkultur, die Zerstörung der griechischen Stadtstaaten während der zwei Generationen dauernden Vorherrschaftskämpfe, der Widerstand gegen die makedonische Macht, die Vereinigung Griechenlands unter makedonischer Führung, die Ausbreitung des Hellenentums durch die Eroberungen Alexanders des Großen und schließlich die Besinnung auf die vergangenen Werte der griechischen Kultur, all das ist doch ein wesentlicher Bestandteil unserer abendländischen Tradition, die von uns allen bewahrt werden muß.

Die Beschäftigung mit der Geschichte Athens und Griechenlands während des fünften und vierten Jahrhunderts ist für die Studenten der Geschischte und vor allem der Alten Geschichte von besonderer Bedeutung, auch wegen des reichen Materials, auf Grund dessen unser Geschichtsbild immer wieder neu geschaffen werden kann, denn die Entdeckung auch des kleinsten Papyrus oder der kleinsten Inschrift, die Neugestaltung oder Neuerklärung auch der kürzesten Schriftquelle zieht eine erneute Untersuchung auch der größten Zusammenhänge nach sich, die es uns erlaubt, ermöglicht und sogar befiehlt, unser Gesamtibild periodisch zu revidieren.

Dazu kommt noch, daß dem Studenten und Forscher die Erfahrung, die er vom Studium dieser Geschichtsperiode gewonnen hat, für die Beschäftigung mit anderen Perioden sehr gut zu statten kommt. Das bezieht sich sowohl auf die Frühzeit, für die wir nur eine dünne schriftliche Überlieferung haben, wie auch auf die spätere Zeit, in der die Schriftquellen versiegen und wir immer mehr auf Inschriften angewiesen sind. In beiden Fällen ist das Studium der klassischen Periode wertvoll, da man von ihm lernt, wie die verschiedenen Quellentypen zu beurteilen sind, auch wenn in einem Falle der eine Typ, im [180] anderen Falle der andere ausfällt oder nur sehr spärlich vorhanden ist.

Unter den Quellen der Geschichte des fünften und vierten Jahrhunderts, wie überhaupt unter den Quellen der abendländischen Geschichte, nimmt das Werk des Thukydides einen ganz besonderen Platz ein. Nicht nur steht Thukydides als intelligenter und gebildeter Zeitgenosse der noch außerdem sich führend an den Ereignissen, die er beschreibt, beteiligte, fast einzigartig unter den Historikern da, sondern er ist auch einer der ganz wenigen, die sich über die Quellen und Methoden der Geschichtsschreibung Gedanken gemacht, diesen Gedanken Ausdruck gegeben und sie dann auch in die Tat umgesetzt haben. So hat er denn in seiner Darstellung des peloponnesischen Krieges scharf zwischen den Reden und den Tatsachen unterschieden und hat in einem besonderen, kurzen und prinzipiellen Abschnitt sich für die Genauigkeit der von ihm beschriebenen Ereignisse und für die Gültigkeit der von ihm berichteten Reden eingesetzt. Das heißt, daß Thukydides in diesen Reden die Ansichten und Einstellungen der führenden Städte und Staatsmänner darlegen wollte, also die Beweggründe aus denen ihre politische Tätigkeit zu erklären war.

Thukydides hat sich also bemüht die subjektiven Meinungen der ausschlaggebenden Griechen objektiv richtig wiederzugeben und hat damit unser Verständnis des peloponnesischen Krieges und den Wert seines Werkes erhöht. Dieses Werk enthält aber nicht nur Tatenberichte und Reden, sondern auch Dokumente, Waffenstillstandsbestimmungen, Friedensverträge und Bündnisse, und die Frage ist berechtigt, ob Thukydides diese Dokumente zu den Tatsachen oder zu den Reden zählte. Da hat man schon seit langem bemerkt, daß die von Thukydides berichteten Ereignisse oft mit den diese Ereignisse bestimmenden Dokumenten nicht ganz im Einklang stehen, daß zum Beispiel

Thukydides das wenigstens teilweise Nichteinhalten von Verträgen berichtet, deren Wortlaut er wörtlich wiedergegeben hatte. Das trifft nicht nur für die drei Vereinbarungen zwischen den Persern und Spartanern zu, die Thukydides im achten Buch zitiert, es gilt auch für den Freundschaftsvertrag Athens und Spartas, der kurz nach dem Frieden des Nikias geschlossen wurde, und auch für diesen Frieden selber, den vorhergehenden Waffenstillstandsvertrag und für das berühmte Bündnis zwischen Athen und Argos, Elis und Mantinea, von dem noch ein Inschriftfragment die Genauigkeit des Thukydideischen Zitats bezeugt. In | allen diesen Fällen dürfen wir annehmen, daß Thukydides sich der Spannung zwischen dem Vertrag und seiner Erfüllung, zwischen dem Dokument und seiner geschichtlichen Wirklichkeit voll bewußt war, und daß er daher die Dokumente nicht zu den Ereignissen, den Tatsachen, den ERGA, sondern zu den Reden, den Beweggründen, den LOGOI zählte.

[181]

Diese Überlegungen haben ein Dokument, einen Vertrag, unberücksichtigt gelassen der im Werke des Thukydides und im peloponnesischen Kriege eine ganz besondere Rolle gespielt hat, den dreißigjährigen Frieden, der im Jahre 446 v. Christi Geburt zwischen Athen und Sparta geschlossen wurde und dessen Bruch vierzehn Jahre später den Anfang des peloponnesischen Krieges markiert. Immer wieder betont Thukydides, daß Athen von Korinth bezichtigt wurde, den Frieden gebrochen zu haben, aber später läßt er die Spartaner zugeben, daß sie die wahrhaft Schuldigen waren. Aber der Friedensvertrag selber wird von dem Historiker weder zitiert noch auch inhaltlich wiedergegeben. Sollen wir annehmen, daß Thukydides erst später im Laufe seiner Arbeit den Wert solcher Dokumente erkannte, oder daß es ihm hier weniger auf den Text ankam, als auf die Interpretation, die die beiden Parteien ihm gaben? Wahrscheinlich ist beides der Fall, und wir dürfen annehmen, daß Thukydides sich erst langsam der Zweideutigkeit der Dokumente bewußt wurde, vielleicht gerade bei Gelegenheit seiner Beschäftigung mit dem dreißigjährigen Frieden.

Diese Annahme führt uns tiefer in unser eigentliches Problem, nämlich die Bewertung und Erklärung einiger historisch überaus wichtiger Dokumente, die Thukydides überhaupt nicht erwähnt hat und deren Echtheit in gewissen Fällen sogar angezweifelt wurde. Die meisten dieser Dokumente gehören soweit sie sich zeitlich festlegen lassen, in die fünfziger Jahre und haben mit dem Übergang der attischen Außenpolitik vom Perserkrieg zur Bundesverwaltung zu tun.

Von diesem Übergang liest man bei Thukydides jedoch nichts, da der Historiker bewußs eine ununterbrochene Entwicklung von den großen Perserkriegen bis zum Ausbruch des peloponnesischen Krieges annimmt und gerade aus dieser Entwicklung heraus den wahren Grund des Krieges herausarbeitet, nämlich Spartas wachsende Angst vor der wachsenden Macht Athens. So sind denn auch die neueren Historiker Thukydides treu und gläubig gefolgt und haben zwischen dem Sieg von Plataea und dem Ausbruch des peloponnesischen Krieges keinen | Bruch der attischen Politik erkennen wollen. Und doch hat man hier wenigstens eine Ausnahme gemacht, die durch die Entdeckung und die Erklärung einer Reihe von Inschriften entstand, den berühmten Tributquotalisten.

[182]

Die erste Reihe dieser Inschriften steht auf einer großen Stele, die im Jahre 454 v. Christi Geburt errichtet wurde, um die Listen der Geldweihungen der mit Athen verbundenen Städte an die Göttin Athena aufzunehmen. Wir können den Anfang dieser Reihe so genau bestimmen, da sie vom ersten Jahr an numeriert wurde und gelegentlich auch durch den jährlichen Archonten datiert wurde. Da die Reihe der attischen Archonten des fünften Jahrhunderts von Diodor gegeben wurde, so kann damit die erste Tributquotaliste dem Jahre 454 zugewiesen werden.

Diese Zahlungen stellen 1/60 der Tributbeiträge dar, auf Grund der von Aristeides anläßlich der Begründung der delischattischen Symmachie veranstalteten Schätzung. Damals mußten die Bündner die Zahlungen für die Aufrechterhaltung der Flotte machen, mit der die Griechen des kleinasiatischen Festlandes von der Perserherrschaft befreit werden sollten und dann die ganze ägäische Welt gegen eine Wiederholung persischer Angriffe geschützt werden sollte. Während der ersten Periode, die mehr als zwanzig Jahre dauerte, befand sich die Bundeskasse auf der Insel Delos, im Jahre 454 wurde sie jedoch nach Athen verlegt. Diodor sowie Plutarch berichten von diesem Ereignis, die Inschriften datieren es sogar aufs Jahr, aber Thukydides sagt kein Wort davon. Da er höchstwahrscheinlich mit den Tatsachen bekannt war, muß man annehmen, daß er sie absichtlich oder unabsichtlich ausließ—entweder weil er sie für unwichtig hielt oder weil sie nicht in seine Gesamtinterpretation jener Geschichtsepoche hineinpaßten. Die modernen Historiker finden sich hier vor einem Dilemma, denn sie behaupten einerseits, daß die Verlegung der Kasse sehr bedautsam war, nehmen aber andererseits zugleich an, daß Thukydides die Ereignisse jener Zeit vollständig und mustergültig berichtete.

Daher wird oft vergessen, daß die uns erhaltenen Inschriften nicht die Tributbeiträge selber verzeichnen, sondern jenen Prozentsatz, 1/60, der daraus der Göttin Athena geweiht wurde. Diese Weihungen, die von den einzelnen Städten gemacht wurden, beweisen zusammen mit ähnlichen Weihungen an Athena selber, an Demeter und an Dionysos, daß wir es hier mit einer panathenischen Organisation zu tun haben, in der die Bundes|städte mit Athen *[183]* nicht nur vertraglich, sondern auch religiös verbunden waren. So wissen wir von Inschriften des fünften Jahrhunderts, daß die mit Athen verbundenen Städte nach Eleusis Getreidespenden machten, an den Dionysien einen Phallos weihten und zu den Panathenäen außer der Geldspende auch eine Kuuh und eine Panhoplie brachten. Solche Beiträge gehen nicht nur über die durch Verträge vorgeschriebenen Tribute weit hinaus, sie haben im Grunde genommen nichts mit einer Bundesgenossenschaft zu tun, sondern kennzeichnen eher eine religiös gerichtete Vereinigung wie die delphische Amphiktyonie. Das heißt, daß wir auf Grund der Tributquotalisten und mit ihnen verbundener Dokumente, ohne direkte Unterstützung der schriftlichen Überlieferung, annehmen dürfen, daß Athen im Jahre 454 v. Christi Geburt sein gegen Persien gerichtetes Kriegsbündnis in eine halbreligiöse Organisation umgestaltet hat.

Daß es sich damals wirklich um eine Neuordnung handelte, beweist ein Dokument, das zwar nicht auf Stein erhalten ist, aber von Plutarch in seinem

Leben des Perikles zitiert wird. Es handelte sich um einen von Perikles eingebrachten Beschluß, ein Psephisma, durch das alle Griechen zu einer Tagung eingeladen wurden, um über die folgenden Punkte zu beraten: (1) die hellenischen Heiligtümer, die von den Persern in 480 und 479 v. Chr. Geburt zerstört wurden; (2) die Opfer, die die Griechen den Göttern versprachen, als sie gegen die Barbaren 479 kämpften und die anscheinend nicht dargebracht worden waren; (3) die ungehinderte Benützung des Meeres und die Bewahrung des Friedens. Um die Einladung auszurichten und das Programm bekannt zu geben, wurden zwanzig besonders beauftragte ältere Männer zu all den Mitgliedern des attischen Bundes und der delphischen Amphiktyonie geschickt. Plutarch berichtet, daß aus dieser panhellenischen Tagung nichts wurde, da sich die Spartaner weigerten, daran teilzunehmen, eine genauere Betrachtung der vorgeschlagenen Programmpunkte zeigt jedoch, daß jeder von ihnen innerhalb der von Athen kontrollierten Organisation tatsächlich ausgeführt wurde.

[184] Die von den Persern verbrannten Heiligtümer sollten nach einem Eid, den die Griechen wohl nach der Schlacht von Plataea geschworen haben, nicht wieder aufgebaut werden, sondern als ein ewiges Denkmal barbarischer Gottlosigkeit als Ruinen bewahrt werden. Dieser Eid ist uns nicht inschriftlich erhalten, aber er wird als Dokument von den attischen Red|nern und von Diodor wörtlich zitiert. Wenn nun so ein Eid 479 geschworen wurde und dann von Perikles anläßlich eines panhellenischen Kongresses die Frage nach den von den Persern zerstörten Heiligtümern diskutiert werden sollte, so könnte man annehmen, daß es sich um die Beseitigung des Eides und den Wiederaufbau der zerstörten Heiligtümer handelte. Diese Annahme wird durch das von Perikles durchgeführte Bauprogramm auf der Akropolis und im attischen Lande bestätigt. Während manche der perikleischen Bauten völlige Neuschöpfungen sind, handelt es sich in anderen Fällen um den Wiederaufbau zerstörter Heiligtümer, sicherlich im Falle des Telesterions zu Eleusis, des Parthenon und des Niketemples. Das heißt, daß der nach Plataea gegebene Eid wenigstens in Athen aufgehoben wurde und daß dort die von den Persern zerstörten Heiligtümer wiederaufgebaut wurden, also der erste Punkt des perikleischen Programmes erfüllt wurde.

Der zweite Punkt betrifft die Opfer, die die Griechen nach Plataea versprachen. Auch von diesen wissen wir von einem Dokument, das von Plutarch zitiert wird, obgleich seine Echtheit von verschiedenen Seiten angezweifelt wurde. Aristeides selber soll nach der Schlacht bei Plataea vor einer Heeresversammlung der Hellenen den folgenden Beschluß beantragt haben: (1) eine jährliche Tagung von Vertretern aus allen beteiligten Staaten in Plataea; (2) Jedes vierte Jahr eine Gedächtnisfeier für die Befreiung von der Persergefahr, die Eleutherien; (3) eine panhellenische Streitmacht von 100 Schiffen, 1,000 Reitern, und 10,000 Fußsoldaten, um den Krieg gegen die Barbaren weiterzuführen; (4) Opfer für die Hellenen von den Plataern dargebracht. Dazu kam noch ein Eid, von dem Plutarch an dieser Stelle nichts sagt, der aber durch Diodor bezeugt ist: niemals mit den Persern Frieden zu schließen, sondern für alle Ewigkeit die Feindschaft gegen die Barbaren von Generation zu Generation weiterzugeben; außerdem gehörte zu dem Eid auch das Versprechen, die von den Persern zerstörten Heiligtümer nicht wieder aufzubauen. Die Verbindung dieses Beschlusses des Aristeides mit jenem des Perikles ist ganz augenscheinlich.

In dem einen wird geschworen die Heiligtümer nicht wieder aufzubauen, in dem anderen beantragt, die Wiederherstellung ins Auge zu fassen, was dann auch später geschah. In dem einen werden Opfer und Feierlichkeiten versprochen, in dem anderen sollen die Opfer erneut diskutiert werden, also scheinen sie nicht dargebracht worden zu sein, und in der Tat zeigt der | Bericht des Thukydides, daß diese Versprechungen wenigstens im fünften Jahrhundert nicht genau eingehalten wurden. Wir können also schon hier an einem guten Beispiel die Spannung zwischen einem gefaßten Beschluß und seiner Ausführung beobachten. Der Beschluß des Aristeides wurde nur zum kleinsten Teil tatsächlich ausgeführt, da innerhalb eines Jahres die attisch-delische Symmachie gegründet wurde, die zwar die jährlichen Tagungen und die gemeinsamen militärischen Streitkräfte von dem Bund von Plataea übernahm, aber den letzteren eben dadurch außer Kraft setzte. Nur der Eid wurde vorderhand eben eingehalten, aber auch er wurde dann vor der Mitte des Jahrhunderts außer Kraft gesetzt. [185]

Der Grund, aus dem der gegen die Perser gerichtete Eid außer Kraft gesetzt wurde, ist natürlich die Beendigung des Perserkrieges und die Friedensverhandlungen mit Persien. Damit kommen wir zu dem letzten Punkte des perikleischen Beschlusses, einen panhellenischen Kongreß zu halten, und damit auch zu dem berühmten Frieden mit Persien, der nach dem athenischen Unterhändler Kallias genannt wird. Perikles beantragte eine Diskussion der Freiheit des Meeres und der Bewahrung des Friedens. Beide Punkte setzen das Ende des sogenannten Ersten Peloponnesischen Krieges voraus, der mit einem fünfjährigen Friedensschluß abgeschlossen wurde. Beide Punkte zielen darauf hin, daß es einer Organisation und eines Aufwandes bedarf, die Freiheit des Meeres und das Fortbestehen des Friedens zu garantieren. Also hatte Perikles anscheinend die Absicht, von den Abgeordneten der Hellenen das Versprechen von regelmäßigen Geldzahlungen zur Aufrechterhaltung des Friedens zu fordern. Wir können jetzt die Weigerung Spartas leicht verstehen: Keine dieser Punkte ging Sparta näher an oder hatte irgend einen unmittelbaren Bezug auf Spartas außenpolitische Ziele, die damals nach wie vor auf die Peloponnes beschränkt blieben. Aber für Athen waren dies Lebensfragen, besonders wenn man bedenkt, daß das Ende der Perserkriege die Auflösung der nur zu diesem Zweck gegründeten Symmachie bedeuten könnte.

Für Athen war demnach die Fortsetzung der Organisation, wenn auch vielleicht unter einem neuen Namen, von ausschlaggebender Wichtigkeit, und als der Kongreß nicht zustande kam, da hat Perikles in diesem Punkte, wie in dem Wiederaufbau der Heiligtümer, auf eigene Faust die delisch-attische Symmachie als Form für seine ursprünglich panhellenischen Pläne benützt. Das können wir daran erkennen, daß die neue Organisation | eben geradeso handelte [186] als ob sie sich der von Perikles vorgeschlagenen Politik widmen wollte. Der Schatz wurde nach Athen überführt und dort verwaltet, die Mitglieder wurden Athen angeschlossen, als ob sie alle attische Kolonisten wären, was doch nur für die Ionier angenommen werden darf und auch für sie eine Fiktion ist. Schließlich hat man den Eindruck, daß der Hauptzweck der Organisation war, Frieden und Freiheit auf dem Meere aufrecht zu erhalten. Vom Krieg gegen Persien ist nach 454 überhaupt nicht mehr die Rede und man kann demnach von einer Friedenspolitik des Perikles sprechen, die mehr als zwanzig Jahre fortgeführt werden konnte; während dieser Zeit kam es zwar zu Aufständen, dem euböischen und dem samischen, die beide von außen her unterstützt wurden, der eine von den Spartanern, der andere von den Persern. Trotzdem endeten beide erfolglos, und es ist keine Übertreibung zu sagen, daß Athen und sein Verbündeten, das heißt der ganze ägäische Raum, von der Mitte der fünfziger Jahre bis zur Mitte der dreißiger Jahre Frieden hatten und genossen. Wir wissen von sehr wenigen Machterweiterungen Athens während dieser Jahre, und die thukydideische These von der Ursache des peloponnesischen Krieges scheint für diese Periode besonders schlecht zu stimmen.

Es hat sich demnach gezeigt, daß eine vorsichtige Verwertung von Dokumenten, ob sie nun auf Stein erhalten sind oder nicht, dem Bericht des Thukydides erweiternd und bereichernd, wenn nicht auch berichtigend, zur Seite treten kann. Es ist dabei erstaunlich aber nicht unerklärlich, daß Thukydides weder die Übergabe des Bundesschatzes von Delos nach Athen, noch auch den Aufruf zu einem panhellenischen Kongreß mit einem Wort würdigt. So eine Entschuldigung läßt sich aber für das Außerachtlassen des Kalliasfriedens nicht machen. Wenn Athen mit Persien Frieden schloß, so hätte Thukydides es berichten müssen, wenn er aber mit Recht diesen Friedensvertrag ausließ, dann kann dieser Vertrag nicht die Bedeutung gehabt haben, die ihm jene Forscher zulegen, die ihn heute für echt erklären—im Gegensatz zu den anderen, die ihn als eine freche Fälschung des vierten Jahrhunderts bezeichnen.

Der Friede des Kallias ist schon in der Antike umstritten gewesen und daher mag er uns als Hauptbeispiel des Verhältnisses zwischen schriftlich oder inschriftlich festgelegtem Dokument und schriftlicher Überlieferung dienen. Thukydides, wie gesagt, hat von einem zwischen Athen und Persien beschlossenen | Frieden nichts zu sagen, er spricht überhaupt nicht von einem Ende der Perserkriege, nachdem er die letzte Fahrt des Kimon nach Kypros erwähnt, die mit dem Tode des großen Atheners endete. Die Datierung dieser Fahrt und von Kimons Tod ist umstritten, aber es scheint mir klar, daß Thukydides diese Ereignisse unmittelbar nach dem Friedensschluß zwischen Athen und Sparta setzt, der seinerseits unmittelbar den Schlachten von Tanagra und Oinophyta folgte. In der ersten wurden die Athener von den Spartanern besiegt, in der zweiten siegten die Athener über die Böoter. Diese zeitliche Verbindung ist noch bestärkt durch die Angabe die wir dem Plutarch verdanken, daß der im Frühjahr 461 v. Chr. Geburt durch Ostrakismos verbannte Kimon nach weniger als fünf Jahren, kurz nach Tanagra, zurückberufen wurde, um über den Frieden mit den Spartanzern zu verhandeln.

Das heißt, daß vor der Mitte der fünfziger Jahre ein Frieden zwischen Athen und Sparta zustande kam und auch der Perserkrieg mit dem Tode des Kimon zu Ende ging. Hier setzt nun unsere andere Überlieferung ein, die durch Diodor und Plutarch, aber auch durch die attischen Redner Isokrates und Demosthenes vertreten wird und die in gerade jene Zeit die Verhandlungen zwischen Persien und Athen setzt, die zum Frieden des Kallias geführt haben sollen.

Bezüglich dieses Friedens besitzen wir zwar nur sehr wenige Angaben, aber alle die sachlichen Bestimmungen. Die ersteren stammen aus Anekdoten, die mehr oder weniger verläßlich sind, aber sich noch ins fünfte oder frühe vierte Jahrhundert zurückführen lassen. Schon Herodot erwähnt zum Beispiel eine Gesandtschaft des Kallias, die sich in Susa aufhielt, und Demosthenes berichtet kennzeichnenderweise, daß dieser Kallias, nachdem er von Persien zurückkehrte, der Bestechung angeklagt und zu einer Geldstrafe verurteilt wurde. Vielleicht ist die Szene der persischen Gesandtschaft in den Acharnern des Aristophanes noch ein Echo dieses Ereignisses.

Die Bestimmungen des Kallias Friedens stammen von einer Inschriftstele, die wohl erst im Jahre 386 v. Chr. Geburt aufgestellt wurde, zusammen mit einer anderen Stele, die den eben geschlossenen, berüchtigten Frieden des Antalkidas enthielt. So konnten diese beiden mit Persien abgeschlossenen Friedensverträge nicht nur von den attischen Rednern sondern auch von den Geschichtsschreibern miteinander verglichen und einander | gegenübergestellt werden. In dem älteren Vertrag gab der Perserkönig seine Kontrolle über die kleinasiatischen Griechenstädte auf, erkannte eine demilitarisierte Zone längs der Westküste Kleinasiens an und versprach, seine Kriegsschiffe vom ägäischen Meere fernzuhalten. Dafür gaben die Athener dem Großkönig das Versprechen, keinen Angriff gegen Persien mehr zu unternehmen. Diesem den Verhältnissen der Mitte der fünfziger Jahre des fünften Jahrhunderts entsprechenden, älteren Vertrage stand der sogenannte Königsfrieden gegenüber, der den umgekehrten Machtverhältnissen um 390 v. Chr. Geburt entsprach. In diesem Vertrage wurden dem persischen König alle die Griechenstädte Kleinasiens überantwortet, und seiner Macht waren keine Grenzen gesetzt. In Griechenland selber bestand der Großkönig auf der Autonomie und Unabhängigkeit jeder einzelnen Griechenstadt, so daß kein Städtebund entstehen und sich seiner Kontrolle erfolgreich widersetzen durfte. Die Spartaner, denen dieser Vertrag auch sehr gelegen kam, wurden die Hüter seiner Bestimmungen. Es unterliegt keinem Zweifel, daß ein Vergleich der beiden nebeneinanderstehenden Dokumente den traurigen Zustand griechischer Schwäche zeigte, die ruhmreiche Größe des attischen Reiches um die Jahrhundertmitte und die verächtliche und verräterische Unterwerfung unter das Joch der Barbaren, die der spartanische Sieg im peloponnesischen Kriege mit sich brachte. Natürlich waren die Athener im Jahre 386 stolz auf ihre Errungenschaften siebzig Jahre früher, und die Spartaner und ihre Freunde waren wütend, diesen alten Vertrag jetzt publiziert zu sehen. Und dieser Verschiedenheit müssen wir die verschiedene Einstellung der schriftlichen Zeugnisse zuschreiben. Athener, wie Demosthenes und Isokrates, priesen den berühmten Frieden des Kallias als eine Großtat attischer Geschichte,

[188]

die allen Griechen Freiheit, Unabhängigkeit und Wohlstand garantierte, während unter spartanischer Kontrolle sogar die Griechen des Festlandes sich dem Willen des Großkönigs beugen müssen. Dieselbe Einstellung dürfen wir auch in dem großen Geschichtswerk des Ephoros voraussetzen, da Diodor, der ihn benützte, nicht nur die Bestimmungen des Kallias Friedens wiedergibt, sondern ihn auch später mit dem Frieden des Antalkidas zu dessen Nachteil vergleicht.

Es gab aber auch Widerspruch. Der chiische Historiker Theopomp, dessen Sympathien auf spartanischer Seite lagen, hat den Wert des Kallias-Friedens direkt angegriffen und behauptet, daß die Athener auf Grund dieses Dokuments sich | stolz Errungenschaften zuschrieben, die nur ihrer Phantasie entstammen. Er bemerkte auch, daß die Stele mit dem Text des Kallias-Friedens erst im vierten Jahrhundert aufgezeichnet wurde und auch daher verdächtig sei. Ein anderer Geschichtschreiber, Kallisthenes, hat dann insofern die Sache zu erklären versucht, daß er annahm, daß der Kallias-Frieden zwar nicht offiziell geschlossen und beschworen wurde, daß aber seine Bestimmungen vom Perserkönig bis zum Anfang des peloponnesischen Krieges eingehalten wurden. Diese Interpretation hat manches für sich, das wir aus anderen Quellen erfahren. Schon erwähnt wurde die Bemerkung des Demosthenes, daß Kallias bei seiner Rückkehr aus Persien der Bestechung angeklagt und zu einer sehr hohen Geldstrafe verurteilt wurde. Es läßt sich schwer vorstellen, daß der von Kallias verhandelte Vertrag damals in Athen in großem Ansehen stand; später mag man ihn aus den Archiven herausgezogen und veröffentlicht haben—und vielleicht nicht ganz mit Unrecht, denn der Perserkönig hatte sich ja mit den Bestimmungen dieses Vertrages seinerzeit einverstanden erklärt. In dieselbe Richtung führt eine Bemerkung des Andokides, der in seiner im Jahr 392 gehaltnen Rede "Über den Frieden" darauf hinweist, daß ein von seinem Vetter kurz vor der sizilischen Expedition mit dem Perserkönig geschlossener Vertrag von den Athenern dadurch sofort gebrochen wurde, daß sie einen aufständigen Satrapen unterstützten. Diese Geschichte zeigt nicht nur, daß damals der Kallias-Frieden als ein Vertrag zwischen Athen und Persien nicht bestand, sondern daß die Athener in ihrer stolzen Überheblichkeit Verträge nicht einhielten, solange sie die Macht hatten, so zu handeln, wie es ihnen paßte. Übrigens erwähnt auch Thukydides, daß beide Parteien sich zu Anfang des peloponnesischen Krieges um Unterstützung durch Persien bemühten—also weder die Athener noch die Spartaner damals einen gültigen Vertrag mit Persien besaßen.

Der Schluß, der aus all diesen Überlegungen gezogen werden kann, ist für das Verständnis des Verhältnisses zwischen inschriftlich erhaltenen Dokument und schriftlicher Geschichtsüberlieferung sehr bedeutsam. Der Friede des Kallias ist ein echtes Dokument, das zu seiner Zeit wenig Bedeutung hatte und gar nicht ratifiziert wurde. Als dieses dann siebzig Jahre später auf Stein verewigt wurde, begann seine eigentliche Geschichte, die wie ein Gelehrter einmal spöttisch behauptete, eher in die Literaturgeschichte als in die politische Geschichte gehört.

Von der Inschrift mit dem Kallias-Frieden ist uns anscheinend kein [190] Bruchstück erhalten, aber von zwei weiteren Dokumenten besitzen wir tatsächlich Steinkopien, deren umstrittene Echtheit und Bedeutung jetzt vielleicht besser verstanden werden kann. Das sind der Eid von Plataea und das Themistoklesdekret.

Herodot hat nichts von einem vor der Schlacht bei Plataea entweder auf dem Isthmos oder in Eleusis oder gar auf dem Schlachtfeld geschworenen Eid zu sagen, Ephoros hat ihn wohl zitiert, da Diodor ihn wiedergibt, und Lykurgus hat ihn in dem Prozeß gegen Leokrates vorlesen lassen, obgleich es nicht ganz sicher ist, ob der in seinem Text zitierte Eid nicht aus Ephoros genommen wurde. Schließlich ist zu Acharnai eine Inschriftstele gefunden worden, auf der nicht nur der Eid, den die Athener vor der Schlacht von Plataea schworen, sondern auch der berühmte und in seiner Echtheit nie angezweifelte Eid der attischen Epheben steht. Und doch gilt der Eid von Plataea auch heute noch weitgehend als eine Fälschung, hauptsächlich weil Theopomp die Historizität auch dieses Ruhmesblattes attischer Geschichte ausdrücklich angezweifelt hat—das heißt behauptet hat, daß die Griechen diesen Eid nicht geschworen haben.

Der Eid besteht, wie er uns schriftlich und inschriftlich erhalten ist, aus drei Teilen, deren erster ein allgemeiner Soldateneid ist, deren zweiter sich ausdrücklich auf die Situation vor der Schlacht bei Plataea bezieht und deren dritter wieder allgemeine Bestimmungen griechischen internationalen Lebens wiedergibt. Der erste und wichtigste Teil enthält den eigentlichen Kampfeseid, das Versprechen im Kampfe auszuharren, den Kameraden zu helfen, den Offizieren zu gehorchen, die Toten zu begraben, Freiheit höher als Leben zu schätzen. So ein Eid sollte und konnte von jeder Armee vor jeder entscheidenden Verteidigungsschlacht geschworen werden, er enthält besonders in dem Teil, der dem Gehorsam gewidmet ist, die Titel spartanischer wie auch athenischer Offiziere, war also für ein gemeinsames spartanisch-athenisches Heer geschaffen, aber sonst hat er so starke Anklänge an die attische Tradition, daß schon Lykurg mit Recht betonte, daß er einem attischen Vorbild folgte. Das heißt natürlich nicht, daß er nicht schon um 479 v. Chr. Geburt verfaßt werden konnte.

Der Theben gewidmete Teil erklärt, daß im Falle eines Sieges Theben dezimiert werden sollte und alle anderen Griechenstädte, die, ohne gezwungen worden zu sein, mit den Barbaren gegen die Griechen zu Felde zogen. Diese Klausel wurde beanstandet, | erstens weil sie nicht ausgeführt wurde und zweitens [191] weil so ein Eid vor der Schlacht doch etwas voreilig war. Doch muß man bedenken, daß wenigstens die Athener nach den Siegen bei Marathon und Salamis kaum mit einer Niederlage bei Plataea rechneten und daß ein tatsächlich unerfüllter Eid eher historisch echt ist als einer der genau eingehalten wurde und der auf Grund der nachträglichen Ereignisse hätte rekonstruiert werden können.

Der dritte Teil des plataïschen Eides ist, wie man schon seit langem erkannt hat, amphiktyonischen Charakters, da er sich dem Frieden unter den griechischen Städten widmet. Keine Stadt soll den Umsturz einer anderen vorbereiten und wenn es zum Kriege kommt soll keine Stadt durch Hunger oder Durst den Gegner zu überwinden versuchen. Der Eid schließt mit einem Anruf an die Götter, diejenigen, die den Eid übertreten, mit Seuchen, Mißgeburten und jedem anderen Elend zu bestrafen. Auch hier klingen die Phrasen der delphischen Amphiktyonie heraus.

Die erste Frage die wir uns stellen müssen, wurde schon in der Besprechung des Kallias-Friedens angedeutet. Damals wurde es klar, daß der Kallias-Frieden bei seiner Veröffentlichung im Jahre 386 politisch und historisch viel wichtiger war als siebzig Jahre früher, als er geschlossen wurde. So dürfen wir jetzt fragen, welche Bedeutung der Eid von Plataea bei seiner Veröffentlichung in den 30er Jahren des vierten Jahrhunderts hatte, nach der Niederlage von Chaironeia, nach der Zerstörung von Theben und dem endgültigen Siege von Mazedonien. Wiederum zeigte ein Vergleich der elenden Gegenwart mit der ruhmreichen Vergangenehit, wie sich die Griechen und besonders die Athener zu jener früheren Zeit benahmen: patriotisch, mutig, aufopfernd und gemäßigt. Jetzt war das alles anders geworden und die Großtat von Plataea stand monumental neben der schmählichen Niederlage von Chaironeia, das nur ein paar Kilometer weiter nördlich liegt.

So können wir verstehen, warum der Eid von Plataea gerade damals von attischen Rednern zitiert und in einem Heiligtum des Kriegsgottes sogar aufgezeichnet wurde, wir können aber auch verstehen, daß athenerfeindliche Griechen, wie Theopomp, großes Vergnügen daran fanden, zu behaupten, daß dieser Eid von Plataea gar nicht von den Griechen geschworen wurde. Diese Behauptung darf man nun nicht so verstehen, wie es namhafte Gelehrte getan

[192] haben, daß die Athener des vierten | Jahrhunderts den Text des Eides entweder frei erfanden, oder aus älteren Dokumenten zusammenstückelten, die mit den Ereignissen vor Plataea nichts zu tun hatten. Der Betrug der Athener bestand nicht in der Fälschung eines Dokuments, sondern darin, daß die Athener einem echten Dokument fälschlich eine Bedeutung gaben, die es ursprünglich nicht besaß. Inwiefern Theopomp in seinem Urteil dadurch beeinflußt wurde, daß der Eid zum Teil nicht eingehalten wurde, läßt sich heute nicht mehr sagen. Der Eid war jedenfalls echt insofern, als er von den Athenern abgefaßt wurde und von den Griechen hätte geschworen werden sollen. Wir haben hier also wieder ein Dokument vor uns, dessen geschichtliche Bedeutung lange nach seiner Entstehung zutage trat, ein Phänomen, das auch sonst in der Geschichte nicht unbekannt ist. Die Frage, die der kritische Forscher bei so einer Gelegenheit immer wieder stellt, bezieht sich auf die Verläßlichkeit der Überlieferung des älteren Dokumentes, das aus dem Archiv geholt wurde, um einer neuen Situation zu dienen; hat der interessierte Herausgeber Änderungen vorgenommen, um das ältere Dokument der neuen Situation anzupassen?

Diese Probleme treten in ganz besonderem Maße bei einem anderen Dokument zu Tage, das wir abschließend kurz betrachten wollen, bei dem berühmten Themistoklesdekret. Es handelt sich um einen athenischen Volksbeschluß, der auf Vorschlag des Themistokles vor der Schlacht bei Salamis gefaßt wurde und der die Evakuierung der Bevölkerung von Athen nach

Troizen, die Verteidigung von Attika durch die griechische Flotte und die Zurückberufung der Ostrakisierten vorsah. Herodotus berichtet zwar im Zusammenhang mit den Orakeln, die den Athenern von Delphi gegeben wurden, daß Themistokles die, "hölzerne Mauer" ($\xi \acute{u}\lambda \iota \nu o \nu\ \tau \epsilon \hat{\iota} \chi o \varsigma$) auf die attische Flotte bezog, daß er schon vorher den Bau athenischer Kriegsschiffe bewerkstelligte und daß damals, d. h. nach dem Orakel, die Athener beschlossen, dem Angriff der Barbaren mit der von der gesamten waffenfähigen Bevölkerung bemannten Flotte entgegenzutreten, zusammen mit all jenen Griechen, die bereit wären Hilfe zu leisten. Von der Evakuierung selber spricht Herodot nur im Zusammenhang mit der tatsächlichen Flucht der Athener, hauptsächlich nach Salamis, und von der Zurückberufung der Ostrakisierten liest man im Herodot nur die Bemerkung, daß der ostrakisierte Aristeides plötzlich im Flottenlager des Themistokles kurz vor der Schlacht erschien, also wohl zurückberufen worden war. | Mit anderen Worten, die Erzählung des Herodot schließt zwar die [193] Maßnahmen des Dekretes nicht aus, nimmt aber darauf kaum Bezug. Andererseits kann man wenigstens von den zwei Hauptpunkten des Dekretes sagen, daß sie nicht ausgeführt wurden, wenn der Bericht des Herodot stimmt. Weder wurden die attischen Frauen und Kinder vor der Schlacht an den Thermopylen nach Troizen gebracht, noch wurde damals die Stadt evakuiert, und die Verteidigung von Attika fand bei Salamis und nicht längs der Ostküste statt. Hier liegt der Schlüssel zum Verständnis dieser und vieler anderen ähnlichen Inschriften: ein Beschluß sagt historisch nur aus, daß er gefaßt wurde, aber nicht daß er ausgeführt wurde.

Die Fragestellung ist demnach falsch, die eine Entscheidung zwischen Herodot und der Inschrift verlangt und dann entweder den Historiker der Ungenauigkeit beschuldigt oder die Inschrift für unecht erklärt. Eher sollte man fragen, ob es irgendeinen Grund gibt, aus dem der angeblich gefaßte Beschluß nicht auch tatsächlich ausgeführt wurde, so daß die von Herodot erzählten Ereignisse auch tatsächlich zutreffen konnten. Hätten wir das Themistoklesdekret als eine Inschrift des Jahres 480 v. Chr. Geburt erhalten, so würden wir weder an ihrer Echtheit noch an der Richtigkeit des herodoteischen Berichtes zweifeln, und wir würden annehmen, daß Themistokles schon einige Monate vor der Schlacht bei Salamis klar erkannte, daß nur eine Evakuierung Athens und eine entscheidende Seeschlacht die Griechen retten könnten und daß diese Überzeugung sich erst später durchsetzte und als richtig erwies.

Nun gehört aber unsere Inschrift nicht ins Jahr 480 v. Chr. Geburt, sondern ins späte vierte oder gar ins dritte Jahrhundert, und sie wurde nicht in Athen, sondern in Troizen gefunden, und während diese Tatsachen machen Forschern die Gelegenheit gab, die Inschrift als eine Fälschung zu bezeichnen, legt sie uns die Verpflichtung auf, ihre späte und abgelegene Veröffentlichung zu erklären. Und das soll abschließend versucht werden.

Schon um die Mitte des vierten Jahrhunderts war das Themistoklesdekret und das des Miltiades so wohl bekannt, daß sie in der Volksversammlung verlesen wurden, um zu zeigen, wie groß der Patriotismus der Athener während der Perserkriege war, im Vergleich mit ihrer furchtsamen Unentschlossenheit gegenüber der Eroberungspolitik Philipp des Zweiten von Mazedonien. Auch von dem Beschluß des Miltiades wissen wir nichts aus Herodot, aber Aristoteles belehrt uns, daß er in den zwei | Worten gipfelte, δεῖ ἐξιέναι "man muß mit der Armee ausziehen"—natürlich nach Marathon. Damals, so sagten die Redner den Athenern, habt ihr eure Stadt im Stich gelassen und habt im Interesse aller Griechen bei Marathon gekämpft und gesiegt, heute bleibt ihr zögernd zuhause und seht zu, während Philipp seine Eroberungen macht. So auch zehn Jahre später, als die Athener dem Angriff des Xerxes allein gegenüberstanden, als sie erkannten, daß keine Landmacht Griechenlands den Eroberungszug der Perser aufhalten konnte. Damals hatten sie den Mut und die Entschlossenheit, ihre Stadt der sicheren Zerstörung durch den Barbaren zu überlassen und sich auf ihre Flotte zu begeben, um von da den Entscheidungskampf zu führen. Heute aber haben sie nicht den Mut Philipp entgegenzutreten, obgleich seine Flotte kleiner und schwächer ist als die der Perser einst war. Den Worten des Miltiades δεῖ ἐξιέναι können die Worte des Themistokles zur Seite gestellt werden, ἐκλιπεῖν τὴν πόλιν "die Stadt aufgeben", τὴν πόλιν παρακαταθέσθαι τῇ ᾿Αθηνᾷ τῇ ᾿Αθηνῶν μεδεούσῃ "die Stadt der Göttin anzuvertrauen, ihr, der Beschützerin von Athen".

Wir können jetzt verstehen, warum die Redner jene Beschlüsse aus den Archiven zogen und nicht den Bericht des Herodot zitierten. Herodot trotz seiner Athenerfreundlichkeit hat natürlich die Perserkriege aus der Perspektive seiner eigenen Zeit gesehen und beschrieben. Dabei stand er den Ereignissen noch nahe genug, um die zeitgenössische Einstellung zu beschreiben, war aber oft schon so sehr in den neuen Gegensätzen zwischen Athen und Sparta befangen, daß er Athen gegen die Anklagen der Spartaner verteidigen wollte. Auch Persien gegenüber hatte Herodot weder Haß noch Verachtung. Im vierten Jahrhundert war all das anders geworden und die Siege der Perserkriege wurden nun mit den Niederlagen gegen Philipp verglichen. Es war nicht anders, als wenn heutzutage man sich der Befreiungskriege gegen Napoleon erinnerte und das Streben nach Einheit und Freiheit von damals auf die heutigen Probleme übertragen wollte. Sicherlich ließen sich Urkunden und Dokumente jener Zeit finden, die heute einen direkten Bezug auf gegenwärtige Probleme hätten und so in dieser Bezugnahme eine größere Bedeutung hätten, als sie je zu ihrer Zeit besaßen.

So stehen die Dokumente und Urkunden und Verträge vor uns, einerseits in völliger Klarheit und wörtlicher Verläßlichkeit, andererseits ohne jene historische Perspektive, die fürs Verständnis der Geschichte so wichtig ist, und gerade wegen dieses | Fehlens der Perspektive sind Dokumente anderen Zeiten und anderen Situationen anpassungsfähig. Dem gegenüber können wir uns nur auf den wahren Geschichtschreiber verlassen, der nicht nur ein Sammler von Daten und Dokumenten ist, sondern der uns erzählt, wie ein gewisses Ereignis zustande kam und sich abspielte. Auch in unserer Zeit, die sich so sehr auf mündliche Berichte, die auf Tonbändern aufgenommen werden, auf Statistiken und objektiven Tatsachen verlassen kann, auch in unserer Zeit ist der Geschichtschreiber unentbehrlich, denn ohne ihn kann keine wirkliche Geschichte geschrieben werden.

24

Review of L. Jeffery, *The Local Scripts of Archaic Greece*

Lilian H. Jeffery. *The local scripts of archaic Greece. A study of the origins of the Greek alphabet and its development from the eighth to the fifth centuries B.C.* Oxford: Clarendon Press 1961. XX, 416 S. 73 Tafeln. 4⁰. (Oxford monographs on classical archaeology, edited by Sir John Beazley and Bernard Ashmole.) *[225]*

Es ist ein Vergnügen, dieses Buch anzuzeigen und zu besprechen, da es in jeder Beziehung eine ausgezeichnete Veröffentlichung ist. Wir mußten auf sie lange warten, aber man kann jetzt leicht verstehen, daß das Schreiben und der Druck schwierig waren: Verständlichkeit ist hier mit Genauigkeit und kritischer Durcharbeitung verbunden, und das Ganze wirkt nicht nur monumental, es ist ein Meisterwerk. Trotz der Größe des Buches und des Werkes tritt die Einfachheit und Bescheidenheit der Verfasserin dem Leser schon im Vorwort entgegen und begleitet ihn bis ans Ende dieses schönen und gut geschriebenen Handbuchs. Es ist teuer (7L 7sh), aber sehr wertvoll, und es wird für viele Jahre das grundlegende Werk über die ältesten Inschriften im griechischen Alphabet bleiben.

Der Hauptteil (S. 66–416) besteht aus einem Handbuch, das alle bekannten und besser erhaltenen Inschriften der Frühzeit einschließlich des fünften Jahrhunderts (mit Ausnahme von Attika, wo wegen der Materialmasse 500 v. Chr. der terminus ante quem sein mußte) nach Landschaften geschieden zusammenfassend und eingehend behandelt und illustriert. Diesem Hauptstück sind zwei Aufsätze vorangestellt, deren einer den Ursprung und die Überlieferung des Alphabets, der andere die technischen Einzelheiten der Inschriften behandelt. Im folgenden gestatte ich mir diese beiden Einführungskapitel (1–65), die Frl. Jefferys Oxforder Doktoratsarbeit von 1951 darstellen, und den sich auf Attika beziehenden Teil des Handbuches besonders eingehend zu behandeln; andere werden andere Teile genauer ins Auge fassen.

Der erste Teil (1–42) versucht vier Fragen zu beantworten: Wo und wann wurde das Alphabet vom Semitischen ins Griechische "übersetzt"? Auf welchen

Wegen wurde die neugeschaffene griechische Schrift in Hellas verbreitet? Wann und von wem wurden die neuen und geänderten Buchstaben geschaffen, und wie ist die Verschiedenheit der griechischen Lokalalphabete zu erklären? Vorerst wird jedoch die wichtige allgemeine Frage behandelt, wie ein schriftloses Volk schreiben lernen kann. Von den zwei Möglichkeiten, durch lange und enge Beziehung zu einem schreibkundigen Volk oder durch die Initiative eines einzelnen Schriftgebers, entscheidet sich die Verf. für die zweite und sagt ganz einfach, daß es einmal einen Griechen gegeben haben muß, der zuerst die fünf Vokale gesondert aussprach und der in voller Kenntnis der phönikischen Sprache und Schrift Griechisch mit phönikischen Buchstaben schrieb. Diese auf wissenschaftliche Forschung begründete Annahme stimmt so sehr mit der alten Überlieferung, die von einem Erfinder spricht, überein, daß wenigstens

[226] ein Leser angenehm berührt und tief beeindruckt war. Dabei be|tont Frl. Jeffery, daß anfänglich weder Änderungen noch Neuerungen zu erwarten sind, sondern daß man annehmen muß, daß der Grieche griechisch mit phönikischen Buchstaben schrieb wie der Tertianer, der eben Griechisch anfängt, Deutsch (oder Englisch) mit griechischen Buchstaben schreibt und so ganz schnell das neue Alphabet lernt.

Nach diesen Vorbemerkungen kommen wir zur ersten Hauptfrage—nach dem "Wo?"—und hier weist die Verf. als Mittelpunkt eine phönikische Faktorei in Griechenland, möglicherweise in Theben, vielleicht zu apodiktisch zurück; Grabungen in Böotien und auf Euböa könnten da manche Überraschungen bringen. Die dorischen Inseln Melos und Thera werden auch zurückgewiesen, und nur Rhodos und Kreta bleiben als griechisce Möglichkeiten übrig; für das letztere spricht die altertümliche Schrift und die phönikischen Funde, für das erste auch die schriftliche Überlieferung. Frl. Jeffery selber entscheidet sich für das von Engländern ausgegrabene Posideion-Al Mina am Orontes, wo griechische Kolonisten von den Phönikern die Schrift übernommen und sie dann in den großen Märkten der damaligen Zeit (Kreta, Rhodos, Korinth, Euboea) weitergegeben haben können. Diese Annahme würden die archäologischen Funde und die schriftliche Überlieferung gleich gut bestätigen.

Als Zeit der Übernahme kommt die protogeometrische Periode des elften und zehnten Jahrhunderts anscheinend nicht in Betracht, und auch das neunte Jahrhundert muß ausscheiden. Vom achten Jahrhundert bleibt nur die berühmte Dipylonkanne, die Frl. Jeffery, nicht als erste, einem der "Phöniker" von Posideion zuweist, der auf Besuch nach Athen kam und zeigen wollte, wie man schreibt. In diesem Zusammenhand kommt auch die Frage nach der Niederschreibung der homerischen Gedichte zur Sprache, und die Verf. behauptet kühn, aber meiner Meinung nach ganz richtig, daß es schwer ist, sich vorzustellen, daß Gedichte solcher Qualität ohne die Hilfe des Schreibens verfaßt und überliefert werden konnten. Eine Entscheidung kann aber weder hier noch im Falle der Beamten- und Siegerlisten gegeben werden, die zwar auch ins achte Jahrhundert hinaufreichen, aber nicht notwendigerweise schon damals aufgezeichnet wurden. So schließt Frl. Jeffery diesen Abschnitt mit dem vorsichtigen, aber ganz glaubwürdigen Schluß, daß die griechische Schrift um 750 v. Chr. entstand.

Die dritte Frage bezieht sich auf den Weg und auf die Weise der Überlieferung der griechischen Schrift, nachdem sie um 750 v. Chr. in einem griechischen Handelsort im Osten, z. B. in Posideion, geschaffen worden ist. Das ganze Alphabet wird nun Buchstabe für Buchstabe vorgeführt und alle frühen Formen werden in natürlicher Reihenfolge aufgezeigt. Die meisten dieser Buchstaben erscheinen in den verschiedenen frühen Denkmälern in recht ähnlichen Formen, besonders wenn man die kleine Anzahl der Denkmäler und ihre oft rohe Zurichtung bedenkt. Doch gibt es hier erstaunliche Ausnahmen, die im einzelnen erwähnt werden müssen. Für die Varianten des Beta scheint es keine einfache Erklärung zu geben, besonders für Korinth, wo das Beta wie ein S aussieht und das Epsilon wie ein B. Dasselbe gilt für Iota, das in manchen Alphabeten (darunter wieder Korinth) wie ein Sigma aussieht; dafür schreibt man in Korinth das Sigma wie ein M. Es scheint, daß die Annahme einer einheitlichen Überlieferung von einem Ausgangspunkt ebenso schwer mit den erhaltenen Denkmälern in Einklang gebracht werden kann wie die Annahme einer zweifachen oder vielfältigen Übernahme. Kirchhoffs Farben sind im Lichte neuerer Forschung stark gebleicht worden, aber sie sind noch erkennbar, und es wäre sehr nützlich, wenn man wüßte, was sie bedeuten. Ein Blick auf die Alphabete von Athen (66) und Korinth (114) zeigt, daß hier Nachbarn zur gleichen Zeit verschieden schrieben.

Schließlich kommt die Verf. kurz auf die vierte Frage nach der Erklärung dieser Verschiedenheiten zu sprechen und a nimmt an, daß die verschiedenen Griechenstädte, zuerst Rhodos, Kreta und die Kykladen, sich ihr Alphabet selber direkt von dem ursprünglichen Übernahmeort in Phönikien geholt haben. Sie schlägt sehr klug vor, daß die lokalen Alphabete von einem Lehrer mitgebracht worden sein können, und daß so seine kleinsten Fehler und Ungenauigkeiten in der Wiederholung festgehalten und sogar vergrößert wurden. Die verschiedenen Aussprachen und Dialekte mögen an anderen Verschiedenheiten | schuld sein, doch kommt man da kaum über Vermutungen hinaus. Schließlich möchte Frl. Jeffery absichtliche Änderungen erlauben, wo Mißverständnisse zu vermeiden waren; so erklärt sie die Unterscheidung von Lambda und Gamma (z. B. in Argos) und von Iota und Sigma (in Rhodos, wo das einfache Iota seinen Ausgang genommen zu haben scheint). [227]

Im Gegensatz zu älteren Theorien, die in Alphabetähnlichkeiten Dialekt- und Stammeszusammengehörigkeit sahen, betont die Verf. ganz richtig, daß Verkehr und Handelsbeziehungen ebenso wichtig sein können, doch muß sie gestehen, daß unser Wissen hier noch in den Anfängen steckt. So schließt dieser erste Teil mit einer Zusammenfassung, die, wenn sie graphisch dargestellt wäre, ein uneinheitliches Netz darstellen würde, dessen sich eine anständige Spinne schämen sollte.

Der zweite Teil ist weniger problematisch und spekulativ und zeigt die Verf. als eine Meisterin des gesamten Materials. Acht lange Seiten sind der

Schriftrichtung gewidmet. Hier liest man, daß die Griechen nicht, wie allgemein angenommen wird, zuerst nur linksläufig, dann nur bustrophedon und schließlich nur rechtsläufig schrieben. Bustrophedon scheint schon vom Anfang an verwendet worden zu sein, wahrscheinlich zuerst, um eine kurze zweite Zeile dem Ende der ersten einfach anzufügen. Da die Mehrzahl der frühesten Inschriften kurz ist, darf doch betont werden, daß die meisten dieser Denkmäler linksläufig anfangen, und diese Schriftrichtung haben die Griechen doch sicher von den Phönikern gelernt. Andrerseits versichert uns die Verf. daß Bustrophedon und Rechtsläufigkeit natürliche Entwicklungen darstellen, die keine besondere Erklärung erforden.

Ebenso ausführlich ist das Schreibmaterial behandelt, von dem man doch einfach hätte sagen können: Stein, Holz Metall, und auch Papyrus, Leder, Wachstafeln, Lindenrinde und Blätter verschiedener Bäume; vielleicht hätte hier nicht nur der etwas mysteriöse Petalismos, sondern auch die Ekphyllophoria genannt werden können, die Stimmzettelabstimmung attischer Behörden wie der Boule. Frl. Jeffery bemerkt erstaunt, daß Tontafeln als Schreibmaterial nur selten benutzt, daß diese Vernachlässigung eines billigen und einfachen Schreibmaterials damit zusammenhängt, daß die Phöniker es eben nicht benutzten. Dabei ist besonders erstaunlich, daß Tontäfelchen in der heroischen Zeit als Schreibmaterial verwendet wurden, und daß demnach die Kenntnis dieses Gebrauchs und sogar die Erinnerung daran verlorengegangen zu sein scheint.

Die Verf. vermutet, daß die ältesten Denkmäler auf Holz aufgezeichnet wurden, und sie führt die Solonischen Axones des frühen sechsten Jahrhunderts als Beispiel an. Sie meint, daß diese Axones rechteckige Balken waren, die waagrecht drehbar in Rahmen verzapft waren; all das muß leider Spekulation bleiben, auch nach Dows ausgezeichneter Veröffentlichung und Wiederherstellung von IG^2 1, 2 (*AJA* 65, 1961, 349–356), da es sich hier um ein beschriftetes Fragment einer Hekatompedontriglyphe handeln kann.

[228] Die Steinschriften werden so erklärt, daß aufrecht stehende unbeschriftete Stelen, wie es solche seit jeher gab, als Schriftträger verwendet wur|den, also die Stele älter als die Inschriftstele war. Diese Stelen wurden ursprünglich senkrecht beschrieben, aber je breiter sie wurden, desto häufiger kam die später gebräuchliche horizontale Schrift vor. Frl. Jeffery bespricht bei dieser Gelegenheit die Bedeutung des Wortes *kyrbis*, das sich manchmal auf den Gegenstand (hauptsächlich Gesetze) und manchmal auf den Schriftträger (Stelen) zu beziehen scheint. Bronzetafeln wurden seit dem sechsten Jahrhundert für besondere Zwecke, wie Verträge und Sakralgesetze, verwendet, aber auch gelegentlich für Weihungen, dann natürlich auf minderwertigem Stein oder auf Holz befestigt. Frl. Jeffery erwähnt, daß das Wort *deltos* für solche Tafeln seit früher Zeit verwendet wurde, aber sie geht nicht noch einen Schritt weiter und bemerkt, daß hier ein phönikisches Vorbild greifbar ist, da *daleth* nicht nur den Buchstaben Delta, sondern auch die Schreibtafel Deltos bezeichnet. Hier haben wir das ursprüngliche Schreibmaterial, und es fragt sich, ob *deltos* nicht das rechteckige Holztäfelchen bezeichnet, in Form vielleicht den mykenischen

Tontafeln nicht unähnlich, auf denen die Linear-B-Rechnungen eingetragen wurden. Leder und Papyrus waren den Griechen des sechsten Jahrhunderts bekannt, und die Kenntnis des Papyrus mag ihnen durch die Phöniker vermittelt worden sein, da das Wort *byblos* in diese Richtung weist; dasselbe gilt auch für das altertümliche Wort *skytale*, das besonders von den Spartanern benützt wurde.

Nach der Schreibrichtung und -methode und dem Schreibmaterial findet man den Inhalt und Gegenstand der ältesten Inschriften kurz, aber gründlich behandelt. Das Fehlen jeglichen Zeichens für das Bestehen schriftlicher Chroniken spricht gegen ihre Existenz, während geschichtlich und politisch bedeutende Namenlisten mindestens seit dem frühen sechsten Jahrhundert inschriftlich bezeugt sind. Inschriften rechtlichen Inhalts sind auch sehr alt, doch darf man hier zuerst von Vorschriften und nur später von Gesetzen sprechen; immerhin reichen die kretischen Beispiele ins siebente Jahrhundert hinauf. Staatsverträge kommen jedoch später auf, und auch die frühesten in Olympia auf Bronzeplatten aufgezeichneten und auf Holzwände oder -tafeln aufgenagelten gehören erst ins späte sechste Jahrhundert.

Die umfangreichsten Gruppen archaischer wie auch späterer Inschriften sind die Grab- und Weihinschriften. Die Verf. betont, daß die Sitte den Namen des Verstorbenen aufzuzeichnen in eine sehr frühe Zeit zurückgeht, bemüht sich aber nicht, diese Zeit zu bestimmen oder der Frage nachzugehen, ob diese Sitte in Griechenland begann; in heroischer Zeit gab es Stelen, aber ohne Inschriften. Frl. Jeffery bespricht die Weihinschriften unter dem allgemeinen Titel "Besitzerinschriften" einschließlich der Graffitos, die einfach den Eigentümer angeben. Sie hebt aber ganz richtig hervor, daß die beschrifteten Weihgeschenke deutlich zeigen, daß die Künstler schreiben konnten und selber viele der Weihinschriften einmeißelten. Hier hätte ich gewünscht, diese Fragen etwas ausführlicher als auf einer kurzen Seite besprochen zu sehen, doch mag das Gesagte anderen genügen. Daß die Athener des frühen fünften Jahrhunderts schreiben konnten, beweist das Ostrakismosgesetz und bestätigen die Ostraka selber, und die Graffiti der Frühzeit erlauben, diese Behauptung auf die Griechen im allgemeinen auszudehnen.

[229]

Der letzte Teil der Einleitung ist der Datierung auf Grund von Buchstabenformen gewidmet, und diese wichtige Erörterung beginnt mit der besonders für die Frühzeit notwendigen Unterscheidung zwischen sorgfältig eingemeißelten und sorglos eingekratzten oder aufgemalten Buchstaben. Auch zeigt ein Vergleich von Stein- und Vaseninschriften, daß die letzteren entwickeltere Buchstabenformen haben. Frl. Jeffery geht so weit anzunehmen, daß die natürlichen Vereinfachungen gemalter Inschriften ton- und richtunggebend wurden und dann von den Steinmetzen übernommen wurden. Diesen Schreibern, besonders den ionischen, die nicht nur die erhaltenen Vaseninschriften, sondern auch die jetzt verlorengegangenen Aufzeichnungen auf Holz, Häuten, Leder oder Papyrus herstellten, muß man dann auch die Entwicklung der Schriftrichtung vom Bustrophedon zur klassischen Rechtsläufigkeit zuweisen.

Mit diesen allgemeinen Betrachtungen schließen diese beiden ersten Kapitel, und man schlägt mit großer Erwartung die nächsten Seiten auf, die die Einzelbehandlungen der frühen Schriftdenkmäler in topographischer und innerhalb der einzelnen Städte in chronologischer Ordnung enthalten.

Wie schon gesagt, darf ich mich auf Attika beschränken, da ich dieses Material besser kenne als irgendein anderes. Hier zeigt sich gleich die Schwierigkeit und die Gefahr der schematischen und systematischen Behandlung eines Gegenstandes, der wie die von Künstlern und Steinmetzen eingegrabenen Formen der Buchstaben dem unbewußten Schönheitssinn und dem Zufall sein Dasein verdankt. Schön und sorgfältig gezeichnete Buchstaben müssen hier mit anderen zusammengestellt werden, in denen der Meißelhieb die Form, nicht die Form den Meißelhieb bestimmte. Und wenn man sich die am Anfang stehende Tabelle mit den chronologisch geordneten untereinander stehenden und sogar numerierten Buchstabenformen ansieht, so bekommt man fast den Eindruck, daß die auf einer waagrechten Linie befindlichen Zeichen auch zusammen auf einer oder der anderen Inschrift standen. Man fragt sich, ob manch ein "kleines Rasenstück" hier nicht ins Herbarium gewandert ist und ob ein Pflanzenbestimmungsbuch das richtige Bild eines Blumengartens gibt. Die Antwort ist klar, aber auch die Feststellung, daß es Frl. Jeffery um Buchstabenformen und nicht um beschriftete Denkmäler zu tun ist, und niemand, der wie ich die Schönheitsliebe und das Kunstverständnis der Verf. kennt, darf ihr vorwerfen, daß sie um des Kleinen willen das Große vernachlässigt hat. Und wenn man sich über die vereinfachten Zeichen der Tabelle beklagen will, da muß man sich gleich die Tafeln ansehen, wo man meistens in guten Photographien die Denkmäler selber betrachten kann; und die den Tafeln beigegebenen Umschriften helfen dem Anfänger den Sinn der Inschriften einigermaßen zu entdecken, während der dem Text nachgestellte Katalog kurze Beschreibungen und maßgebende Verweise enthält.

[230] Niemand, der Kunst und Handwerk kennt, wird eine geradlinige und ebenmäßige Entwicklung der Schrift erwarten, und ein Blick auf die ersten vier Tafeln zeigt sofort das fast völlige Fehlen jeder Einförmigkeit. Das reiche Material der Vaseninschriften wie auch der Bronzeinschriften mußte beiseite gelassen werden, da es zwei eigene Bücher ausmachen würde, Bücher, die leider noch ungeschrieben sind, aber erst geschrieben werden können, nachdem die notwendigen Vorarbeiten gemacht sind. Auch die Münzinschriften bleiben unberücksichtigt, doch betont die Verf. ganz richtig, daß sie mit den Vaseninschriften und nicht mit den Steininschriften verglichen werden müssen; wenn sie aber sagt, daß Kraay "argues strongly" für eine Datierung der ersten attischen Eulen ins letzte Viertel des sechsten Jahrhunderts, da möchte man doch "wrongly" hinzufügen. So müssen wir uns mit einigen wohlbekannten Proben der ältesten und berühmtesten Vaseninschriften begnügen, von denen die älteste auf der Dipylonkanne von der Verf. als eine von einem Fremdling angefertigte Schriftprobe ausgeschieden wird; ich darf hier auch darauf hinweisen, daß Frl. Jeffery im Anfang des zweiten Hexameters Reste einer Alphabetreihe entdeckt, wie sie auch bei anderen frühen Schreibversuchen

vorkommen. Von der Mitte des siebenten Jahrhunderts an kann man aber die attische Schrift in Dipintos und Graffitos verfolgen, und das hier gesammelte Material ist außerordentlich bedeutsam, da es sonst noch nicht zusammenfassend vorgelegt wurde.

Wir wenden uns den Steininschriften zu, zu deren Erforschung Frl. Jeffery selber so viel beigetragen hat.

Da haben wir am Anfang ein kleines Bruchstück mit einer frühen metrischen Inschrift (Nr. 2 = IG^2 1,484), mit der man nichts anzufangen weiß und die nicht einmal attisch zu sein braucht. Dasselbe gilt von der Inschrift auf naxischem Marmor (Nr. 7 = IG^2 1,672), die man eher mit inselionischen Denkmälern vergleichen sollte. So kommt man zu Nr. 8 (= IG^2 1,997) und ist erstaunt, diesen unebenen und schlecht behauenen, aber wohl beschrifteten Stein als Teil einer Reliefstele des siebenten Jahrhunderts bezeichnet zu finden. Und warum soll denn die einfache Inschrift Κεραμὸς στέλε 'Ενιάλο θυγατρ[ὸς τô - - -]ο ein Hexameter sein? Ich glaube, daß die Inschrift wohl später sein kann und auf der Basis einer Porosstele stand. Dieses Denkmal gehört zusammen mit einigen anderen (Nr. 12. 13. 15. 17. 21) zu einer Gruppe, die durch die mit Nr. 15 und 17 verbundenen Skulpturen und durch die sichere Zuweisung von Nr. 18 als Gründungsweihung der Panathenäen vom Jahre 566 v. Chr. ins zweite Viertel des sechsten Jahrhunderts gehört. Dieser Gruppe weist die Verf. auch die attische Inschrift auf der Stele von Sigeion (72) zu, man ist jedoch enttäuscht, von dieser im Britischen Museum befindlichen Inschrift nur eine Kopie einer alten Zeichnung auf Tafel 71 zu finden.

Frl. Jeffery teilt die Denkmäler in Gruppen, deren jede ungefähr 25 Jahre umfaßt, und sie gesteht, daß diese Einteilung "etwas willkürlich" ist; sie hätte sie auch als ungenau bezeichnen können. Wäre es da nicht besser gewesen anzunehmen, daß sich im großen und ganzen der Inschriftstil wie der Stil der Kunst und der Dichtkunst nach Generationen scheiden läßt, und daß sich dieser Stil von Generation zu Generation mit dem allgemeinen Geschmack ändert? So gehören die obenerwähnten Inschriften mit anderen zu einer Frühgruppe, die der Solonischen Zeit (bis ca. 560 v. Chr.) entspricht. Dann kommen die Denkmäler der Tyrannenzeit, ob sie nun von Freunden oder Fienden des Peisistratos und seiner Söhne errichtet wurden. Hierher gehören die ganz vorzüglichen Grabinschriften des Tettichos (Nr. 19), des Chairedemos (Nr. 20) und ein anderes auch von Phaidimos gefertigtes Grabmal (Nr. 23), wie auch eine Reihe von anderen Inschriften (Nr. 25. 29. 30. 31. 32. 35). Schleißlich berücksichtigt die Verf. noch eine dritte | Gruppe, die die späte Tyrannenzeit und die ersten Jahrzehnte der Demokratie umfaßt und daher nicht völlig von ihr beschrieben werden kann, da das Buch für Athen plötzlich und willkürlich um 500 v. Chr. abbricht.

Von besonderer Bedeutung für diese dritte Gruppe sind die Weihung des Nearchos (S. 75 erwähnt, aber nicht in den Katalog aufgenommen), der Altar des jüngeren Peisistratos (Nr. 37, dessen Datierung selber umstritten ist), die öffentliche Weihung nach dem Sieg über Böoter und Chalkider (Nr. 43) und die schon ins fünfte Jahrhundert gehörende für Kallimachos (IG^2 1, 609). Hier kann es keinem Zweifel unterliegen, daß wir nicht ein Bild, sondern zwei Strömungen wahrnehmen können. Die eine weist in die archaische Zeit zurück (IG^2 1, 609), die andere nimmt den Stil der Frühklassik vorweg (Nr. 37), genau wie die Skulptur und Vasenmalerei dieser Periode.

Die vorstehenden Bemerkungen sind kritisch, aber nicht kritisierend. Wenn man die Denkmäler des archaischen Athen liebt, sieht man sie nicht gerne seziert und katalogisiert, und ich weiß, daß Frl. Jeffery auch so fühlt. Aber das Werk mußte getan werden, nicht nur für Attika, dessen Inschriften ja wohl bekannt sind, sondern für alle Landschaften, und das Einzeldenkmal wurde da gelegentlich etwas stiefmütterlich behandelt. Der Gesamteindruck ist jedoch monumental und sehr befriedigend. Einzeldenkmäler mögen ein bißchen horizontal oder vertikal verschoben, Neufunde mögen eingefügt, Verbesserungen hier und da gemacht werden, aber der große Bau wird für lange dastehen und uns alle beherbergen.

25

Leagros

I. VICTOR STATUES OF THE FIFTH CENTURY IN ATHENS [155]

Among the Attic votive offerings of the sixth and fifth centuries there are dedications made by victorious athletes. The sixth century is represented by the capital of a Doric column which supported a bronze bowl, by fragmentary inscriptions on the lips of five bronze bowls, by an inscribed bronze discus, and by a stone jumping-weight with a dedicatory inscription.[1] The inscription from Eleusis, *I.G.*, I², 803, engraved on the lip of a marble basin is wrongly restored as referring to a victory in a gymnastic contest.[2] We have no certain evidence of statues of athletes set up in that period on the Akropolis. It may be mentioned, however, that some of the statues of horses and horsemen and their inscribed bases, found on the Akropolis, could be understood as dedications of victors in the horse race. We may even suggest that the statue of Rhombos, the moschophoros, was a dedication of a victor who won an ox.[3] Yet there is no certain evidence for these assumptions.

There are several instances which show clearly that in the first half of the fifth century the dedications of victor statues became popular in Athens. We recognize here the increasing Peloponnesian influence and we find just at the beginning of this period, about 500 B.C., the first examples of the use of the

Hesperia 8 (1939) 155–164

1. Doric capital: *I.G.*, I², 472; *S.E.G.*, 1, no. 8; W. Kroll, *R.E., Suppl.* 4, col. 16, 38ff. Another bronze bowl was supported by the base with the inscription *I.G.*, I², 464, but the preserved fragments of this inscription, coming from two different faces of the base, do not indicate that the dedication was made by an athlete after his victory. Bronze bowls: *I.G.*, I², 401–404, 406. Bronze discus: *I.G.*, I², 445; J. Jüthner, *Jahreshefte*, 29, 1935, pp. 32ff. Jumping-weight: *I.G.*, I², 802; Hampe and Jantzen, *Jahrbuch*, 52, 1937, *Olympiabericht*, 1, pp. 82ff. For the whole group compare W. H. D. Rouse, *Greek Votive Offerings*, pp. 149ff.
2. It can rather be restored to [--- εὐχ]όμενος παι[δον?---]ἀνέθεκ[εν---]; cf. *Jahreshefte*, 31, 1938, Beiblatt, col. 51.
3. Payne and Young, *Archaic Marble Sculpture*, pls. 2ff.; W. H. D. Rouse, *Greek Votive Offerings*, p. 151, note 11; A. Mommsen, *Feste der Stadt Athen*, pp. 98f.; *I.G.*, II², 2311, 72ff.

Peloponnesian bronze technique in Athens. At the same time we find Peloponnesian sculptors in Athens (this evidence is taken from the preserved artists' signatures), and the development of the severe style in Attic art is partly due to the same influence.[4] We may suggest that the statue of Hipparchos, son of Charmos, was a victor statue. This statue was melted down when Hipparchos was ostracised in 487/6 B.C. The στήλη ἀτιμίας was set up not before that year.[5]

[156] The following list contains inscriptions which presumably | belong to dedications made by victors in gymnastic contests (except for nos. 2 and 8);[6] the dedications of victors in musical contests are here omitted.

1. Column with the dedicatory inscription of Kallias, son of Didymias, erected after a victory he won as a boy in the Great Panathenaia soon after or soon before 480 B.C.[7] It is a tempting assumption that his column bore one of the preserved marble statues, the so-called Kritios boy, which belongs to the same time and represents a young athlete. The Kritios boy, however, is not the only Kouros found on the Akropolis and we may assume that some of the other Kouroi also belonged to dedications made by victorious athletes (cf. Payne and Young, *Archaic Marble Sculpture*, pls. 97ff.). For the political activity of Kallias, see no. 5.

2. Bronze statue of Kallias, son of Hipponikos, the base of which is preserved with the inscription *I.G.*, I², 607. This base has recently again been identified as belonging to the statue of Aphrodite dedicated by Kallias and made by Kalamis.[8] We accept, however, the objections made by F. Studniczka.[9] Furthermore we may suggest that the dedication was made by Kallias after his victories in the horse race in Olympia (Schol. Aristophanes, *Clouds*, 64). We know of several other victors in horse races who dedicated mere statues of themselves (e.g., Pausanias, 6, 1, 6f.). If the monument could be dated prior to 480 B.C., we would understand why its existence is not recorded in our literary tradition.

3. Bronze statue of Epicharinos as a hoplitodromos, a work of the artists Kritios and Nesiotes (perhaps mentioned in *I.G.*, II², 1500, 12). Though the inscription is mutilated, we may assume that it did not contain any reference to the fact that it was after a victory when Epicharinos dedicated the monument.

---

4. Cf. C. A. Robinson, Jr., *A.J.A.*, 52, 1938, p. 455; S. Lauffer, *Ath. Mitt.*, 62, 1937, pp. 84f.
5. Cf. Lykourgos, *In Leocratem*, 117; H. Friedel, *Der Tyrannenmord*, pp. 39f.; F. Schachermeyr, *R.E.*, 19, col. 155, 32ff.
6. Cf. C. Wachsmuth, *Die Stadt Athen*, I, p. 603, note 5; A. Furtwängler, *Ath. Mitt.*, 5, 1880, pp. 26ff.; W. W. Hyde, *Olympic Victor Monuments*, pp. 26f.; H. Friedel, *Der Tyrannenmord*, p. 36 (add here the reference to Pliny, *Nat. Hist.*, 34, 17).
7. *I.G.*, I², 608 + 714; *Jahreshefte*, 31, 1938, Beiblatt, cols. 47ff.
8. W. Judeich, *Topographie²*, p. 74, note 8; G. P. Stevens, *Hesperia*, 5, 1936, pp. 451ff. See also Harrison and Verrall, *Mythology and Monuments*, p. 386; C. H. Weller, *Athens and Its Monuments*, pp. 254f.; J. Six, *Jahrbuch*, 30, 1915, p. 77 and Fig. 3; K. Freemann, *Greece and Rome*, VIII, 1938, pp. 22ff.
9. *Abh. Sächs. Ges. Wiss.*, 25, 1907, Abh. 4, pp. 54ff., including Fig. 11; G. Lippold, *R.E.*, 10, col. 1534, 4ff.; H. Swoboda, *R.E.*, 10, coll. 1615ff.

A preserved bronze statuette gives us an impression of the statue of Kritios and Nesiotes.[10]

4. Marble statue of Phayllos, from Kroton, who joined the Athenian navy in the battle of Salamis and who was ὀλυμπιονίκης (?) and three times victor in the | Isthmian games.[11] The cutting on the top of the base, mentioned by H. G. Lolling, *Catalogue*, no. 212, is situated on the left end of the preserved upper surface, indicating that our fragment belongs to the right half of the base. Thus the restoration of the epigram (see note 9) is incorrect since it assumes the lack of only two letters on the left side. For marble statues of victors, see W. W. Hyde, *Olympic Victor Monuments*, pp. 324ff. Since Phayllos's name occurs long before 480 B.C. as a love-name on Attic vases (cf. R. Lullies, *R.E.*, 19, col. 1904, 50ff.), and as his dedication can be dated about 470 B.C., he must have spent a good deal of his life in Athens. The monument on the Akropolis was a private dedication made by Phayllos himself. The crew of the ship he lead in the battle of Salamis consisted of citizens of Kroton who resided at that time in Greece (Pausanias, 10, 9, 2). Thus we may doubt whether the public dedication of Kroton in Delphi was erected in honor of Phayllos.[12]

[157]

5. Statue of Kallias, son of Didymias, the famous pancratiast. The inscription, published as *I.G.*, I², 606, records his many victories, among them the one in the Panathenaic games for which he dedicated the statue mentioned above (No. 1). This second dedication can be dated about 445 B.C. and after this date he may have been ostracised.[13] We may suggest that his ostracism took place in connection with Perikles' successful fight against the opposition in 443 B.C.

6. Statue of a victor signed by the painter and sculptor Mikon. The inscription, published as *I.G.*, I², 534, can be dated, in spite of the Ionic alphabet, prior to 440 B.C. We have, however, no certain evidence that this dedication was made by a victor. We know that Mikon was famous for his victor statues, but the dedicatory inscription has not yet been restored in a satisfactory way (cf. G. Lippold, *R.E.*, 15, cols. 1557f.).

7. Statue of the son of Kallaischros who was twice victor in the Isthmian and Nemean games, *I.G.*, I², 829. While the other monuments, listed above, are dedications made to the goddess Athena and were found on the Akropolis, this inscription is a dedication to the Twelve Gods and was found in Salamis.[14]

---

10. *I.G.*, I², 531; Harrison and Verrall, *Mythology and Monuments*, pp. 406f.; C. H. Weller, *Athens and Its Monuments*, p. 262, Fig. 160; G. Lippold, *R.E.*, 11, col. 1915, 35ff.; A. Schober, *Jahreshefte*, 30, 1937, Beiblatt, cols. 215ff. The date proposed by E. Löwy, *Sitzunsber. Ak. Wien*, 216, Abh. 4, 1937, pp. 8f., is too late.
11. *I.G.*, I², 655; M. N. Tod, *Greek Hist. Inscr.*, p. 26, no. 21; a new restoration was proposed by F. Hiller and reported by H. Pomtow, *R.E.*, Suppl., 4, col. 1204, 52ff.; Robinson and Fluck, *Greek Love-Names*, pp. 167ff.; W. W. Hyde, *A.J.P.*, 59, 1938, pp. 407f.; H. E. Stier, *R.E.*, 19, coll. 1903f.
12. Cf. H. Pomtow, *R.E.*, Suppl., 4, coll. 1204f.; H. E. Stier, *R.E.*, 19, col. 1903, 38ff.
13. Cf. *Jahreshefte*, 31, 1938, Beiblatt, cols. 48f.; T. L. Shear, *Hesperia*, 7, 1938, p. 361.
14. Cf. L. Bürchner, *R.E.*, 1 A, col. 1832, 30ff.; S. Solders, *Die aussersstädtischen Kulte*, p. 71; O. Weinreich in Roscher, *Myth. Lex.*, 6, col. 781, no. 18.

Though we have another inscription, found in Salamis, where the Twelve Gods are mentioned (*C.I.G.*, I, 452; other references in note 4), we may doubt whether this inscription does not belong to the sanctuary of these deities in the Agora in Athens. If the reading τοι δοδεκαθεοι is correct we may remember that the sanctuaries of the Twelve | Gods in Kos and Crete were called δωδεκάθεον.[15] The assumption that *I.G.*, I², 829 belongs to Athens, made already by F. Hiller, does not clash with the fact that an inscription containing a list of prizes given in a gymnastic contest, *I.G.*, I², 846, was also found in Salamis. We may even suggest that this inscription, too, belongs to Athens and to the Panathenaic games.

The dedication of the son of Kallaischros was set up in the second half of the fifth century. Though the name Kallaischros occurs in several instances within the period, we may suggest that it was the famous Kritias, son of Kallaischros, who made the dedication to the Twelve Gods. His statue may have been removed after 403 B.C., and that could explain the fact that there were no victor statues in Lykourgos's time in the Agora of Athens (*In Leocratem*, 51).

8. Bronze four-horse chariot dedicated by Pronapes, son of Pronapides, after more than three victories, among them one each in the Nemean, Isthmian, and Panathenaic games.[16] The letter forms of the inscription illustrated by N. Kyparissis, Δελτ. 'Αρχ., 11, 1927/28, p. 133, no. 8, fig. 6, indicate that the monument belongs to the fifth century. The two preserved names are the first part of a pentameter, since there is uninscribed space on the left side of the inscription, and we may assume that the dedicatory inscription consisted only of a pentameter (cf. *I.G.*, I², 661 and p. 205; Aristotle, 'Αθ. Πολ., 7, 4). The preserved slab has anathyrosis on the back and on the right lateral face. The anathyrosis on the right lateral face and the four dowel holes on the top can be seen on two photographs which show the slab still built in the north part of the west door of the Parthenon (cf. N. Balanos, *Les Monuments de l'Acropole*, pls. 140a and 141b; the piece in question is the fourth large slab from the top). The four dowel holes on the top belong to the front hoofs of two bronze horses. Thus we may assume that the whole base consisted of six slabs, two on the smaller front face, three on the lateral faces. The left front slab is preserved. A similar base, consisting of six slabs, bore the δούριος ἵππος, dedicated by Chairedemos, son of Euangelos (cf. G. P. Stevens, *Hesperia*, 5, 1936, pp. 460f., fig. 14). The front length of our base can be restored to three meters. Thus the size of the chariot was the same as that of the chariot dedicated by the Athenians after their victory over the Boeotians and Chalkidians at the end of the sixth century and renewed in 446 B.C.[17] | For the name Pronapes, N. Kyparissis has

---

15. Paton and Hicks, *Inscriptions of Cos*, no. 43; A. Maiuri, *Nuova Silloge Epigrafica di Rodi e Cos*, p. 141; M. Guarducci, *Inscriptiones Creticae*, I, pp. 69ff., no. 13.

16. *I.G.*, II², 3123. This monument is included in our list because of its historical significance and because it is the only chariot, so far known, which was dedicated by a victor and set up on the Akropolis.

17. To the inscription, *I.G.*, I², 394 I, belongs the fragment E.M. 12410. The site where this monument was located has been determined by L. B. Holland, *A.J.A.*, 28, 1924, pp. 77 and

already referred to *I.G.*, I^2, 400 Ia, a public dedication made by the Knights when Lakedaimonios, Xenophon, and Pronapes were their generals. The preserved base bore the statues of a horse and of a man standing beside the horse (cf. C. Anti, *Bulletino Communale*, 47, 1921, pp. 90ff.). The monument was set up presumably in 446 B.C., and if it was ever moved and put on one of the west antae of the Propylaea, that evidently did not happen in the fifth century since the antae were not built to support statues.[18] The general of the Knights in 446 B.C., Pronapes, may be identical with the dedicant Pronapes, son of Pronapides. The name is rare and both were Knights. *I.G.*, I^2, 400 Ia shows the Ionic eta which also occurs in *I.G.*, II^2, 3123; for the earlier letter forms of *I.G.*, I^2, 400 Ia we may compare the Koronea epigrams.[19] Furthermore it was perhaps the same Pronapes, son of Pronapides, successful general and victor in several chariot races, who was between 468 and 466 B.C. one of the accusers of Themistokles. He is mentioned together with Leobotes and Lysandros in the eighth letter of Themistokles (G. Niessing, *De Themistoclis epistolis*, pp. 40ff.), and we may assume that the author of this letter had his knowledge from Krateros (U. Kahrstedt, *R.E.*, 5 A, col. 1694, 11ff.). The full name was Pronapes, son of Pronapides, Prasieus, and this name as well as all the other evidence indicates that he belonged to the aristocracy (cf. K. Mras, *Wiener Studien*, 55, 1937, pp. 78ff.). He was born before 490 B.C., and his first political activity may have been the accusation of Themistokles. He was Knight and victor in several chariot races. In 446 B.C. he served as general of the Knights together with Lakedaimonios, son of Kimon, and his dedication on the Akropolis may be dated about 435 B.C. We know of still another bearer of the name Pronapes in the fifth century (J. Kirchner, *P.A.*, 12251), and it may be suggested that our Pronapes was identical with the father of Amynias whom we know from allusions in comedy.[20] These passages together with the scholia give us a picture of the personality of Amynias which, though not favorable, agrees with the impression we obtain from the life of Pronapes, provided we realize that his son was an impoverished aristocrat still proud of his noble birth. Of this kind of nobleman Aristophanes made more fun than of any class in Athens, save the radical democrats.[21] Since Amynias was appointed delegate to Thessaly he must have had some connection with Kleon and his party (E. Meyer, *Gesch. d. Alterthums*, 4, pp. 366ff.). His | mission to Pharsalos was a   *[160]*

402, and L. Weber, *Phil. Woch.*, 53, 1933, cols. 331f. The propylon itself was reconstructed in drawings by G. P. Stevens, *Hesperia*, 5, 1936, pp. 474ff. (cf. R. Stillwell, *A.J.A.*, 42, 1938, pp. 432f.); Stevens also discussed the chariot problem on pp. 504ff. of the same paper.

18. Cf. W. Judeich, *Topographie*^{2}, pp. 80, 229; R. P. Austin, *The Stoichedon Style*, p. 64; P. Graindor, *Athènes de Tibère à Trajan*, pp. 5ff.; G. W. Elderkin, *Problems in Periclean Buildings*, pp. 10f.; G. Lippold, *R.E.*, 13, col. 2293, 35ff.; K. I. Gelzer, *Die Schrift vom Staate der Athener*, pp. 103f.

19. Kyparissis and Peek, *Ath. Mitt.*, 57, 1932, pp. 142ff.; K. Reinhardt, *Hermes*, 73, 1938, pp. 234ff.

20. J. Kirchner, *P.A.*, 737 and Aristophanes, *Knights*, 570. I consulted the commentaries to each passage in the editions of T. Mitchell, W. Ribbeck, F. H. M. Blaydes, T. van Leeuwen, W. T. M. Starkie, Hall and Geldart, B. B. Rogers.

21. Cf. A. W. Gomme, *Cl. Rev.*, 52, 1938, pp. 97ff.

failure and we may suppose that it belongs to the year 424 B.C. when the Penestai in Thessaly were unable to stop Brasidas on his way to Thrace.[22] The scholiast to *Clouds*, 31, asserts that his proper name was not Amynias but Ameinias, and Aristophanes changed the name because of a law forbidding mockery of the archon in comedy. Whether or not we believe in the existence of this law, the identity of Amynias, son of Pronapes, with the archon of 423/2 B.C. has so far not been accepted.[23] Yet to suppose that the scholiast invented the whole story is not likely and it would fit very well into the career of Amynias.[24] We may add that the pronunciations of ει and υ in the later fifth century in Athens were much alike.[25]

II. LEAGROS

We should bear in mind these examples of dedications made by victorious athletes in Athens in approaching the proper subject of this paper, the dedication of Leagros, son of Glaukon, to the Twelve Gods set up outside of their temenos in the Agora of Athens.[26]

The question what kind of statue was dedicated by Leagros has not yet been discussed. We have still another dedication to the Twelve Gods (no. 7 in our list) which was made by a victorious athlete, perhaps by the aristocrat Kallias, son of Kallaischros. We have already discussed the possibility that, though found in Salamis, it was once part of a monument which stood in the Agora of Athens. We obtain, in any case, from this inscription evidence that victor statues were dedicated to the Twelve Gods. It may be suggested that another dedication to the Twelve Gods, *I.G.*, II², 4564, was set up by a victor as indicated by the crown which was engraved | below the inscription.[27] The

---

22. Cf. F. Hiller, *R.E.*, 6 A, col. 121, 20ff.; F. Miltner, *R.E.*, 19, cols. 494f. Add here Aristophanes, *Wasps*, 1271ff. and the scholion to this passage.

23. Cf. E. Kalinka, *Die pseudoxenophontische 'Αθηναίων πολιτεία*, p. 14; K. I. Gelzer, *Die Schrift vom Staate der Athener*, pp. 71 and 128ff.

24. J. Beloch, *Gr. Gesch.*, II², 2, p. 390, in his archon list accepted Amynias as the archon eponymous of the year 423/2 B.C.

25. Cf. E. H. Sturtevant, *The Pronunciation of Greek and Latin*, pp. 124 and 135; Robinson and Fluck, *Greek Love-Names*, p. 172f.

26. T. L. Shear, *Hesperia*, 4, 1935, pp. 355ff.; B. D. Meritt, *Hesperia*, V, 1936, pp. 358f., no. 2. I am indebted to Professor T. L. Shear for giving me the opportunity to use the part of the excavator's diary which concerns the sanctuary of the Twelve Gods.
In the following reports or articles the Leagros base or the temenos of the Twelve Gods is mentioned or discussed: H. G. G. Payne, *J.H.S.*, 54, 1934, pp. 185f.; G. Karo, *Arch. Anz.*, 49, 1934, col. 128, Fig. 3; T. L. Shear, *A.J.A.*, 39, 1935, p. 177; P. Lemerle, *B.C.H.*, 59, 1935, p. 249; J. F. Crome, *Ath. Mitt.*, 60/61, 1935/36, pp. 306f.; F. Schachermeyr, *R.E.*, 19, col. 188, 44ff.; J. Kirchner, *R.E.*, 19, col. 191, 35ff.; O. Weinreich in Roscher, *Myth. Lex.*, VI, cols. 772ff., 780, no. 16; M. N. Tod, *J.H.S.*, 57, 1937, p. 169; E. Löwy, *Sitzungsber. Ak. Wien*, 216, Abh. 4, 1937, p. 9; W. Dörpfeld, *Alt-Athen*, 1, pp. 67ff., 131; Ch. Picard, *Rev. Arch.*, 9, 1937, pp. 283f.; O. Kern, *Religion der Griechen*, 2, p. 93; 3, p. 321; Robinson and Fluck, *Greek Love-Names*, pp. 132dd.; T. L. Shear, *C.W.*, 31, 1938, p. 76; Ch. Picard, *R.E.G.*, 51, 1938, p. 94; A. Raubitschek, *Jahreshefte*, 31, 1938, Beiblatt, col. 44; Flacelière, J. Robert, and L. Robert, *R.E.G.*, 51, 1938, p. 421, no. 52; B. D. Meritt, *Hesperia*, 8, 1939, p. 64, no. 21.

27. The date given for this inscription by O. Weinreich in Roscher, *Myth. Lex.*, VI, col. 845, 60ff., is too late. Thus, the reference he gives for his interpretation of Tyche as the Thirteenth God

assumption, however, that this base bore a victor statue would clash with the assertion of Lykourgos (*In Leocratem*, 51) that no victor statues stand in the Agora of Athens. Furthermore, the significance which the cult of the Twelve Gods obtained in Olympia (O. Weinreich in Roscher, *Myth. Lex.*, VI, cols. 781ff.) shows clearly enough that these deities were connected with the gymnastic contests. N. Gardiner, *Greek Athletic Sports*, p. 74, referred to a passage in Herodotos (II, 7) recording the distance from the altar of the Twelve Gods in Athens to Olympia; there is, however, an inscription preserved, *I.G.*, II^2, 2640, which indicates the distance from the altar of the deities in Athens to the harbor.[28]

It seems, therefore, safe to suggest that Leagros's dedication to the Twelve Gods was a victor statue set up after a victory he won in one of the panhellenic games, perhaps in the Panathenaia. The inscription itself does not indicate that it belongs to a victor monument, but we find the same omission in nos. 2 and 3 of our list and in K. Purgold, *Inschriften von Olympia*, no. 143. We may assume either that such omission was customary in this early time[29] or, as is probable, that the statue itself sufficiently indicated the purpose of the dedication. This interpretation is certainly right for the statue of Epicharinos and for the chariot set up by Gelon, and it can also be applied, as we shall see, to the statue of Leagros.

A further evidence for this interpretation may be found in a vase painting (Plate 9) which shows the dedication made by Leagros and contains the love-name Λέαγρος καλός.[30] The platform upon which the young athlete stands cannot be understood as a βῆμα, though there was no difference between the βῆμα and the statue base either in vase painting or in reality. A βῆμα is, however, no place for an athlete to stand. But obviously appropriate for an athlete are the ἀγάλματ' ἐπ' αὐτᾶς βαθμίδος ἑσταότ'.[31] Therefore, the vase painting may be interpreted as representing the statue of a young victor, standing on a rectangular base, with a laurel crown on his head. His right foot is advanced; in his right hand he holds a javelin, in his left an aryballos and a sponge or a discus-bag.[32] The similarity between this painting and the statue which must have stood on the Leagros base is obvious if we notice the preserved dowels and dowel holes which indicate the posture of the statue (see the illustration in *Hesperia*, 4, 1935, p. 357, Fig. 14). We may even suggest that the small hole near the left front corner on the top of the base once received the end of the javelin held by the athlete in his right hand.

[163]

---

is not valid. Cf. A. Greifenhagen, *Röm. Mitt.*, 52, 1937, pp. 238ff. Another dedication to the Twelve Gods, *I.G.*, II^2, 2790, was found so far from the Agora that it can be doubted whether it actually belongs to this sanctuary of the deities.

28. Cf. R. L. Scranton, *A.J.A.*, 42, 1938, pp. 529f.

29. It is also significant that none of the inscriptions listed above specifies the dedication as aparche or dekate; cf. Hampe and Jantzen, *Jahrbuch*, 52, 1937, *Olympiabericht*, 1, pp. 79f.

30. *C.V.A.*, Baltimore, fasc. 2, III I, pls. 5 and 6, pp. 13f.

31. Pindar, *Nemean*, 5, 3f.

32. The latter interpretation was given by K. McKnight Elderkin, *Harv. Stud. Cl. Phil.*, 35, 1924, pp. 119ff. Neither interpretation is satisfactory, but I cannot give any other.

The bearded man standing in front of the statue has always been interpreted as a trainer.[33] This interpretation cannot be maintained if the scene is the Agora of Athens and not a gymnasium. The bearded man is crowned with an ivy wreath and this ornament is not in keeping with the interpretation of this figure as an instructor.[34] We have, however, the inscription both words of which begin near the head of the bearded man indicating that they refer to him. Leagros himself is standing before the statue which he dedicated to the Twelve Gods. The javelin in the right hand of the statue indicates that the victory was won in the pentathlon.[35]

If our interpretation of the vase painting on the cup in Baltimore is correct, we may include it in E. Langlotz's list of paintings which show Leagros in different states of his youth.[36] Our vase painting is certainly earlier than the cup in Brussels. The cup in Baltimore was attributed to the Kiss Painter and dated ca. 500 B.C. The date, however, which was proposed for the inscription of the Leagros base is the decade 490–480 B.C.[37] Yet Miss G. M. A. Richter with whom I had the opportunity to discuss the question told me that, "though stylistically the vase has been dated about 500 B.C., an absolute date as late as 490 B.C. would perhaps be possible if we suppose that the painter retained an old-fashioned style."[38] The date of the Leagros base must be discussed together with the date of the earlier structure of the precinct of the temenos since the base was set against the structure's west face at a time when the early precinct was still standing. The date of the earlier building, as I was informed by the excavators, goes back to the sixth century. The second structure was erected when the statue was already removed, but it is dated still in the fifth century. Since the disappearance of the statue can be connected with the devastation of Athens by the Persians we can conclude that only the first structure belongs to the time before 480 B.C. This assumption explains the good preservation of the front face of the base, in comparison with the top which was worn when used as a floor.[39] We also understand why the victory of Leagros, whose descendants were famous in the fifth century, is not recorded in our literary tradition.[40] If the first structure is rightly | assigned to the time before 480 B.C., it must have been a part of the parapet built around the altar dedicated by

[164]

---

33. See, however, the doubts expressed by P. Hartwig, *Meisterschalen*, p. 43.
34. For instructors with wreaths, see F. Hauser, *Jahrbuch*, 10, 1895, p. 110; Robinson and Fluck, *Greek Love-Names*, pp. 169f.
35. Cf. W. W. Hyde, *Olympic Victor Monuments*, pp. 164f.; J. Jüthner, *R.E.*, 19, cols. 524ff.
36. *Zur Zeitbestimmung*, pp. 48ff.; Robinson and Fluck, *Greek Love-Names*, pp. 132ff.; see the additions made by A. Rumpf, *Gnomon*, 14, 1938, p. 456.
37. B. D. Meritt, *Hesperia*, 5, 1936, p. 357: 490–480 B.C.; E. Löwy, *Sitzungsber. Ak. Wien*, 216, Abh. 4, 1937, p. 9: not long before the death of Leagros in 464 B.C.; Robinson and Fluck, *Greek Love-Names*, p. 134: about 485 B.C. It may be noted that the theta with a circle in the middle does not occur in the Hecatompedon and Marathon inscriptions as Robinson and Fluck assert. The Leagros inscription is not metrical, not even a bad hexameter, as the same authors suggest.
38. A similar answer was given me by Miss D. K. Hill.
39. Notice here the horizontal striation of the right and left vertical margins which occurs also on the base with the Marathon epigram; cf. J. H. Oliver, *Hesperia*, 2, 1933, no. 11.
40. Cf. B. D. Meritt, *Hesperia*, 8, 1939, no. 15.

Peisistratos, son of Hippias, when he was archon eponymous. His archonship has generally been dated in some year immediately before 512/1,[41] but new evidence suggests a date as late as 497/6 B.C. (B. D. Meritt, *Hesperia*, 8, 1939, no. 21). There may have been, however, an earlier altar of the Twelve Gods on the same place, though the preserved structure belongs to the building of Peisistratos.[42] Shortly after the building of the temenos of the Twelve Gods, Leagros must have made his dedication. We do not know whether there existed any connection between Leagros and the family of the tyrants, but, if so, it certainly was not close. We do not know anything of Leagros's father, Glaukon. He was certainly not the Glaukon whom J. Kirchner, *R.E.*, 7, col. 1402, 7ff., put in the first place in his list of the bearers of this name.[43] It is, however, remarkable that Leagros was elected general immediately after the trial of Themistokles which happened between 468 and 466 B.C. when the aristocratic influence again increased in Athens. On the other hand the author of the eighth letter of Themistokles assumes that Themistokles and Leagros were friends.

Soon after Leagros's statue was dedicated it was illustrated on the cup now in Baltimore. The vase painting belongs, therefore, to the class of vases with representations of statues (K. Schefold, *Jahrbuch*, 52, 1937, pp. 30ff.).

Our knowledge of Leagros's life is increased. The celebration of his youth on the vase paintings of the late sixth century will now be understood as a result of his activity as an athlete, and the fact that he was allowed to set up his dedication by the altar of the Twelve Gods shows his political significance which led to his στρατηγία in 465/4 B.C.

Indeed almost all of the victor monuments which we have listed above were dedicated by men who took an active part in political life. Now we understand the passage in Lykourgos (*In Leocratem*, 51) which states that, in contrast to other Greek cities, there were no victor statues in the Agora of Athens.[44] Thus the statue of the pancratiast Autolykos, set up near the prytaneion,[45] is said to have been erected long after his death in commemoration of his political merits (cf. G. Lippold, *R.E.*, 12, cols. 1994f.). We may understand all these victor statues as the forerunners of the honorary statues of the fourth century.[46]

41. An even earlier date was proposed by J. J. E. Hondius, *Hermes*, 57, 1922, pp. 475ff.: ca. 525 B.C.; F. Cornelius, *Die Tyrannis in Athen*, p. 10, note 1: before 514 B.C.; O. Kern, *Die Religion der Griechen*, 3, p. 321: 523/2 B.C., or somewhat later.
42. Cf. Herodotos, 6, 108, 2; and the discussions of W. Aly, *Klio*, 11, 1911, p. 21; W. Judeich, *Topographie*², pp. 64, 350, note 3; J. F. Crome, *Ath. Mitt.*, 60/61, 1935/6, p. 306.
43. Cf. W. S. Ferguson, *Hellenistic Athens*, pp. 201 and 212; J. Beloch, *Griech. Gesch.*, IV², 2, p. 457; F. Jacoby, *Frag. Gr. Hist.*, 2, no. 80. Pythermos, no. 2.
44. Misunderstood by W. W. Hyde, *Olympic Victor Monuments*, p. 26.
45. A building which was far from the Greek Agora; cf. E. Vanderpool, *Hesperia*, 4, 1935, p. 470, note 4; W. Dörpfeld, *Alt-Athen*, 1, p. 36.
46. Cf. W. H. D. Rouse, *Greek Votive Offerings*, p. 269.

26

Two Monuments Erected after the Victory of Marathon

[53] Before the battle of Marathon the Polemarch Callimachus of Aphidnae promised in the Demos's name to sacrifice to Artemis Agrotera, if victory should fall to Athens, as many goats as there were Persians left dead upon the field.[1] After the battle in which Callimachus himself died, the Athenians kept his promise and sacrificed every year 500 goats, for they had not goats enough for a single sacrifice of 6,400—the number of Persians who remained on the battlefield. Another vow which Callimachus made in his own name we know only from the preserved inscription on the monument erected on the Acropolis after his death:[2]

[Καλίμαχος μ' ἀν]έθεκεν 'Αφιδναῖο[ς] τἀθεναίαι·
ἄν[γελον ἀθ]ανάτον, hοὶ ο[ὐρανὸν εὐρὺν] ἔχοσιν· |
[hὸς στέσας πολέ]μαρχο[ς] 'Αθεναίον τὸν ἀγο͂να·
τὸν Μέ[δον τε καὶ h]ελένον θά[νε δόλιον ἐμαρ·]
παισίν 'Αθεναίον Μα[ραθο͂νος ἐν ἄλσει ἀμύνον].

It has been customary to suppose that the column with the dedication of Callimachus was crowned by a statue of Hermes,[3] since the indication ἄγγελος ἀθανάτων seemed to be suitable only for this god. Yet it is hard to understand why a general should promise before a battle to dedicate to the goddess Athena a statue of Hermes. It would obviously be more likely for Callimachus to promise to dedicate, in case of a victory, a statue of Nike, and, though we have no reference to Nike as ἄγγελος ἀθανάτων, it is a phrase which accurately

American Journal of Archaeology 44 (1940) 53–59

1. H. Berve, *Miltiades*, p. 86, note 1, doubts without reason the credibility of this report (Xenophon, *Anabasis* 3, 2, 12) without referring to the dedication of Callimachus, *IG.* i², 609.
2. *IG.* i², 609; Ad. Wilhelm, *Anz. Akad. Wien* 1934, p. 111; J. Kirchner, *Imagines*, no. 17; R. P. Austin, *The Stoichedon Style*, p. 6, note; cf. *AJA.* 43, 1939, p. 711.
3. E. Löwy, *Sb. Akad. Wien* 216, Abh. 4, 1937, p. 3, even considered this monument a dedication to Hermes.

Figure 26.1. Reconstruction of the dedication erected in honor of Callimachus.

describes her many well-known functions as "Botin der Götter."[4]

Among the archaic Nike statues of the Acropolis there is one, no. 690,[5] which was long ago connected with the "Persian horseman," no. 606, and considered a dedication for the victory of Marathon; it was even supposed to have stood upon a column.[6] Yet for these proposals there has been lacking the definite proof which is now at hand. The plinth of this statue fits into a rectangular cutting in the abacus of an Ionic capital which I was able to join together from several fragments.[7] The diameter of the column crowned by this capital can be determined from the preserved underside of the capital itself, and this diameter corresponds to the diameter of the column with the inscription of Callimachus. It can, therefore, be assumed that Nike 690 belongs to the dedication of Callimachus. The total height of the monument was | more than 12 feet; its reconstruction in plaster has not yet been undertaken, but the drawing in Figure 26.1 may give some impression of the dedication.[8]

[55]

The date given for the Nike statue according to its style was the end of the archaic period, and this dating is now confirmed by the statue's connection with the Callimachus inscription recording a dedication made presumably in 489 B.C.

The Nike of Callimachus is very closely related to the Kore dedicated by Euthydikos. The strands of hair across the forehead of the Euthydikos Kore and the strands at the back of the head of the Nike are rendered in the same technique, deeply cut, precise, metallic.[9] Similar, too, are the necks of the two statues, their plump shoulders and their breasts, soft, round, and full; the himation-folds across the right arm, the heavy folds below the breast, the small folds of the skirts tightly drawn across the legs,[10] and the vigorously modeled knees—all are similarly executed in Nike and in Kore. The similarities are such that we can conclude that both Nike and Kore are derived from the same workshop, perhaps even the work of the same sculptor.[11]

The ends of the Kore's himation must be restored in a manner similar to that in which they are preserved on the Nike. In spite of the many similarities between the two statues there are some differences to be reckoned with; the

---

4. Bernert, *RE.* 17, col. 290, 16; W. D. Rouse, *Greek Votive Offerings*, p. 142.
5. E. Langlotz in H. Schrader, *Die archaischen Marmorbildwerke der Akropolis*, pp. 122ff., no. 77.
6. F. Studniczka, *JdI.* 6, 1891, p. 248; E. Petersen, *AM.* 11, 1891, p. 383.
7. The largest fragment of the capital was published several years ago by H. Möbius, *AM.* 52, 1927, p. 166. There are now four other fragments A.M. 3776, 3820, and θ 312; cf. the report by W. Züchner, *AA.* 1936, p. 329; *Bull. Bulgare* 12, 1938, pp. 168ff.
8. The drawing is illustrated in the report of P. Lemerle, *BCH.* 61, 1937, p. 443. Cf. E. Blegen, *AJA.* 42, 1938, p. 302; G. M. Young, *JHS.* 58, 1938, p. 217; R. Hampe, *Antike* 15, 1939, pp. 168ff.
9. Notice that the Nike's hair is gathered into a knot on her neck; we find the same hairdress on a bronze statuette from the Acropolis (De Ridder, no. 788, Fig. 294) belonging to the third decade of the fifth century.
10. Notice the depression running down the center of the middle folds of the himation; in the small folds of the skirt across the left legs of the two statues notice the ridge lying below the cutting of the fold.
11. Associated with this group is a third statue, Kore no. 628, Payne and Young, *op. cit.*, pl. 96, 4, the relationship of which with the Euthydikos Kore is obvious. The Kore may be dated after 480 B.C.

most striking of these is the arrangement of the folds on the back of the Nike's skirt, which cannot be compared with any part of the Kore's drapery; these folds, rhythmically composed, resemble vase-paintings of Euthymides and show the last affected period of the archaic style. This is an earlier element of style. Contrasted with it is the motion of the Nike, which differs from the motion of the earlier Nike statues not only in its direction to the right instead of to the left, but above all in the departure from the old "Knielaufschema." Although the relationship between Nike and Kore is so close, the former sculpture belongs to the severe style, the latter to the late archaic. Thus, if our assumption is right, that these two statues belong to the same artist, we must recognize this sculptor as one of the founders of the severe style. In his work we can see the characteristics of his great contemporaries, a sturdy conservatism paradoxically mixed with vigorous intimations of the classicism which was to follow. Another sculptor of the period was Euenor, born in Ephesos, who made two Korai in the old style, but later, after 480, the Athena no. 140, the earliest representation of the goddess in the severe style.[12] Most of the famous sculptors on the mainland of Greece in this period worked in bronze and only in the East was the high standard of marble sculpture maintained. May we assume that the artist of Nike 690 came from the East, as did Euenor? If so, we would say he came from Paros. The style of the Kore resembles Eastern sculptures of the beginning of the severe style. As to the inscription of the Callimachus dedication, the shape of the letter phi is the same as that in the Archermos signature.[13] The blossom on the side of the Ionic capital can be compared only with two capitals in Samos.[14] But we do not know the name of the artist who made the Nike and no artist's signature has yet been successfully connected with the Euthydikos base.[15]

[56]

The dedication of the Polemarch Callimachus, erected in his name by the Demos, must have been destroyed in 480. It was not reconstructed after 479. There is no reference in literature of its existence; the recovered and reconstructed monument takes its place beside the record in Xenophon of the vow to Artemis Agrotera.

Of all the monuments erected in Attica after the victory of Marathon it seems that not a single one remained preserved in later antiquity except the stelae upon the tomb in Marathon, of the existence of which we know from Thucydides and Pausanias.[16] But again a recovered monument can give us more definite information concerning a historical incident than the information Herodotus has left us.

Pausanias says that the dead of Marathon were buried not in the public

---

12. H. Schrader, *Arch. Griech. Plastik*, pp. 36, 91; K. Schefold illustrated, *JdI*. 52, 1937, p. 46, Fig. 8, a vase-painting representing this monument; cf. P. Lemerle's report in *BCH*. 61, 1937, p. 434; G. M. Young, *JHS*. 58, 1938, p. 217; Ch. Picard, *REG*. 52, 1939, pp. 115 and 125f.
13. Ad. Wilhelm, *Anz. Akad. Wien* 1934, p. 115; J. Kirchner, *Imagines*, no. 13; *JOAI*. 31, 1938, Beibl., col. 28.
14. Still unpublished.
15. F. Winter, *AM*. 13, 1893, p. 123, note 1.
16. Thuc. 2, 34; Paus. 1, 29, 4 and 32, 3.

cemetery in Athens[17] but on the battlefield itself. It was this information which caused some to misinterpret the fragments of a cenotaph found in Athens,[18] and not until the significance of the technical details of the fragments was considered[19] did it become possible to understand the nature of these fragments.[20] But now it is obvious that we have two pieces of the base on which stood a stele with the names of the 192 men slain in the battle of Marathon. This base had engraved on its front two epigrams which run according to an almost certain restoration:[21]

I. Ἀνδρῶν τῶνδ' ἀρετέ[ς λάμφσει κλέος ἄφθιτον] αἰεί[:]
[εὔτολμοι Πε]ρ[σōν, hοὶ στόρεσαν δύναμιν]·|
ἔσχον γὰρ πεζοί τὲ[ν ἄλκιμον Ἀσίδος hίππο]ν˙
hελλά[δα μ]ὲ πᾶσαν δούλιο[ν ἐμαρ ἰδεν].

II. Ἐν ἄρα τοῖσζ'ἀδά[μαντος ἐνὶ φρεσὶ θυμός], hότ'αἰχμὲν
στέσαμ πρόσθε πυλῶν ἀ[ντία τοχσοφόρον],|
ἀνχίαλομ πρέσαι β[ολευσαμένον δ'ἐσάοσαν]
ἄστυ βίαι Περσōν κλινάμενο[ι στρατιάν].

There is no doubt that the first epigram refers to the battle of Marathon and the historical interpretation is certain. The second epigram was engraved later than the first; | this fact was seen long ago and is not to be doubted. The historical situation referred to in the second epigram has not yet been recognized, mainly because it seemed unreasonable to assume a fight in Phalerum without having any literary tradition of this event.[22]

A new examination of the old fragment *IG*. i², 763 showed that there was in the upper surface of the base a cutting, a part of which is still preserved. This cutting is 0.17 m. from the front edge of the stone, its depth is 0.07 m., and its preserved length 0.125 m. (see Plates 10 and 11). The left corner of this cutting, though damaged, can be clearly recognized, and we see that it must have been only a small cutting, located in the extreme right corner of the front of the base. It is, therefore, obvious that this cutting cannot belong to the first epigram, or to the stele engraved with the names of the men who fell in the battle of Marathon. According to Oliver's restoration of the | monument the

[57]

[58]

---

17. A. S. Arvanitopoulos, Ἐπιγραφική, p. 119, tries without success to prove that there is a reference in Pausanias to the cenotaph in the Kerameikos.
18. E. Löwy, *Sb. Akad. Wien* 216, Abh. 4, 1937, p. 7.
19. Ad. Wilhelm, *Anz. Akad. Wien* 1935, p. 83, points to the fact that we owe this new epigraphical method, Ferguson's "Architectural Epigraphy," to American scholars; cf. *JHS*. 53, 1933, p. 134; S. Lauffer, *AM*. 62, 1937, p. 82.
20. J. H. Oliver, *Hesperia* 5, 1936, pp. 225ff.
21. Ad. Wilhelm, *Anz. Akad. Wien* 1934, p. 94; for full bibliography see M. N. Tod, *JHS*. 57, 1937, p. 175; E. Löwy, *Sb. Akad. Wien* 216, Abh. 4, pp. 6f.; P. Friedländer, *Studi Ital. di Fil.*, N.S. 25, 1938, p. 93, note 2.
22. Ad. Wilhelm, *loc. cit.*, p. 100; A. S. Arvanitopoulos, Ἐπιγραφική, p. 118, proposes that the second epigram refers to all combatants, but since this assumption is certainly wrong, he is not able to use his good idea—mentioned before by P. Maas, *Hermes* 70, 1935, p. 236—that the second epigram speaks about an event which happened in Phalerum.

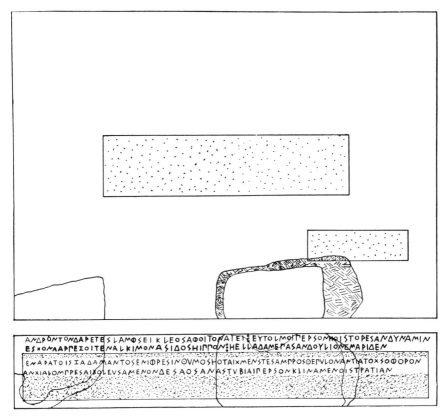

Figure 26.2. Reconstruction of front face and top of Marathon base.

total front length of the base was 1.05 m., but according to Wilhelm's restoration of the inscription the total length was not less than 1.27 m., and this measure is applied in our drawing in Figure 26.2. The great stele had, therefore, a breadth of 0.7 m., the smaller one of 0.3 m. If our drawing is correct it can be seen that the smaller stele was inserted in such a way that it did not conceal the large one. Thus we may assume that it was put in after the large stele was already standing in its place. This second stele can be connected with the second epigram, which was also a later addition.

The first epigram refers to men who had fallen in a battle and so does the second. The connection of the two epigrams and the two stelae on the same base shows that the two events to which they belong must have been connected locally as well as temporally. Herodotus does not mention any battle immediately after Marathon, but says that the Persian fleet sailed around Sunium and tried to land in Phalerum (6, 116). The Greeks, informed of the purpose, and aware of the danger, hurried from Marathon back to Athens and camped in the Heracleum in Kynosarges. This place was outside the town near the famous Gymnasium.[23] They arrived in Phalerum before the Persians did, for if they had stayed at the Heracleum the Persians could have landed without any impediment. After the Greek army arrived in Phalerum the Persians turned their ships and went back to Asia; that is all Herodotus reports. According to this statement it has been supposed that Datis did not dare to make an attack and did not even try to land when he saw the Greeks stationed along the shore. As a matter of fact, literary tradition does not allow further conclusions. The second epigram tells us that the men whose memory it celebrates stood, their spears in their fists, before the gates of their town, on the shore of the sea, in front of the bowmen. They forced the Persian expedition, which wanted to burn their town, to turn back, and saved their homes. Let us combine these facts with the report of Herodotus. The Persians brought their ships close into shore and attempted to land under the cover of their bowmen, but the Greeks stood firm, and the Persians, unable to effect a landing, were compelled to withdraw their fleet. It is probable enough that Persian arrows were responsible for some casualties among the Greeks.

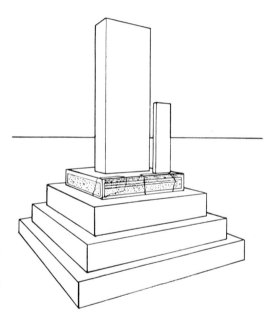

Figure 26.3. Reconstruction of monument with Marathon epigram. (Adapted from *Hesperia* V, 1936, p. 227, Fig. 1.)

After the men who fell in the battle of Marathon were buried in the Soros, the Athenians decided to erect a cenotaph in the Kerameikos. Later they were mindful of the men who fell in Phalerum, engraved their names on a second stele inserted near the other on the same base, and put the second epigram below the first (see Figure 26.3). We have still another fragment of an early

23. There was a main road leading from Athens through Kynosarges to Phalerum, for the Spartan Anchimolius, who died in a battle in Phalerum, was buried on the road to Athens near the Heracleum; cf. W. Judeich, *Topographie*², p. 422.

Attic public grave monument which was found some years ago in the Agora excavations.[24] The second line of this small | fragment contains the name of [59] Darius, indicating that the monument refers to a battle between Athens and Persia which took place before 486 B.C. As we already have the grave monument of the dead of Marathon, we must look for another battle between Athenian and Persian troops in the lifetime of Darius. That happened in 498, when twenty Athenian ships took part in the Ionian revolt (Cf. C. A. Robinson, *AJP.* 60, 1939, pp. 232f.). After the burning of Sardis the united Ionian army retreated toward Ephesus and was here defeated by the Persians.[25] The Athenians under Melanthius were ordered home and left the Ionians. The reason for this sudden departure was the reaction in Athens after the report of the burning of Sardis reached the town. Peisistratus, son of Hippias, was elected Archon eponymous for the following year.[26] Those who died in this battle in Ephesus were buried in the Kerameikos and a fragment of the base of the grave monument is preserved. The first line can be restored according to *IG.* i^2, 945, 1.6[27] in the following way:

[σόματα τόνδ' ἀνδρôν ∪∪ —hυπ]|ϵδέχσατο γα[îα· - - -

The repeated devastation of Athens by the Persians destroyed this monument as well as the monument erected for the dead of Marathon and the one containing the dedication of Callimachus. Neither Herodotus saw them, nor anyone else later. The details of the incident at Phalerum passed into oblivion,[28] and the defeat of the Greek army at Ephesus was even doubted.[29]

24. J. H. Oliver, *Hesperia* 4, 1935, no. 15; cf. *JOAI.* 31, 1938, Beibl., col. 44. The similarity between this inscription, which can be dated in 498 B.C., and the inscription on a fragment of the base which bore the statues of Harmodius and Aristogeiton (B. D. Meritt, *Hesperia* 5, 1936, no. 1) leads us to suggest that the latter fragment belongs to the monument made by Antenor and set up not earlier than 488 B.C. This date had already been proposed by P. Corssen (see the reports in *PW.* 23, 1903, coll. 350f. and *AA.* 1903, p. 41; H. Friedel, *Der Tyrannenmord,* pp. 34ff.), but has not been accepted in recent publications; W. Judeich, *Topographie*^{2}, p. 68, note 8; A. J. B. Wace, *An Approach to Greek Sculpture,* p. 17; B. D. Meritt, *Hesperia* 5, 1936, p. 357. Further evidence which may support Corssen's argument is the great influence which members of the tyrant's family exerted in Athens until the battle of Marathon. It is hard to believe that before that time the statues of Harmodius and Aristogeiton stood in the Agora of Athens.

25. Herod. 5, 102f. As further evidence for the fact that Athenian troops took part in the Ionian revolt we may refer to Pausanias 1 29, 5, "... καὶ δευτέραν εἰς τὴν νῦν 'Ιωνίαν ἐστράτευσαν,..."

26. B. D. Meritt, *Hesperia* 8, 1939, no. 21 and p. 164, note 1; cf., however, G. Welter, *AA.* 1939, cols. 29ff.

27. H. T. Wade-Gery, *JHS.* 53, 1933, p. 77; see also p. 99, line 7 of the Eurymedon epigram.

28. We could, however, refer to Aristophanes, *Wasps* 1079ff., a passage which contains a description of the events in Phalerum as well as a description of the battle of Marathon (Cf. H. L. Crosby, *Classical Studies Presented to Edward Capps,* p. 75).

29. J. Beloch, *Gesch.* ii, 1, p. 11, note 1.

27

An Original Work of Endoios

[245] In the last twenty years scholars have begun to use the evidence gained from a comparison of Attic vase-paintings with the contemporary sculptural remains for the stylistic interpretation and for the more accurate dating of both classes of monuments.[1] The epigraphical evidence, however, has been utilized only in a very few instances.[2] The famous Potter Relief from the Akropolis (Plate 12) is a good example of a monument for the interpretation of which the threefold evidence may be employed to good advantage.[3] As sculpture, the relief is one of the masterpieces of late archaic art; the vase held in the hand of the potter is a faithful copy of contemporary kylikes; the inscription reveals both in its text and in its lettering the artist of the monument.

THE NEW FRAGMENTS

To the seven fragments which were found in 1887 two more can be added now, but no photographs showing their place in the relief are as yet available.[4] The restored drawing (Figure 27.1) is based on two separate photographs and a drawing, all much smaller in scale.

American Journal of Archaeology 46 (1942) 245–253

1. The pioneer work was done by Gottfried von Lücken, "Archaische griechische Vasenmalerei und Plastik," *AM.* 44, 1919, pp. 47–210, and by Ernst Langlotz, *Zur Zeitbestimmung der strengrotfigurigen Vasenmalerei und der gleichzeitigen Plastik*, 1920.
2. A characteristic example of the way in which epigraphical evidence has been treated are E. Löwy's two papers *Zur Datierung Attischer Inschriften*, 1937, and *Der Beginn der rotfigurigen Vasenmalerei*, 1938; see the review in *AJA.* 43, 1939, pp. 710–713. Quite different is W. B. Dinsmoor's "The Correlation of Greek Archaeology with History" in *Studies in the History of Culture*, pp. 185–216.
3. See W.-H. Schuchhardt in H. Schrader, *Die archaischen Marmorbildwerke der Akropolis*, pp. 301–302, and pl. 176; compare C. Picard, *Manuel d'Archéologie grecque, La sculpture* ii, pp. 22–23, 902.
4. This combination should be added as no. 19 to the list published in *AJA.* 45, 1941, p. 70. See the summary in *AJA.* 46, 1942, p. 123.

One of the newly added fragments (EM 6520) has been known for almost fifty years. It was first published by H. G. Lolling, who thought that it belonged originally to an archaic Ionic capital and that the inscription was of later, probably Roman, date.[5] R. Heberdey made a drawing of this fragment (Figure 27.2), and he, too, thought that it was part of an Ionic capital. The fragment joins the top of the Potter Relief on the right side and contains the

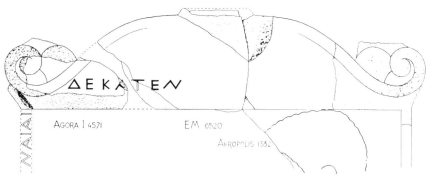

Figure 27.1. Top of Potter Relief.

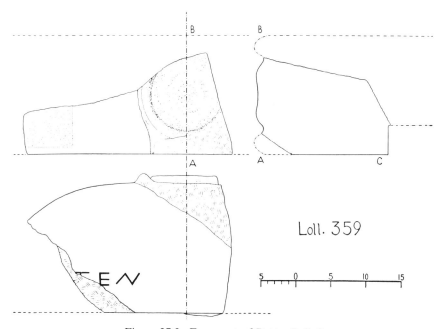

Figure 27.2. Fragment of Potter Relief.

5. Κατάλογος τοῦ ἐν 'Αθήναις 'Επιγραφικοῦ Μουσείου i, no. 359.

central part of the architrave, with the cutting for an akroterion on its top. The shape of the cutting, an annular depression, indicates that the akroterion was of bronze. It may have been in the form of a bronze vase, an appropriate decoration for the offering of a potter.[6] The relief was crowned not only by a central akroterion, but also by two lateral finials made of marble and set into square cuttings. The socket for the right akroterion is preserved on top of the Potter Relief,[7] and these lateral akroteria may have consisted of some kind of sculptured ornament.

[247] The second of the newly added fragments comes from the American excavations in the Athenian Agora (Plate 13).[8] Since the fragment was found in
[249] surface soil, it may | have been thrown down the north slope of the Akropolis during the great excavations in the eighties, rather than in antiquity. The restored drawing (Figure 27.1) may seem to indicate that the two new fragments do not actually join, but an examination of the fragments themselves might prove that the one from the Agora fits the other piece and that it preserves traces of the socket for the left akroterion. The whole crowning member of the relief was certainly decorated with a painted ornament, but its restoration must remain quite hypothetical, since only faint traces are visible on the small photograph now available (Plate 13).

The inscription previously known to have run along the left margin of the relief[9] is now shown by the new fragments to continue on the left side of the crowning member. This fact recalls the suggestion made by F. Studniczka[10] that the inscription may have begun near the very bottom of the margin, and that the letters thought to be part of the dedicator's name belong to that of his father. It now seems likely that the stonecutter engraved the last word of the dedicatory inscription above the relief because there was no space left on the margin itself.[11]

Various restorations of the dedicator's name have been suggested; they all

---

6. Several inscribed bases of early Attic dedications have on top circular sockets which may have received stone vases; the inscriptions are published as *IG*. i², 543, 600, 643, and 671. The names of the dedicators of these monuments are those of potters or painters: Aischines, Kephalos, Smikros, Xenokles; see *Bull. de l'Inst. Arch. Bulgare* 12, 1939, p. 144, note 5.

7. See H. G. Lolling, *op. cit.*, no. 281; G. Dickins, *Catalogue of the Acropolis Museum* i, p. 272, no. 1332.

8. It was discovered on March 3, 1937, north of the modern street, in Sector OA, directly below the north slope of the Akropolis; the inventory number of the fragment is I 4571. The following information is provided by the filing card: Part of the top and of the left side is preserved; below the inscribed face appears to be the top of a recess. The fragment seems to be the upper left corner of a stele or monument with an elaborate system of decorative mouldings. Much red color is still preserved in the inscription.

9. *IG*. i², 718.

10. Reported by E. Langlotz, *op. cit.*, p. 92.

11. This conclusion, however, is not cogent, as is shown by another dedicatory inscription from the Akropolis (*IG*. i², 607; see *Hesperia* 8, 1939, p. 156, no. 2) which begins at a considerable distance from the left edge but turns around the corner at the end of the line. Moreover, the position of δεκάτεν above that part of the relief which had no carving may have special significance. For Langlotz assumed (see W.-H. Schuchhardt, *op. cit.*, p. 301) that this part of the relief contained the painting of a stand with vases upon it, signifying that the tithe (δεκάτη) was derived from the sale of such objects.

agree in the assumption that the potter who made this impressive offering must have been one of the masters whose name is known from his signatures.[12] Of the first preserved letter only the lower part of a slanting stroke remains (Plate 15). This stroke may belong to one of the following letters: alpha, gamma, kappa, mu, rho, sigma, chi. The uneven spacing of the letters in the preserved part of the inscription makes it impossible to determine the exact number of letters which may be restored between the lower end of the margin and the first preserved letter: [ca. 12] \ ιος: ἀνέθεκεν [:τἀθεναίαι]|δεκάτεν.[13]

THE DATE [250]

The dating of the Potter Relief is based on the style of the relief itself, on the letter forms of the inscription, and on the shape of the kylix in the hand of the potter. The profile of the kylix with its offset rim has been interpreted in different ways by A. Furtwängler and by H. Payne.[14] Furtwängler compared the kylix both with preserved vases and with representations of kylikes on red-figured vase paintings.[15] Payne, however, declared that this type of kylix went out of fashion between 530 and 520 B.C., and he therefore suggested this decade as the probable date of the Potter Relief.[16] In a recent study of the shapes of Attic kylikes H. Bloesch connected the cup on the Potter Relief with one particular class of kylikes, the earliest example of which is dated at the end of the sixth century[17]

The letter forms of the inscription have only once been used for the dating of the whole monument.[18] E. Löwy declared that the forms of the letters alpha, epsilon, theta, and nu date the relief in the Kimonian period, ca. 460 B.C., and that the four-stroke sigma would indicate an even later date.[19] It is hardly necessary to point out that Attic inscriptions of the two decades preceding 450 B.C. have an entirely different appearance. The letter forms of the inscription on the Potter Relief, which Löwy took as evidence for a date near the middle of the fifth century, are rather characteristic for inscriptions which are either

---

12. See A. Furtwängler, *Aegina*, p. 495.
13. Another possible restoration is [ca. 12] \ ιος: ἀνέθεκεν [:ho κεραμεὺς]|δεκάτεν.
14. A. Furtwängler, *op. cit.*, pp. 494–495; H. Payne and G. M. Young, *Archaic Marble Sculpture from the Acropolis*, p. 48, and pls. 129–130.
15. The cup illustrated on an amphora of the Berlin Painter clearly shows that this type of kylix was well known at the beginning of the fifth century; see J. D. Beazley, *Der Berlin Maler*, pl. 14 a.
16. The kylikes which Payne compared with the cup on the Potter Relief are apparently the so-called Droop Cups, the Attic origin of which, asserted by E. A. Lane (*BSA*. 34, 1936, p. 152), seems to be assured, since Beazley discovered the signature of Nikosthenes on one of them (*JHS*. 55, 1935, p. 81). It may be worth mentioning that the workshop of Nikosthenes was probably taken over by the potter Pamphaios (see *RE.*, s.v. Nikosthenes) and that [Παμφ]αῖος is one of the possible restorations of the dedicator's name on the Potter Relief.
17. *Formen attischer Schalen*, p. 144.
18. The remarks made by H. Lechat (*Sculpture attique avant Phidias*, pp. 365–366) and E. Langlotz (*op. cit.*, p. 93) may be disregarded here because they are too general.
19. *Sb. Akad. Wien* 217, Abh. 2, 1938, p. 95; see above note 2.

Ionic or influenced by Ionic writing.[20] Of particular interest may be a comparison of the inscription on the Potter Relief with the preserved signatures of the sculptor Endoios.[21] One of these inscriptions (*IG*. i^2, 978), although written in the old Attic alphabet, may have been engraved by the same hand as the inscription on the Potter Relief (Plate 14);[22] the Endoios signature agrees with the inscription on the Potter Relief in the forms of the letters epsilon and nu, and particularly in the deeply cut grooves of the strokes. Both the Archermos inscription (see note 20) and the Endoios signatures belong to the last quarter of the sixth century, and the Potter Relief may be assigned to the end of the same period.

Only a few words may be added concerning the style of the relief itself. Payne [251] compared the folds of the potter's himation with those of the seated figures of the Siphnian treasury,[23] and this frieze has been attributed to Endoios by Rumpf.[24] Schuchhardt went so far as to suggest that the sculptor of the Potter Relief also made the reliefs on the so-called ballplayer base;[25] and this base was found together with the base carrying the signature of Endoios (*IG*. i^2, 983). The badly preserved painting on this statue base, representing a seated man, is the only original work of Endoios which is known so far.[26] There exists a certain similarity between this painting, the Potter Relief, and the Athena statue from the Akropolis commonly identified as the work of Endoios which is mentioned by Pausanias in his description of the Akropolis.[27] Both the potter and the Athena are rather heavily built, and their bodies are well modeled. The great liveliness of the Athena statue is achieved by the artist through the unsymmetrical position of the legs which indicates a departure from the static character of most of the archaic seated figures.[28] Similarly, the potter is not represented in exact profile, but his body is slightly turned toward the spectator, his right shoulder is visible, and his breast is rendered in a foreshortened way; his left leg seems to be slightly drawn back.[29] The legs of the chair on which

20. See the Archermos inscription illustrated by J. Kirchner, *Imagines*, pl. 6, no. 13.
21. *IG*. i^2, 492, 978, 983; see *JOAI*. 31, 1938, Beiblatt, cols. 62–68; A. Rumpf, *La Critica d'Arte* 14, 1938, pp. 44–45.
22. The top of the illustration shows the right part of *IG*. i^2, 978 (photograph of the back of a squeeze), and the lower part shows the inscription on the two new fragments of the Potter Relief (photograph of the back of a squeeze).
23. *Op. cit.*, p. 48, note 2.
24. *Loc. cit.*, pp. 42–45.
25. *Op. cit.*, pp. 301–302.
26. Rumpf (*loc. cit.*, p. 41) and Langlotz (in H. Schrader, *op. cit.*, p. 111) date this painting near the middle of the sixth century. But the stylistic evaluation of the painting, as given by Rumpf, seems to be based solely on a restored drawing, since the stone itself never did show many traces (see the photograph illustrated in *JHS*. 45, 1925, p. 167, Fig. 2), and the painting is now practically gone. The type of pedestal to which the painted base belongs seems to be Ionic in origin; see *Bull. de l'Inst. Arch. Bulgare* 12, 1939, pp. 148–149, and compare J. Mosel, *Studien zu den beiden archaischen Reliefbasen*, pp. 4, 25, and 48.
27. See E. Langlotz in H. Schrader, *op. cit.*, pp. 109–111, no. 60, and pl. 85. It seems that Endoios was known for his representations of seated figures; see Pausanias vii, 5, 9.
28. See the fine description given by H. Payne, *op. cit.*, pp. 46–47.
29. Quite similar is the relief in the Ince-Blundell collection which S. Casson aptly compared with the painting on the base signed by Endoios (*JHS*. 45, 1925, p. 172, Fig. 4); see B. Ashmole, *Catalogue of the Ancient Marbles at Ince Blundell Hall*, p. 96, note 2.

the potter is seated are similar to those of the throne of the Athena statue.

The preceding observations suggest a date for the Potter Relief at the end of the sixth century. Most closely related to the style both of the relief and of the inscription are Ionic works and, in particular, monuments connected with the Ionic-Attic sculptor Endoios.

THE SIGNATURE

Two fragments of the artist's signature are preserved on the Potter Relief. Lolling had already observed that an inscription was engraved also on the right margin of the relief, and he copied carefully the scant traces of the letters.[30] Dickins went further and suggested that these letters are part of the signature of the

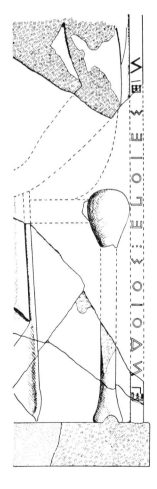

Figure 27.3. Signature on the Potter Relief.

---

30. *Op. cit.*, no. 281 b.

artist;[31] neither Payne nor Schuchhardt noted the existence of this second inscription. The preserved letters are engraved from right to left on the right margin of the relief, and the inscription begins near the bottom and ends near the middle of the margin.[32] | Both the beginning and the end of the signature are preserved; the end is illustrated in Plate 16. The total length of the inscription and the distance between the four preserved letters permit an estimate of the number of letters that originally composed the inscription. There were approximately fifteen, and this number tallies well with the assumption that the inscription contained only the name of the artist and the verb ἐποίησεν (Figure 27.3). It is to be noted that the inscription was damaged in ancient times by the addition of strokes which make the reading very difficult.[33] These additional strokes are evidently not parts of normal letters.[34] If they are disregarded, the signature may be read as beginning with epsilon, followed by one of the three letters: gamma, mu, or nu. The whole inscription ended in epsilon-nu, which may easily be restored to [---ἐποίεσ]εν. There remain approximately seven letters for the name of the artist.[35] Among the few names beginning with epsilon-nu,[36] the name of Endoios suggests itself, and the whole signature may be completed to Ἔν[δοιος ἐποίεσ]εν.

ENDOIOS

A. Rumpf, in his monograph on Endoios, has attributed to this artist, among other monuments, not only the Rampin horseman[37] and the peplos kore,[38] but also the frieze of the treasury of Siphnos at Delphi.[39] These attributions

---

31. *Op. cit.*, p. 272. S. Casson could not have been aware of this observation when he suggested (*Technique of Early Greek Sculpture*, p. 144) that the dedicator of the relief "may conceivably have carved the relief himself."

32. This is supporting evidence for the assumption that the dedicatory inscription began at the lower end of the left margin.

33. Lolling must have noticed that some of the preserved strokes do not belong to the original inscription, because he gives the distance between the letters as 43 millimeters, which does not take into account the added strokes. Surprisingly similar is the damage done to the artist's signature on the Siphnian frieze, and Rumpf restored (*loc. cit.*, pp. 44–45) in this inscription the name of Endoios; for another signature of Endoios which was destroyed intentionally in antiquity, see *JOAI.* 31, 1938, Beiblatt, cols. 66–68, and compare W. B. Dinsmoor, *op. cit.*, pp. 187 and 198.

34. An Ionic gamma is inserted between the first and second letters, and this gamma is engraved from left to right, while the original inscription is written from right to left. Between the last two letters of the signature (Plate 16) three vertical strokes are inserted; two are engraved right over the horizontal bars of the epsilon, while the third, standing between the letters epsilon and nu, may easily be taken for an iota. It is clear, however, that the signature, or any other conceivable word here used, cannot end with the letters epsilon-iota-nu, and that the combination epsilon-iota would have been written in that period simply as epsilon.

35. This name began with epsilon or eta, but eta may be ruled out, since most names beginning with eta have a rough breathing, and a rough breathing would probably have been written in that period.

36. The second letter was most likely a nu.

37. See W.-H. Schuchhardt in H. Schrader, *op. cit.*, pp. 212–225, no. 312.

38. See E. Langlotz in H. Schrader, *op. cit.*, pp. 45–48, no. 4.

39. Rumpf did not discuss in this connection kore no. 602 which has been attributed to Endoios or to his studio; see *JOAI.* 31, 1938, Beiblatt, col. 65, note 11; E. Langlotz in H. Schrader, *op. cit.*,

are based, so it seems, on each other, rather than | on external evidence. The attribution of the Siphnian frieze is of special significance, because Rumpf compares one of the figures of the frieze with the seated Athena from the Akropolis, and because there is preserved the signature of that artist who made the part of the frieze which Rumpf attributed to Endoios. Though the name of the artist is almost entirely lost, Rumpf does not hesitate to restore the name of Endoios.[40]

The literary tradition knows Endoios as a sculptor of marble statues, as well as of statues of wood and of ivory. The statue base from the Kerameikos proves that he was a painter, and the Potter Relief shows that his skill was great also in the carving of reliefs.

---

pp. 50, no. 7, and 111. This combination would support Rumpf's attribution of the Siphnian frieze to Endoios, while it may weaken the alleged connection of Endoios with the peplos kore and with the Rampin horseman; compare C. Picard, *op. cit.* ii, pp. 899–900.

40. It is worth mentioning that the formula of the signature on the Siphnian frieze (ὁ δεῖνα τάδε καὶ τὄπισθεν ἐποίε) is similar to that of Endoios's signature on *IG*. i², 983 ("Ενδοιος καὶ τόνδ' ἐποίε). It is quite possible that a similar signature was inscribed on the base of the statues of the Charites and Horai in Erythrai. Pausanias says (7, 5, 9) that Endoios's authorship of the cult image may be deduced from the marble statues of the Charites and Horai which stood outside the temple; the signature may have been: Ἔνδοιος τάσδε καὶ τὄνδον ἐποίε.

28

The New Homer

[317] The epigram on the plinth of the statue representing a personification of the Iliad (above, pp. 62–65) is of considerable archaeological and historical importance.[1]

> Ἰλιὰς ἡ μεθ' Ὅμηρον ἐγὼ καὶ πρόσθεν Ὁμήρ[ου]
> Παροτάτις ἵδρυμαι τῶι με τεκόντι νέω[ι].

The pentameter indicates that the statue of the Iliad was set up next to a statue of her "new" father. The first line also speaks of two Homers, one older and one younger than the Iliad herself. The older Homer must be the famous poet of the Iliad, but it was the new Homer next to whose statue stood that of the Iliad.

This interpretation of the epigram permits at once the identification of the "New Homer" with C. Iulius Nikanor who was known as the *Neos Homeros*. Most of the evidence concerning this remarkable man (who was a friend and contemporary of the emperor Augustus) was assembled more than ninety years ago.[2] Some new evidence has recently come to light, and, while a careful study of it will be necessary, a preliminary account of it may be submitted here in order to indicate the significance of the new discovery and to stimulate further study of this man.[3]

Hesperia 23 (1954) 317–319

---

1. The statue of the Iliad has been discussed repeatedly; reference may be made to P. Graindor, *Athènes sous Hadrien*, pp. 262–266 (with bibliography). The artist's signature preserved on the statue representing the Odyssey is now published as *I.G.*, II², 4313.
2. See Karl Keil, *Rh. Mus.*, 18, 1863, pp. 56–70. For recent treatments of Nikanor, see P. Graindor, *Athènes sous Auguste*, pp. 8–10, 168, 176; A. Stein, W. Kroll, and C. Wendel, *R.E.*, s.v. Nikanor, nos. 18 and 22; J. Day, *An Economic History of Athens under Roman Domination*, p. 149; S. Accame, *Il Dominio Romano in Grecia dalla guerra Acaica ad Augusto*, pp. 178–179; Th. Ch. Sarikakis, *The Hoplite General in Athens* (Diss. Princeton, 1951), Athens, 1954, pp. 73–74.
3. I am grateful to H. A. Thompson for having discussed with me the archaeological implications of the association of the Iliad monument with Nikanor. In the original publication of the Iliad base [*Hesperia* 23 (1954) 64] its association with the Library of Pantainos was suggested. This appears to

The Attic inscriptions in which Nikanor's name appears are the following:

I. *I.G.*, II², 3785 (a = E. M. 3124, identified by Grimaldi): inscribed base of an honorary statue of C. Iulius Nikanor, the son of Are(i)os. This inscription suggests the identification of Nikanor with the son of the philosopher Areius Didymus who was a friend of the emperor Augustus (Suetonius, *Divi Augusti Vita*, 89, 1), and of Maecenas (Aelianus, *Varia Hist.*, 12, 25).

II. *I.G.*, II², 3786–3789: a series of inscribed statue bases honoring Iulius Nikanor, the New Homer and the New Themistokles. Two of these monuments (3786 and 3787) were set up on or near the Akropolis, one (3788) in the Peiraieus, and one (3789) in Eleusis. These identical inscriptions assure the identification of Nikanor with Nikanor the New Homer, a native of Hierapolis, mentioned by Stephanus Byzantius, s.v. Ἱεράπολις. In one of these inscriptions (3788) the designation of Iulius Nikanor as the New Homer and the New Themistokles is completely preserved, while in the others it has been erased; this erasure itself supports the identification of Nikanor with the man of this name mentioned by Dio Chrysostom, 31, 116.

III. *I.G.*, II², 1723; S. Dow, *Hesperia*, 3, 1934, pp. 162–167 (with Fig. 11 on p. 163): archon list in which Iulius Nikanor, the New Homer (and the New Themistokles), is mentioned as Hoplite General. This inscription is dated by Dow (p. 166) "at the earliest late in the reign of Augustus." Notice that the laudatory epithet ("the New Homer") was added to Nikanor's name in this public document; notice also that the epithet was not erased.

[318]

IV. *I.G.*, II², 1069 + (?) 1119 + E. M. 5245 (apparently unpublished) + I 175 + I 1059 + I 6132 + I 6387 (unpublished fragments from the Agora). A second copy of the same inscription was set up in Eleusis: *I.G.*, II², 1086 + A. N. Skias, Ἐφ. Ἀρχ., 1895, col. 121, no. 34 (already associated with *I.G.*, II², 1086 by Skias, but apparently excluded from publication in *I.G.*). In the Attic decree (*I.G.*, II², 1069), Nikanor is mentioned with his epithets (not erased) and as Agonothetes of the Augustan Games and as Hoplite General; for the combination of these two offices, see J. H. Oliver, *The Athenian Expounders of the Sacred and Ancestral Law*, p. 85. The association of *I.G.*, II², 1069 (which seems to be lost) with the other six Athenian fragments is suggested (a) by the occurrence of one significant word (κεχειροτονημένον in line 8) in the first line of the small fragment from Eleusis published by Skias ([κε]χειροτο[νημένον]), and (b) by the occurrence on the same fragment (line 5) of the name of Iulius Nikanor. Moreover, the larger fragment from Eleusis (*I.G.*, II², 1086) has an erasure in line 14 in which the epithets of Nikanor can be restored since the corresponding line of the Athenian fragment (*I.G.*, III, 47, line 14; the letters of this line are omitted in *I.G.*, II², 1119) allows the reading and restoration: καὶ ὁ Ἰού|[λιος ...]. This document, the text of which has still to be studied, evidently is concerned with real estate on the island of

involve some chronological discrepancy inasmuch as Nikanor, the New Homer, was active in the time of Augustus whereas the library was erected in the reign of Trajan. The difficulty can scarcely be resolved without additional evidence. One can conceive, however, that a statue of the New Homer, dating from his life time or soon after, was set up a good deal later in the Library of Pantainos, and that the statues of the Iliad and the Odyssey, whose style suggests a date in the second century, were then added so as to make up a group.

I am also grateful to Fr. W. M. A. Grimaldi, S.J., and to E. Vanderpool who examined most of the inscriptions mentioned in this report.

Salamis, and thus confirms once more the identification of this Nikanor with the Nikanor mentioned by Dio, ὃς αὐτοῖς καὶ τὴν Σαλαμῖνα ἐωνήσατο.

Attention may be called to the literary problem connected with these inscriptions and with the statue of the Iliad. The official designation of a man as the New Homer and the erection of his statue with that of the Iliad are puzzling. It seems that Nikanor | was the author of a poem which could be identified or equated with the famous Iliad of Homer. The suggestion may be made that he was the author of the *Ilias Latina.*[4]

A final suggestion may be made to explain Nikanor's second epithet: the New Themistokles. Since the most noteworthy deed of the famous Themistokles was his victory at Salamis, one should expect that Nikanor, too, was the victor of a battle of Salamis. It so happens that Augustus staged in 2 B.C. a naumachia between the Persians and the Athenians (Ovid, *Ars Am.*, 1, 171–2) in which the Athenians won (Cassius Dio, 55, 10, 7). Graindor rightly connected this event with the sham battles at Salamis which the Athenian ephebes staged during the Empire (*Athènes sous Auguste*, pp. 128–9). It is now possible to confirm this connection and to see in Nikanor the man who "won" the battle of Salamis in 2 B.C., and who established the commemorative celebrations of the ephebes at Salamis. This explains satisfactorily his second title: the New Themistokles.[5]

---

4. See C. Hosius, *Gesch. d. Röm. Lit.*, 4th ed., pp. 505–508.

5. Attention may be called to the repeated references to monuments and to sanctuaries on Salamis in the much discussed inscription *I.G.*, II², 1035 (especially lines 30–34); for the Augustan date of this document, see H. S. Robinson, *A.J.A.*, 47, 1943, pp. 298–299, and note 20 (with bibliography).

29

Demokratia

In publishing the relief stele with the "Law against Tyranny," Meritt pointed out[1] that the relief represented the seated, bearded Demos being crowned by Demokratia standing behind his throne. This interpretation is confirmed by the peculiar and repeated reference in the law to "Demos and Demokratia" (lines 8–9, 13, 16–17) which are within the context surely synonymous terms.[2] The relief suggests clearly that the Demos representing all the people of Athens is honored by and thus especially attached to one particular form of government, Demokratia.

This close association in word and picture of Demos and Demokratia on a monument of 337/6 B.C. recalls the famous painting of Theseus, Demos, and Demokratia which, according to Pausanias (1, 3, 3), clearly indicated that Theseus established democracy for the Demos. It is not too far fetched to assume that the relief, in spite of its poor workmanship,[3] represents the figures of Demos and Demokratia from the famous painting of Euphranor (Pausanias, 1, 3, 4; Pliny, *Nat. Hist.*, 35, 129).[4] Theseus himself could be supplied on the other side of the seated Demos who, in the painting even more than in the relief, faced the spectator. At any rate, the relief with its representation of Demos and Demokratia and the law with its repeated employment of these two abstract nouns in a somewhat metaphorical meaning both reflect the same spirit as the painting.

Pausanias himself indicates (1, 3, 3) that there was in addition to the painting another tradition which presented Theseus as the founder of democracy, and he contrasts this account with other popular and poetic traditions which

---

1. *Hesperia*, 21, 1952, pp. 355–359, No. 5; see *S.E.G.*, 12, 87; 15, 95; Al. N. Oikonomides, *Polemon*, 6, 1956/57, Symmeikta pp. 28–36.
2. See, however, M. Ostwald, *T.A.P.A.*, 86, 1955, p. 120, note 93.
3. H. A. Thompson, *Hesperia*, 22, 1953, p. 53, pl. 20, a.
4. See J. H. Oliver, *Demokratia*, pp. 107–108, 164–166, pl. II.

consider Theseian Athens a monarchy rather than a democracy. No doubt, the nature of the former tradition is political, and this is indicated by the further point that the democracy which Theseus established lasted until the rise of the tyrant Peisistratos; F. Jacoby suggested[5] that Pausanias relied here on the Atthis of Philochoros (who may have been using here, as elsewhere, the work of Androtion; see below).

The political tradition of Theseus as founder of Athenian democracy is old,[6] but it seems that in the years immediately preceding 340 B.C. several authors reinterpreted this tradition to mean that Theseus founded a moderate democracy. Ruschenbusch, | who attributed this innovation to Androtion,[7] called attention to the remark in Isokrates' *Panathenaicus* of ca. 340 B.C. (5, 148) that the Theseian democracy lasted for no less than a thousand (*sic*) years until the age of Solon and the tyranny of Peisistratos; Isokrates is here evidently following the same tradition as Pausanias. Mention may also be made of a passage in the speech against Neaira (Demosthenes, 59, 75) in which the speaker described the moderate democracy of Theseus in considerable detail and in constitutional terms. This concept of Theseus as a moderate democrat had only a limited success, for in Theophrastos (*Characters*, 26, 6) we find Theseus again as the champion, the founder, and the first victim of radical democracy.[8]

The evidence so far assembled indicates that there was a considerable interest in Demokratia at Athens around 340 B.C.[9] How much older the personification of Demokratia is one cannot tell, except for recalling that the tomb of Kritias is said to have had a representation of Oligarchia and Demokratia (Scholion on Aeschines, 1, 39 = Diels, *Frag. d. Vors.*, 88 A 13), and that a series of Attic ships from the beginning of the fourth century on bore the proud name of Demokratia.[10]

It has been suggested, most recently by J. H. Oliver,[11] that the Athenians celebrated the anniversary of the restoration of democracy in 403 B.C., which fell on the 12th of Boedromion, by a sacrifice to Demokratia. The evidence for this suggestion is tenuous but telling. In a list of public sacrifices of the years 332/1 and 331/0 (*I.G.*, II^2, 1496; add *Hesperia*, 9, 1940, pp. 328–330, no. 37), one reads and restores (combining lines 131–132 and 140–141) "from the sacrifice to Demokratia made by the strategoi – – –." We do not know whether such sacrifices were also performed in the preceding and following years, but the time of the year, between the Eleusinia (Metageitnion) and the

---

5. On *F. Gr. Hist.*, 328 F 19; see also his comments on 239 A 20.
6. Isokrates, *Helen*, 32–37, and *Panathenaicus*, 126–129; E. Ruschenbusch, *Historia*, 7, 1958, pp. 415–418.
7. *Op. cit.*, pp. 408–415.
8. *Op. cit.*, pp. 408–409, and my comments in *Bull. d'arch. et d'hist. dalmate*, 56–59, 2, 1954–1957 (*Mélanges Abramic*, 2), pp. 50–51.
9. See also the general survey by J. A. O. Larsen, *Cl. Phil.*, 49, 1954, pp. 1–14.
10. F.'Miltner, *R.E.*, Suppl. V, s.v. *Seewesen*, col. 948, lines 49–51; add *I.G.*, II^2, 1623, line 326; see also the vocatives "Oh Democracy" in Aristophanes, *Acharnians*, line 618, and *Birds*, line 1570, and A. Debrunner, *Festschrift für E. Tièche*, pp. 20, 23, note 9.
11. *Demokratia*, pp. 105–106.

Asklepieia (17th of Boedromion) strongly supports the assumption that the sacrifices to Demokratia were performed on the anniversary of the liberation of Athens and of the restoration of democracy. This combination is supported by the appearance of Demokratia (or of a torch-holding woman who was later so identified) on the tomb of Kritias and by the naming of an Athenian ship Demokratia (*I.G.*, II^2, 1606, lines 59–60) which was already an "old" ship in 374/3 B.C.

L. Deubner[12] called attention to a round altar from the Akropolis (Plate 18; *I.G.*, II^2, 4992; see also the earlier publications *C.I.A.*, II, 1672 and III, 165) which | was dedicated to Athena Demokratia and for which he suggested tentatively the date of the restoration of democracy in 89/8 B.C.[13] D. Bradeen kindly examined this round altar which was hollowed out on top[14] and carried on its cylindrical face an inscription of three (not two) lines, since traces of a first line are still visible on the stone (Plate 18). One may assume that this was a joint altar (like *I.G.*, II^2, 4983) of some deity and Athena to which the name of Demokratia was added later, but possibly by the same hand. Whether the name on the moulding is that of the archon Herodes (of 59/8 B.C.[15]) one cannot tell but it seems to be contemporary with the main inscription. The letter forms suggest that this altar of Athena and Demokratia was erected in the Augustan period, and J. H. Oliver has assembled some evidence to show that there may have been a connection between Demokratia and Augustus.[16]

[240]

Of even greater importance for our understanding of the cult of Demokratia in Athens is the inscription on two seat blocks in the Theater of Dionysos (Plate 19; *I.G.*, II^2, 5029a[17]) for the priest of Demokratia, the priest of Demos and Charites, and for a priest whose designation can no longer be read. D. Bradeen examined the stones and took a squeeze of the inscription which belongs, to judge from the letter forms, in the third century before Christ. Adolf Wilhelm had already called attention[18] to the dedicatory inscription now published as *I.G.*, II^2, 4676 which he himself had restored as a dedication made by either Eurykleides or Mikion in his capacity as priest of Demos and the Charites.[19] This inscription which was also recently examined by D. Bradeen is engraved on two adjoining thin plaques (ca. 0.08–0.10 m. thick) of which only fragments of the right one are preserved (Plate 17). The dedicatory inscription itself was engraved on a smooth band along the upper edge of the plaques. Only the title of the dedicator is preserved; he was priest of Demos and the Charites. Below the dedicatory inscription were three citations symmetrically arranged; of these only part of the middle one (at left of the

---

12. *Attische Feste*, p. 39.
13. See also W. S. Ferguson, *Hellenistic Athens*, p. 444, note 2.
14. For this type of altar, see *Hesperia*, 28, 1959, p. 76, no. 3.
15. M. Thompson, *Hesperia*, 10, 1941, p. 224, note 74; S. Dow, *Hesperia*, Suppl. 8, 1949, p. 117, lines 55 and 127.
16. *Demokratia*, pp. 158–169.
17. Oliver, *Demokratia*, p. 107.
18. *S.E.G.*, 1, addenda to no. 45 on p. 137.
19. *Beitrage*, pp. 76–81; see Oliver, *Demokratia*, p. 106.

preserved plaque) and the one on the right are preserved, mentioning the priest of Demos and Charites and the priest of Ptolemy Euergetes and Berenike. The establishment of the cult of Demos and Charites is one of the features of the restored democracy under Eurykleides and Mikion and belongs to the same occasion as that of the cult of Ptolemy and Berenike.[20] Unfortunately, the nature of the dedication *I.G.*, II², 4676, cannot be determined with certainty but it should be noted that the preserved fragments were found in the theater of Dionysos and that the citations are like those on the seats of the theater. Combining the evidence of the two inscriptions (*I.G.*, II², 4676 and 5029a) one may restore the priest of Demokratia as the missing left entry of *I.G.*, II², 4676, and the priest of Ptolemy Euergetes and Berenike as the missing right entry of *I.G.*, II², 5029a, and assume that the older cult of Demokratia was combined with that of Demos and Charites, when the latter was established in or shortly before 222/1[21] (*I.G.*, II², 844, lines 38–39). This combination would provide us not only with the knowledge of the continuation of the cult of Demokratia in the third century B.C., but also with the possible location of the cult in or near the later sanctuary of Demos and Charites and Aphrodite Hegemone, that is, just behind the building in which Euphranor's painting was kept and near the place where the statue base of Demokratia was found (see below).[22]

It appears, therefore, that there was a cult of Demokratia in Athens, beginning perhaps as early as the restoration of democracy in 403 B.C., certainly of considerable importance during the third quarter of the fourth century, still active toward the end of the third and even during the first century B.C. The cult had a priest who may have been during the second century B.C. also the priest of Demos and Charites (and was as such mentioned in many ephebe inscriptions of that period), and the location of the cult was in the northwest corner of the Agora, on the southern side of the dromos leading from the Kerameikos to the Agora.

One more piece of evidence completes the picture of the personification of Demokratia in fourth century Athens; the statue of Demokratia, looking perhaps not unlike the relief picture of Demokratia on the stele with the "Law against Tyranny" of 337/6.[23] The existence of this statue has been known for a long time since one of the ephebe decrees of 106/5 was to be set up next to Demokratia (*I.G.*, II², 1011, line 62), but not much attention has been paid to this formerly isolated reference.[24] More recently a fragment was found of an honorary decree by the "Volunteer Guard"[25] for Demetrios Poliorketes, which contains the provision that a statue of Demetrios should be erected in the Agora

20. Ferguson, *op. cit.*, pp. 237ff., especially pp. 238, 242, note 3; R. E. Wycherley, *Testimonia*, pp. 59–61 (add *I.G.*, II², 5029a); Oliver, *Demokratia*, pp. 106–107.
21. For the date, see W. B. Dinsmoor, *The Athenian Archon List*, pp. 161–162.
22. For the topography, see *I.G.*, II², 2798; Wycherley, *op. cit.*, pl. IV, location 13; W. Judeich, *Topographie*², p. 363; G. Roger Edwards, *Hesperia*, 26, 1957, p. 343, no. 27.
23. M. Lang, *The Athenian Citizen*, Fig. 29.
24. Waser, *R.E.*, s.v. Demokratia.
25. See also *Hesperia*, 17, 1945, pp. 114–116, no. 68, and *I.G.*, II², 1209, both referring surely to the same organization and probably to the same event.

next to Demokratia.[26] Even so, Wycherley[27] did not recognize the statue of Demokratia as a separate Agora monument, and he felt unable to say anything about its location, its date, and the occasion of its erection.

In fact, the statue base of Demokratia has been known since Fourmont [242] copied it in the church of Nikolaos Blassarou, right across from the sanctuary of Demos and Charites.[28] Later, A. Boeckh published the inscription from Fourmont's copy in the C.I.G., I, 95:

"Athenis apud Nicolaum βλασταροῦ," ex Fourmontianis.
Secundus versus maioribus litteris scriptus.

I....ΒΑΣ
...ΜΟΚΡΑΤ..
ΕΓΙΝΙΚΟΚΡΑΤΟΥΣΑ[ΡΧΟΝΤΟΣ
ΑΝΕΘΗΚ]ΕΝΣΤΕΦΑΝΩΘΕΙΣΑΥΓ[Ο ---
ΑΡΕΤ]ΗΣΕΝΕΚΑΚΑΙΔΙΚ[ΑΙΟΣΥΝΗΣ

Unfortunately, during republication in C.I.A., II, 1156 and in I.G., II², 2791, the inscription was further mutilated by the omission of the first two lines which were thought to belong to a different and separate inscription. Actually, this is the dedication of the Boule of the year 333/2 whose members were crowned by the Demos (probably before their terms of office was over),[29] and who in turn erected a dedicatory statue of Demokratia. Similarly, the Boule of 357/6 dedicated its statue on a base of Eleusinian stone to the Twelve Gods, and the Boule of 332/1 erected its dedicatory statue also on a base of Eleusinian stone probably to Hephaistos.[30] The interpretation suggested here of I.G., II², 2791, which is now lost, has been reached by the confident association with it of an inscribed fragment of Eleusinian stone (E.M. 3913) which is here published (Plate 20) with the kind permission of the director of the Epigraphical Museum, Dr. Markellos Mitsos; for the description I am indebted to Miss Anna Benjamin. The stone is broken at the left and at the back, while part of the smooth top, of the right side, and of the bottom are preserved.

Height, 0.34 m.; preserved width, 0.20 m.; preserved thickness, 0.35 m.; the inscription is engraved between neat guide lines.

333/2

[vacat ca. 5..]|[ca. 7] ΒΑΣ [ca. 5] vacat ca. 5
[Δη]μοκρατ[ί]α
['Η βουλὴ ἡ] ἐπὶ Νικοκράτους ἄ[ρ]χοντος

26. W. Peek, Ath. Mitt., 66, 1941, pp. 221–227, and, with improved restorations, Ad. Wilhelm, Wiener Jahresh., 35, 1943, pp. 157–163.
27. op. cit., p. 210, no. 696.
28. A. Mommsen, Athenae Christianae, p. 126, no. 126; I. N. Travlos, Πολεοδωμικὴ 'Εξέλιξις τῶν 'Αθηνῶν, pl. XII, no. 93.
29. J. Kirchner, Ath. Mitt., 29, 1904, p. 250.
30. I.G., II², 2790 and 2792; for the use of Eleusinian stone, L. T. Shoe, Hesperia, Suppl. 8, 1949, pp. 342–343, 351–352.

[ἀνέθηκ]εν στεφανωθεῖσα ὑπ[ὸ] τοῦ δήμου
[ἀρετ]ῆς ἕνεκα καὶ δικ[αι]οσύνης·

The first line cannot now be restored with confidence and is therefore left unrestored; it may have contained an artist's signature.

It was undoubtedly next to this statue that that of Demetrios Poliorketes was to be set up[31] and that the stele *I.G.*, II^2, 1011 was to be erected, but there is no other mention of this statue of Demokratia in either our epigraphical or our literary tradition. There was, however, a statue of Demokratia in the Bouleuterion of Pergamon and next to it a bronze tablet was to be set up.[32]

It may seem idle to speculate why the Boule of 333/2 chose to erect a statue of Demokratia, at a time when Alexander was establishing "democracies" in the Greek cities of Asia Minor. In this very year, Alexander wrote personally to the Demos of the Chians (Dittenberger, *Sylloge*^{3}, 283) asking them to establish a democracy, and Arrian concluded his account of these activities by observing (*Anabasis*, 1, 18, 2) that Alexander ordered the cities to dissolve the oligarchies and to establish democracies. Toward Athens, Alexander showed considerable friendship; after the victory at Granikos, he dedicated three hundred panoplies to Athena on the Akropolis (*Anabasis*, 1, 16, 7) and during the archonship of Nikokrates (333/2, *Anabasis*, 2, 11, 10 through 24, 6), he spared the life of Iphikrates the Younger out of friendship for the city of Athens (2, 15, 4). It is doubtful whether this evidence has any bearing on the erection of the statue of Demokratia by the Boule of 333/2 B.C. It is significant, however, that during the following two years, 332/1 and 331/0, public sacrifices were made by the generals to Demokratia (*I.G.*, II^2, 1496, lines 131–132 and 140–141; see above pp. 224–225).

No more can be gained from a glance at the known inscriptions of this year: *I.G.*, II^2, 337 (decrees of the first and second prytanies), 338 (first prytany), 339 (second prytany; see *S.E.G.*, 16, 54), 408 (also second prytany, as B. D. Meritt has noted in his copy of the *Corpus*), 340 (undated), 341, 4594a, and *Hesperia*, 9, 1940, pp. 59–66, no. 8, referring to some officials and to the ephebes who served during this year and were honored, as the Boule, for their service. Of some interest is the dedication in Eleusis of a silver phiale by the Boule of this very year 333/2; see *I.G.*, II^2, 1544, lines 49–50.

In closing, we may recall the ringing phrases of an Athenian orator in which Vittorio De Falco[33] recognized Demades' well-known reply to Alexander: ἡμῖν δ' ἀπόρθητός ἐστιν ἡ δημοκρατία· ὁμονοοῦμεν πρὸς ἀλλήλους, τοῖς νόμοις ἐνμένομεν, καρτερεῖν ἐν τοῖς δεινοῖς ἐπιστάμεθα, τὴν τῆς ἐλευθερίας τάξιν οὐκ ἐνκαταλείπομεν.

---

31. *Ath. Mitt.*, 66, 1941, p. 226.
32. Dittenberger, *Sylloge*^{3}, 694, lines 30–31.
33. *Demade Oratore*^{2}, pp. 49–50, no. 83; see also Al. N. Oikonomides, Πλάτων, 9, 1956, p. 123, no. 22.

30

The Mission of Triptolemos

A study of the Mission of Triptolemos seems to us an appropriate offering to honor Homer A. Thompson; as the Attic Hero spread the knowledge of agriculture to civilize the world, so has our friend, the scholar, generously given the fruit of his work on Attic monuments to enlighten and inform our world.

The connection between artistic representations of Triptolemos and the political history of Athens, hinted at by C. Dugas,[1] has been more fully developed by J. W. Day,[2] who reached the conclusion that "during the fifth century Triptolemos was employed as a symbol of Athens' civilizing mission at the head of her imperial amphictyony."

The mission of Triptolemos was a favorite topic in Athens even before the time of Kimon and Perikles. Dugas, who was concerned almost exclusively with representations on Attic vases, listed fifteen black-figured examples; Grossman and Bindokat added four and three (two of them fragments) respectively.[3]

Studies in Athenian Architecture, Sculpture and Topography Presented to Homer A. Thompson, Hesperia: Supplement 20 (1982) 109–117 [with Isabelle Raubitschek]

1. "La mission du Triptolème d'après l'imagerie Athènienne," *MélRome* 62, 1950, pp. 7–31, esp. p. 14.

2. *The Glory of Athens: The Popular Tradition as Reflected in the* Panathenaicus *of Aelius Aristides*, Chicago 1980, pp. 15–39, esp. p. 24. J. P. Barron is going to include a chapter on the Eleusinian cult in his forthcoming book on "Athenian propaganda," and M. Sakurai is preparing an English version of her article on "The Eleusinian Cults and the Development of Athenian Democracy."

3. B. Grossman, "Eleusinian Gods and Heroes in Greek Art," diss. Washington University, 1959, pp. 67–77 and 81–92; A. P. Bindokat, "Demeter und Persephone in der attischen Kunst des 6. bis 4. Jahrhunderts," *JdI* 87, 1972, pp. 78–157. A list of the black-figured vases is given here in the order in which they will be mentioned in the first part of this essay: D refers to Dugas (see footnote 1 above), pp. 23–24, nos. 1–3 and p. 31, nos, 103, 104; G refers to Grossman, pp. 67–77 and 81–82; B refers to Bindokat.

    1. Brussels A 130; *CVA*, Brussels 1 [Belgium 1], 7 [20]:1; *ABV*, p. 308, no. 82 (Swing Painter); D 2; G AP 18.

    2. Göttingen J 14; *ABV*, p. 309, no. 83 (Swing Painter); D 1; G AP 20, 39.

[110] A grouping of these vases according to chronology and iconography permits a better understanding of their representations. The earliest (nos. 1 and 2) show Triptolemos sitting on a country cart,[4] a bearded man holding ears of grain in his hand and giving instructions to mortals about the cultivation of grain. These vases are attributed to the Swing Painter and dated ca. 540–520 B.C. Both show the cart above the ground, floating by levitation.[5]

On vases of the Leagros Period, the cart was rendered more majestic by transforming the stool into a thronelike seat and by adding occasionally wings to carry it over land and sea. The wings are probably related to the winged cart of Dionysos made for him by Hephaistos.[6] Dionysos sits on a winged cart on the black-figured amphora in Compiègne (no. 3), while Triptolemos, on the other side of the same vase, has a wingless cart and a seat with arm rests. The only black-figured representation of Triptolemos on a winged cart is on an amphora by the Edinburgh Painter (no. 4; see H. R. W. Smith, "From Farthest West," *AJA* 49, 1945, pp. 470–471, fig. 4:1); there the hero is not associated with Dionysos.[7]

Triptolemos appears with Demeter on a fragmentary black-figured amphora in Reggio,[8] where, identified by name, he is a member of a divine procession accompanying Demeter; but here he is not on his mission. On the amphoras

---

3. Compiègne 975; *CVA*, Compiègne [France 3], 10 [108]:4, 7; *ABV*, p. 331, no. 13 (Priam Painter); D 12; G AP 17.
4. Beverly Hills (Dr. Marion Prinzmetal); *ABV*, p. 478, no. 2, *Paralipomena*, p. 217 (Edinburgh Painter); D 13.
5. Vatican 385; *ABV*, p. 374, no. 195 (Leagros Group); D 7; G AP 21.
6. Lenormant Collection (formerly); D 5; G AP 16.
7. Würzburg 197; D 4; G AP 19.
8. Bologna, Mus. Civico 5; *CVA*, Bologna 2 [Italy 7], 14 [313]:4; D 11.
9. San Simeon 5503; D 9.
10. Hamburg; D 10.
11. Providence, Rhode Island School of Design 25.083; *CVA*, Providence 1 [USA 2], 10 [63]:1a; D 3.
12. Budapest 50.732; B V 53.
13. Munich 1539 (J 543); *CVA*, Munich 8 [Germany 37], 398 [1816]:4, 402 [1820]:2, 412 [1830]:2; D 8; G AP 37.
14. Acropolis 675; *ABV*, p. 377, no. 236, *Paralipomena*, p. 171, no. 9 (near Chiusi Painter); D 6; G AP 23.
15. Athens, National Museum 430 (CC 967); *ABV*, p. 587, no. 1 (Beldam Painter); D 104; G AP 40.
16. Athens market; *ABV*, p. 518 (Haspels, *ABL*, no. 64: Theseus Painter); G AP 38.
17. Prague, National Museum 1867; *ABV*, p. 708, no. 19 (Emporium Painter); G AP 22.

4. See H. L. Lorimer, "The Country Cart of Ancient Greece," *JHS* 23, 1903, pp. 132–151.
5. For illustrations of no. 2, see A. B. Cook, *Zeus*, 1, Cambridge 1914, p. 213, fig. 156; P. Jacobsthal, *Göttinger Vasen*, Berlin 1912, pl. 5.
6. This is suggested by the red-figured kylix in Berlin (2273) by the Ambrosios Painter (*ARV*², p. 174, no. 31; see Cook, *op. cit.*, p. 216, fig. 159), and the hypothesis is strengthened by the appearance of Hephaistos in such a cart on a kylix in Florence; see E. Simon, *Götter der Griechen*, Munich 1969, pp. 223–224, fig. 209; M. Robertson, review of F. Brommer, *Hephaistos*, Mainz am Rhein, *AJA* 84, 1980, p. 104.
7. Other amphoras with Dionysos on one side and Triptolemos on the other are nos. 5 (C. Albizzati, *Vasi antichi dipinti del Vaticano* [n.p., n.d.], pl. 54, no. 385), 6 (Cook, *op. cit.*, p. 214, figs. 157a and b), 7 (E. Langlotz, *Griechische Vasen in Würzburg*, Munich 1932, p. 34, pl. 51, no. 197).
8. Reggio, Mus. Nazionale 4001; *ABV*, p. 147, no. 6 (manner of Exekias); H. Metzger, *L'imagerie athénienne*, Paris 1965, p. 8, pls. I and II.

of the Leagros Period (nos. 8–14), the attentive women on either side of Triptolemos cannot be securely identified as Demeter and Kore; for no. 10, see R. Pagenstecher, "Zu unteritalischen Terracotten," *AA* [*JdI* 23], 1917, p. 107, fig. 37. No. 13 has been assigned in the *CVA* to "near the Chiusi Painter and the Antiope group."

The Mission of Triptolemos on three black-figured lekythoi reflects the already established red-figured scenes with snakes as an aid in drawing the chariot (nos. 15–17). These vases belong to the fifth century; two of them show the hero alone (nos. 15 and 16) and the third, the hero flanked by two unidentified women.

The iconography of these vases shows the policy of Peisistratos and his sons, who fostered the cultivation of grain and of the vine and who promoted the cults of Dionysos and Demeter.[9] The City Eleusinion may have been founded alteady in the age | of the tyrants.[10] The association of Demeter and Triptolemos *[111]* is implied by the scenes on the black-figured vases (nos. 1–17) which depict the Mission. The mission is not mentioned in the Hymn to Demeter which is not only earlier but also purely Eleusinian.[11] The elevation of Triptolemos as the hero who uses a "chariot of the gods" may belong to the time of the establishment of the City Eleusinion, which should have taken place according to the vases and the excavations in the reign of Peisistratos. The poet Onomakritos, a friend of his son Hippias, is said to have composed poems and oracles about Triptolemos which he attributed to Mousaios to give them greater authority.[12] Thus the vase paintings enrich our knowledge and understanding of the establishment of the Eleusinian cult in the city of Athens by showing that the story of Triptolemos was added at that time, possibly taken from Argos, the home of Peisistratos's second wife, Timonassa.[13]

The transformation of Triptolemos from an instructor of the Athenian farmers in the art of agriculture into a hero, charged by Demeter to spread the knowledge of farming throughout the world, took place, according to the vases illustrating this mission, between 510 and 480 B.C., when renewed activity is attested both in Eleusis and in the Eleusinion in Athens.[14] This propaganda effort with its emphasis on the Mission of Triptolemos beyond the borders of Attica may be connected with the claims of the newly founded Athenian democracy to be the mother city of all the Ionians.[15]

The skyphos painted by Oltos and found on the Akropolis is unfortunately

9. See A. Andrewes, *The Greek Tyrants*, London 1956, pp. 113–114.
10. See H. A. Thompson, "Activities in the Athenian Agora: 1959," *Hesperia* 29, 1960, pp. 334–338; H. A. Thompson and R. E. Wycherley, *The Athenian Agora*, XIV, *The Agora of Athens*, Princeton 1972, pp. 152–153; R. E. Wycherley, *Stones of Athens*, Princeton 1978, pp. 71–72.
11. See N. Richardson, *The Homeric Hymn to Demeter*, Oxford 1974, pp. 194–196 and 301–302 with notes on lines 476–482.
12. See Herodotos, 7.6.3; Pausanias, 1.14.2; 22.7; 8.37.5.
13. See Aristotle, *Athenaion Politeia*, 17; Pausanias, 1.14.2.
14. See *IG* I^2, 5 (818 is another copy) = *IG* I^3, 5; L. H. Jeffery, "The Boustrophedon Sacral Inscriptions from the Agora," *Hesperia* 17, 1948, pp. 86–111; J. Travlos, *Pictorial Dictionary of Athens*, New York 1971, pp. 198–199; Wycherley, *op. cit.*, pp. 71–72; K. Clinton, "IG I^2 5, The Eleusinia, and the Eleusinians," *AJP* 100, 1979, pp. 1–12, esp. note 13 on pp. 4–5.
15. See Herodotos, 1.147; 5.97; 9.97.

so fragmentary that the composition of the three figures is not certain:[16] Demeter, holding ears of grain, sits on a throne; Triptolemos must be in his chariot of which only the footrest is preserved; a second female figure, elaborately dressed in embroidered chiton and mantle, is shown frontally. Details now missing from the skyphos by Oltos may be seen on fragments of a contemporary phiale by the Euergides Painter, also from the Akropolis.[17] Boardman observed that this painter shared "Epiktetos' admiration for Hipparchos."[18] It seems that the Athenians of the early democracy continued the interest of the tyrants in the Eleusinian cult. On neither vase is the head of Triptolemos preserved; we cannot tell, therefore, whether he was already beardless and long haired as he appears in the canonical scenes of the fifth century.

[112]

After the Persian Wars representations of the Mission of Triptolemos on vases became frequent, perhaps as an acknowledgment of the aid given by the Eleusinian deities to the Athenians both at Marathon (Herodotos, VIII.65) and at Salamis (Plutarch, *Themistocles*, 15), perhaps in gratitude for Demeter's gifts to the Athenians and, through the Athenians, to the world.[19] It was at that time, in 468 B.C., that Sophokles' *Triptolemos* was performed and favorably received. Although only fragments are preserved (A. C. Pearson, *Fragments*, 2, Cambridge 1917, pp. 239–242), the play must have dealt with the Mission of Triptolemos since some of the fragments contain references to details seen on the vases: the dragons which pull the chariot (fragment 596) and Demeter's charge to Triptolemos (fragment 597). The transmission of this message to certain unspecified nations, to which Dionysios of Halikarnassos refers (*Antiquitatum romanarum* 1.12.2), may have been part of Sophokles' play; an inscription of ca. 460 B.C. concerning the proper conduct of the Mysteria refers in a fragmentarily preserved passage[20] to "these cities" meaning the cities who share the cult of Demeter and benefit from her gifts. Later in the fifth century, the allies of the Athenians and the Greeks in general were requested and invited to contribute wheat and barley, the "fruits of the earth," to the sanctuary in Eleusis, certainly in return for the benefactions received from Demeter and from the Athenians through the Mission of Triptolemos ($IG\ I^2, 76 = IG\ I^3, 78$).

The strongest indication of the importance of the Mission of Triptolemos and of the benefactions of Athens is given by the vases of the period between the Persian and Peloponnesian Wars. Twelve of the 245 vases attributed by Beazley to the Berlin Painter (ARV^2, pp. 196–214) illustrate the Mission of

---

16. E 13; Graef-Langlotz, *Die antiken Vasen von der Akropolis zu Athen*, Berlin 1931, II, ii, p. 41, pl. 39, 449a–d; ARV^2, p. 66, no. 135.
17. Graef-Langlotz, *op. cit.*, p. 11, no. 147, pl. 6; ARV^2, p. 89, no. 19.
18. *Athenian Red Figure Vases, the Archaic Period*, London 1975, p. 60.
19. In the "Ephebic Oath" (M. N. Tod, *A Selection of Greek Historical Inscriptions* II, Oxford 1948, no. 204; P. Siewert, "The Ephebic Oath in Fifth-Century Athens," *JHS* 97, 1977, pp. 102–111), which surely goes back to the Persian Wars, the young men swore by the fruits of the earth, and in the *Panegyricus* of Isokrates (which repeats details of the funeral orations which also go back to the Persian Wars) the gifts of Demeter to the Athenians and to the world are emphasized (28–29).
20. $IG\ I^2$, 6 and 9 (*SEG* 10, 6 and 26, 2), lines 36–43 = $IG\ I^3$, lines 36–43; copies of this text were set up in Eleusis and in the Eleusinion in Athens.

Triptolemos. The earliest of these[21] shows the hero standing next to his chariot which is now winged for the longer trip outside Attica. He holds the Eleusinian scepter and Demeter hands him ears of grain; Persephone and Hades appear on the reverse. Two fragmentary vases from the same period probably echo the same scene, as suggested by the reconstruction offered by G. Schwarz.[22]

The missionary aspect of Triptolemos's travels was illustrated even more clearly by showing him pouring a libation before his departure, evidently an adaptation of the "warrior's departure," an invention of the Kleophrades Painter.[23] The scene first appeared on the Berlin Painter's hydria in Copenhagen (2696: ARV^2, p. 210, no. 181) of his middle period. Triptolemos is seated in his winged chariot holding out a phiale | toward Kore, who is about to pour a libation from an oinochoe in her right hand while she holds a flaming torch in her left. The same scene is repeated on a hydria in Graz from the painter's late period.[24] To this period belong also two stamnoi in the Louvre (G 371 = ARV^2, p. 208, no. 158 and C 10798 = ARV^2, p. 208, no. 159) with Demeter, Triptolemos and Kore in a multifigured scene, perhaps not independent of Makron's earlier skyphos.[25] [113]

The invention of the multifigured scene, including Triptolemos with Demeter and Kore, may be credited to Makron, who showed this group among six other figures on his famous skyphos in the British Museum.[26] Here all the characters are labeled: Triptolemos, a beardless youth with long hair, myrtle crowned, with chiton and mantle, seated in a chariot equipped with wings and snakes to carry him over land and sea; behind him regal "Demetre" with embroidered chiton and mantle, a polos on her head, a flaming torch in her right hand, three ears of grain in her left. In front of Triptolemos is "Pherephatta" about to pour a libation from the oinochoe in her right hand into the phiale held out to her by Triptolemos, while in her left she holds a flaming torch. Others present include Triptolemos's parents "Eumolppos" and "Eleusis" as well as the gods Zeus, Poseidon, and Amphitrite. Of the 354 vases attributed to Makron, this skyphos of ca. 480 B.C. is the only one which shows Triptolemos and which depicts him in a friezelike composition. One must, however, resist the temptation to claim that the vase painter was inspired by and copied a contemporary mural. This caution is also suggested by a vase painted by an unidentified member of Brygos's workshop which was found in the same tomb

---

21. Karlsruhe 68.101, *Paralipomena*, p. 344, no. 131*bis*.

22. "Zwei eleusinische Szenen auf einem Kelchkrater des Berliner Malers in Athen," *AA* [*JdI* 86], 1971, pp. 178–182: ARV^2, p. 211, no. 201 and p. 205, nos. 119 and 120.

23. See E. Simon, *Opfernde Götter*, Berlin 1953, p. 71.

24. Universität G 30: see G. Schwarz, "Eine Hydria des Berliner Malers in Graz," *ÖJh* 50, 1972/73, pp. 125–133. On some vases, the Berlin Painter shows Triptolemos alone, but the wings on the chariot indicate that he is on his mission: Vienna 3726 = ARV^2, p. 205, no. 113; Dresden 289 = ARV^2, p. 201, no. 69 (both vases are of the middle period); Robinson Collection = ARV^2, p. 203, no. 97 (of his late period).

25. B. Philippaki has observed a close connection in the manufacture of stamnoi in the workshops of the Berlin Painter and of Makron (*The Attic Stamnos*, Oxford 1967, pp. 44 and 151).

26. E 140 = ARV^2, p. 459, no. 3; see Simon, *op. cit.* (footnote 6 above), pp. 111–112, fig. 105.

at Capua as the Makron skyphos,[27] and which contains an eight-figure scene consisting of Triptolemos in a snakeless winged chariot, stretching out his phiale to a draped woman who holds a flower in her right hand. She stands in a building, indicated by two Doric columns; in this building yet another woman is standing and Hades is sitting on a throne. Behind Triptolemos is another woman and behind her a warrior with a phiale, a woman with two torches and a winged female holding both oinochoe and phiale. This ambitious composition, rather ineptly drawn, owes nothing either to Makron's vase painting or to a hypothetical mural in the Eleusinion.[28]

During the following period, when Athens had become rich and powerful, the Mission of Triptolemos gained in popularity to judge by its appearance on [114] vase paintings | by distinguished artists working in the best workshops, for instance by the Altamura Painter and by his "younger brother," the Niobid Painter. Among the six Triptolemos pictures of the Altamura Painter there is one in Munich[29] which recalls a vase by the Berlin Painter (see footnote 21 above). Here the hero is standing behind his chariot while an Eleusinian goddess is facing him, holding an oinochoe in her right hand. Two of the eight Triptolemos vases attributed to the Niobid Painter have similar two-figure scenes,[30] while the canonical three-figure scene, with Triptolemos between Demeter and Kore, appears on a number of vases painted by the two artists.[31] On other vases, the two painters include the three figures in multifigured compositions.[32] The Niobid Painter alone increased the number of figures in

27. J. D. Beazley, "The Brygos Tomb at Capua," *AJA* 49, 1945, pp. 153–158. This cup is now in Frankfurt's Städelsche Kunstinstitut V/7 = *ARV²*, p. 386; see also M. Wegner, *Der Brygosmaler*, Berlin 1973, pp. 53–56.

28. A fragmentary pelike in the Getty Museum in Malibu is probably to be dated to the same ripe archaic period. It shows Triptolemos and another figure on one side, with a Dionysiac scene on the other. M. Robertson, who will publish this vase, attributes it to the Pan Painter.

29. 2383 = *ARV²*, p. 591, no. 23; K. Vierneisel generously sent us a picture.

30. A neck amphora in Leiden (PC 76 = *ARV²*, p. 605, no. 59) and a Nolan amphora in the British Museum (E 274 = *ARV²*, p. 604, no. 53) on which the chariot and the oinochoe are omitted (*CVA*, British Museum 3 [Great Britain 4], 13 [178]: 2a–c).

31. Most pleasing is the scene on a kalyx-krater by the Altamura Painter in Lyons (E 120 = *ARV²*, p. 591, no. 24, illustrated by C. Dugas, "Le Peintre d'Altamura au Musée de Lyon," *JHS* 71, 1951, p. 58, pls. 25, 26); more crowded are the four figures on another kalyx-krater in Leningrad (639 = *ARV²*, p. 591, no. 15, illustrated by A. A. Peredolskaya, *Krasnofigurnye attischeskie vazy*, Leningrad 1967, pp. 146–147, no. 170, pl. 112:1). The Niobid Painter has two vases with three figures: a stamnos in Lugano (*Paralipomena*, p. 395, no. 41 *ter*, illustrated in Münzen und Medaillen A.G., *Auktion* 34, Basel 1967, no. 165, pl. 54) with a shape which is favored by the Altamura Painter; see Philippaki, *op. cit.* (footnote 25 above), pp. 73 and 153. This stamnos shows Triptolemos wearing only a mantle which leaves his right shoulder bare. The other vase is a kalpis in the Metropolitan Museum of Art (41.162.98 = *ARV²*, p. 606, no. 80, illustrated in *CVA*, Fogg Museum, Gallatin Collection 1 [USA 8], 56 [404]: 1); its representation reverts to earlier details in showing the dress of Triptolemos as a chiton with mantle and in having snakes added to the chariot, as on the British Museum skyphos by Makron (see footnote 26 above), although the vase is assigned to the late work of the Niobid Painter by T. B. L. Webster, *Niobidenmaler*, Leipzig 1935, p. 22, no. 50.

32. The Altamura Painter used a crowded six-figure composition only once, in a miniature frieze on one side of the neck of his name vase, the volute-krater in London, British Museum E 469 = *ARV²*, p. 589, no. 1. The Niobid Painter employed five figures on the body of a bell-krater in Perugia (*ARV²*, p. 603, no. 34), and possibly on a kalyx-krater from Lokroi in Reggio (*ARV²*,

such a scene to eight on two kalyx-kraters. The one in the Louvre (G 343 = ARV^2, p. 600, no. 17), possibly in imitation of the Altamura Painter's name vase in the British Museum (see above, footnote 32), shows the Mission on one side of the neck, while on the krater from Spina (Ferrara T13 = ARV^2, p. 602, no. 24) it occurs on the lower of the two zones on side B. As on the Altamura Painter's krater in Lyons (see above, footnote 31), Demeter and Persephone stand in front of Triptolemos's chariot which is equipped with both wings and snakes; the other five figures are depicted as participants in Eleusinian rites by the bunches of grain or by the torches which they hold in their hands. The scene on the reverse shows a Dionysiac thiasos, suggesting once again that Athens was the donor of both grain and wine.[33]

The continued popularity of the Mission is shown by the great number of vases on which it was represented between 475 and 425 B.C.[34] Five of these were decorated by Polygnotos, a successor of the Niobid Painter, who was not so much influenced by wall paintings as his master and his own followers have been assumed to have been.[35] It is clear, however, that Polygnotos imitated the Niobid Painter in his representations of the Mission of Triptolemos.[36] These representations, in turn, may have inspired a member of Polygnotos's group, the Kleophon Painter,[37] to compose the picture on the volute-krater in the Stanford University Museum of Art, 70.12. It seems, therefore, that there was an unbroken tradition from the Berlin Painter to the Kleophon Painter, a

[115]

p. 603, no. 40, illustrated in *NSc*, 1917, p. 154, fig. 60) where two figures to the right of Triptolemos may be restored to balance the two on his left behind his chariot.

33. D. Feytmans claims ("Une représentation inusitée du départ de Triptolème," *AntCl* 14, 1945, pp. 285–295) that on a pelike in Brussels (R 235 = ARV^2, p. 1121, no. 11) the youth standing in a chariot drawn by horses is Triptolemos, although he has no attributes identifying him. Since the women standing in front of the chariot hold, on one side, two torches, and, on the other, an oinochoe and a phiale, the scene is surely Eleusinian, possibly echoing representations of the Mission. Under one handle Herakles is seated, under the other Dionysos enthroned.

34. Dugas, *op. cit.* (footnote 1 above), counted 36 in the second quarter of the fifth century B.C. (pp. 26–28) and 16 in the third quarter (pp. 28–29). To these should be added the oinochoe in the National Museum, Athens 1545 with a two-figure libation scene; see E. Buschor in FR III, p. 262.

35. See P. E. Arias and B. B. Shefton, *A History of 1000 Years of Greek Vase Painting*, New York 1962, p. 366; C. M. Robertson, "Attic Red-Figure Vase-Painters," *JHS* 85, 1965, p. 97.

36. None of the chariots on the five vases has snakes, and they all float, except for no. 1.

1. Stamnos in Florence, 75748 = ARV^2, p. 1028, no. 8; this may be the latest.
2. Kalyx-krater in Duke University, Art Museum = *Paralipomena*, p. 442, no. 27 *bis*. The four figures are identified by inscriptions; the one behind Kore is "Hekate." The scene on the back, three mantled youths, has no connection with the Mission.
3. Neck amphora in London, B.M. E 281 = ARV^2, p. 1030, no. 36; here, the hero is clad only in a mantle as on the Niobid Painter's stamnos in Lugano (see above, footnote 31).
4. Neck amphora in Cannes, private collection = ARV^2, p. 1031, no. 37 = *Paralipomena*, p. 442.
5. Stamnos in Capua, 7529 = ARV^2, p. 1028, no. 7. Triptolemos is clad only in a mantle (see no. 3). The three draped women on side B continue the Eleusinian scene since one carries an Eleusinian torch, another a scepter.

37. See Arias and Shefton, *op. cit.* (footnote 35 above), p. 368, on the relationship between the two painters. The argument for the identification of the painter of the Stanford vase will be presented elsewhere.

tradition probably independent of sculpture, relief, and wall or panel painting.

The fragmentary relief showing the Mission of Triptolemos which was found in the Eleusinion in the Agora (S 1013) seems to be later than most of the representations on vases, although it is proof that the worship of Triptolemos on his Mission took place during the fifth century in the sanctuary which also contained a "temple" of the hero.[38] The great relief from Eleusis in the National Museum of Athens is strictly Eleusinian (see above, footnote 11) and does not seem to have influenced the representation of the Mission of Triptolemos on vases which were painted in Athens.[39]

The Stanford krater shows on Side A (Plate 21) a unique scene of the Mission of Triptolemos consisting of five figures. On the left is Demeter (Plate 22) sitting on the "agelastos petra" which was located next to the well Kallichoron in the sanctuary in Eleusis;[40] she is identified by an inscription which is now difficult [116] to read. In her left | hand she holds a torch and in her missing right hand she probably held out a phiale into which the maiden who is standing in front of her is pouring a libation out of an oinochoe. This maiden, labeled "Parthenos," holds a torch in her left hand. The actual mission scene (Plate 21) shows Triptolemos, Persephone and a second woman (possibly Hekate). Triptolemos wears a crown of ears of grain, and he is holding others in his left hand; his right hand is stretched out holding a phiale for the libation. His winged chariot equipped with two snakes is already afloat. The young woman standing in front of him is labeled "Kore"; she may be the same person as the "Parthenos" standing in front of Demeter, or the figure called "Parthenos" may refer to the well called Parthenion which is an old name for Kallichoron.[41] Kore pours the libation from the oinochoe in her right hand while her left hand holds the Eleusinian torch reversed toward the ground. Of the label for the figure behind her only the letter E is preserved; it may have been Hekate; she stands quietly holding an Eleusinian scepter in her right hand. Evidently, the vase painting shows two Eleusinian scenes, the arrival of Demeter in Eleusis and her reception by the "Parthenos" Kallidike, the daughter of Keleos (*Hymn*, line 145), and the departure of Triptolemos, sent off by Demeter in gratitude for the reception she had received in Eleusis.

The other side of the krater (B, Plate 23) shows two pairs of two figures each. Dionysos (the first four letters of the name are still visible) stands quietly in right profile; he holds a thyrsos in his right hand and a kantharos in his left. The young woman to the right of the god stands frontally and holds an oinochoe in her right hand and an Eleusinian torch in her left; she is about to pour wine into the kantharos held by Dionysos. Behind the god in a separate scene stands

---

38. See T. L. Shear, *Hesperia* 8, 1939, pp. 207–211, fig. 9; Travlos, *op. cit.* (footnote 14 above), pp. 198–201, fig. 262, to which there is no reference in the text.
39. See R. Lullies, *Greek Sculpture*, London 1960, p. 80, figs. 172, 173; G. E. Mylonas, *Eleusis and the Eleusinian Mysteries*, Princeton 1961, pp. 192–193.
40. See Apollodoros, 1.5.1; G. E. Mylonas, *The Hymn to Demeter and Her Sanctuary at Eleusis*, St. Louis 1942, p. 69.
41. See *Hymn to Demeter*, line 99, and the comments by Richardson, *op. cit.* (footnote 11 above), pp. 326–327.

a frontal Papasilenos, white haired, partially bald and bearded; he leans on a staff held in his right hand and carries a wineskin, decorated with an ivy garland, over his left shoulder; he is labeled Pom(po)s. The young woman to the left of Papasilenos is named but her name cannot be restored with confidence;[42] she stands in right profile and plays the lyre held in her left hand; her bent right wrist indicates the use of the plectrum. The use of the lyre, along with the dignified poses of all four figures, indicates that the scene is no orgy but a solemn religious procession with Eleusinian overtones.

The two sides of the krater present good illustrations of the benefactions of Demeter and Dionysos which are praised by Teiresias in the *Bacchai* of Euripides (lines 271–286).

The scenes on the front and back are ingeniously connected by the figures under the handles. Running away from the Triptolemos scene on Side A and toward the lyre player on Side B is a vigorous satyr (Plate 24), a thyrsos over his right arm and a spotted animal skin over his left shoulder. More unexpected is the Pan (so labeled) under the other handle (Plate 22), who, also wearing an animal skin, looks back at Demeter on the "agelastos petra" while running toward the Dionysiac scene. Although no connection is attested between Pan and either Dionysos or Demeter for Greek art of the fifth century B.C.,[43] the appearance of this god of open spaces on the same vase with the deities of grain and of wine may be intended to emphasize the essentially agricultural milieu of the setting as a whole. The political-historical note is again echoed when we remember that Pan saved the Athenians in the Persian Wars by frightening their enemies (Herodotos, 6.105, 8.65; Pausanias, 1.28.4).

The Stanford volute-krater contains the most detailed representation of the Athenian benefactions to the world, a claim which was the very basis of a contemporary decree regulating the offering of the First Fruits (Aparchai) in the sanctuary of Eleusis and ordering sacrifices to the goddesses and to Triptolemos (lines 36–38).[44] This claim of having benefited all mankind is also stressed by Plato in the *Menexenos* (238 A–B) the dramatic date of which belongs to the same period, before the death of Perikles. According to Isokrates (*Panegyricus*, 28–29) the Athenian benefactions were a standard topic of the public funeral orations, the Epitaphioi, which contained a patriotic version of Athenian history. It was to this tradition that the conservative Dadouchos Kallias, the son of Hipponikos, referred when he delivered his speech in Sparta (Xenophon, *Hellenika* 6.3.6): Sparta was to make peace with Athens because the Spartans were the first beneficiaries of the gifts bestowed by Triptolemos.

During the last quarter of the fifth century B.C., when Athenian power and authority over the allies began to wane, the Mission of Triptolemos was less

42. See Münzen und Medaillen, A.G. *Auktion 40*, Basel 1969, no. 108, pp. 65–67, pls. 45, 46: (---)AIT(---)A ∨ E.
43. Our negative results in a search for such a connection were affirmed by H. Walter (*per litt.*) in June 1980; cf. idem, *Pans Wiederkehr*, Munich 1980.
44. *IG* I², 76 = *IG* I³, 78; see R. Meiggs and D. Lewis, *A Selection of Greek Historical Inscriptions*, Oxford 1969, no. 73; a copy of this decree was to be set up in the Eleusinion in Athens, and a small fragment of this copy has been found.

often depicted on Attic vases.[45] In the fourth century B.C., the vase paintings of this scene correspond more closely to the dedicatory reliefs which show Triptolemos along with many other Eleusinian deities, including Dionysos;[46] the Mission itself has become mythical history, and it had lost its political relevance after the downfall of the Athenian Empire. The religious and cultural significance of the Mission of Triptolemos continued into the Hellenistic and Roman periods just as the cultural mission of Athens was cherished in later times.[47]

45. Dugas (*op. cit.* [footnote 1 above], pp. 29–30, nos. 88–91) cites only five vases of this period, and only one has been included in *ARV²*, p. 1315, no. 2a.

46. For example, the kalyx-krater by the Telos Painter, *ARV²*, p. 1427, no. 37, and the skyphos by the Marsyas Painter, *ARV²*, p. 1475, no. 8, both omit the libation scene.

47. See A. Alföldi, "Frugifer Triptolemos im ptolemäisch-römischen Herrscherkult," *Chiron* 9, 1979, pp. 553–606. G. Schwarz is preparing a monograph on the representations of Triptolemos in Hellenistic and Roman Art.

31

The Eleusinian Spondai
(IG I³, 6, lines 8–47)

The original purpose of the Olympic *Ekecheiria* was to promote concord [263] (*homonoia*) and peace (*eirene*), or so it was claimed by the champions of Olympic Panhellenism at the end of the Peloponnesian War and during the period of the Common Peace (*Koine Eirene*) at the beginning of the fourth century B.C., by Gorgias, Hippias, Lysias, Isocrates. We have a detailed description of the working of the Olympic Truce during the Games of 420 B.C. and during the preceding ones of 424 B.C. (Thucydides, 5, 49–50; for the earlier conflict at Lepreon, see 5, 31). It seems that in 424 B.C., the Spartans were putting a garrison into Lepreon after the Eleans had proclaimed the Olympic Truce and before their heralds had reached Sparta and announced it; for this (alleged) violation, the Spartans were fined two thousand minae (two for each soldier), and since they refused to pay or to agree to any compromise, they were excluded from the next Games in 420 B.C. This meant that no Spartan could participate under his own name as a Spartan and one who was discovered having had his victory announced in the name of the Boeotians was officially flogged. The Eleans feared, said Thucydides, that the Spartans would enter using military force, but they did not, and all the Greeks participated except the athletes from Lepreon and Sparta. This story makes it clear that the Olympic Peace was obligatory and that it was announced, that the Eleans were acting on behalf of the Greeks in arranging and conducting the games and in enforcing the truce. The story suggests, moreover, that the Ekecheiria was announced to entire national groups (comparable to the *Ethne* of the Amphictyony) and not to individual cities.

Turning from the Olympic "Peace" to the Eleusinian "Peace", we notice both differences and similarities. The *Spondai* were not only announced but also concluded by Athenians from the Eleusinian families of the Eumolpidai and of the Kerykes who were appropriately called *Spondophoroi* (*IG* II², 1672,

Philia Epē, Bibliothēkē tēs ēn Athēnais Archaiologikēs Hetaireias 103 (Athens, 1986), 2:263–265 [with Mariko Sakurai]

line 4); these went to the Κοινόν of the Aitolians (M. N. Tod, *GHI*, II, 137), to the Phocians (Aeschines, 2, 133—134), and to the Islands (*IG* II², 1672, line 4), but apparently not to any one individual community. An earlier document of the second quarter of the fifth century B.C. records the establishment of the *Spondai* (*IG* I³, 6, lines 8–47).

```
           σ]πονδὰς εἶν-                         ὰς δὲ σπονδὰς
         [αι] τοῖσι μύστ-                         εἶναι ἐν τεῖσ-
     10  [εσιν]καὶ το[ῖς]                     30  ι πόλεσιν hό[σ]-
         [ἐπ]όπτεισιν [κ]-                       αι χρῶνται τό-
         [αὶ τ]οῖς ἀκολ[ο]-                       ι hιεροῖ καὶ 'Α-
         [ύθ]οισιν καὶ [χ]-                       θεναίοισιν ἐ-
         [ρέ]μασιν τὸν [ὀ]-                       κεῖ ἐν τεῖσιν
     15  [θ]νείον καὶ ['Αθ]-                  35  αὐτέσι πόλεσ-
         [ε]ν[δ]ίοισιν [h]ά-                      είζοσι μυστε-
         πασιν· ἄρχε[ν] δ-                        ρίοισιν τὰς [σ]-
         ὲ τὸν χρόνο[ν] τ-                        πονδὰς εἶνα[ι]
         ὸν σπονδὸν [τ]ὸ                      40  τὸ Γαμελιῶνο-
     20  Νεταγειτνιό-                             ς μενὸς ἀπὸ δ[ι]-
         νος μενὸς ἀπ[ὸ]                          [χ]ομενίας κα[ὶ]
         διχομενίας [κ]-                          τὸν 'Ανθεστε[ρ]-
         αὶ τὸν Βοεδρ[ο]-                         [ι]ῶνα καὶ τὸ 'Ελ-
         μιῶνα καὶ τὸ [Π]-                    45  αφεβολιῶνος
     25  υανοφσιῶνος                              μέχρι δεκάτε-
         μέχρι δεκάτε-                            ς hισταμένο.
         ς hισταμένο· τ-                                vacat
         ιν· τοῖσι δὲ ὀλ-
```

"There is to be a treaty (*Spondai*) between the foreign initiates (*Mystai* and *Epoptai*), their company and their possessions, and all the Athenians." After the duration of the treaty is accurately defined, it is further stated that "the treaty is to be in force in all the cities which frequent the sanctuary and it is to apply also to the Athenians living in these very cities." Finally, the treaty is said to apply also to the Lesser Mysteries, and the duration of this second treaty is also given with precision.

This passage has been understood to mean that "heralds were sent out in advance to all the cities that recognized the festival inviting them to declare a truce" (see H. W. Parke, *Festivals of the Athenians*, 61, referring to G. E. Mylonas, *Eleusis*, 244). We believe that the phrase σπονδὰς εἶναι rather refers to the validity of a treaty in the same sense in which Thucydides uses the words, 5, 18,3: ἔτη δὲ εἶναι τὰς σπονδὰς πεντήκοντα 'Αθηναίοις...καὶ Λακεδαιμονίοις; also 5, 49,3: 'Ηλεῖοι δὲ τὴν παρ' αὐτοῖς ἐκεχειρίαν ἤδη ἔφασαν εἶναι. We would have here a mutual guarantee and protection of foreign visitors to the Eleusinian Mysteries which would cover also the Athenians residing in the cities whence visitors go to Eleusis. There is no mention of heralds or *Spondophoroi* in this inscription (which is not completely preserved) nor in the completely preserved decree regulating the *Aparchai* (*IG* I³, 78). It could be that the mission of sacred heralds was started later, perhaps in the early fourth century B.C.

Considering the historical circumstances which form the background of the new organization of the Mysteries ca. 470 B.C., Mylonas himself has called already attention to their growing popularity, saying simply that "everyone knew that Demeter gave the victory of the Greeks in the battle of Salamis" (see *Eleusis*, 107 and I. K. Raubitschek, *Hesperia*, Supplement 20, 1982, 112 [*supra*, pp. 232–233]). The story told by Herodotus about Dikaios the Athenian (8, 62) culminates in the observation that any Athenian and any Greek who so desired could be initiated in the Mysteries. This may have been possible already before the Persian Wars, but after the wars the Athenians made a great effort to present the Mysteries as a panhellenic gathering comparable and even superior to the Olympic Games because of their promise to the initiates of a good life after death. | Since the center of the new panhellenic alliance was located on the island of Delos, the Athenians put the Mysteries of Eleusis forward as a panhellenic festival with its center in Attica. The institution of the *Spondai* in imitation of those of Olympia would encourage the attendance, and the Mysteries themselves were expected to surpass the athletic competitions at the other panhellenic Games because they provided not only the spectacle of the *Dromena* but the Mysteries themselves with their promise of an everlasting happy life.

IV

POETS, LITERATURE, AND HISTORIOGRAPHY

32

Das Denkmal-Epigramm

Die zwei Fragen die ich erörtern möchte, ohne sie beantworten zu können, *[3]* beziehen sich auf den Ursprung des Epigramms und auf das Verhältnis zwischen Epigramm und zugehörigem Denkmal. Beide Fragen sind archäologisch, historisch, und epigraphisch eher als literarisch, und dieser Abstand von den anderen Themen der Tagung fordert eine kurze Erklärung.

Das Epigramm ist eine Aufschrift in metrischer Form, die sowohl von einem Gedicht wie von einer Aufzeichnung verschieden ist. Mit dem Gedicht hat das Epigramm die Form gemeinsam, unterscheidet sich aber von ihm dadurch dass es eine *Inschrift* ist die eng und einzigartig mit einem Denkmal verbunden ist, um dieses Denkmals willen geschaffen wurde und ein wesentlicher Teil dieses Denkmals ist. Als Inschrift ist ein Epigramm von einer Aufzeichnung nicht verschieden, doch unterscheidet es sich von ihr dadurch, dass die Aufzeichnung ohne direkten Bezug auf das Schreibmaterial steht von dem sie gelöst gedacht werden kann. Die uns bekannten Gesetze, Beschlüsse, und Verträge des Altertums sind uns zwar in den meisten Fällen auf Stein oder Bronze erhalten, sie verlieren aber wenig ihrer Bedeutung wenn sie auf Papyrus oder gar Papier übertragen werden. Die echten Epigramme jedoch, aus welchen Gründen sie immer von ihren Denkmälern gelöst in alten oder neuen Sammlungen geschrieben oder gedruckt erhalten sind, führen ein Schattendasein, auch wenn sie von Meisterdichtern stammen.

Es geht hier um das Epigramm wie es vor der Zeit des Simonides auf den Grabmälern und Weihgeschenken der Griechen stand und über das wir aus der schriftlichen Überlieferung sehr wenig erfahren. Mit Simonides scheint das literarische Epigramm im eigentlichen Sinne anzufangen und | vom 5. Jh. an *[4]* wird es zu einer Kunstform, deren Verhältnis zur früheren Elegie das nächste Thema dieser Tagung sein wird. Daher soll diese Beziehung heute nur ganz

Fondation Hardt, *Entretiens sur l'antiquité classique* 14, *L' Epigramme Grecque* (Vandœvres—Genève, 1969) 3–36 kurz zur Sprache kommen und nur von der Seite des Denkmals her erörtet werden.

Die Frage nach dem Ursprung einer Kulturform, sei es nun in der Kunst, in der Philosophie, oder in der Literatur, birgt in sich die Entscheidung zwischen Schöpfung und Entwicklung, und diese schwierige Entscheidung lässt sich auch bei der Frage nach dem Ursprung des Epigramms nicht vermeiden. Im allgemeinen ist es zwar richtig dass Schöpfung und Entwicklung weder in der göttlichen Welt noch in der menschlichen Kultur wirkliche Gegensätze darstellen; jede Betrachtung zeigt dass Entwicklung einsetzt sobald Schöpfung stattfand, dass die eine ohne die andere weder denkbar noch erfahrungsgemäss fassbar ist. In diesem Sinne ist es berechtigt von der Schöpfung und weiteren Entwicklung des griechischen Alphabets, der attischen Tragödie, des griechischen Portraits zu sprechen, und das meinen wir ja auch wenn wir die Frage nach dem Ursprung einer Literaturform stellen. Natürlich sind wir uns dessen bewusst dass es keine Schöpfung *ex nihilo* gibt, das heisst dass der Erfinder entweder geformtes oder ungeformtes Material vor sich hatte aus dem er in neuartiger Verbindung und ingeniöser Veränderung etwas schuf das vorher nicht existierte, das einen bestimmten Zweck erfüllte und das dann weiterhin von anderen entwickelt wurde.

[5] Was das Epigramm betrifft, so scheint die Sache verhältnismässig einfach zu liegen da wir es nicht mit einem flüchtigen Gedanken oder einem geflügelten Wort zu tun haben, auch nicht mit einer momentanen Handlung, wie beim Epos, der Chorlyrik, dem Drama, sondern mit einem festen dreidimensionalen Gegenstand dessen Bestehen oder Nichtbestehen keinem Zweifel unterliegt, der von allem Anfang an als Denkmal geschaffen wurde, das heisst, um Bild und Wort zu verewigen. Während das Verhältnis zwischen mündlicher Überlieferung und schriftlicher Festlegung beim Epos und auch bei der Chorlyrik immer problematisch bleiben wird und wir kaum je beweisen werden können, dass es sich bei den frühesten Aufzeichnungen nur um Gedächtnishilfen für Dichter und Sänger handelte, so liegt die Sache beim Epigramm ganz anders. Mit ihm fängt die Literatur an, denn das Epigramm wurde aufgeschrieben um gelesen zu werden, nicht nur von Zeitgenossen, sondern besonders von Nachkommen und späteren Generationen. Grabmäler gab es schon vorher aber diese bestanden aus Hügeln und Stelen; auf sie beziehen sich eine Reihe von Homerstellen und solche Grabstelen sind uns aus dem zweiten und dem frühen ersten Jahrtausend bekannt. Die im Epos gebrauchten Worte *Tymbos, Sema, Stele* beziehen sich demnach nicht nur auf Denkmäler der Zeit Homers sondern auch auf die der früheren Heroenzeit über die er dichtet. Dabei ist das *Sema* mit dem *Tymbos* identisch und die Stele stand wohl auf dem Grabhügel (*Il.* 7, 86; 11, 371; 16, 457; *Od.* 11, 75; 12, 14–15). Von einer Inschrift ist hier nicht die Rede, auch nicht von einem Klagelied das bei der Errichtung des Grabmals gesungen wurde.

Es gibt aber eine Stelle in der *Ilias* (7, 81–91), die ein neues Element einführt, das so eng mit dem Grabepigramm verbunden ist dass wir diese Verse genauer betrachten müssen. Hektor spricht hier von dem Zweikampf zwischen ihm und einem erlesenen Griechenhelden und nimmt verständlicherweise an, dass er in so einem Kampfe der Sieger sein würde. Die Waffen des erschlagenen Gegners will Hektor der Leiche abnehmen, zum heiligen Ilion bringen und am Tempel des Ferntreffers Apollo aufhängen—aber die Leiche wird er bei den Schiffen abgeben damit die Achäer dem Toten am breiten Hellespont ein *Sema* aufwerfen können. "Und," sagt dann Hektor, "einmal wird einer der spätergeborenen Menschen sagen, wenn er in seinem Schiffe vorbei | fährt: Dies ist das *Sema* eines Mannes der vor langem starb, den einmal in noblem Einzelkampfe der treffliche Hektor erschlug. Das wird einer einmal sagen, aber mein Ruhm wird niemals vergehen."

Hans-Martin Lumpp, ein Schüler von Hommel aus Tübingen, hat in einem ausgezeichneten Aufsatz, der in den *Forschungen und Fortschritten* für 1963 (Bd. 37, Heft 7, S. 212–215) erschien, diese Homerstelle mit einem Grabepigramm aus Korfu (Peek, *G.V.* 73) zusammengestellt und auf die ausserordentliche Ähnlichkeit dieser Epigramme hingewiesen. Und so zeigt auch ein Vergleich weitgehende Übereinstimmungen. Bei Homer liest man:

ἀνδρὸς μὲν τόδε σῆμα πάλαι κατατεθνηῶτος
ὅν ποτ᾽ ἀριστεύοντα κατέκτανε φαίδιμος Ἕκτωρ.

Und auf der korkyräischen Stele steht:

Σᾶμα τόδε ᾽Αρνιάδα Χάροπος τόνδ᾽ ὤλεσεν Ἄρες
βαρνάμενον παρὰ ναυσὶν ἐπ᾽ ᾽Αρράθοιο ῥοαῖσι
πολλὸν ἀριστεύ[[τ]]οντα κατὰ στονόεσ(σ)αν ἀϝυτάν.

Eng verbunden damit sind zwei weitere Epigramme (Peek, *G.V.* 1224 und 321) eines von Athen, schon aus dem sechsten Jahrhundert:

Στῆθι καὶ οἴκτιρον Κροίσο παρὰ σῆμα θανόντος
ὅν ποτ᾽ ἐνὶ προμάχοις ὤλεσε θόρος Ἄρες,

das andere aus Thisbe in Böotien wohl auch aus dem sechsten Jahrhundert, dessen Pentameter die Worte enthält:

ὅς ποτ᾽ ἀριστεύον ἐν προμάχοις [ἔπεσεν].

Diese enge Beziehungen zwischen der Iliasstelle und erhaltenen griechischen Steinepigrammen des späten 7. und 6. Jh. legt die Frage nahe, ob der Dichter schon Grabepi|gramme vor sich hatte oder ob sich die Epigramme auf die Homerstelle beziehen, oder ob das enge Verhältnis zwischen beiden anders zu erklären ist. Am unwahrscheinlichsten scheint mir die Annahme zu sein, dass Homer ganz unbewusst das Grabepigramm erfand und es dann zufällig in seinen Gesängen Generationen später entdeckt und auf Stein aufgezeichnet wurde. Aber die erste Möglichkeit muss ernsthaft erwogen werden, nämlich die, dass dem Dichter jener Stelle schon Grabepigramme vorlagen, auf die er sich bezog. Man könnte behaupten, dass der spätergeborene Schiffer, der das Grabmal des von Hektor getöteten Griechen betrachtet, gar nicht wissen

konnte, dass der Mann vom "glänzenden Hektor" erschlagen worden war—
wenn er es nicht auf einer Inschrift las die auf dem *Sema* stand.

Trotzdem scheint mir diese bestechende Annahme abwegig zu sein auch wenn es sich herausstellen sollte, dass die Iliasstelle einen späteren Zusatz darstellt, was ich nicht glaube. Die zweifache Betonung des Sprechens und nicht des Lesens: καί ποτέ τις εἴπῃσι und ὥς ποτέ τις ἐρέει, zeigt doch dass der Dichter nicht an eine Inschrift dachte, sondern an einen Ausspruch, und diese Idee, der auch Werner Peek zustimmt, möchte ich etwas weiter ausführen.

Schon Friedländer hat in seiner schönen Einleitung zum elegischen Epigramm (*Epigrammata* 65–70) die einfache und überzeugende Feststellung gemacht, dass es gut möglich ist dass die inschriftlich erhaltenen Grabgedichte den mündlich gesungenen Trauerliedern nachgebildet sind, dass das Grabepigramm mit der Grabelegie eng verbunden ist. Friedländer sieht in den längeren Epigrammen eine Verknüpfung von sachlichen Angaben und lyrischen Reflexionen, und er dehnt diese Betrachtung vom Grabepigramm auf das Weihepigramm aus. Diese allgemeinen Beobachtungen lassen sich an Einzelbeispielen gut belegen.

[8] Eines der ältesten und längsten Epigramme stammt von einem runden Grabbau in Korfu, der leider bis heute nicht | ordentlich veröffentlicht ist. Die einzeilige Inschrift läuft in sechs Hexametern unter dem Rande des Sockels auf dem einst vielleicht die Statue eines Löwen lag. Der Text (Peek, *G.V.* 42) muss noch immer in der alten Umzeichnung gelesen werden.

Υἱοῦ Τλασίαϝο Μενεκράτεος τόδε σᾶμα
Οἰανθέος γενεάν· τόδε δ'αὐτôι δᾶμος ἐποίει·
ἐς γὰρ πρόξενϝος δάμου φίλος· ἀλλ' ἐνὶ πόντοι
ὤλετο, δαμόσιον δὲ καϙὸν [∪∪—∪∪—∪]
Πραξιμένες δ'αὐτôι γ[αία]ς ἀπὸ πατρίδος ἐνθὸν
σὺν δάμοι τόδε σᾶμα κασιγνέτοιο πονέθε.

Man hat oft bemerkt, dass dieses Epigramm kein grosses dichterisches Meisterwerk ist, aber gerade darum zeigt es uns noch klarer die Verbindung zwischen Totenrede in gebundener Sprache und Inschrift. Man kann sich leicht vorstellen dass diese sechs Verse, die alles wesentliche über den Toten aussagten, über seinem Grabe gesprochen wurden und dann auch auf dem Grabmal aufgezeichnet wurden. Die Inschrift verewigt demnach ein Ereignis das sich anlässlich der Begräbnisses oder anlässlich der Fertigstellung des Grabdenkmals zugetragen hat. Das Gedicht konnte gesprochen werden ehe es aufgeschrieben wurde.

Ein ähnliches Beispiel unter den Weihepigrammen wurde in Krissa bei Delphi gefunden und ist leider auch nur in alten Zeichnungen erhalten. Die Lesung ist klar mit Ausnahme eines Wortes das aber leider von grosser Bedeutung ist (*S.E.G.* 15.351).

Τάσδε γ' Ἀθαναίαι δρα[χμ]ὰς Φανάριστος ἔθηκε
Ἥραι τε, hος καὶ κênos ἔχοι κλέϝος ἄπθιτον αἰϝεί.

Es mag sich hier um eine Weihung eines Bündels von Spiessen handeln wie sie als Geld noch im siebenten Jahrhundert ver|wendet wurden und die hier, wie auch in der später zu erwähnenden Inschrift von Perachora, Drachmen, d.h. eine Handvoll, genannt werden; doch muss zugegeben werden, dass die alte Zeichnung δραϝεος bietet was unverständlich ist. Die Weihung wurde von Phanaristos an Athena Pronaia und an Hera gemacht, der ja auch Pheidon um jene Zeit eine ähnliche Weihung in Argos stiftete. Am Schluss wird der Zweck erwähnt; Phanaristos wollte immer unvergänglichen Ruhm haben, so wie auch Hektor in der *Ilias* (7, 91) sagte τὸ δ' ἐμὸν κλέος οὔποτ' ὀλεῖται. Das Gedicht, das hier auf dem Stein steht, auf dem die Drachmen aufgestellt waren, hätte anlässlich der Weihung laut gesprochen worden sein können, das heisst, dem Epigramm geht auch hier ein Denkspruch voraus der durch die Inschrift verewigt wurde; über seine Formulierung wird noch später etwas zu sagen sein.

[9]

Es ist demnach zulässig anzunehmen, dass das Denkmal-Epigramm ursprünglich eine schriftliche Fixierung eines Ausspruchs enthält, der anlässlich eines Begräbnisses oder einer Weihung getan wurde. Das heisst dass der in gebundener Rede abgefasste Grabspruch oder Weihspruch älter als die ältesten Inschriften sein kann, die ja nur in ein paar Ausnahmefällen über das siebente Jahrhundert zurückgehen.

Einer dieser Ausnahmefälle bestätig die hier vorgelegte Überlegung. Es handelt sich um den berühmten Becher des Nestor dessen Text ich hier mit neuen Lesungen und Ergänzungen von Albrecht Dihle vorlege die er selber gesondert besprechen wird (*S.E.G.* 19.621).

Νέστορος: ε[ἰμ]ὶ: εὔποτ[ον]: ποτέριο[ν]·
hὸς δ'ἀ(π)ὸ (τό)δε [πίε]σι: ποτερί[ο]: αὐτίκα κένον
hίμερ[ος: ha ιρ]έσει: καλλιστε[φά]νο: Ἀφροδίτες.

Wir haben hier zwei Inschriften vor uns die sich beide auf denselben Gegenstand beziehen aber verschiedenes von ihm aussagen. Zuerst kommt ein Trimeter der einer einfachen | Besitzerinschrift ähnelt; dieser folgen zwei Hexameter die behaupten, dass wer aus diesem Becher trinkt, den wird augenblicklich Begierde nach der schönbekränzten Aphrodite ergreifen. So ein in gebundener Rede gemachter Ausspruch könnte vom Besitzer des Bechers laut gemacht werden und brauchte gar nicht eingeritzt zu werden. Wie bei den vorher besprochenen Grab- und Weihsprüchen handelt es sich auch hier um ein erst durch die Aufzeichnung zum Epigramm gemachtes Gedicht. Der besondere Wert dieses Gedichtes liegt jedoch nicht nur darin dass es ein fast unabhängiger Ausspruch ist, sondern auch in seinem Alter, denn das Gefäss wird noch vor 700 v. Chr. datiert und gehört somit mit der Inschrift auf der Dipylonkanne zu den ältesten griechischen Inschriften. Übrigens ist sie der Form nach eng der Dipyloninschrift verwandt, die auch mit dem Relativum hός anfängt und keinen sicheren Hinweis auf das Gefäss auf dem sie eingekratzt wurde enthält. Während man im Falle der schon lange bekannten Dipyloninschrift (*S.E.G.* 22.83) im Zweifel sein kann ob das Epigramm überhaupt etwas

[10]

mit dem Gefäss auf dem es steht zu tun hat, zeigt die Inschrift auf dem Nestorbecher was dieses Verhältnis wirklich war. Man kann sich vorstellen dass jemand mit dem Becher in der Hand die zwei Hexameter aussprach die dann aufgeschrieben wurden; und das gleiche könnte auch bei der Dipylonkanne der Fall gewesen sein. Damit scheint erwiesen zu sein dass es ursprünglich Kurzgedichte gab, dichterische Aussprüche, die auf Grabmäler, Weihgeschenke und andere Gegenstände aufgezeichnet wurden als man der Schreibkunst kundig wurde. Obgleich es sich hier um Epigramme handelt wie sie uns von der späteren Tradition des 5. und 4. Jh. wohl bekannt sind, ist mit diesem Typus der Reichtum der archaischen Denkmalepigramme nicht erschöpft, denn das wichtigste Element des frühen Epigramms wurde noch nicht erwähnt, das Epigramm als ein Ausspruch des Denkmals.

[11] Die erste Zeile des Nestorepigramms gibt uns ein einfaches Beispiel eines redenden Epigramms: "Ich bin des Nestors wohl zu trinkender Becher." Warum ist hier und in vielen anderen ähnlichen Fällen, die Friedländer gesammelt hat (S. 162–165, 177), ein Trinkgefäss redend eingeführt? Die meisten dieser redenden Inschriften hat neulich Mario Burzachechi sorgsam zusammengetragen und eingehend besprochen (*Epigraphica* 24 [1962]); leider kann ich mich seiner Hauptthese nicht anschliessen, nämlich dass hinter den sprechenden Inschriften sprechende Gegenstände stehen, die als lebendig betrachtet werden müssen.

Um mit dem wohl ältesten Beispiel, dem Trimeter des Nestorbechers anzufangen, so ist die Ergänzung von ἐ[ἰμ]ί zwar nicht gesichert aber durch viele spätere Beispiele höchst wahrscheinlich gemacht. Um den Unterschied zwischen Νέστορος τόδε ποτήριον und Νέστορος εἰμί ποτήριον zu verstehen, müssen wir annehmen, dass der Sprecher in dem einen Fall Nestor selber ist, in dem anderen Fall er den Becher sagen lässt was er selber nicht sagen kann, da er nicht immer dabei ist wenn der Becher bewundert und verwendet wird.

Diese Übertragung der Botschaft vom ursprünglichen Sprecher auf den Gegenstand über den der Ausspruch gemacht wird lässt sich auch beim Grabmal erkennen, das zwar zu einer bestimmten Zeit von einem Freund oder Verwandten dem Verstorbenen errichtet wurde, das aber darüber hinaus den am Grabe gemachten Spruch späteren Generationen durch die Inschrift vermitteln soll. Die objektiven Feststellungen, wie wir sie an einigen Beispielen, nämlich den Grabepigrammen des Arniades und Menekrates, beobachtet haben, machen den subjektiven Angaben die vom ursprünglichen Sprecher dem Grabmal in den Mund gelegt werden schon früh Platz. Dabei ist es gar nicht so sicher dass die Grabstatue, sei es nun eine Sphinx, ein Löwe, oder gar ein Menschenbild eher als sprachfähig betrachtet wurde als eine einfache Stele oder ein Becher, wie der des Nestor.

[12] Eines der ältesten gesicherten Beispiele ist das auf Thasos gefundene Grabmal des Glaukos, des Sohnes des Leptines, der in den Gedichten des Archilochos (fr. 115 Lasserre) erwähnt wird und somit vielleicht noch ins siebente Jahrhundert gehört. Was dieses grosse Denkmal einst trug, lässt sich nicht mehr feststellen, da nur das rechteckige Fundament erhalten ist, eine Stufe deren

Quadern aus gewöhnlichem Kalkstein bestehen mit Ausnahme von zwei Marmorblöcken die an entsprechenden Stellen in die Längsseiten eingelassen waren und von denen einer die folgende Inschrift trägt (*S.E.G.* 18.338):

Γλαύϙω εἰμὶ μνῆ-
μα τῶ Λεπτίνεω· ἔ-
θεσαν δέ με ωἱ Βρέντ-
εο παῖδες.

Beide Sätze sind vom *Mnema* selber gesprochen das wohl das ganze Denkmal und nicht gerade eine sprechende Tierfigur darstellt. Die ungenannten Söhne oder Mannen des Brentes lassen also das Denkmal den Grabspruch weitergeben, der nicht nur den Namen des Verstorbenen sondern auch die Identität der Errichter des Denkmals angibt. Diese sprechen durch die Inschrift zu dem Betrachter des Denkmals und teilen ihm alles mit was ihnen wesentlich erscheint. So wird die Schrift eine Übertragung und Fortführung der Rede noch ehe sie das gesprochene Wort völlig ersetzt und ein Eigenleben erhält. Eine Spur dieser ursprünglichen Vorstellung ist noch in mancher unserer Anzeigen erhalten die in unserem Namen Einladungen aussprechen und Hochzeiten oder Todesfälle bekannt geben. Von einer Lebendigmachung ist in keinem dieser Fälle die Rede.

Manchmal hat man jedoch den Eindruck, dass der Sprecher Worte einer Grabstatue in den Mund legt die von ihr und nur von ihr gesprochen werden können. Aber das ist eine dichterische Fiktion und es ist nicht erstaunlich, dass unser ältestes und bestes Beispiel als ein Gedicht Homers bezeichnet wird (Peek, *G.V.* 1171).

[13]

1 Χαλκῆ παρθένος εἰμί, Μίδου δ' ἐπὶ σήματος ἧμαι.

2 Ὄφρ' (vel ἔστ') ἂν ὕδωρ τε νάῃ καὶ δένδρεα μακρὰ τεθήλῃ

3 ἠέλιος δ' ἀνιὼν φαίνῃ λαμπρά τε σελήνη,

4 καὶ ποταμοὶ πλήθωσι, περικλύζῃ δὲ θάλασσα

5 αὐτοῦ τῇδε μένουσα πολυκλαύτῳ ἐπὶ τύμβῳ

6 σημανέω παριοῦσι, Μίδας ὅτι τῇδε τέθαπται.

Es unterliegt keinem Zweifel, dass das hier vorliegende Gedicht in den Mund der Statue eines weiblichen Wesens gelegt wird, wohl einer Sphinx deren Bronzebild das Grabmal des Midas krönte (siehe L. Weber, *Hermes* 52, 1917, 543–544). Historisch betrachtet hat George Huxley (*Greek, Roman and Byzantine Studies* 2 [1959] 94–96) es wahrscheinlich gemacht dass der grosse König Midas, um 700 v. Chr. enge Beziehungen zur Griechenstadt Kyme hatte, aber diese Verbindung gibt uns keinen Beweis dafür, dass das vorliegende Gedicht schon zu dieser Zeit geschaffen und auf Stein oder Bronze eingetragen wurde. Plato zitiert die ersten und letzten zwei Zeilen (*Phaedr.* 264 c) und Diogenes Laertius gibt an (1.6.89) dass das Gedicht nach einem Zitat des Simonides von Kleoboulos von Lindos und nicht von Homer stammt. Da sich die Verse des Simonides anscheinend auf unser Epigramm beziehen, ist dieses, so wie wir es vor uns haben, schon im späten 6. Jh. bekannt gewesen (siehe aber Weber, *loc. cit.* 536–545). Schon von Plato wurde betont dass das Gedicht sowohl von vorn wie von hinten gelesen werden konnte (οὐδὲν διαφέρει αὐτοῦ πρῶτον ἢ ὕστατόν τι λέγεσθαι). Das trifft aber für sechs untereinander stehende Hexameter nicht zu, und daher hat man schon seit langem die zwei mittleren Verse die Plato zwar nicht wiedergibt die aber durch das Zitat des Simonides beglaubigt sind, als späteren Zusatz weggestrichen. Es wurde aber, soviel ich weiss, die Frage gar nicht aufgeworfen wie denn dieses Gedicht tatsächlich aufgezeichnet war, das heisst wie die Verszeilen auf dem Inschriftträger verzeichnet waren. Die einzige Angabe die wir darüber besitzen steht in der *Vita Homeri*, die dem Herodot zugeschrieben wurde (11), wo es heisst dass Homer nach dem Tode des Midas auf Ansuchen der Verwandten ποιεῖ τὸ ἐπίγραμμα τόδε τὸ ἔτι καὶ νῦν ἐπὶ τῆς στήλης τοῦ μνήματος τοῦ Γορδίεω ἐπιγέγραπται (στίχοι τέσσαρες). Wenn man in dem Wort στήλη einen verlässlichen Hinweis auf die Form des Grabmals sehen darf so würde es sich wohl um eine Säule handeln wie beim Säulendenkmal des Xenvares (Peek, *G.V.* 52), das noch gleich zu besprechen sein wird. Die Sphinx würde also auf einem ionischen Kapitell gesessen haben, was auch für Kleinasien die passende Säulenbekrönung wäre, und wie wir sie von der Säule der Naxier in Delphi und von anderen Weih- und Grabdenkmälern der Zeit um 600 v. Ch. kennen. Freilich aus dem frühen 7. Jh. sind uns keine ionischen Säulen bekannt. Was den Platz der Inschrift selber betrifft, so könnte sie entweder auf dem Abacus des Kapitells, oder in den Kanneluren oder auf dem Postament gestanden haben, ich glaube aber dass man hier eine Entscheidung treffen kann die der Form des Gedichtes gerecht wird. Dieses besteht augenscheinlich aus drei Teilen die miteinander so verbunden sind dass die mit ὄφρα oder ἔστε eingeleiteten drei Zeilen sowohl auf das vorhergehende κεῖμαι wie auf das folgende μένουσα bezogen werden können, und dass die erste oder die letzten beiden Zeilen (in umgekehrter Ordnung) den Anfang machen können. Diese Überlegung legt die Vermutung nahe dass das Epigramm so in drei oder vier Kanneluren eingeschrieben wurde dass der erste Hexameter in einer stand, die nächsten drei in der links oder rechts danebenstehenden Kannelur, und die letzten beiden Hexameter in der darauf folgenden Kannelur (oder den beiden folgenden Kanneluren). Steht nun der Betrachter vor der Säule so kann | er entweder mit der linken oder der rechten Zeile zu lesen anfangen und er erhält in jedem Fall ein vollständiges Gedicht. Dass es sich hier um ein sehr altertümliches Denkmal gehandelt haben muss ist ganz klar, obwohl es keineswegs sicher ist dass es noch aus dem frühen 7. Jh. stammt (siehe X.I. Καρούζου, Ἐπιτύμβιον Χρηστοῦ Τσούντα 561–562).

Die Frage, ob ein Grabdenkmal eine geraume Zeit nach dem Tode des Mannes errichtet wurde dessen Andenken es ehrt, wurde schon anlässlich des sich auf Glaukos, den Sohn des Leptines, beziehenden Epigrammes erwähnt, und sie muss auch im Zusammenhang mit dem Midasepigramm erwogen werden. Während diese Beispiele ziemlich unsicher sind, besitzen wir noch Reste des Grabmals des Archilochos das erst eine oder gar zwei Generationen nach seinem Tode errichtet wurde. Das erst vor fünf Jahren in Paros gefundene und noch unveröffentlichte ionische Kapitell gehört wohl der ersten Hälfte des 6. Jh. an und kann mit früharchaischen samischen Kapitellen wie mit dem berühmten Kapitell der Naxiersäule verglichen werden. Auf der Oberseite befindet sich nun eine tiefe Einarbeitung die zur Aufnahme einer Grabfigur und zwar einer Sphinx diente wie auch sonst archaische Sphingen mit Zapfen in Kapitelle eingelassen wurden. Der Inhaber dieses Grabmals das wie gesagt nicht lange nach 600 v. Ch. anzusetzen ist, wird in einem Epigramm genannt, das um 400 v. Chr. in den wagrechten Teil des Voluten canalis eingetragen wurde, nachdem die Verzierung abgemeisselt worden war. Dieses Epigramm zeigt nun dass es sich um das Grabdenkmal des Archilochos handelt (Daux, *B.C.H.* 85, 1961, 846–847):

᾽Αρχίλοχος Πάριος Τελεσικλέος ἐνθάδε κεῖται
τὸ Δόκιμος μνημήϊον ὁ Νεοκρέωντος τόδ᾽ ἔθηκεν.

Ursprünglich wollte man annehmen dass das Epigramm in die Zeit seiner Eintragung gehört und dass das hoch|archaische ionische Kapitell daher für das Grab des Archilochos wiederverwendet wurde. Aber Kontoleon, der diesen Vorschlag während der dem Archilochos gewidmeten Tagung der Fondation Hardt machte (*Archiloque*, Entretiens 10, 45–46; Χαριστήριον εἰς A. K. ᾽Ορλάνδον 415–416), hat mir jetzt brieflich versichert, dass er bereit sei anzunehmen, dass das Kapitell zum ursprünglichen Grabdenkmal des Archilochos gehörte, das erst im sechsten Jahrhundert errichtet wurde, und dass das Epigramm ursprünglich auf einem anderen Teil des Denkmals stand und anlässlich der Errichtung des Archilochium (Arist. *Rh.* 1398*b* 11) um 400 v. Chr. auf dem Kapitell selber aufgezeichnet wurde. [16]

Wir sind davon ausgegangen die "sprechenden" Epigramme zu erklären und haben gesehen dass im allgemeinen das Denkmal den Ausspruch seines Errichters in dessen Worten wiedergibt, dass aber ein Dichter die Worte eine Grabfigur selber sprechen lassen kann. Eines der besten und vielleicht das früheste Zeugnis dieser Art ist das dem Homer zugeschriebene Midasepigramm das der Grabstatue einer Sphinx in den Mund gelegt wird. Anderseits gibt es Denkmäler in denen die Inschrift nicht von einem Abbild einer lebendig gedachten Figur sondern von einem ganz unlebendigen Pfeiler oder einer unlebendigen Säule gesprochen gedacht wird. So steht auf einem hocharchaischem dorischen Kapitell das auf Korfu gefunden wurde das einfache Epigramm (Peek, *G.V.* 52).

Στάλα Ξενϝάρεος τοῦ Μείξιός εἰμ᾽ ἐπὶ τύμοι.

Die Verwendung des Wortes στάλα wie des Ausdruckes ἐπὶ τύμοι (τύμβῳ) zeigt den Einfluss der epischen Sprache, wie wir sie zum Beispiel in *Il.* 11, 317 ausgedrückt finden: στήλη κεκλιμένος ἀνδροκμήτῳ ἐπὶ τύμβῳ; oder in *Od.* 12, 14: τύμβον χεύαντες καὶ ἐπὶ στήλην ἐρύσαντες. Es unterliegt wohl keinem Zweifel, dass der Zusammenhang zwischen Epos und | Epigramm nicht nur formal und sprachlich ist, sondern auch sachlich sein muss. Es mag dahingestellt bleiben ob Homer hier auf eine heroische Sitte anspielt oder sich nur auf die [17] zu seiner Zeit übliche Sitte bezieht, dass aber das homerische Grab dem archaischen entspricht ist sicher und für unser Verständnis des Urprungs der archaischen Kultur aus der Wiedererweckung der Heroenzeit von grosser Bedeutung. Zugleich sehen wir in dem Grabmal des Xenvares nicht nur die heroisch-epische Tradition sondern auch die künstlerische Neuschöpfung der archaischen Zeit die sich nicht nur in dem auf Stein eingetragenen Epigramm zeigt, sondern in dem prächtigen dorischen Kapitell, eines der ältesten und schönsten seiner Art, so schön dass es allein als ἄγαλμα zu bezeichnen ist und keiner weiteren Zutaten bedurfte. In so einem Denkmal hat der heroische Geist einen neuen Ausdruck gefunden der dann für die Jahrhunderte der Klassik massgebend wurde.

Trotzdem dürfen wir uns nicht vorstellen dass die Säule mit dem schönen dorischen Kapitell als lebendig gedacht wurde und dass deshalb das Epigramm in der ersten Person abgefasst ist. Die Säule spricht nicht, sie lässt sich lesen, sie vermittelt dem Beschauer etwas das ihm sonst nicht mitgeteilt werden kann, das Epigramm wird zum Teil des Denkmals und wird daher so ausgedrückt wie es dem Künstler natürlich scheint. Wenn ein Bildhauer oder ein Maler ein paar Jahre später ein Bild des Hermes schafft so spricht dieses Bild auch direkt zum Beschauer und wenn der Künstler nicht sicher ist dass seine Darstellung ganz klar ist und dass seine Mitteilung durch das Bild vermittelt wird, dann setzt er ein paar Worte hinzu wie Ἑρμῆς εἰμὶ Κυλλήνιος (Kirchner, *Imagines*[2] 6) die nicht mehr und nicht weniger sagen als das Bild selbst aber es noch verdeutlichen und daher den Beschauer ebenso direkt ansprechen wie das Bild selber.

[18] Diese Erklärung des sprechenden Epigrammes die auf den schönen Betrachtungen von Chrestos Karouzos (*Aristo|dikos* 33–38) aufbaut, darf noch durch ein weiteres Beispiel bestätigt werden das mit dem schon vorher erwähnten Weihepigramm des Phanaristos aus Krissa zu verbinden ist, da es sich auch hier um die Weihung von Bündelgeld handelt. In den Ausgrabungen des Heraheiligtums in Perachora wurde eine Reihe von hochaltertümlichen Inschriften gefunden, die von den Ausgräbern mit Überzeugung noch ins dritte Viertel des 7. Jh. gesetzt wurden und die jedenfalls in die Zeit um 600 v. Chr. gehören in der die Einführung gemünzten Geldes durch Pheidon von Argos stattfand. Auf einer dieser Porosblöcke steht nun das Epigramm (*S.E.G.* 15.193):

Δραχμὰ ἐγό, Ἥρα λευϙ[όλενε, κεῖμαι ἐν αὐ]λᾶι

Die Ergänzung ist zwar nicht sicher, wenn aber ἐν αὐ]λᾶι angenommen wird so kann dazwischen nur ein Wort wie κεῖμαι fehlen. Wir können hier bei aller Kürze das wesentlich Neue des Epigramms erkennen worin es über die Aufzeichnung eines mündlich gemachten Ausspruchs hinausgeht und seine eigene Form und raison d'être findet—es ist ein wesentlicher Bestandteil des Denkmals auf dem es steht, das ohne die Inschrift entweder unverständlich oder schwer verständlich ist. Dafür hat auch das Epigramm selber an Unabhängigkeit verloren und kann kaum ohne das Denkmal auf dem es steht gedacht oder verstanden werden.

Diese zwei Weihinschriften von Krissa und Perachora erlauben uns den Ursprung des Weihepigramms selber ganz kurz zu besprechen. Auch hier sind wir berechtigt auf dieselbe Iliasstelle (7, 82–83) zurückzugehen die schon anfangs im Zusammenhang mit dem Grabepigramm erwähnt wurde. Sagt doch Hektor dass er nach dem Siege über seinen griechischen Gegner

τεύχεα συλήσας οἴσω προτὶ Ἴλιον ἱρήν,
καὶ κρεμόω προτὶ νηὸν Ἀπόλλωνος ἑκάτοιο.

Hektor hätte somit die Waffen des getöteten Griechen zum Apollotempel [19] getragen und sie dort aufgehängt. Es handelt sich hier zweifellos um eine Weihung wie wir sie vom 7. Jh. an durch zahlreiche Beispiele in vielen griechischen Heiligtümern belegen können. Besonders in Olympia wurden viele Waffen gefunden die alle wohl von Weihungen wie der des Hektor stammen. Inschriften tragen diese Waffen von Olympia erst seit dem 6. Jh. (E. Kunze, VIII. *Bericht, Olympia*, 83–110), während sich auf kretischen Waffen des 7. Jh. schon Inschriften finden, die den Namen des Siegers und somit der Weihunden aber nicht des Besiegten enthalten. In der früheren Zeit waren die Waffenweihungen und alle anderen Weihungen namenlos, und als die ersten Weihinschriften auftraten, hatten sie einen anderen Charakter als die ältesten Grabinschriften. Diese zeigen ganz klar die Verbindung mit dem Begräbnis und der damals gesprochenen oder gesungenen Totenklage, und diese elegische Stimmung bleibt den Grabepigrammen stets erhalten. Das Weihepigramm stammt jedoch aus dem Gebet das nach der Weihung an die Gottheit gerichtet wurde. Auch das kennen wir schon von Homers *Ilias* wo Chryses in seinem Gebet an Apollo darauf hinweist dass er in der Vergangenheit viele Opfer dargebracht hat und daher jetzt um die Erfüllung seines Wunsches bittet. So erklärt es sich auch, dass die meisten Weihinschriften auf die Weihung als in der Vergangenheit liegend zurückverweisen, aber das Gebet im Optativ aussprechen.

Als Beispiel mag uns das älteste erhaltene Weihepigramm dienen das nach dem Stil der Bronzestatuette auf der es eingetragen wurde ins frühe sechste oder gar ins späte siebente Jahrhundert datiert wird (Jeffery, *Local Scripts of Archaic Greece*, 94, no. 1):

Μάντικλος μ' ἀνέθεκε ϝεκαβόλοι ἀργυροτόχσοι
τᾶς δεκάτας· τὺ δὲ Φοῖβε δίδοι χαρίϝετταν ἀμοι[βάν].

Das Weihgebet war so abgefasst, dass der Weihende erklärte, er habe etwas [20] dem Gott geweiht, und dann seinen Wunsch hinzufügte. In der Aufschrift hat dann das Weihgeschenk die Rolle des Weihenden übernommen und es erklärt seinem Wunsche gemäss: mich hat der Weihende geweiht, dann wendet es sich in seinem Namen an die Gottheit: Phoibus, mögest Du ihm seinen Wunsch erfüllen.

Die berühmte, von Mantiklos dem Apollo geweihte Statuette wird auf Grund ihres Stiles und der Sprache und Form ihrer Inschrift für böotisch gehalten, und diese Zuweisung erinnert uns an die drei berühmten Epigramme die Herodot im Heiligtum des Ismenischen Apollo in Theben sah und kopierte (5.

59–61). Sagt er doch von ihnen dass sie in kadmeischer Schrift geschrieben waren und dass die Καδμήια γράμματα den ionischen Buchstaben recht ähnlich seien (πολλὰ ὅμοια ὄντα τοῖσι Ἰωνικοῖσι). Das heisst, dass sie im altertümlichen böotischen Alphabet abgefasst waren. Die Inschriften selber standen auf drei Dreifüssen, die Weihungen der mythischen Zeit waren:

a. Ἀμφιτρύον μ' ἀνέθεκε νεὸν ἀπὸ Τελεβοάον.

b. Σκαῖος πυγμαχέον με ϝεκαβόλοι Ἀπόλλονι
νικάσας ἀνέθεκε τεῒ περικαλλὲς ἄγαλμα,

c. Λαοδάμας τρίποδ' αὐτὸς ἐϋσκόποι Ἀπόλλονι
μουναρχέον ἀνέθεκε τεῒν περικαλλὲς ἄγαλμα.

Wenn diese Inschriften auch nicht aus der Heroenzeit stammen, so waren sie doch in den Tagen der Herodot schon hochaltertümlich und gehören demnach noch ins frühe sechste oder späte siebente Jahrhundert. Von einer Fälschung kann kaum die Rede sein, wohl gaben aber die Inschriften eine Tradition wieder, die in eine noch frühere Zeit zurückreichte und sich an die
[21] hochaltertümlichen Drei|füsse wohl geometrischer Zeit anknüpfte. Als man dann Inschriften aufzuzeichnen anfing, war es natürlich dass man die drei Dreifüsse mit Inschriften versah die ihre überlieferte Bedeutung klar machten. Es ist nicht einmal sicher, dass Herodot die Sachlage nicht so verstand, denn hier wie an anderer Stelle (5. 58) wo er von Φοινικήια γράμματα spricht, die auch den ionischen ähnlich waren, lässt er erkennen dass er diese Übernahme der Schrift von den Phönikern nicht in die heroische Zeit setzt, sondern zwei- oder dreihundert Jahre vor seiner eigenen Zeit als die Griechen von den Phönikern die Schrift lernten (οὐκ ἐόντα πρὶν Ἕλλησι, ὡς ἐμοὶ δοκεῖ). Obgleich in der Form von Weihinschriften, und daher dem Weihgeschenk sozusagen in den Mund gelegt, sind diese Inschriften zugleich auch Beischriften die die Bedeutung des Gegenstandes auf dem sie stehen erklären.

So dürfen noch ein paar Bemerkungen zu den ältesten metrischen Beischriften unsere Betrachtungen über das Denkmalepigramm abschliessen. Beischriften haben den einzigen Zweck eine bildliche Darstellung zu erklären, und wenn man bedenkt dass die archaische Kunst noch nicht die später geläufige mythologische Kunsttypologie entwickelt hat, so ist es verständlich dass sie sich so oft wie möglich der Beischriften bediente. Solche einfache Beischriften sind uns von der frühen Vasenmalerei wohlbekannt und auch die ihr gleichzeitigen Schildbänder, die um 600 v. Chr. anzusetzen sind, zeigen gelegentlich Namensbeischriften.

Ein gutes Beispiel aus dem frühen 6. Jh. findet sich auf einem Gefäss das wir allgemein arrhyballos nennen und das vor einigen Jahren in Korinth gefunden wurde (*SEG* 14. 303; *AJA* 69, 1965, Tafel 56) und in Darstellung und Inschrift ganz einzigartig ist. Links von dem Flötenspieler steht sein Name Πολύτερπος, und in einer Schlangenlinie um den springenden und die drei stehenden Tänzer steht der Vers

Πυρϝίας προχορευόμενος αὐτὸ δέ ϝοι ὄλπα.

Die ersten zwei Worte "Pyrrhias vortanzend" beziehen sich zweifellos auf [22] den springenden Jungen, und auch das Ende des Hexameters muss sich auf ihn beziehen, nach Ausweis des αὐτὸ δέ das an die wiederholte Verwendung von αὐτός im Epigramm des Menekrates erinnert. Die Frage ist nur ob der Vasenmaler, das von ihm angefertigte Gefäss eine Olpe nannte und in dem Verse auf die von ihm dargestellte Szene aus dem Leben des Pyrrhias hinwies und ihm die Olpe widmete, oder ob wir die Möglichkeit erwägen sollen, statt ὄλπα, μολπά zu lesen um anzunehmen, dass die Molpe dem Pyrrhias gilt. Jedenfalls haben wir hier ein treffliches Beispiel für Namens- und Versbeischriften die Szenen korinthischer Künstler erklärend zur Seite standen.

In jene Zeit gehört auch die berühmte Lade des Kypselos die von dessen Nachkommen in Olympia gestiftet wurde (Pausanias 5.17.5). Von Kypselidenweihungen wissen wir auch sonst; eine Goldschale mit der Inschrift Κυψελίδαι ἀνέθεν ἐξ Ἡρακλείας (SEG 14. 302) ist ins Museum von Boston gelangt; auf eine andere spielte schon Platon an (*Phaedr.* 236 b), sowie andere Schriftsteller der klassischen Zeit. Sie wurde in den Platonscholien zitiert ist uns aber erst im Lexikon des Photios (s.v. Κυψελιδῶν ἀνάθεμα) erhalten (Geffcken 36):

Εἰ μὴ ἐγὼ χρύσεος σφυρήλατός εἰμι κολοσσὸς
ἐξώλης εἴη Κυψελιδᾶν γενεά.

Das Gedicht drückt in poetischer Umschreibung aus dass die Kypseliden eine aus Gold gehämmerte grosse Statue geweiht haben und daher den Gott um die Erhaltung ihrer Herrschaft bitten, ein Gedanke der völlig in die Überlieferung archaischer Weihungen gehört aber nicht der Formelsprache der anderen Weihepigramme folgt. Die berühmte Lade hatte anscheinend keine Weihinschrift die dem Pausanias bemerkenswert genug erschien um sie wiederzugeben, aber es stan|den auf ihr viele in Verse gefasste Beischriften die [23] Pausanias mit grosser Mühe abschrieb da sie in altertümlicher Schrift geschrieben und oft in Bustrophedon-Ordnung aufgemalt waren (5.17.6): γράμμασι τοῖς ἀρχαίοις γεγραμμένα καὶ τὰ μὲν ἐς εὐθὺ αὐτῶν ἔχει, σχήματα δὲ ἄλλα τῶν γραμμάτων βουστροφηδὸν καλοῦσιν Ἕλληνες. Zweifellos hat sich also Pausanias sehr für die Inschriften interessiert und sie so gut er konnte abgezeichnet. In der Regel gaben die Beischriften nur die Namen der dargestellten Personen und manchmal hat der Künstler auch nicht einmal den Namen hinzugefügt; so sagt dann auch Pausanias enttäuscht (5.17.9–10) ταύτης τῆς γυναικὸς ἐπίγραμμα μὲν ἄπεστι ἥτις ἐστί—und weiter unten—τὸ δὲ ὄνομα ἐπὶ τῇ 'Αλκήστιδι γέγραπται μόνη. Das zeigt dass Pausanias alle Beischriften als ἐπιγράμματα bezeichnete. Andrerseits war die Beischrift bei leicht zu erkennenden Helden wie Herakles nicht nötig (5.17.11): ἅτε δὲ τοῦ Ἡρακλέους ὄντος οὐκ ἀγνώστου τοῦ τε ἄθλου χάριν καὶ ἐπὶ τῷ σχήματι, τὸ ὄνομα οὐκ ἔστιν ἐπ' αὐτῷ γεγραμμένον. Tod und Schlaf (Θάνατόν τε καὶ Ὕπνον) hätte Pausanias auch ohne Beischriften erkannt, συνεῖναι δὲ καὶ ἄνευ τῶν ἐπιγραμμάτων (5.18.1).

Uns interessieren natürlich vor allem die Versinschriften die alle auf der zweiten und vierten der fünf Bilderzonen stehen, die wohl die Hauptzonen des ganzen Kunstwerks waren. Pausanias sieht hier einen Mann und eine Frau die erst durch den Hexameter identifiziert werden (5.18.2): τὰ δὲ ἐς τὸν ἄνδρα τε καὶ γυναῖκα ἑπομένην αὐτῷ τὰ ἔπη δηλοῖ τὰ ἑξάμετρα· λέγει γὰρ δὴ οὕτως·

α. Ἴδας Μάρπησσαν καλλίσφυρον, ἄν οἱ Ἀπόλλων
ἅρπασε, τὰν Εὐανοῦ ἄγει πάλιν οὐκ ἀέκουσαν

Das nächste Bild das einen Vers beigeschrieben hat zeigt Medea auf einem Thron mit Jason und Aphrodite (5.18.3) γέγραπται δὲ καὶ ἐπίγραμμα ἐπ᾽ αὐτοῖς·

β. Μήδειαν Ἰάσων γαμέει, κέλεται δ᾽ Ἀφροδίτα.

[24] Die anderen Epigramme mögen schnell zitiert werden ohne ihren Zusammenhang zu geben (5.18.4–19.6), doch darf darauf hingewiesen werden dass wir in fast allen Fällen bewusst beschreibende und erklärende Sprüche vor uns haben; das wird dadurch angedeutet, dass in fünf Fällen die Hauptfigur mit dem hinweisenden Fürwort οὗτος oder ὅδε bezeichnet wird, was zwanglos zu den berühmten beschreibenden Epigrammen der hellenistischen und der späteren Zeit überleitet. Aber die Sprache der Epigramme ist episch und die epischen Formeln sind leicht zu erkennen. Der Verfasser der Sprüche war eben entweder ein Rhapsode oder zum mindesten mit der epischen Tradition wohl vertraut. In derselben Hauptzone wie die Idas- und Medeaepigramme stehen auch die folgenden zwei:

γ. Λατοΐδας οὗτος τάχ᾽ ἄναξ ἑκάεργος Ἀπόλλων
Μοῦσαι δ᾽ ἀμφ᾽ αὐτόν, χαρίεις χορός, αἷσι κατάρχει.
δ. Ἄτλας οὐρανὸν οὗτος ἔχει, τὰ δὲ μᾶλα μεθήσει.

Auf der vierten Zone sieht Pausanias das Bild der Dioskuren die Helena wegführen während Aithra auf dem Boden liegt; dazu sagt er (5.19.3): ἐπίγραμμα δὲ ἐπ᾽ αὐτοῖς ἔπος τε ἑξάμετρον καὶ ὀνόματός ἐστιν ἑνὸς ἐπὶ τῷ ἑξαμέτρῳ προθήκη·

ε. Τυνδαρίδα Ἑλέναν φέρετον, Αἴθραν δ᾽ ἑλκέτον Ἀθάναθεν.

Die Frage erhebt sich ob wir hier einen Zusatz zum Hexameter vor uns haben wie Pausanias angibt und Hommel durch Beispiele belegt hat (*Rh. Mus.* 88, 1939, 198–206; Friedländer, *Epigrammata* zu 54 e), oder ob ein Hexameter und ein weiteres ergänzendes Wort aus einem grösseren Zusammenhang geschnitten wurden.

ζ. Ἰφιδάμας οὗτος τε Κόων περιμάρναται αὐτοῦ.
η. Οὗτος μὲν Φόβος ἐστὶ βροτῶν ὁ δ᾽ἔχων Ἀγαμέμνων.
[25] θ. Ἑρμείας ὅδ᾽ Ἀλεξάνδρῳ δείκνυσι διαιτὴν
τοῦ εἴδου Ἥραν καὶ Ἀθάναν καὶ Ἀφροδίταν.
ι. Αἴας Κασσάνδραν ἀπ᾽ Ἀθαναίας Λοκρὸς ἕλκει.

Die Bedeutung all dieser Versbeischriften liegt in der abschliessenden Bemerkung des Pausanias (5.19.10), dass der Künstler der Lade unbekannt ist und auch jemand anders die Epigramme verfasst haben mag, dass aber ihre Hauptvorlage das Epos des Korinthers Eumelos war, was man besonders an seinem delischen Prosodion sehen kann: τῆς δὲ ὑπονοίας τὸ πυλὺ ἐς Εὔμηλον τὸν Κορίνθιον εἶχεν ἡμῖν, ἄλλων τε ἕνεκα καὶ τοῦ προσοδίου μάλιστα ὃ ἐποίησεν ἐς Δελφούς. Diese Stelle wurde allgemein als ein Zeugnis für Eumelos als Verfasser der Epigramme der Kypseloslade aufgefasste und abgelehnt, zuletzt auch von Bowra (Cl. Q. 13, 1963, 147), aber Pausanias spricht nur von der ὑπόνοια und nimmt wohl an, dass der Künstler der Lade und der Verfasser der Epigramme, ob sie nun identisch waren oder nicht, sich auf das Werk des Eumelos, des einheimischen epischen Dichters bezogen. Die Bedeutung dieser Überlegung für die Kunst Korinths und seines Einflussgebietes kann hier nicht weiter verfolgt werden, aber die bezeugte Verbindung der Epik des 8. und 7. Jh. mit der Epigrammatik des 7. und 6. Jh. darf hier ganz kurz verfolgt werden.

Es fällt sofort auf dass viele der ältesten Epigramme aus dem korinthischen Kreis stammen, aus Megara und Böotien, aus Delphi und Perachora, und vor allem aus Korfu, und dies sind gerade auch die Epigramme die viele homerische Formeln zeigen und im allgemeinen der homerischen Tradition folgen. Während Paul Friedländer das Vorbild Homers und der Epik für die frühe Epigrammatik stark betont hat, wurden von verschiedener Seite neulich Einwände gemacht, zuerst von Werner Peek (Πείρατα Τέχνης, Wiss. Zt. der Universität Halle, 4, 1954/55, 2. Heft, 235–237), | der Friedländers Homereinfluss für übertrieben hielt, dann von T. B. L. Webster (Glotta 38, 1960, 251–263), der zwischen homerischem Wortgut und unhomerischer Dichtersprache unterscheiden wollte, und schliesslich von James Notopoulos (Hesperia 29, 1960, 194–196), der behauptete dass sich die Versinschriften jener Frühzeit nicht auf das Epos beziehen, sondern auf die mündliche Überlieferung, auf die die Rhapsoden selber zurückgreifen. Jede dieser Ansichten hat sehr viel für sich und ist auf Beobachtungen begründet deren Verlässlichkeit sich nicht leugnen lässt. Das Epigramm verdient aber auch aus sich selbst erklärt zu werden. Homerische Überlieferung, mündliche Tradition, zeitgenössische Poesie, Grab- und Weihkult, all diese Elemente haben zur Formung des Epigramms beigetragen, sie waren sozusagen das Rohmaterial aus dem das Denkmalepigramm geschaffen wurde, aber die Schöpfung selbst darf nicht mit der Summe der Elemente des Geschaffenen verwechselt werden, und dieser Schöpfung des Denkmalepigramms müssen wir jetzt in der Diskussion versuchen näherzukommen.

[26]

DISCUSSION

M. Pfohl: Im Anschluss an Herrn Raubitscheks Hinweis, dass es im geistigen Bereich keine Schöpfung *de nihilo* gebe, möchte ich auf die phönizischen Vorbilder der griechischen Inschriften verweisen. Die Forschung hat bisher meist nur die Formen und Verwendungsweisen der Buchstaben berücksichtigt,

[27]

kaum die Sprachformeln und die Kompositionsformen der inschriftlichen Texte. Nun zeigen aber die Inschriften des 8. und des 7. Jh. bereits ein ausgeprägtes Formular, und wir müssen daran denken, dass die Griechen mit dem Alphabet auch die Idee, Inschriften auf Grabsteine (weniger auf Weihdenkmäler) zu setzen, von den Phöniziern übernommen haben. Wenn wir nun den phönizischen Bestand betrachten, werden wir erkennen, dass gewisse griechische Formeln Übersetzungen phönizischer Modelle sind. Ich mache darauf aufmerksam, dass derlei Forschung erst noch betrieben werden muss, was jetzt um so eher möglich ist, als genügend phönizische Dokumente zur Verfügung stehen. Ich weise u.a. auf die phönizische Grabinschrift aus der ersten Hälfte des 9. Jh. hin, die man jüngst auf Cypern fand und die Dikaios in seinem Museumsführer von Nicosia 1953 veröffentlicht hat. Auf den ganzen Komplex haben bisher nur Paul Friedländer, *Epigrammata*, in einer kurzen Anmerkung und Eduard Norden, *Aus altrömischen Priesterbüchern*, hingewiesen.

[28]

M. Labarbe: M. Raubitschek a soulevé le difficile problème du rapport entre la littérature écrite et la littérature orale. Au point de vue formel, l'épos a certes influencé l'épigramme archaïque (comme il a influencé les premières productions du lyrisme). Mais la tradition épique ne suffit pas à expliquer l'apparition, au VIIe siècle, de ce genre indubitablement nouveau, caractérisé à la fois—sans parler du mètre—par sa présentation écrite et par | un singulier effort de concision. Je ne puis croire que les premières épigrammes aient été la simple transcription, la simple perpétuation par l'écriture, de vers récités lors d'une inhumation ou d'une dédicace. Car, dans cette hypothèse, on ne voit pas pourquoi le poète aurait renoncé à l'étirement de l'idée, à la prolixité, au délayage même, si fréquents et si compréhensibles dans le style oral jusque-là traditionnel. On l'imagine mal disant deux ou quatre vers devant un auditoire, puis se taisant. Non, ce qu'il faut admetree, c'est qu'il composait expressément *pour le monument*. Le procédé qui consiste à faire parler celui-ci ("Ich-Rede") ne s'accommode pas d'une autre explication; par surcroît, il est remarquable que, dans les dédicaces, on ait l'aoriste ($ἀνέθηκε$), et non point le présent.

M. Raubitschek: Das Verhältnis zwischen Epos und Epigramm ist so zu verstehen, dass Leute, vielleicht Rhapsoden, die die Sprache des Epos kannten, die kurzen Gedichte verfassten, welche auf Grab- und Weihdenkmälern und auf Gegenständen wie dem Nestorbecher aufgeschrieben wurden.

M. Dihle: Der Unterschied zwischen mündlicher und schriftlicher Dichtung besteht bekanntlich darin, dass der mündliche Dichter Formeln, der schriftliche konzipierende Wörter als Elemente seiner Poesie verwendet. Die Einheit des epischen Stiles von der mündlichen zur schriftlichen Epoche beruht darauf, dass der schriftlich erfindende Dichter seine Ausdrücke aus alten und neuen Wörtern nach den Modellen der alten Formeln bildet. Im Fall der frühen Inschrift vom Nestorbecher kann man bereits Indizien für die schriftliche Erfindung erkennen: zwei Mal verwendet der Verfasser, der doch mit einem in den erhaltenen Epen nicht belegten aber nach homerischer Konvention gebildeten Ausdruck καλλιστεφάνου 'Αφροδίτης in epischer Tradition steht, das Wort ποτήριον, das in der ionischen Prosa und bei den Lesbiern—hier sicherlich als Entlehnung aus der Umgangssprache—zuerst vorkommt und dem epischen Vokabular fremd ist. (Bei Homer gibt es κύπελλον, δέπας und κοτύλη, also mehrere Bezeich|nungen zur Auswahl.) Wo man derartige Einzelwörter in die epische Kunstsprache eindringen sieht, liegt der Schluss auf die schriftliche Entstehung des Textes nahe.

Die Fähligkeit dessen, der die zur Wiedergabe grosser Zusammenhänge entwickelte Sprachtechnik des Epos erlernt hat, gelegentlich auch konzis und pointiert mit Hilfe eben dieser Kunstsprache zu formulieren, ist durch die Gnomen bei Homer und vor allem bei Hesiod reichlich bezeugt. Wer eine Gnome wie solche der Ὑποθῆκαι Χείρωνος prägen konnte, vermochte auch eine kurze Grabschrift in epischer Sprache zu erfinden.

M. Gentili: Le locuzioni epiche sono più frequenti nelle iscrizioni per persone morte in guerra e che si sono distinte per atti di valore. È interessante, sotto questo profilo, rilevare la differenza tra l'iscrizione per Arniada morto combattendo presso le navi sulla riva dell' Aratto e l'epigramma sepolcrale da Corcira per Menecrate. Nella prima gli epicismi sono più frequenti, e il motivo è evidente: per commemorare l'*arete* di un combattente, l'autore poteva disporre largamente del repertorio formulare e lessicale epico. Per il secondo invece, che commemora i meriti politici di un prosseno di Corcira, la lingua epica non offriva una identica diponibilità: πρόξενϝος non è omerico, come non è omerico δαμόσιον κακόν che compare per la prima volta in Solone (3,26 D.).

Inoltre per ciò che concerne il grosso problema dei rapporti tra iscrizioni ed epos, già a suo tempo posto in chiari termini dal Wackernagel e ora ripreso in una direzione totalmente opposta a quella del Leumann da Hoekstra (*Mnemos.* 1957, 193–225) e Ruijgh (*L'élément achéen dans la langue épique*, Assen 1957), io credo che in primo luogo si debba chiarire a quale epos ci si riferisce. Se ci si vuole riferire solo all'epos omerico, le obiezioni sono molte: perché pensare a un adattamento ai dialetti locali di forme omeriche divulgate da rapsodi ionici quando storicamente è più verisimile ammettere la presenza di antiche tradizioni epiche orali (e anche scritte) in regioni della Grecia continentale quali | l'Argolide e la Beozia? Bisognerebbe supporre insostenibilmente (l'obiezione è stata ben formulata de Hoekstra) che in quelle regioni fossero cadute in oblio antiche leggende di dei, di re e di eroi per essere più tardi reintrodotte da rapsodi ionici. E se consideriamo il problema sotto l'altro aspetto del rapporto poesia-uditorio, c'è ancora da chiedersi se nel VII e nel VI secolo la società della Grecia continentale di lingua dorica, tessalica, beotica ecc. fosse in grado di capire la lingua dell'Omero recitato dai rapsodi ionici, quando le iscrizioni destinate ad essere lette dal pubblico di qualsiasi ceto sociale e quindi di diversa cultura erano scritte nella lingua epicorica. Infine questa presunta omericità culturale costringerebbe a considerare ognuno degli autori di epigrammi, nella maggior parte dei casi mediocri letterati locali, come un esperto traduttore dal greco di Omero al greco del suo paese. Dunque è più ragionevole ammettere con

Hoekstra che in alcuni centri della Grecia continentale, che si trovarono isolati in seguito all'invasione dei Dori, lo stile orale si sia mantenuto sotto forme locali adattate ai dialetti parlati, in sostanza un adattamento paragonabile alla ionizzazione avvenuta in Asia Minore.

M. Labarbe: L'emploi des dialectes locaux n'a pas exclu le recours à la langue épique. C'est que celle-ci, depuis des siècles, était la seule langue de la poésie et que le genre nouveau de l'épigramme exigeait des concessions à la tradition. Il ne faut pas croire à de réelles difficultés de compréhension dans les régions non ioniennes du monde grec: les lieux d'origine des plus récents poètes du *Cycle* (Stasinos de Chypre, Hagias de Trézène, Eugammon de Cyrène) indiquent assez quelle diffusion l'épos avait fini par connaître. Sa langue artificielle était une κοινή, de même qu'il y avait, pour le lyrisme choral, une κοινή dorienne que pratiquèrent les Ioniens Simonide et Bacchylide, le Béotien Pindare.

M. Raubitschek: Das Verhältnis zwischen der gemeingriechischen epischen Dichtersprache und den oft in epichorischen Dialekten geschriebenen
[31] Epigrammen ist so zuverstehen, dass | die Verfasser der Epigramme sich der lokal gebräuchlichen Aussprache und Ausdrücke bedienten.

M. Dihle: Ich möchte mir die dialektische Differenzierung der frühen, an der epischen Tradition orientierten Versinschriften etwa folgendermassen erklären: Das grosse Epos wurde bis ins 6. Jh. hinein mündlich gepflegt, denn die Herstellung von Texten, die viele tausend Verse umfassten, überstieg die technischen Möglichkeiten des 7. Jh. (vgl. D. L. Page, *Archiloque*, Entretiens Hardt 1963, S. 120 ss.). Diese Pflege geschah durch das Medium einer in ganz Griechenland einheitlichen, mit keinem gesprochenen Dialekt identischen Kunstsprache, die, wie Hesiod trotz oder gerade wegen seiner kleinen Anzahl von Boiotismen zeigen kann, überall verstanden wurde. In dem Masse, in dem aus der epischen Sprachtradition, aber mit z.T. neuen Wörtern und für aktuelle Zwecke kurze Texte wie Grab- oder Weihepigramme *schriftlich* konzipiert wurden, lag es nahe, sie dem epichorischen Dialekt anzupassen. Dass es echte lokale Varianten der Sprache des grossen Epos gegeben habe, dafür fehlt jeder Anhalt. Ganz entsprechend sind die Ionismen der Elegie und die Mehrschichtigkeit des Dialektes der frühen Chorlyrik zu erklären. Lyriker wie Alkman und Terpander erwarben ihren Ruhm in Sparta, stammten aus dem äolischen Osten und schöpften aus Homer—und alle diese Elemente finden sich in ihren schriftlich konzipierten Texten.

M. Gentili: Non comprendo a quale κοινή si riferisca il prof. Labarbe. Certamente si può parlare di una κοινή formulare, ma non di una κοινή linguistica di dialetto ionico. Basteranno due esempi, il primo, addotto da Hoekstra dall' iscrizione beotica in esametri (Frie. 35) nella quale l'espressione formulare δίδοι χαρί ϝετταν ἀμοιβάν ha *un solo* riscontro in Omero (*Od.* 3, 58 δίδου χαρίεσσαν ἀμοιβήν). Ancora, in un' iscrizione tessalica in esametri (Pe. 217, Frie. 32) si legge nel primo verso νηπία ἐόσ᾽ ἔθανον καὶ οὐ λά[β]ον ἄνθος ἔτ᾽ ἥβας. È possibile pensare che ἄνθος ἥβης che ricorre una sola volta in *Il.* 13, 484 (καὶ δ' ἔχει ἥβας ἄνθος, | ὅ τε κράτος ἐστὶ μέγιστον), ma non è infrequente nell'elegia arcaica (Tirteo 7,28; Solone 12,1; Mimnermo i,4; 2,3 D.) e compare in Esiodo (*Th.* 988), derivi dall'epopea omerica? [32]

M. Giangrande: Geschriebene Poesie, d.h. Gedichte, die für die Öffentlichkeit berechnet sind, ist *ipso facto* etwas Künstliches, das von der gesprochenen Umgangssprache wesentlich verschieden ist. Abweichung vom homerischen Sprachgebrauch in den frühen Epigrammen war unvermeidlich, sie lag in der Natur der Sache. Jeder epichorische Dichter bediente sich zwar eines schon konstituierten Gemeinguts (ich meine den homerischen Wortschatz). Abweichungen und "Neuerungen" sind aber nichts Unerwartetes. Die Frage ist nur, ob sie als "slips of the pen" zu betrachten sind, die als solche den Heimatdialekt des Dichters verraten, oder—was ich für wahrscheinlicher halte—als absichtliche Abweichungen nach dem Prinzip der *imitatio cum variatione*. Ein Prinzip, das in der griech. Literatur zu allen Zeiten galt. Um in der innerepischen Tradition zu bleiben, so haben nicht nur die Alexandriner (wie man zu glauben pflegte), sondern schon Antimachos absichtliche Änderungen des homerischen Wortlauts beim Anspielen auf Homer vorgenommen.

M. Pfohl: Es ist bezeichnend, dass das archaische Hexameterepigramm gerade dort eine besondere Blüte entfaltete, wo eine lokale epische Tradition vorliegt, z.B. in Korinth und Kerkyra, in Böotien, auf Delos. Auf die Existenz lokaler epischer Schulen schlossen u.a. auf Grund der epigrammatischen Dokumentation Paul Friedländer und J. A. Notopoulos (*Hesperia* 29, 1960, S. 195f.).

Bezüglich der Länge von Epigrammen der Frühzeit erinnere ich an den überraschenden Neufund eines attischen Epigramms elegischen Metrums von 6 Versen (F. Willemsen: *AM* 78 (1963) S.141/145(II), Taf. 72f.)

M. Robert: A-t-on un seul texte permettant de supposer que, dans un enterrement grec, on prononçait devant la tombe un | éloge du défunt? La coutume de l'*épitaphios logos* pour un groupe de guerriers morts est autre chose. Il n'y a point de tel texte à alléguer. La coutume paraît caractériser Rome, opposée à la Grèce, et on ne peut la supposer gratuitement pour expliquer l'origine d'un genre. Un texte comme celui de Lucien, *De luctu*, montre l'absence d'un éloge individuel dans les funérailles grecques. [33]

M. Luck: Die beiden Verse *Ilias* VII, 89–90 erinnern so stark an ein Grabepigramm, dass man im ersten Augenblick erstaunt ist. Aber wenn man genauer hinsieht, fallen die Unterschiede ins Auge; 1. πάλαι κατατεθνηῶτος ist vom Standpunkt der Nachwelt ausgesprochen, viele Jahre nach dem Tode. So etwas kann auf einer echten Grabschrift nicht stehen. 2. Eine echte Grabschrift preist den Toten; hier wird der Mann verherrlicht, der ihn getötet hat, als wäre es eine grosse Ehre, von Hektor erschlagen zu werden. 3. Der Name des Toten fehlt; er hat überhaupt keine Bedeutung. Gewiss, die Form ist epigrammatisch; das klingt wie ein Epigramm; aber wenn es ein Epigramm sein soll, dann müsste man es eher als eine ganz besondere Art von Weihgedicht bezeichnen, keinesfalls als Grabschrift.

M. Raubitschek: Das von Hektor zitierte Epigramm ist zwar vom typischen Grabepigramm insofern verschieden, dass es eine dichterische Umgestaltung darstellt, aber gerade darum setzt es die Normalform voraus; dem homerischen πάλαι entspricht das später so oft verwendete (und missverstandene) ποτέ.

M. Pfohl: Bei H.-M. Lumpps Frage: "Verwendet Homer hier bereits eine geprägte Form der Grabinschrift oder hat er diese Inschriftenform erst angeregt?" mag man sich für die erste Möglichkeit entscheiden, wenn man die nicht seltenen "epigrammatischen" Stellen bei Homer und unsere Überlieferung von Hexameterepigrammen aus homerischer Zeit selber in Erwägung zieht. Hinter diesem Problem aber steht letztlich die Frage nach dem Zeitpunkt der Übernahme des Alphabetes aus dem Osten.

[34] Die Ilias-Stelle VII 87–89 enthält einen Hinweis darauf, dass das Wesen auch des aufgeschriebenen Epigrammes in der "Mündlichkeit" besteht. Durch das laute Lesen des Wanderers am Grabe lebt der Nachruhm jeweils neu auf.

M. Gentili: Chiedo al prof. Raubitschek la sua opinione sulle tesi contrastanti del Preger e di Leo Weber a proposito dell'epigramma per Mida. Si tratta davvero di uno "Stein-Epigramm"? Esso è certo anteriore a Simonide che nel carme a Cleobulo di Lindo polemizza con il pensiero espresso nei versi 3–5 dell'epigramma. Ma Simonide dice λίθον δὲ καὶ βρότεοι παλάμᾳ θραύοντι. Perciò è da supporre che il primo verso dell'epigramma χαλκέη παρθένος εἰμί sia stato tramandato in una redazione diversa. Cioè, come ha suggerito lo Snell, in luogo di χαλκέη παρθέμος εἰμί doveva essere qualcosa come παρθένος εἰμὶ λίθοιο. Non si può pensare che con λίθος Simonide abbia inteso riferirsi soltanto alla stele, perché nell' epigramma è la statua che afferma la sua perennità. E ancora qualche altro dubbio: se l'epigramma fu realmente inciso, come Simonide ha potuto conoscerlo? Attraverso una tradizione orale o scritta? Se scritta, bisognerebbe postulare almeno all' inizio del V secolo l'esistenza di una raccolta di epigrammi epigrafici trascritti dalla pietra. E il defunto fu proprio il re della Frigia, come attesta Platone (*Phaedr.* 264 c)? L'autore è Cleobulo di Lindo, attribuzione attestata da Simonide? Ma l'iscrizione arcaica era anonima; dunque l'assegnazione a Cleobulo non è antica ma dovrebbe connetersi con l'aneddotica che fiorì intorno ai Sette Sapienti, in particolare intorno a Cleobulo. Inoltre, se l'epigramma proviene, come è stato supposto, da Cuma eolica, la forma linguistica tradisce una normalizzazione nella tradizione letteraria. Tutti questi dubbi e queste ipotesi inducono a credere che l'epigramma per Mida era l'esemplare di una serie di epitafi fittizi, non reali.

M. Raubitschek: Das Verhältnis zwischen Midasepigramm und dem "Lied" des Simonides könnte so zu verstehen sein, dass sich Kleobulos auf das
[35] Midasepigramm bezog—vielleicht in einem | gnomischen Zusammenhang—und Simonides dem Kleobulos widersprach—vielleicht ohne den Namen des Midas zu kennen.

M. Luck: Ich habe bisher die Erläuterungen Platons im Phaidros zur kyklischen Form des Midas-Epigramms so aufgefasst, dass man mit jedem beliebigen Vers anfangen kann, also nicht nur 1, 2, 3, 4 und 4, 3, 2, 1 sondern auch 2, 3, 4, 1 usw. Das setzt aber voraus, dass die in der *Homer-Vita* und bei Diogenes Laertios überlieferten Verse 2a und b unecht sind; der Text von Beckby enthält sie z.B. nicht. Dann sehe ich aber nicht ein, warum die vier Verse nicht rings um eine stattliche Säule geschrieben sein konnten, auf der die bronzene Sphinx sass? Das wäre dann ein ähnlicher Fall wie die Menekrates-Inschrift (Text Nr. 3). Man konnte mit irgend einem Vers beginnen und, um das Monument herumgehend, die Inschrift zu Ende lesen. Interpunktionen, wie sie auch z.B. auf der Menekrates-Inscrift sichtbar sind, sorgten dafür, dass man nicht mitten in einem Vers begann. Die Buchstaben müssen von königlichem Format gewesen sein.

M. Pfohl: Ich weise auf die Interpretation hin, die H. Fränkel in *Dichtung und Philosophie des frühen Griechentums* dem Midasepigramm hat zuteil werden lassen (Stolz auf das neue Material der Bronze). Dazu passt eine frühe syrakusanische Inschrift, in der der Künstler selbstbewusst die Verarbeitung eines neuen Materials hervorhebt.

M. Gentili: Credo molto dubbia la datazione proposta dal professor Raubitschek per l'epigramma esametrico 'Ἀρχίλοχος Πάριος, κτλ. Il termine μνημήϊον è attestato per la prima volta in un'iscrizione del V secolo (*IG* I² 1037 = 78 Peek) e in Pind., *Pyth*. 5,49.

M. Pfohl: Im Hinblick auf die Datierung des Archilochosepitaphs erinnere ich daran, dass im Gesamtvokabular der Grabinschriften vom 7. bis 5. Jh. v. Chr. in meiner Publikation *Greek | Poems on Stones*, Vol. I (Leiden, 1967) das Wort nur einmal in einer attischen Inschrift des 5. Jahrhunderts begegnet.

M. Giangrande: Man sollte sich davor hüten, Urteile über Echtheit und dgl. zu fällen, solange eine eingehende Untersuchung des Sprachgebrauchs der archaischen Epitaphien nicht vorliegt. Der Gebrauch und die etwaige Frequenz von Formeln, Wortgruppen und dgl. sind noch nicht abschliessend geklärt worden: Erst nachdem eine solche konkrete Untersuchung (an Hand vom Parallelhomer und solchen Werken) gemacht worden ist, werden wir imstande sein, mit Angaben und nicht mehr mit Eindrücken zu operieren.

33

Review of F. Solmsen, *Hesiod and Aeschylus*

[70] *Hesiod and Aeschylus.* By Friedrich Solmsen. ("Cornell Studies in Classical Philology," Vol. XXX.) Ithaca, N. Y. Cornell University Press, 1949, Pp. ix, 230. $3.00.

"Whereas a Homeric tag is frequently introduced merely as ornament, Hesiod is usually cited for his substance" (M. Hadas).[1] Solmsen is primarily concerned with the substantial contributions made by Hesiod and with their influence on Solon and Aeschylus. He knows, however, that "the ambiguity inherent in the much-used word 'influence' should not deceive us" (165). Accordingly, he speaks more often of the Attic poets' *use* of Hesiod's works. And yet, from Solmsen's book, one gets the impression that "the development of Greek poetry moves with the gradual but inevitable power of an incoming tide." William C. Greene, who made this observation in his review of Solmsen's book (*AJP* 71 [1950] 316), added: "Something of this impression may be felt also in the movement of thought of Solmsen's book." Whether one uses the term "development" or "influence," the book is an expression of the belief in the poetic and philosophic tradition of classical Greek literature which, if not inevitable, is definitely meaningful and significant. This does not imply, however, that Solmsen is unaware of the unique originality of the individual poet. In fact, the first part of his book is devoted to a careful examination of "the results of Hesiod's own speculations" (vii) and to a clear distinction "between what he himself creates and what he merely passes on" (5). Even in the second part ("Solon and Aeschylus"), which concentrates on the Hesiodic elements in the works of the two Attic poets, Solmsen retains his interest "in discerning the individual—or the Athenian—element in the attitude of Solon and Aeschylus" (105–106). For, as Greene pointed out (*loc. cit.*), "only a just appreciation of the debt [to Hesiod] can enable us to measure the originality of Aeschylus."

"Hesiod's Zeus is Homer's Zeus—Homer's Zeus and something more" (7).

Classical Weekly 45 (1951) 70–72

1. *History of Greek Literature* (New York 1950) 34.

Solmsen tries to make clear what this "something more" is, starting with the observation that Hesiod's "own message ... consigned much that Homer had told ... to the realm of lies" (8). In his emphasis on Hesiod's orginality, Solmsen would probably not agree with Kurt von Fritz's observation (*Rev. of Religion* 11 [1946/47] 259): "A careful reader of Hesiod can hardly escape the conclusion that he failed to unite these many logically inconsistent stories in one scheme, not because he was aware of a deeper significance behind the seeming inconsistencies, but on the contrary, because he was not aware of them."

The careful modesty and exacting objectivity of Solmsen's scholarship make it difficult and yet imperative for a reviewer to state boldly and clearly what in Solmsen's opinion were Hesiod's main contributions and his own message. As I see it, Solmsen emphasizes Hesiod's concept of the three generations of Gods: Uranos-Kronos-Zeus; and he discovers in this concept more than a mere genealogical and systematic order. Hesiod is credited with a conscious awareness of the conflict, physical as well as moral, between the old and the new orders, the latter represented by Zeus. In describing the government of Zeus, Hesiod distinguishes, but does not separate, the two sources of Zeus's power, physical force and justice, represented by Kratos and Bia on the one side and, on the other, by Dike and her sisters, Eunomie and Eirene; see J. Kühn, *Würzb. Jahrb.* 2 (1947) 259–294 (on Eris and Dike). Finally, Hesiod was concerned with the origin and meaning of violence and injustice. "We here see Hesiod as the predecessor of Solon, a first | defier of corrupt justice." H. T. Wade-Gery, to *[71]* whom we owe this statement (*Phoenix* 3 [1949] 90) added the following fine characterization of Hesiod's concept of justice: "From Hesiod through Solon to Aeschylus and Euripides, the Nightingale [*Erga* 202–212] was a real power in Greek opinion and behaviour, and the Hawk had to listen. With the Macedonian, and the Roman, things changed. The Hawk professed benevolence, and the Nightingale ate out of his hand. Aratos was court poet to Antigonos, Virgil to Augustus. They both professed to imitate Hesiod, but it was a strange imitation." It is for his insistence on Truth and Justice that Hesiod has been called "the first Presocratic," not only by Wade-Gery (*Phoenix* 3 [1949] 81), but also by H. Diller in a good article devoted to Hesiod and the beginning of Greek philosophy in *Antike und Abendland* 2 (1946) 140–151. F. J. Teggart, *JHI* 8 (1947) 45–77, sees in Hesiod the champion of progress.

Solmsen's interpretation of Hesiod is both convincing and attractive. And yet, although it is based exclusively on Hesiod's work itself, one feels that it received its stimulus and direction from the classical tradition of which Hesiod himself is a source. If Solon and Aeschylus, to mention only those authors whom Solmsen examined in detail, had not known Hesiod or had used his work in a different way, Solmsen's interpretation would have followed a different path. As Solmsen himself says of C. Gottfried Hermann and F. G. Welcker (216–217, note 147), "it is difficult to resist the impression that their solutions reflect their respective temperaments," one might say of Solmsen's account of Hesiod that it reflects the respective temperaments of Solon and Aeschylus." This, however, is no criticism, but praise, for Solmsen has revealed to us the originality of Hesiod not as he, as a modern scholar, sees it, but as it was seen by two of the greatest representatives of classical poetry, Solon and Aeschylus.

The second part of Solmsen's book contains, accordingly, an examination of the Hesiodic elements in the works of Solon and Aeschylus. Solmsen begins with Heraclitus's statement "The teacher of most men is Hesiod," but uses it in an utterly un-Heraclitean spirit. His general introduction on the Hesiodic tradition (103–106) is a masterpiece of precision and conciseness.

At the risk of doing injustice to the richness and delicacy of Solmsen's interpretation, I consider it essential to stress his key observations. Hesiodic in Solon are the appeal to the Muses (a formal element), the emphasis on unjust wealth, and the triple concept of Eunomie, Eirene, and Dike. Solon pays little attention to Hesiod's praise of hard work, his concept of Hope is different, and his Athenian political experience and activity produced certain significant alterations in his Hesiodic borrowings, especially concerning unjust wealth, Justice (Dike), and Lawfulness (Eunomie); on Hope see not only K. von Fritz. *Rev. of Religion* 11 (1946/47) 257–259, but also F. Martinazzoli, *SIFC* 21 (1946) 11–22. Solmsen's discussion of Eunomie may be supplemented, by those interested in the historical sequel, with Larsen's brilliant interpretation of the Cleisthenian idea of Isonomia which occupies an intermediary position between the Solonian Eunomie and the Periclean (or post-Periclean?) Demokratia (*Essays in Political Theory Presented to George H. Sabine* [Ithaca, N. Y. 1948] 1–16).

Aeschylus's debt to Hesiod is discussed at greater length, and it was evidently larger than Solon's. To put it crudely, Aeschylus took the plot for his Prometheus plays from Hesiod; see K. Reinhardt, *Aischylos als Regisseur und Theologe* (Bern 1949) 29–39. He also took from Hesiod the concept of the Trilogy as such, derived from the three divine generations of Uranos-Kronos-Zeus. Solmsen claims that the three generations of the Oresteia and of the Oedipus Trilogy owe their compositional unity to the model of Hesiod. I suspect that this suggestion will have to be examined critically, perhaps with some consideration given to Elderkin's startling discovery of the Achilles and Theseus Trilogies (or Tetralogies) on the François Crater (*Art in America* 33 [1945] 29–33).

As for the Prometheus play itself, Solmsen has assembled not only the Hesiodic passages, but also those which are either critical of the received story or simply different from it. Solmsen indicates that Aeschylus was offended by and critical of Hesiod's treatment of Zeus, and he does not join E. Vandvik in claiming that Hesiod's and Aeschylus's treatment of Prometheus and especially of Zeus did not differ markedly. Vandvik's book, *The Prometheus of Hesiod and Aeschylus* (Oslo 1943) appeared too late to be thoroughly examined by Solmsen (124, note 1); see also Vandvik's more recent article in *Symb. Osl.* 24 (1945) 154–163; I. Macciotta, *Dioniso* 10 (1947) 83–101; A. Lesky, *Anz. f. d. Altertumsw.* 1 (1948) 99–108 (a thoughtful and exhaustive critical bibliography). For Vandvik, Prometheus was a cheat and a villain who "boasts of benefits which are of a dubious quality" (*The Prometheus of Hesiod and Aeschylus* 30) while "the impression of the injustice and brutality of Zeus" (*ibid.* 4) is false and was not intended | by Aeschylus; W. A. Irwin (*Journal of Religion* 30 [1950] 91) thinks that "Prometheus is Man himself," comparing him once more with Job. Solmsen, however, points out that the reconciliation between Prometheus and Zeus was un-Hesiodic, and was the original contribution made by Aeschylus, perhaps the very message which the Attic poet tried to convey with his Trilogy. J. A. Davison's historical interpretation of the Prometheia in *TAPA* 80 (1949) 66–93 shows merely the timeliness of Aeschylus's treatment, while the precise identifications (Prometheus = Protagoras, Cronos = Cimon, Zeus = Pericles or Ephialtes) at best apply only to the action of the preserved play. If Plato's Protagoras is historical, as Davison assumes (*op. cit.* 74, note 17), the dependence of Aeschylus's Prometheus on that of Protagoras (or vice versa) is more important than any historical allusions found in the play.

The final chapter of Solmsen's book is devoted to an examination of the *Eumenides*. Inasmuch as this play deals with the conflict between the old and the new gods, Solmsen sees Hesiod's influence at work, and he is probably right. And yet the old gods of Aeschylus are not those of Hesiod, nor is the recognition and reconciliation of the Erinyes at all Hesiodic. See R. P. Winnington-Ingram, *JHS* 68 (1948) 130–147. After all the evidence is examined—and Solmsen's examination is thorough and unprejudiced—the differences between Aeschylus's treatment and that of Hesiod are more significant than the similarities. It appears, therefore, and this is perhaps Solmsen's main thesis, that Aeschylus and also Solon owed to Hesiod not only those ideas and concepts which they borrowed from him, but also those in which they differed from him. By setting himself against the Hesiodic tradition, by criticizing, altering, and adding to it, Aeschylus actually followed the lead of the earlier poet and became part of the classical tradition.

34

Erga megala te kai thomasta

[217] On a souvent dit que les mots d'Hérodote qui servent de titre à cet article se rapportent aux actions humaines dont il est question dans sa description de tout ce qu'il avait vu et appris. Cette interprétation ne semble pas justifiée par certains passages de l'œuvre même d'Hérodote. Cet auteur juge (1, 93) qu'il n'y a pas beaucoup de choses remarquables ($\theta\omega\mu\alpha\tau\alpha$) en Lydie. Il ne décrit qu'un seul monument ($\xi\rho\gamma o\nu$) de ce pays, le tombeau d'Alyatte, qui, dit-il, est le plus grand monument qu'il connaisse, sauf ceux d'Égypte et de Babylonie. En effet, dans son livre sur l'Égypte, nous trouvons mentionnés et décrits plus d'édifices et de monuments que dans tout le reste de son œuvre. Au commencement de ce livre (2, 35), on trouve un passage très caractéristique. Hérodote y dit que l'Égypte a plus d'ouvrages admirables ($\theta\omega\mu\alpha\sigma\iota\alpha$) que le reste du monde et des monuments ($\xi\rho\gamma\alpha$) qui sont plus remarquables que tous les autres et que, pour cette raison, il lui faut parler plus abondamment de l'Égypte ($\tau o\upsilon\tau\omega\nu$ $\epsilon\iota\nu\epsilon\kappa\epsilon\nu$ $\pi\lambda\epsilon\omega$ $\pi\epsilon\rho\iota$ $\alpha\upsilon\tau\eta s$ $\epsilon\iota\rho\eta\sigma\epsilon\tau\alpha\iota$). Mais la meilleure preuve qui se puisse citer, pour montrer qu'Hérodote désirait faire entrer dans son œuvre la description des plus grands monuments, se trouve ailleurs. En expliquant pourquoi il donne en détail l'histoire de Samos (3, 60), il allègue que cette île contient trois des plus remarquables monuments ($\xi\xi\epsilon\rho\gamma\alpha\sigma\mu\epsilon\nu\alpha$) de la Grèce, c'est à savoir: la conduite d'eau faite par Eupalinos, le môle et le sanctuaire d'Héra. Un quatrième passage peut être, en outre, mentionné; celui (7, 24) qui concerne la construction du canal à travers la péninsule de l'Athos. Ce canal est appelé un $\xi\rho\gamma o\nu$, entrepris par Xerxès, qui désirait montrer sa puissance et en laisser le souvenir ($\mu\nu\eta\mu o\sigma\upsilon\nu\alpha$ $\lambda\iota\pi\epsilon\sigma\theta\alpha\iota$).

Si nous essayons d'avancer une interprétation personnelle du passage d'Hérodote cité au début de ces pages, nous devons en chercher la justification [218] dans d'autres parties de son exposé. Car cette | phrase est, non sans raison, considérée comme étant essentiellement une déclaration de principe pour l'ensemble de l'œuvre. Hérodote mentionne seulement quelques-uns des

---

Revue des Etudes Anciennes 41 (1939) 217–222. Traduit de l'anglais par Robert Fawtier.

monuments élevés en terre grecque et se rapportant à l'histoire grecque. Ceci explique peut-être pourquoi il n'a pas été encore suggéré que cet historien aurait essayé de faire place dans ses recherches aux monuments aussi bien qu'aux événements et aux coutumes.[1]

Voyons, rapidement, les monuments élevés en Grèce et par des Grecs. Il nous faut citer l'offrande d'Arion au cap Ténare, décrite par Hérodote (1, 24) comme un petit monument (οὐ μέγα), et les statues de Cléobis et Biton à Delphes, une offrande des Argiens (1, 31).[2] Hérodote n'a pas vu les autres statues attiques et de meilleure qualité de la même période, parce que, de son temps, elles étaient déjà sous la terre. Nous pouvons énumérer ensuite la statue d'or d'Apollon à Thornax (1, 69), dont l'or fut fourni aux Spartiates par Crésus, et le cratère de bronze envoyé par les Spartiates à Crésus (1, 70), qui, intercepté, fut offert au sanctuaire d'Hèra à Samos. Les aventures de ce cratère rappellent celles du trépied des Sept-Sages.[3] Hérodote nous signale, en outre: deux autres vases, l'un, présent votif des Samiens (4, 152), l'autre, offrande de Pausanias (4, 81); la peinture dédiée par Mandroclès dans le sanctuaire d'Hèra (4, 88) et les colonnes élevées par lui sur le Bosphore (4, 87).

Il est question de tous ces monuments dans les quatre premiers livres. Dans les livres suivants, qui traitent d'affaires ne concernant que les Grecs, on ne trouve que peu d'indications sur des œuvres d'art. Le récit ininterrompu des événements ne permet pas de telles digressions et Hérodote présume que ses lecteurs connaissent les monuments de la Grèce péninsulaire. Dans les derniers livres, nous trouvons seulement une description détaillée des anciennes inscriptions de l'Isménion (5, 59) et une courte note sur la construction du sanctuaire delphique par les Alcméonides (5, 62). Les statues de Damia et d'Auxésia sont mentionnées (5, 82–83), mais seulement à propos de la guerre entre Athènes et Égine et, de même, Hérodote présume que tout le monde a vu le monument funéraire d'Anchimolios près de l'Héracleion (5, 63); | il ne [*219*] cite que comme preuve de l'invasion laconienne en Attique. Une raison semblable explique la mention du quadrige dédié après la victoire sur les Béotiens et les Chalcidiens (5, 77) et celle de la fondation du sanctuaire de Pan sur l'Acropole (6, 105). Hérodote raconte plutôt longuement une histoire sur la statue d'or d'Apollon qui avait été volée par les Perses à Délion en Béotie (6, 118); mais cette mention n'est qu'une incidente dans l'histoire. Il y a néanmoins trois passages dans lesquels Hérodote parle de monuments pour eux-mêmes. Le premier concerne l'inscription des morts des Thermopyles (7, 228); le second, les offrandes votives après Salamine (8, 121–122), et, le troisième, celles qui suivirent la victoire de Platées (9, 80). Ni Hérodote ni Thucydide ne parlent du groupe des tyrannicides; ceci est d'autant plus remarquable que tous deux s'occupent avec détails de l'acte d'Harmodios et d'Aristogiton et

---

1. Cf. le commentaire de Stein dans son édition d'Hérodote; W. W. How et J. Wells, *A Commentary on Herodotus*. J. E. Powell, *A Lexicon to Herodotus*, p. 141, renvoie au préambule, s.v. *Deed, Act, Achievement*.
2. Cf. aussi 8, 27, où deux autres grandes statues sont mentionnées.
3. Diogène Laërce, 1, 28–29; Plutarque, *Solon*, 4, et *Conv.*, 13; Athénée, 495 D.

déclarent que ce meurtre ne fut pas le début de la démocratie athénienne (Hér., 6, 123; Thuc., 1, 20, 2; 6, 54, 1). Nous pouvons, par conséquent, supposer que cette raison explique le silence d'Hérodote sur ce monument.

Si nous comparons ce que dit Hérodote, dans son œuvre, des monuments dédiés soit par des Grecs, soit par des monarques d'Asie et d'Égypte dans les sanctuaires grecs, nous nous rendons compte qu'il signale peu de monuments grecs et nous découvrons quel principe a présidé au choix des monuments cités dans son œuvre. Nous trouvons une énumération détaillée des offrandes votives de Gygès, de Midas (1, 14) et d'Alyatte (1, 25) au dieu de Delphes, et une référence à la construction par Alyatte du sanctuaire de Milet (1, 19–20). Hérodote s'intéresse beaucoup aux offrandes votives de Crésus à Delphes (1, 50, 92) et nous pouvons rapprocher les consécrations des monarques perses dans des sanctuaires grecs du fait que les Perses n'avaient point d'images de leurs dieux (1, 131). Il y avait aussi des offrandes égyptiennes dans ces sanctuaires. Hérodote mentionne les broches envoyées par la Samienne Rhodopis (2, 135), l'offrande votive de Ladikè à Cyrène (2, 181) et celles d'Amasis dans beaucoup de sanctuaires grecs (2, 183). Tous ces passages montrent que la taille du monument et aussi la valeur de la matière employée sont la raison qui guide Hérodote dans son appréciation. Mais ses contemporains pensaient-ils comme lui?

[220] Avant de poursuivre cette idée, voyons rapidement quels sont les plus grands monuments d'Asie et d'Égypte indiqués par Héro|dote. Nous avons déjà parlé de l'imposant tombeau d'Alyatte (1, 93). On peut y ajouter la statue d'or travaillée au marteau mentionnée par lui à cause de la matière même (7, 69). Dans le sanctuaire de Zeus (Bélus), à Babylone, Hérodote admire l'or et sa valeur (1, 183) et, dans sa description du sanctuaire d'Héraclès à Tyr (2, 44), il conserve le même point de vue. Dans bien des passages concernant les monuments de l'Égypte, il insiste sur la quantité, les dimensions, la valeur (2, 63, 91, 106, 110, 111, 121, 130, 138, 141, 143, 153, 155, 172, 173; 3, 37). Il convient, d'ailleurs, d'ajouter que c'est souvent l'étrange, le curieux et l'incroyable qui l'intéresse le plus. Relatons seulement à ce sujet les édifices et ouvrages en bois de la Scythie (4, 108), la statue d'Aristée (4, 15) et un haut-relief taillé dans le rocher en Perse (3, 8); mais bien d'autres passages cités plus haut trouveraient aussi leur place ici.

Hérodote attribue aux nations civilisées d'Asie et d'Égypte une part importante dans le développement de la civilisation grecque. Aussi observe-t-il que bien des héros grecs viennent d'Orient. Son opinion sur l'importance de l'Orient a dû être engendrée par son examen des monuments d'Asie et d'Égypte. Il dit, en effet, que la construction des sanctuaires et l'art de faire des statues et des reliefs ont été inventés en Égypte et empruntés par les Grecs aux Égyptiens (2, 4). L'habituelle représentation égyptienne du dieu Pan lui semble avoir été adoptée en Grèce (2, 46). Les hermès seuls sont dit inventés à Athènes, ou plutôt empruntés par les Athéniens à un peuple non grec—mais aussi non asiatique ou africain—les Pélasges (2, 51).

Hérodote n'a pas une très haute opinion de l'individualité de l'art grec. A son avis, les artistes sont des ouvriers, donc des gens des basses classes (2, 167). Il ne cite dans toute son œuvre qu'un artiste, Théodore de Samos (1, 51; 3, 41). S'il parle une fois de Dédale (7, 170), ce n'est pas comme d'un artiste. Pouvons-nous croire que son attitude au sujet de l'art ne lui était pas particulière, mais qu'elle était aussi celle qui régnait en Grèce et chez les Athéniens à cette époque? L'esthétique, à laquelle nous devons la connaissance des noms des premiers artistes grecs, n'était pas encore née; les monuments élevés en Attique avant 480 avaient tous été détruits par les Perses et, sauf de rares exceptions, étaient ensevelis quand les Athéniens commencèrent la restauration de l'Acropole. Mais bien des monuments votifs, élevés après 480, furent aussi enfouis quand Périclès entreprit de construire le Par|thénon. Furtwängler dit *[221]* que les Athéniens de cette époque avaient peu de respect pour les œuvres d'art. Il me semble pourtant qu'ils estimaient fort, au contraire, celles de ces œuvres dont la matière avait une grande valeur. Car nous voyons que la gloire du plus célèbre artiste du ve siècle est fondée sur ses deux grandes statues de divinités à Olympie et à Athènes; ces deux statues imposantes ne nous donneraient pas l'impression d'une valeur artistique intrinsèque avec leur masse d'or et d'ivoire. Nous apprenons aussi, par les comptes des trésoriers attiques et par les inscriptions relatives aux deux Victoires en or, que la valeur de la matière avait une grosse importance dans les œuvres d'art même au ve siècle. Les offrandes de prix et les images des dieux formaient l'essentiel du trésor d'Athèna; car Périclès, d'après un passage significatif de Thucydide, dit que l'or des monuments votifs et des statues des dieux peut être utilisé pour des fins militaires. On doit, à ce propos, signaler un passage d'Hérodote (1, 164), où nous lisons que les Phocéens, au temps de leur embarquement, emportèrent les images de leurs dieux et toutes les offrandes votives des sanctuaires, à l'exception des bronzes, des marbres et des peintures. Nous sommes fondés à supposer que les Phocéens n'ont pas uniquement emporté des statues de bois, mais surtout celles d'or et d'argent.

Les observations précédentes suggèrent que, dans l'opinion des Grecs du ve siècle, la valeur d'une œuvre d'art dépendait de sa taille et de la matière employée. Si nos sources littéraires ne nous fournissent rien touchant la qualité de ces œuvres—au sens où nous l'entendons—pouvons-nous croire qu'une telle appréciation n'a pas existé? Cette conclusion serait contredite par les monuments existants et par les principes d'esthétique en usage depuis le ive siècle. Car, tout comme nous, l'Antiquité considéré les vi^e et ve siècles comme l' "acmé" de l'art grec et, pour la plus grosse part, de l'art européen.

La soudaine croissance économique et la grande opulence de quelques cités grecques ont, je crois, rendu possible la production d'un certain nombre d'œuvres d'art originales. Un passage d'Hérodote sur la richesse de Siphnos (3, 57) et ce qui a survécu du trésor de Siphnos à Delphes viennent à l'appui de cette opinion. On peut y ajouter la construction du temple de Delphes par les Alcméonides et le fait que l'art grec s'est développé dans les plus riches sanctuaires, Delphes, Olympie et finalement Athènes. L'art | d'Égine prospéra *[222]* au temps de la puissance économique de cette île; la décadence survint au ve

siècle. Mais c'est l'art attique du Ve siècle qui nous fournit notre meilleure preuve. Toutes les grandes constructions de Sparte appartiennent à une période ancienne et, au passage significatif de Thucydide (1, 10) rappelant que le pouvoir politique et l'influence de Sparte furent hors de proportion avec les monuments de cette ville (voir aussi 1, 84), nous pouvons ajouter comme commentaire que la puissance économique de Sparte fut petite. D'un autre côté, bien des monuments du Ve siècle sont toujours debout à Athènes, des œuvres d'art du VIe siècle remplissent le musée de l'Acropole; bien des musées européens contiennent des copies d'œuvres d'art du Ve siècle et la masse des vases attiques est immense. La grande importance de la céramique attique des VIe et Ve siècles se montre unique, non seulement par la quantité de vases existants, mais surtout par la haute qualité de leur travail. Et, cependant, seuls, Solon (ϵἰς ἑαυτόν, 2, 47) et Critias (frag. 1), l'un contemporain des débuts de la peinture attique sur vases, l'autre de sa fin, nous parlent de potiers et de poterie;[4] mais Hérodote, dans un passage célèbre (5, 88), nous dit qu'Égine défendait sa céramique en interdisant l'importation des vases attiques. Après les guerres médiques, Athènes devint la capitale d'un grand empire et ses chefs eurent à leur disposition des sommes considérables (Aristophane, *Paix*, 605, 616).

En résumé, pour en revenir au préambule d'Hérodote, nous avancerons que cet historien, qui connaissait les grands centres de vie humaine en Asie et en Égypte, était d'avis que les grands monuments doivent figurer dans la composition de l'histoire générale d'une nation, parce qu'ils montrent, tout comme ses actions et ses coutumes, la véritable structure d'une société.

4. Cf. aussi Aristophane, *Nuées*, 470; *Paix*, 202.

35

Herodotus and the Inscriptions

The discovery of the now famous "Themistocles Decree" (Jameson, *Hesperia* [59] 29 (1960), pp. 198–223) has renewed our interest in the literary and epigraphical evidence pertaining to the Persian Wars; see the critical treatments by Habicht, *Hermes* 89 (1961), pp. 1–35, and Moretti, *Riv. di Fil.* 38 (1960), pp. 390–402. It appears that there existed in Athens, in addition to the account of Herodotus, a series of popular and patriotic stories some of which were based on documents which themselves may not have been made public until the fourth century before Christ.

In what follows, no attention is paid to the anecdotes which are known only from the literary tradition of the fourth century; the tentative suggestion may be made, however, that the Atthidographers from Hellanicus to Philochorus included many of these stories as well as many of the documents listed below; see Jacoby's remarks in *Atthis*, pp. 196–215, and in the introductions to the individual Atthidographers in *F. Gr. Hist.*; see also my remarks in *Charites, Studien z. Altertums Wissenschaft* 234–239 [*supra*, pp. 146–152], and Biliński, *L'antico oplite corridore di Maratona* (1959) (rejected by Moretti, *Athenaeum* 38 (1960), pp. 154–156).

THE SALAMIS EPIGRAM

The so-called Marathon epigrams (*SEG* 16.22 with the criticism of Amandry, *Deutsche Beitr. z. Altert.* 12/13 (1960), pp. 1–8, Peek, *Hermes* 88 (1960), pp. 493–498, and Pritchett, *Univ. of Cal. Publ. in Class. Arch.* 4/2 (1960), pp. 167–168) could not be properly understood until Pritchett used the *Persians* of Aeschylus to explain the emphasis on footsoldiers and ships in the first epigram. Only here, in the play (lines 435–471), and in Plutarch's *Aristides* (9) is the fighting under Aristides on Psyttaleia put side by side with the naval battle, while the account of Herodotus is short and without details (8.74 and 95). There must have been an old tradition in Athens to which the play and the epigram belong and upon which Plutarch was able to draw; the second copy of the first epigram (*SEG* 16.139) shows that this tradition was alive in the fourth century.

The second epigram which was surely later inscribed has not yet been convincingly connected with any of the battles of the Persian Wars; Wade-Gery's suggestion that the poem speaks of "sea-girt gates" points to Thermopylae rather than to Marathon.

[60]
THE THEMISTOCLES DECREE

Against the authenticity of this fourth century inscription from Troizen (part of the monument described by Pausanias 2.31.10) internal and external arguments have been advanced (see above). It must be remembered that decrees of this type were not inscribed on stone, and comparable material from the literary tradition (including the speeches in Thucydides) has still to be examined. Externally, it is clear that Herodotus refers to a decree such as this (7.144.3) but that his narrative does not bear out the provisions of the preserved decree. On the other hand, the number of Athenian ships of which Herodotus speaks (8.1 and 14: $127 + 20 + 53 = 200$) is the same as that in the inscription, and the sudden decision to flee to Troizen, Aegina, and Salamis (8.41.1) must have been preceded by certain preparations, as Herodotus (8.40.1) intimates and the inscription describes.

The essential fact is, however, that the account of Herodotus does not follow the provisions of the decree: a) The evacuation of women and children began later and was not confined to Troizen (8.41.1). b) The priests did not stay behind but only the poor and the treasurers (8.52.2). c) It was not just citizens and resident aliens who manned the fleet but also Plataeans and Chalcidians (8.1), and the fleet was not divided into two sections of a hundred ships each. Thus, most of the provisions which can be checked were not carried out. This throws suspicion against the historical relevance rather than against the genuineness of the inscription; most decrees and treaties order things to be done but do not record whether these things were done.

On the other hand, there is the amnesty of which we hear also from Aristotle (22.8) and Nepos (*Aristides* 1-2). Its purpose, according to Plutarch (*Aristides* 8 and *Themistocles* 11), was to prevent the desertion of dissatisfied Athenians to Xerxes; the decree speaks instead of the desire for national unity. The story told by Nepos that Aristides fought at Salamis before he was freed from the punishment of ostracism is explained by the decree which orders the ostracized to return to Salamis and to await there further decisions. Thus, this provision of the decree was carried out, but confirmation of it does not come from Herodotus who mentions the return (8.79) but not the recall of Aristides; see, however, Andocides 1.107, Aristotle, Plutarch, and Aristides (46, p. 248 Dindorf).

When Demosthenes (19.303) referred to the fact that Aeschines had quoted in his speech the decrees of Miltiades and Themistocles and the Ephebic Oath, neither the two orators nor their audiences questioned the genuineness of these documents; nor should we.

THE OATH OF PLATAEA

The Ephebic Oath and the Oath of Plataea were inscribed together on a fourth century stele (*SEG* 16.140) and were quoted together by Lycurgus at about the same time, ca. 330 B.C. (*Against Leocrates* 76–82; see also Diodorus 11.29.2). The former is generally considered genuine, the latter is generally condemned, mainly because Herodotus does not mention it (7.132.2 and Thucydides 3.52.2, 63.3, 64.2–3 are not precise enough) and because Theopompus says (*F. Gr. Hist.* 115 F 153) "the Hellenic Oath, which the Athenians say the Greeks swore before the battle against the barbarians at Plataea, is falsely reported" (or attributed: *katapseudetai*). Since we know from Herodotus (9.86.1) that the Greeks did not plan to subject the Thebans to the punishment threatened by the oath ("tithe"), it seems clear that they either did not swear it or did not keep it. Hence, Lycurgus's claim of Athenian origin for the oath (80), and Theopompus's insistence that it was an Athenian invention show that we are again dealing with a problem of historicity and not of genuineness. | It may be [61] remarked in passing that the oath shares with the Themistocles decree, with the Covenant of Plataea, and with the famous statement on the assessment of Aristides (Thucydides 1.96.1) the use of the word "barbarian" as referring to the Persians.

THE COVENANT OF PLATAEA

This agreement which was sponsored by Aristides (Plutarch, *Aristides* 21.1) has been condemned as a forgery because it is not mentioned by Herodotus (except perhaps in 9.86.1 and 106), Thucydides (except perhaps in 3.58.4, 68.1, 71.2–4), or Isocrates (14, *The Plataicus*), and because it evidently did not have any effect on the events of the following years. It does, however, anticipate the original compact (also drafted by Aristides) of the Delian League (see *ATL* 3, pp. 225–233), and its oath formula is in part preserved (Diodorus 9.10.5); see my remarks in *TAPA* 91 (1960) [*supra*, pp. 11–15].

THE PEACE OF KALLIAS

In spite of recent attacks (e.g. by R. Sealey, *JHS* 80 (1960), pp. 194–195), the peace of Kallias (mentioned also in Diodorus 9.10.5) is a genuine document; see Andrewes, *Historia* 10 (1961), pp. 15–18. Demosthenes makes clear (19.273–274) that the Athenians did not think much of the vaunted peace at the time when it was concluded; later when it was renegotiated (Andocides 3.29) it became immediately obsolete through the support Athens gave to a rebellious satrap. After the treaty was published (we do not know when) it suddenly was turned into an important historical document. Its historical validity was rightly challenged by Theopompus (*F. Gr. Hist.* 115 F 153–154) and Callisthenes (Plutarch, *Cimon* 13; see Pearson, *The Lost Histories of Alexander the Great* (1960), p. 30, note 42) but its genuineness is assured also by its inclusion in Craterus's collection (*F. Gr. Hist.* 342 F 13).

In conclusion it may be affirmed that not a single one of the documents listed here can be considered a forgery although some of them are known to us only in the form in which they were published in the fourth century; see the parallel study by Graham, *JHS* 80 (1960), pp. 94–111 with Jeffery's additions (*Historia* 10 (1961), pp. 139–147). The genuineness of these documents does not assure, however, their historical relevance. After all, not only the biased anti-Athenian Theopompus (*F. Gr. Hist.* 115 F 281 and 306) but also Herodotus (6.123.2), Thucydides (1.20.2), and Aristotle (6.2–3 and 18.5), in their references to the "popular" tradition, take issue with stories which are commonly believed but not necessarily true. As students of history, it is as important for us to know what people believed to have happened as what in truth did happen but may not have been known or believed at the time. In fact, the former is more easily ascertained and it is more relevant to the understanding of the historical tradition as it existed in the past.

36

The Speech of the Athenians at Sparta

Concerning the speech of the Athenians at Sparta (1.72–78) Gomme and de Romilly[1] have already called attention to most of the questions which have puzzled me and which I should like to discuss, although I may not be able to answer them. The problems concern the occasion of the speech, its content, its intention, and its effect.

The occasion for the presence of the Athenian embassy in Sparta has not been explained either by Thucydides (1.72) or by any of his ancient and modern readers. The embassy happened to be there already and it was concerned with other matters, that is, it was not sent to reply to the charges made by Sparta's allies. Later, namely after the allies of the Spartans had departed, the ambassadors | of the Athenians returned home after having completed their negotiations.

The authoritative tone of the speech delivered by the Athenian embassy is indicated not only by the words used (which after all may be those of Thucydides) but also by the summary[2] with which the historian introduces the speech and which is bound to be a reasonably accurate account. This means that the Athenians present at Sparta were competent and responsible and could

The Speeches in Thucydides, ed. P. A. Stadter (Chapel Hill, N.C., 1973) 32–48

1. A. W. Gomme, *A Historical Commentary on Thucydides* (Oxford, 1945), 1:252. Full discussions of this passage are given by Gomme, pp. 233–46 and 252–55, E. Schwartz, *Das Geschichtwerk des Thukydides* (Bonn, 1929), pp. 102–16, J. de Romilly, *Thucydide et l'impérialisme athénien* (Paris, 1947), pp. 205–29, H.-P. Stahl, *Thukydides: Die Stellung des Menschen im geschichtlichen Prozess* (Zetemata 40, Munich, 1966), pp. 43–54. Since this manuscript was submitted to the editor, I have been able, through the kindness of Peter Herrmann, to consult L. Reich, "Die Rede der Athener in Sparta" (Ph.D. diss., Hamburg University, 1956). It is a good and detailed study which reaches the startling conclusion "dass die Reden in Sparta, wie wir sie lesen, in Wirklichkeit nie gehalten worden sind, sondern dass sie eine freie Schöpfung des Thukydides sind, der seine Leser mit der *alêthestatê prophasis* des Peloponnesischen Krieges bekannt machen wollte."

2. Or preamble, as Westlake names them in his article in this book.

take it upon themselves to state their city's position without fear of repudiation at home or disbelief in Sparta.

My first suggestion is the claim that the Athenian embassy described by Thucydides is the very one which was sent according to Plutarch (*Per.* 30) at the motion of Pericles with the purpose of denouncing the Megarians for having cut down the trees of the "sacred field" between Eleusis and Megara, and of justifying the so-called Megarian Decree which was Athens' reply to the sacrilege.

It has been said[3] that this embassy cannot have taken place so early because the Athenian herald Anthemocritus died or was killed in Megara and the Athenians replied to this alleged criminal act of impiety with the famous decree of Charinus which provided for "two invasions annually" into Megarian territory. It is claimed accordingly that this deterioration of relations between Megara and Athens could not have taken place while negotiations between Sparta and Athens were still going on. This argument ignores the fact that the Athenian charges against the Megarians were religious and not political and therefore not subject to negotiations between Athens and Sparta. The matter had become political when the Athenians passed the so-called Megarian Decree which excluded the Megarians from the Attic markets. To explain the political punishment for a religious offense Pericles evidently asked an embassy to be sent to Megara and Sparta, and it was at this point that the Athenian ambassadors appeared in Sparta, and that Anthemocritus died or was killed in Megara. This time, the Athenian response was unpolitical: "enmity without truce and declaration" (*aspondos kai akêryktos echthra*), "death to any Me|garian found in Attica," and "twice every year" an "invasion" of the Megarid. That this was a demonstration and not war is made clear by the fact that it was to take place twice a year and was not to be announced or to be concluded by a truce.

[34]

There is another link between the speech of the Athenian embassy reported by Thucydides and the decree of Pericles according to which an embassy was sent to Sparta and Megara. Plutarch remarks that the decree contained "a reasonable and humane justification" (*eugnômonos kai philanthrôpou dikaiologias echomenon*) which the herald was to convey. These adjectives seem to be eminently fitted to describe the speech of the Athenians in Sparta. Notice the two main themes: we are worthy to rule and we rule worthily.

To sum up, we may assume that the Athenian embassy which happened to be in Sparta when Sparta's allies made their first accusations against Athens had come at the request of Pericles in order to explain to the Spartans the Megarian Decree, and that this embassy was charged with giving a reasonable and humane justification of Athenian policy.

The next question concerns the relationship of the speech as given by Thucydides in his own words, the preamble (1.72), and the words put by him into the mouth of the embassy (1.73–78).

---

3. W. R. Connor, *American Journal of Philology* 83 (1962): 231.

The purpose of the speech summarized by Thucydides is two-fold: not to make a defense against the accusations raised by the cities but to show in general that the Spartans should not quickly make a decision but look at the situation at greater length. In the second place, the Athenians want to show the Spartans their power and to remind the older ones of something they knew and to tell the younger ones something of which they were ignorant; in this way they believed they would persuade both groups to be quiet rather than to make war. One could easily claim that Thucydides' summary of what the Athenians plan on saying does not quite agree with what they do say, or at any rate that the summary does not suggest the kind of speech quoted in direct oration.[4]

This argument may be carried a step further. Gomme already observed[5] that Archidamus's speech seems to ignore completely the | preceding speech of the Athenians, and we may add that the speech of Sthenelaïdas refers to it in a peculiar way (1.86.1). The ephor says three things about it. First, they spent a long time praising themselves but did not claim that they did not do harm to our allies and to the Peloponnesus. This criterion, that they did not do harm (*hôs ouk adikousi*), recurs in the Plataean Debate (3.52–54) where the Spartans ask the prisoners "whether they had done anything good for the Spartans and their allies" (3.52.4) which is rephrased by the Plataeans "that they did no harm" (*ouk adikeisthai*, 3.54.2). Evidently, this is a standard argument of the Spartans.

The second point made by the ephor is also interesting: if they were noble when fighting against the Medes, but are now bad toward us, they deserve double punishment (*diplasias zêmias*) because they turned out bad having been good. This is exactly what the Thebans tell the Spartans about the Plataeans who had recounted their noble deeds during the Persian Wars (3.67.2). A good record, the Thebans say, ought to help those who are being injured, but for people who behave shamefully it should double their punishment (*diplasias zêmias*). The third point to which Sthenelaïdas replies (1.86.4) occurs also both in the summary (1.72.1) and in the speech itself (1.78.1): take time over your decision; let no one tell us how we must make our decision. All these are conventional responses which illustrate the famous remark of Thucydides about his speeches (1.22.1): "as each seemed to me to say *ta deonta* concerning the various circumstances." On the other hand, the next sentence of the speech of Sthenelaïdas is decisive (1.86.5): "vote for war." If he did not say that, Thucydides was lying; but he did say it, and Thucydides refers to this part of the speech with the words (1.22.1) "keeping as close as possible to the entire *gnômê* of what was really said."

The speech of Sthenelaïdas, and perhaps all the speeches of Thucydides, consists of two parts: what the man really said which has to be absolutely reliable—"vote for war" and what Thucydides thought he was rather bound to

---

4. See de Romilly, *Thucydide*, pp. 224–27.
5. Gomme, *Commentary on Thuycydides*, 1:252.

say—"the Athenians are unreliable hypocrites."

We may now return to the speech of the Athenians and ask whether the summary (1.72) contains what they really said, the speech itself (1.73–78) what [36] they were bound to say. If this is true, | the summary contains what Thucydides heard them say (1.22.1, *autos êkousa*) or heard that they said (1.22.1, *emoi apangellousin*) or thought that they really meant (1.22.1, *tôn alêthôs lechthentôn*), namely, do not rush making your decision, and remember or consider how we gained our power and how great it is. The speech itself (1.73–78) should then contain not only these two points, which in a way it does, but also others which the Athenians were bound to make because they represent the view which they had of themselves and especially, to judge from the historical situation, the views held by Pericles. This speech merits, therefore, our closest attention.

The first and the main theme of the speech of the Athenians is the justification of their empire: "we do not hold what we possess unreasonably (*apeikotôs*)" they say at the beginning (1.73.1) and "we are worthy (*axioi*) of the empire we hold" they say at the conclusion of their account of their achievements during the Persian Wars (1.75.1). Similarly, Euphemus begins his speech at Camarina, more than fifteen years later, with the same observation (6.82.1). "It is necessary as well to state that we hold the empire reasonably (*eikotôs*)," and he closes his argument concerning the Athenian actions during the Persian Wars with the statement (6.83.1), "therefore we are worthy (*axioi*) to rule."

The similarity between these two statements is so great that one has seen in them two expressions of one and the same attitude without appreciating the contrast between the idealism of the earlier speech and the realism of the later.

This is most clearly shown by the fact that Euphemus disclaims during the Sicilian Expedition what the Athenians assert before the outbreak of the war. "We do not speak in fine phrases (*kalliepoumetha*)" says Euphemus (6.83.2)—echoing the famous words of the Athenians at Melos (5.89). "We are not using fine phrases (*onomata kala)"*—such as "we rule reasonably (*eikotôs*) because we overthrew the barbarian singlehanded" (or "we rule justly [*dikaiôs*] because we put down the Mede," 5.89); but before the beginning of the war, the Athenians in their speech at Sparta used precisely this argument when they said (1.73.4) "we say that at Marathon we alone fought in the forefront against the barbarians," an argument which the Spartans had turned [37] against the Athenians | in their speech at Athens before the battle of Plataea (Hdt. 8.142.2), "you stirred up this war, when we wanted nothing of it and the contest began for your territory," and which had been first suggested by the closing remarks of Miltiades' speech to Callimachus (Hdt. 6.109.6): "If you come over to my opinion, your country will be free, and the first state in Greece."

Euphemus refuses to use not only the argument based on the glory of Marathon but also the more striking one: we rule "because we risked our lives for the freedom of these men more than for that of all, and for our own." The meaning of this claim becomes clear from a passage in the speech of the Athenians at Sparta (1.73.4–74.3) which culminates in the assertion "therefore we state that we conferred on you no less than we received." The reason why the Athenians dare assert that they did more (they actually say "no less") for the Peloponnesians than for themselves was the fact that Athens had been captured but the Athenians continued to fight for the freedom of those Greeks who had not yet been overrun by the Persians. The full force of this point is made by the Athenians in their remarks closing this first argument (1.74.4): "But if we had yielded to the Mede, like the others, out of fear for our territory, or if we had refused afterwards to embark in our ships (like people already defeated), your naval inferiority would have made a sea fight unnecessary, and the enemy's objectives would have been obtained at leisure." The very same argument is made more forcefully in that famous passage in which Herodotus declares (7.139.5): "If a man should now say that the Athenians were the saviors of Greece, he would not miss the truth." How closely Thucydides follows Herodotus in the very words he uses will be pointed out presently; at the moment it suffices to become aware that Herodotus knows that his argument is liable to envy (*epiphthonon*, 7.139.1) while Euphemus glibly declares, after saying that he would not use the two claims mentioned above (Th. 6.83.2), "all are free from envy (*anepiphthonon*) who provide for their own safety."

It has become clear from this comparison of the speeches of the Athenians in Sparta (1.73–78) and of Euphemus in Camarina (6.82–87) that they represent two different stages in the development of the claim of the Athenians to be entitled to rule: in the first speech the claim is based on virtuous conduct, in the second on power. We notice in passing that in this respect the Euphemus speech agrees with the Athenian position in the Melian Dialogue (5.85–113) and with that of Alcibiades in his two great speeches in Athens (6.16–18) and in Sparta (6.89–92). On the other hand, the speech of the Athenians at Sparta (1.73–78) agrees with the famous praise of Athens which Herodotus must have written about the same time that the Athenians delivered their speech (Hdt. 7.139). The two statements agree not only in sentiment but also in content and may indeed represent the view which the Athenians held of their position in the world on the eve of the outbreak of the Peloponnesian War.

[38]

The Athenians, at the beginning of their speech (1.73.2), not unlike Thucydides himself (1.20.1), refuse to talk about the distant past because of the unreliability of mere hearsay. But they consider it necessary to speak of the Persian Wars although the events are well known and bothersome to hear again and again; the Athenians do not admit (as Herodotus does, 7.139.1) that what they have to say may be *epiphthonon*, but they later complain (1.75.1) that they do not deserve to be viewed invidiously (*epiphthonôs diakeisthai*). After Marathon, the famous decision to embark (*esbantes es tas naus pandêmei*) is mentioned (1.73.4) in words recalling the Themistocles Decree (*eisbainein eis tas ... naus*).[6] The battle of Salamis kept Xerxes from attacking and destroying the individual cities of the Peloponnesus which could not have defended themselves and each other against the great Persian fleet. The same points are

---

6. R. Meiggs and D. Lewis, *A Selection of Greek Historical Inscriptions* (Oxford, 1971), no. 23, ll. 13–14.

made by Herodotus (7.139.3: "taken city by city") and in Herodotus by Mnesiphilus (8.57.2: "they will scatter each to his city"), but the argument is different because Herodotus emphasizes the importance of the Athenian decision neither to flee nor to surrender but to fight. The Athenians speaking at Sparta assert that the fleet was the decisive part of the Greek armed forces (1.74.1), a view already expressed by Themistocles to Eurybiades, Hdt. 8.62.1: "The ships are the decisive factor of the war." They further assert that they made the three most useful contributions to the success of the battle: the largest

[39] number of ships (a little | less than half of the four hundred; see Hdt. 8.42–48), the most intelligent general, and the most unflinching enthusiasm; only the first and the third of these three contributions are repeated by Euphemus (6.83.1). The most intelligent (*xynetôtaton*) general, namely Themistocles, is not mentioned either by Euphemus or by Herodotus in his praise of Athens (7.139), although Herodotus later (8.124) reports that Themistocles was acclaimed as most intelligent (*sophôtatos*) and honored by the Spartans on account of his intelligence (*sophiê*) and cleverness (*dexiotês*).[7]

To sum up, the first and major theme of the speech of the Athenians, the justification of their empire, is presented in the same spirit as the famous praise of Athens by Herodotus, but the two passages do not directly depend on each other. They represent what Thucydides called (1.22.1) *peri tôn aiei parontôn ta deonta malista eipein* and thus reveal to us the view which the Athenians held of themselves at this important moment of their history.

The second theme of the speech of the Athenians is their claim that they exercise their rule with moderation (1.76.4: *metriazomen*). It should be recalled that Thucydides did not refer to either theme in his own summary of the speech (1.72.1). He says that the Athenians wanted to show the power (*dynamis*) of their city and from this statement one would expect a speech like the one which Pericles is said to have delivered later (1.140–144). Instead, to use the summary given by the Athenians themselves (1.73.1), they say " we want to make clear that we do not hold what we possess unreasonably, and that our city is noteworthy." Not a word about *dynamis* here! The two themes, however, are well defined, first the justification of empire, which has already been discussed, and second the claim that Athens is noteworthy (*axia logou*).[8]

This difference between Thucydides' own summary, in which the *dynamis* is emphasized, and the speech itself with its strong stress on the *axion* is the more striking because of the repeated statement made by Thucydides that the Spartans entered the war "because they were afraid the power of the Athenians would grow" (*epi meizon dynêthôsin*, 1.88), and that the truest cause of the war was

[40] that "the Athenians were becoming great" (1.23.6). Perhaps | we should connect *alêthestatê prophasis* (1.23.6) and *tôn alêthôs lechthentôn* (1.22.1) and assume that what was "really" said was not necessarily actually said. Since Thucydides asserts that the "most true explanation" was "least manifest in word"

---

7. See de Romilly, *Thucydide*, pp. 207–9.
8. De Romilly noted this contrast, ibid., pp. 224–29.

(*aphanestatê logô*, 1.23.6), it is clear that the difference between *dynamis* and *axion* was consciously employed by the historian as if he wanted to say: the Athenians may have spoken of their merits and their worthiness but they really revealed their *dynamis*, and it was fear of this power which made the Spartans decide on war (1.88). This means that Thucydides expected his reader to recognize and to understand the difference between his estimate of the true intention of the Athenian speech ("to show the power of the city") and the arguments that may have been used by the Athenians ("to reveal the justice of the city"). In closing this argument, we may once more glance briefly at Plutarch's description of the motion made by Pericles which directed an Athenian embassy to proceed to Sparta (*Per*. 30). The embassy was to present a "reasonable and humane justification" of Athenian conduct, and that is exactly what the Athenians accomplished in their speech at Sparta.

Considering now more carefully the way in which the Athenians developed the second theme of their speech (1.75–77), we are struck at once by the fact that they cover the same period which Thucydides described in great detail immediately following his account of the conference at Sparta, in his famous Pentecontaetia (1.89–117). We shall therefore be first concerned with the agreement and the differences between the speech and the narrative of the historian. There is, to start with, full agreement between Thucydides (1.94–96,1) and the Athenians (1.75.1–2) concerning the manner in which Athens took over the command of the Greeks after the departure of Pausanias: *ou biasamenoi*, "not forcing" (1.75.2) corresponds to *hekontôn*, "willing" (1.96.1). The name of Pausanias is not mentioned by the Athenians although Thucydides gives as reason for the takeover "the hatred of Pausanias" (1.96.1) and tells later in detail (1.128–135) the story of the misfortune of Pausanias. The following account of the development of the command is both novel and at variance with the story told by Thucydides himself: the Athenians declare that they were forced (1.75.3) by fear, honor, advantage (*hypo deous, epeita kai timês, hysteron kai ôphelias*) to increase their power. The precise Greek words which are used [41] merit close attention because they are not easy to understand and have been misunderstood by the commentators.[9] There has been reluctance to accept the three "causes" which led to the development of Athenian power as more or less clearly defined chronological stages and to assume that it was first fear (presumably for the Persians and their possible retaliation), then the desire for honor and prestige (being at the head of a great alliance such as Greece had never seen before), and finally the economic advantage (which came from the accumulation of the tribute and other imperial income) which led the Athenians on to ever greater power and preeminence. The reluctance to accept this interpretation is derived from passages in the work of Thucydides himself. In the first place, in the very same speech, the Athenians reverse the order of the causes and assert (1.76.2) that they were overcome by honor, fear, and advantage (*timês kai deous kai ôphelias*) and during the Sicilian expedition, the Athenian

---

9. See, however, Stahl, *Thukydides*, p. 47.

Euphemus declares at Camarina (6.83.4) that the Athenians admit to holding the command because of fear (*dia deos*)—this fear surely does not apply to the Persians. It is thought, therefore, that the three causes were meant to be operative during the whole period, and this interpretation was encouraged by the disassociation of *to prôton* from *epeita* and *hysteron*, but above all because the narrative of the Pentecontaetia as given by Thucydides does not contain or even suggest two internal divisions which would produce a period during which the Athenians acted out of fear of Persian counterattacks, another period during which Athens enjoyed the prestige of leadership of a great alliance, and a third period during which advantage (perhaps mutual advantage) held the alliance and the empire together. And yet, the mere mention of such periods makes it possible to identify them within the course of events described by Thucydides. The first period comes to an end with the battle at the river Eurymedon which brings the "Persian Wars" as such to a conclusion (1.100.1) and after which only the Egyptian expedition and, not independent of it, Cimon's last compaign in Cyprus occurred (1.104; 109–110; 112.1–4), neither of which was dictated

[42] or caused by fear. The second period begins with the revolt of | Thasos which, in contrast to the revolts of Naxos and Carystos, had nothing to do with the pursuit of the Persian Wars, and this period comes to an end with an event which Thucydides did not mention but which falls between the death of Cimon (1.112.4) and the battle of Coronea (1.113), namely the transfer of the Treasury of the League from Delos to Athens which made the empire for the first time profitable to the Athenians. It seems, therefore, reasonable to suppose that the Athenians refer in fact to three periods during which they increased and maintained their power, namely the war against Persia, the control over virtually all of Greece (except the southern Peloponnesus), and the economic organization and centralization (if not unification) of the Aegean area (as known from the Tribute Lists and Documents). Since this is evidently not the concept which Thucydides himself presented in his account of the Pentecontaetia, it must be a view held by others, possibly including Pericles himself.

The next sentence introduces a new consideration which is so directly connected with the concept of advantage (*ôphelia*) that some editors attach it to the preceding argument. "Furthermore, it did not seem safe" say the Athenians (1.75.4), "that people hated by the majority (sc. of the allies), after some had even already revolted and had been struck down, when you were no longer friendly in the same way (as before) but suspicious and at odds (with us), that such people (namely we) should run the risk of letting them (i.e., the allies, or it, i.e., the empire) go." This clumsy but literal translation is to make clear that the Athenians do not refer here to the situation before the outbreak of the Peloponnesian War, but to an earlier moment when the question of dissolving the alliance actually must have come up. This is indicated by the imperfect *edokei*, by the reference to some revolts which had been crushed (Thasos and Euboea, 1.100–101 and 114) while relations with the Spartans were strained and the revolting allies (if permitted to secede) would turn to the Spartans. Since none of this happened during the Thirty Years Peace (as shown clearly during the revolt of Samos), it appears that the Athenians explained why they could not dissolve the alliance when it had fulfilled its original purpose: it was of advantage (*ôphelia*) to them to keep it and it was unsafe (*ouk asphales*) to give it up. The Athenians conclude that | nobody can blame people who look [43] to their own advantage when they are in great danger. It may be that Plutarch (*Per.* 17.1) refers to the same situation when he reports that before Pericles moved the Congress Decree (i.e., before the transfer of the Treasury) "the Lacedaemonians began to be offended by the increase in power of the Athenians."

The following chapter (1.76) gives the explanation and the justification of the observations which have just been discussed (1.75) and which certainly convey a note of apology: we were forced (*katênankasthêmen*) to acquire and increase our power, and it was not safe for us to give it up. Moreover, you Spartans do the same in the Peloponnesus, and you would have done the same in Greece if you had kept the command (1.76.1). Thus what we are doing is perfectly natural and in line with the universal law which affirms that the weaker is constrained by the stronger (1.76.2). In fact, we think this to be a justified attitude and you thought so, too, until you discovered that it was advantageous to use the argument of righteousness (*tô dikaiô logô*) which has never kept anybody from making use of his power if he had any (1.76.3).

Professor Andrewes has called our attention to the fact that the doctrine here announced is characteristic of Thucydides and that the historian subscribed to it himself.[10] He pointed out, moreover, that "the Athenians' statement at Sparta goes far beyond any topical need to answer the Corinthians or advise the Spartans." Actually, this was not the intention of the Athenians nor did they identify themselves with this rather sophistic point of view which they do make their own after the Peace of Nicias and especially in the Melian Dialogue. What the Athenians here (1.76) really say is not "that might is an inescapable fact, however little we like or approve of it" (Andrewes) but that although this is so we are people worthy to be praised because while being so human as to exercise power we are more just than our power would require (*dikaioteroi ê kata tên hyparchousan dynamin*, 1.76.3). In fact, we believe that if others were to take our place they would show whether or not we are acting moderately (*metriazomen*). This argument of the Athenians reveals clearly that their claim to be | noteworthy (1.73.1) does not rest on the justification of their power but [44] on the moderate use they have made of it.

Unfortunately this moderation has brought the Athenians not praise but ill repute; the one would have been deserved, the other is unfair (1.76.4). What follows is a demonstration rather than an example of this observation, as Geoffrey de Ste. Croix has shown convincingly.[11]

The area in which the Athenians chose to demonstrate their moderation is the administration of justice as a whole and not only the conduct of commercial

---

10. *Proceedings of the Cambridge Philosophical Society* 186, n.s. 6 (1960), pp. 5–6: empire as a fact of nature. See, however, Stahl, *Thukydides*, p. 50.
11. *Classical Quarterly* 55, n.s. 11 (1961): 95–100.

cases, as has sometimes been thought. We would expect that political, military, economic, and financial control were issues of far greater importance than the administration of justice, but we must realize that in civilized society, then as now, the settlement of disputes of whatever kind is the mark of good government, depending on whether force (*biazesthai*) or arbitration (*dikazesthai*, 1.77.2) is employed. This means that the Athenians identify the essential issue and argue it on principle—without using any details. To be sure, both the author of the "Constitution of the Athenians" preserved among the works of Xenophon[12] and Aristophanes, especially in the *Wasps*, show clearly that the Athenian love of court trials had not gone unnoticed, and Professor Turner has commented on the meaning of *philodikein* as it occurs in the speech of the Athenians (1.77.1) and elsewhere.[13] It is, therefore, natural and justified that the Athenians should complain that they are thought to be litigious if they substitute the use of law for the use of force. "If in lowering ourselves, we arrange for trials under treaty provisions in cases against our allies and on the basis of legal equality in our own courts—we are called litigious." This translation of 1.77.1 differs from that offered by Geoffrey de Ste. Croix[14] in considering the two phrases *en tais xymbolaiais ... dikais and par' hêmin ... nomois* as parallel so that the Athenians point out that in accepting treaty provisions with the weaker allies they are lowering themselves (*elassoumenoi*) and they are in fact doing the same by treating their weaker allies according to the same laws as themselves when they come to Athens. Justice, as the Athenians proclaim at Melos (5.89), is determined on the basis of equal necessity but the superior do what they can and the weak yield. If the Athenians had acted before the outbreak of the Peloponnesian War according to this principle, there would and could have been neither treaties nor legal equality (*homoioi nomoi*) in any dealings between them and their allies. But in fact the Athenians lowered themselves, they made themselves weaker in order to treat with their weaker allies on the basis of equality (*apo tou isou*, 1.77.3 and 4). Instead of being praised for this moderation, they are hated by those whom they treat justly, and they are attacked by others.

[45]

This point which is succinctly stated in one sentence (1.77.1) is then elaborated rather extensively in the following paragraph. Other people exercising power elsewhere behave less moderately toward their subjects than we do, but nobody criticizes them. For if one uses force, one does not have to administer justice, too (1.77.2). Our allies are used to just treatment and therefore they complain whenever they are dissatisfied with their position or our decisions. But, of course, they would have to admit that the weaker has to yield to the stronger (1.77.3). People are more enraged if they are treated unjustly than if violence is done to them. In the first case, an equal is thought to have taken advantage of them, in the latter a superior to have forced them (1.77.4). They suffered more under Persian rule and endured it, but our rule seems now to be harsh;

12. [Xenophon], *Resp. Ath.* 1.16; H. Frisch refers in his commentary (p. 223) to Thucydides 1.77.
13. *Classical Review* 60 (1946): 5–7.
14. *Classical Quarterly*, p. 99.

rightly so, for present burdens are always heavy to bear (1.77.5). After these general observations, the Athenians turn directly to the Spartans and say (1.77.6): "If you were to put us down and to rule in our stead, you would quickly change the goodwill which you now receive because of the fear of us; just as you showed this when you for a short time held the command against the Persians—you will experience similar things also at the present time. For your legal conventions are quite different from those of others, and moreover none of you who goes abroad either adheres to these conventions or to those current in the rest of Greece."

The question forces itself upon us whether this last argument can have been made before the final defeat and before the establishment of the hated Spartan hegemony, both of which were predicted by the Athenians. Gomme feels that the open reference to Pausanias's conduct "would make the prophecy an easy one"; de Romilly[15] disagrees; I wonder whether Thucydides himself could have expressed this view already during the "Ionian War" of 412 B.C.

To sum up, the Athenians have demonstrated in their speech not only that they have a right to rule according to what we would call the "law of history" but that they exercise their rule moderately and justly in spirit; this argument is much more like that of the Funeral Oration than of the First Pericles Speech or any of the speeches made during the war. It shows Athens at her best, and it gives a true picture of Periclean Athens before the conduct of the war demoralized men, people, and policies.

The peroration (1.78) returns to the main purpose of the speech as it was stated at the beginning both by Thucydides (1.72.1: "do not decide hastily") and by the Athenians (1.73.1: "lest too easily you decide wrongly"). "Take time over your decision," say the Athenians (1.78.1), and it is to this wise counsel that the Spartan ephor Sthenelaïdas angrily replies (1.86.4): "Let nobody tell us that we should take counsel." This repeated use of the word *bouleuomai* calls to our mind the contrast between *euboulia* and *aboulia*, good counsel and bad counsel, which not only the Athenian politicians but also the poets explored, not only Sophocles in the *Antigone* but also Euripides in the *Medea*. Not satisfied with the general advice, the Athenians once more explain in detail what in their opinion should be considered by the Spartans: do not be persuaded by other people's opinions and accusations; in turn, Sthenelaïdas urges the opposite course of action (1.86), and Thucydides himself admits that to a certain extent the Lacedaemonians were persuaded by the argument made by their allies when they declared that the treaty had been broken (1.88). Secondly, the Athenians point to the unpredictability (*ho paralogos*) of war, an argument which Gomme calls "a commonplace in Thucydides." It certainly is not a view attributed by Thucydides to Pericles. Not only does Pericles not mention it in his First Speech which is a call to war, but he implicitly denies it in his last speech when he says (2.64.1) that the plague was an event which went beyond what was expected and that it was the only thing which went

---

15. De Romilly, *Thucydide*, pp. 221–23.

beyond expectation. In fact, Thucydides praises the foresight (*pronoia*) of Pericles with regard to the war (2.65.6). Evidently, the Athenians in their speech at Sparta are made to emphasize the uncertainty and unpredictability of war, either because they really thought so (and thus disagreed with Pericles) or because they want the Spartans to think so and to avoid war.

Finally, the Athenians put forth the paradox of war (1.78.3): in war you act first and think afterward when you have suffered misfortune. Therefore, the Athenians imply, think first and you won't vote for war. But we are not caught in any such mistake, conclude the Athenians (1.78.4), and when we say "we" we mean not only ourselves but, as far as we can see, you, too. And we say unto you, do not dissolve the treaty as long as we both are still free to choose good counsel (*euboulia*), do not break the oaths, but resolve differences by arbitration according to the agreement. This attitude was still maintained by Pericles in the beginning of his First Speech (1.140.2) when he says that it had been agreed that differences should be settled by arbitration and at the end when he reaffirms (1.144.2) his willingness to arbitrate according to the treaty. Pericles adds, "we shall not start the war but resist those who do" (*archomenous de amynoumetha*); the Athenians at Sparta say virtually the same thing as a parting shot. "Taking the Gods by whom we have sworn the oaths as our witnesses, we shall try to defend ourselves against you if you start the war (*amynesthai polemou archontas*), and we shall do this wherever you choose to act."[16]

The two main questions, then, are: Did the Athenians deliver a speech and what did they say? Was the speech which is attributed to them provocative and was it meant to be provocative? The first question has been answered by Kagan with disarming candor: "if we are to deny his simplest statement of fact, we must give up any hope of dealing with the history he purports to describe."[17]

[48] When it comes to the question what the Athenians ac|tually said, we have to decide between Thucydide's summary of the speech and the speech quoted by him; the former adhered as closely as possible to what was "really" said and what was understood by the Spartans; this means that the Athenians conveyed by their speech the greatness of their power while they actually used arguments based on their virtue and moderation. Thus the speech itself contained what the Athenians actually said and what Thucydides thought the situation demanded of them to say. To answer the second question, the speech was provocative to the Spartans but it was not meant by the Athenians to be provocative, and Thucydides makes this difference clear by emphasizing in his summary that the Athenians displayed their power, but in the speech itself that they emphasized their moderation and justice.

16. See the remarks on this and similar "stings in the tail" by H. D. Westlake, *Greek, Roman, and Byzantine Studies* 12 (1971): 500.
17. D. Kagan, *The Outbreak of the Peloponnesian War* (Ithaca, N.Y., 1969), pp. 293–300, esp. p. 293. See, however, the opposite view expressed by Schwartz, *Geschichtswerk des Thukydides*, p. 105, and by Stahl, *Thukydides*, p. 43.

The speech of the Athenians at Sparta contains a moral justification of Athenian democracy. A comparison of the speech of the Athenians at Sparta with the speeches of the Athenian generals at Melos and of Euphemus at Camarina shows clearly the difference between the cynicism of an Alcibiades and the idealism of a Pericles. This means that we possess in the speech of the Athenians at Sparta an authentic statement on the glory and virtue of the Athenian Empire in the days of Pericles.

37

Andocides and Thucydides

[121] There is little doubt that Thucydides knew of Andocides; he may even have known him. And yet, he did not mention him by name even in the one passage[1] in which he clearly refers to him. One wonders whether this was written after Andocides delivered (and published) the speech[2] in which he revealed the famous scene in prison,[3] an early case of pleabargaining. There is a slight but perhaps not insignificant link between the accounts given by Andocides and Thucydides; both authors emphasize the calming effect the revelations of Andocides had on the hysterical atmosphere prevailing at that time in Athens.[4] Thucydides may not have believed all that Andocides had said, but he surely agreed with him that he had saved the situation at that time. It is, therefore, the more remarkable that Thucydides kept from his readers the identity of this man, if he knew it.[5]

Less than ten years after the speech on the Mysteries (1), Andocides delivered (and presumably published) his oration on Peace with the Lacedaemonians (3) which shows clearly, I think, that he had read Thucydides. At the very beginning of the speech (3.2), he invited his audience to look back at their peaceful relations with the Lacedaemonians in the past, because one must use the past as containing sure signs ($\tau\epsilon\kappa\mu\dot\eta\rho\iota\alpha$) of the things to be in the future ($\pi\epsilon\rho\grave{\iota}$ $\tau\hat{\omega}\nu$ $\mu\epsilon\lambda\lambda\acute{o}\nu\tau\omega\nu$ $\check{\epsilon}\sigma\epsilon\sigma\theta\alpha\iota$). This is exactly what Thucydides claimed to be his great contribution:[6] those who wish to observe the past and what will once again happen in the future ($\tau\hat{\omega}\nu$ $\mu\epsilon\lambda\lambda\acute{o}\nu\tau\omega\nu$ $\check{\epsilon}\sigma\epsilon\sigma\theta\alpha\iota$) "will consider my work

---

Classical Contributions: Studies in Honor of M. F. McGregor (Locust Valley, N.Y., 1981), 121–123

1. Thuc., 6.60.2.
2. Andoc., 1.
3. Andoc., 1.48–69.
4. Thuc., 6.60.3 and 5 = Andoc., 1.51, 56, 58, 59, 66, 68, with the repeated mention of suspicion, $\dot{\upsilon}\pi o\psi\acute{\iota}\alpha$.
5. This point has been neglected by R. Seager, *Historia* 27 (1978) 223.
6. Thuc., 1.22.4.

sufficiently useful." Umberto Albini[7] has called attention to the passage in Thucydides (un concetto tucidideo) and to other similar passages in the same speech of Andocides[8] and in the later works of other orators. Special consideration deserves a statement by Lysias[9] because it reflects the same attitude to the utility of historical knowledge and employs the same terminology as do Thucydides and Andocides; unfortunately, the speech which may be dated "soon after"[10] 403 B.C., is considered by Kenneth Dover[11] a hypothetical defense, "a paradigm of the defence speeches which Lysias wrote for men who remained in the city."[12] The speaker urges the court to keep in mind[13] what happened (γεγενημένων) under the Thirty in order to take better counsel (ἄμεινον... βουλεύσασθαι). After giving the pertinent historical evidence, he adds[14] the advice: using the examples of previous events (τοῖς πρότερον γεγενημένοις παραδείγμασι χρωμένους) you must take counsel concerning what will happen in the future (χρὴ ... | βουλεύεσθαι περὶ τῶν μελλόντων ἔσεσθαι). [122] George Kennedy, in his discussion of Andocides 3,[15] referred to a passage in Aristotle's *Rhetoric*[16] which looks like a paraphrase of the statements in the orators, and which he considered "derived from sophistic rhetoricians." If this were true, Thucydides would be addressing himself to political orators and would be telling them that his work will not only be useful to them as a reliable record of past events which in some way will again take place in the future, but that in fact it will be a lasting possession (κτῆμα ἐς αἰεί—a storehouse) rather than a beautiful composition to be heard and enjoyed but once.

The realization that Andocides and Lysias, and of course Isocrates,[17] accepted Thucydides' concept of the utility of history confirms the traditional interpretation of the historian's famous statement.[18] Even more important is the application of this concept by Andocides to the issue at hand, namely to the peace with the Lacedaemonians. The orator chose the same period of Athenian history to show that the Athenians prospered whenever they were at peace with the Spartans[19] as Thucydides had chosen[20] to reveal the Athenians as daring and aggressive and the Spartans as quiet and conservative—just as the Corinthian had described them.[21] Evidently, the same events can be made

---

7. Umberto Albini, ed., Andocide, *De pace* (Florence 1964) 55.
8. Andoc., 3.29.
9. Lys., 25.21–23.
10. Donald Lateiner, "Lysias and Athenian Politics" (unpublished Stanford dissertation 1971) 79.
11. Kenneth Dover, *Lysias and the Corpus Lysiacum* (Berkeley 1968) 189.
12. Lateiner (above, note 10), 85.
13. Lys., 25.21.
14. Lys., 25.23.
15. George Kennedy, *AJP* 79 (1958) 41.
16. Arist., *Rh.* 3.16.11 = 1417b.
17. C. H. Wilson, *Greece and Rome* 13 (1966) 56.
18. See now W. R. Connor, *CJ* 72 (1977) 289–298, and R. Leslie, *ibidem* 342–347.
19. Andoc., 3.3–9.
20. Thuc., 1.89–117.
21. Thuc., 1.70.2–3.

to teach different lessons, and I think that Andocides was quite aware of this, and so was Isocrates after him.[22]

Of all the pleas for peace of the classical period,[23] Andocides alone develops the theme that Athens should make peace with Sparta because the Athenians benefited greatly in the past whenever they were at peace with the Lacedaemonians. Considering the novelty of this argument, it is not surprising that Andocides was less accurate than persuasive; he did not win his argument, and most modern readers have rejected his account because it is very much at variance with that of Thucydides and Ephorus.[24] Only Wesley Thompson has made a valiant and on the whole successful attempt "to show that Andocides' account is not really so confused and erroneous as scholars have imagined."[25] The two worst mistakes are the statements[26] that it was Miltiades the son of Cimon (and not Cimon the son of Miltiades) who brought about the peace which closed the war which raged in Euboea (and not in Boeotia). Thompson suggested[27] that Andocides reversed by mistake the "War in Euboea" and the "War because of Aegina," while Johan Henrik Schreiber defended the "Euboean War."[28] The confusion of Cimon with his father is the more perplexing because the orator Aeschines repeated the entire passage of Andocides[29] including this obvious mistake.[30]

Aeschines defended the Peace of Philocrates by reviewing the events of Athenian history since the victory of Salamis in order to show that whenever Athens was at peace, the Athenian democracy was well off. This historical summary begins earlier and ends later than the account of Andocides, but for the period covered by Andocides Aeschines follows him very closely, adding nothing and using often the very same words.[31] The emphasis of Aeschines is, however, on peace in general, and on Athenian democracy in particular, while Andocides stresses the benefits which the Athenians enjoyed whenever they were at peace with the Lacedaemonians.

Although Aeschines may have used Andocides directly, both orators follow in substance the Athenian popular tradition which Thucydides already sought to correct. This is indicated by a number of significant agreements between Andocides and Thucydides. Both mention the possession of Megara and Pegai,[32] the Five Years' Peace,[33] the building of the walls,[34] and the Thirty Years'

22. Isoc., 1.34; 2.35; 4.141; 6.59, all mentioned by Albini, and 8.11.
23. Ar., *Ach.* and *Pax*; Xen., *Hell.* 6.3 (Kallias's speech); Isoc., *De Pace*.
24. Diodorus; see now H. D. Westlake, *Phoenix* 31 (1977) 325–329.
25. Wesley Thompson, *TAPA* 98 (1967) 489.
26. Andoc., 3.3.
27. Thompson, 488.
28. Johan Henrik Schreiber, *SymbOslo* 51 (1976) 28.
29. Andoc., 3.3–9.
30. Aeschines, 2.172–176.
31. See Thompson, 483, note 2.
32. Thuc., 1.103.4 = Andoc., 3.3.
33. Thuc., 1.112.1 = Andoc., 3.4.
34. Thuc., 1.107.1 and 108.3 = Andoc., 3.5.

Peace;[35] what is more striking, there are several correspondences between the second speech of Pericles[36] and the speech of Andocides[37] showing that both relied here on the popular and patriotic tradition concerning the greatness of Athens: the treasure of coined silver on the Acropolis,[38] the numbers of horsemen, archers and triremes,[39] and the revenue from the tribute.[40] Thompson, who examined these details carefully, came to the conclusion that Andocides used the book of Hellanicus,[41] and Schreiber also offers Hellanicus as a source for the "popular version of Athenian history."[42] I think both Andocides and Hellanicus relied on what Aristotle called the Populars (δημοτικοί), and that Thucydides' criticism of Hellanicus[43] would apply equally well to the accounts given by Andocides and Aeschines.

One of the reasons for attributing the "ancient tradition" represented by Andocides (and Theopompus) to Hellanicus was Thompson's decision that this account must be the work of a chronographer who made "a serious effort to order the events of the Pentekontaetia in a chronological sequence,"[44] but he did not consider for a moment that Andocides and Theopompus may have the correct story of Cimon's ostracism, return and death. And yet, many years ago, I tried to show that Thucydides agreed with the story,[45] and Schreiber has gone a long way in the same direction.[46]

The preceding sketchy remarks are offered to a friend of forty years who has taught us a great deal about the history of the very period treated by Thucydides and Andocides, the Pentekontaetia. My modest conclusions are: Andocides knew the history of Thucydides, he accepted Thucydides' view of the utility of history, but he offered a summary of Athenian history which followed the very same popular tradition which Thucydides had set out to correct.[47]

35. Thuc., 1.115.1 = Andoc., 3.6.
36. Thuc., 2.13.
37. Andoc., 3.
38. Thuc., 2.13.3 = Andoc., 3.7 and 8.
39. Thuc., 2.13.8 = Andoc., 3.7 and 8.
40. Thuc., 2.13.3 = Andoc., 3.9.
41. Thompson (above, note 25) 483–490.
42. Schreiber (above, note 28), 29 and especially in a still unpublished monograph entitled "Kimon-Studien."
43. Thuc., 1.97.2.
44. Thompson, 489.
45. A. E. Raubitschek, *Historia* 3 (1955) 370–380; *AJA* 70 (1960) 37–38.
46. Johan Henrik Schreiber, *SymbOslo* 52 (1977) 29–36.
47. I have benefitted greatly from the criticism generously offered to me by Bob Connor, David Lewis, Lionel Pearson, Wesley Thompson; how fortunate to have such friends!

38

Die sogenannten Interpolation in den ersten beiden Büchern von Xenophons griechischer Geschichte

[315] Der vorliegende Bericht wird mit um so größerem Zögern vorgelegt, als er nicht nur eine allgemein gemachte Annahme in Frage stellt, sondern auch die Warnung des Altmeisters unserer Studien ignoriert, der vor ein paar Monaten erklärte (K. v. Fritz, *The Position of Classical Studies in Our Time*, FIEC Sitzung, Paris, August 1972, S. 2), daß von all den Theorien der letzten 150 Jahre, die sich mit der Komposition von Xenophons *Hellenika* befaßten, die letzte weniger begründet sei als die älteste; Herr von Fritz hat allerdings mein Manuskript erst nachträglich eingesehen und versichert mir jetzt brieflich: "Ihre chronologischen Betrachtungen haben mir im allgemeinen eingeleuchtet." Ihm und F. W. Michel, der ebenfalls das Manuskript mit Zustimmung las, sei auch hier mein Dank ausgesprochen.

Es wird allgemein angenommen, daß der Text des Xenophon durch eine Reihe verschiedener Zusätze ergänzt und zum Teil verstellt wurde, und diese sogenannten Interpolationen sind in den letzten Ausgaben von J. Hatzfeld (1949) und E. Delebecque (1964) durch kleineren Druck ausgeschieden und wurden von Hatzfeld in einem besonderen Anhang ausführlich behandelt (S. 153–158).

Es soll im folgenden versucht werden, den Text des Xenophon ohne die Annahme von Interpolationen zu verstehen, wobei wir uns dessen bewußt sein müssen, daß einige dieser Zusätze, wie die von Archonten, Ephoren, Olympiaden, und Olympioniken als anachronistisch und zum Teil als historisch falsch betrachtet wurden. Andrerseits habe ich vergeblich nach Antworten auf die Fragen gesucht, warum, zu welchem Zweck, wann und von wem diese Zusätze gemacht wurden und warum sie gerade mit dem Ende des Peloponnesischen Krieges aufhören, und ich konnte mich mit G. Jachmanns Zurechtweisung (*Klio* 35, 1942, 79), daß das "einfältige" Fragen sind, nicht befriedigen.

Mit dem Hauptproblem, der Chronologie, hat sich D. Lotze neulich zweimal eingehend beschäftigt (*Philologus* 106, 1962, 1–13, und: Lysander und der peloponnesische Krieg, *Abh. Leipzig, Phil.-Hist. Kl.* 57/1, 1964, 72–98), und H. Baden hat diese Frage auch nochmals kurz und bündig behandelt (*Untersuchungen zur Einheit der Hellenika Xenophons*, Hamburg 1966, 49–58).

Die erste Schlüsselstelle ist wohl der Anfang des zweiten Kapitels (1,2,1), mit dem ein neues Jahr beginnt, das durch die 93. Olympiade mit dem Sieg des Euagoras von Elis in dem neubegründeten Wettbewerb mit dem Zweigespann und des Eubotas von Kyrene im Wettlauf, durch das Ephorat des Euarchippos und das Archontat des Euktemon (408/7) bestimmt wird. Daß es sich hier nicht um ein im Spätsommer anfangendes Jahr handelt, zeigt die erste Episode, in der Xenophon berichtet, daß Thrasyllos am Anfang des Sommers nach Samos fuhr (1,2,1–13). Dann erzählt Xenophon die Ereignisse, die sich noch im selben Jahre in Lampsakos zutrugen (1,2,15–17), und fügt kurze Bemerkungen über Koryphasion und Trachis (1,2,18) und über die Wiedervereinigung von Medien und Persien (1,2,19) hinzu. All das soll sich nach Xenophon im Jahre 408 zugetragen haben, während die Herausgeber dafür das Jahr 410 oder 409 annehmen. [316]

Da die letzten Ereignisse, die noch von Thukydides beschrieben wurden, in den Sommer von 411 gehören und Xenophons *Hellenika* dort anfangen, nämlich am Beginn des Winters (1,1,2), so mußte das erste Kapitel dieses Werkes den zwei Jahren 410 und 409 gewidmet sein, also, nach der thukydideischen Zählung, dem Ende des 21., dem 22. und dem Anfang des 23. Jahres des Krieges. Daß Xenophons Geschichtsdarstellung mit dem Jahr 410 anfängt, wird auch von Diodor (oder seiner Quelle) bestätigt (13,42,5), wenn er behauptet, daß Xenophons *Hellenika* 48 Jahre umspannten (410–362); verwunderlich bleibt jedoch Diodors Bemerkung (ebd.), daß Thukydides die Ereignisse von 22 Jahren berichtete, was nicht mit den Angaben des Thukydides (8,60,3 und 109,2) übereinstimmt: Er schloß vor dem Ende des 21. Jahres.

Vielleicht ist es am besten, mit den am Ende des ersten Kapitels erwähnten Ereignissen anzufangen und zu zeigen, daß diese wohl im Jahre 409 stattgefunden haben können.

Die kurz berichtete Eroberung von Selinunt und Himera (1,1,37) wird von Diodor (13,54–62) fürs Jahr 409/408 erzählt, obwohl natürlich der Anfang des karthagischen Feldzuges schon früher (13,43–44) berichtet wurde. Jedenfalls kann man nicht behaupten, daß Xenophons Angaben chronologisch falsch sind.

Dasselbe trifft auf die Wahl des Kratesippidas als Nauarch (1,1,32) zu, die von Diodor in seinem Bericht über das Jahr 409/408 angeführt wird (13,65,3), und auf die Verhandlungen zwischen Hermokrates und Pharnabazos (1,1,31), die ins gleiche Jahr des diodorischen Berichtes gehören (13,63,2). Somit steht nichts der Annahme im Wege, daß die ganze auf Hermokrates bezügliche Stelle (1,1,27–31) ins Jahr 409/408 gehört und Diodor 13,63,1–6 entspricht.

Wir dürfen uns jetzt dem ersten Teil des ersten Kapitels (1,1,1–26) zuwen|den, um zu überprüfen, inwieweit die dort erwähnten Ereignisse in die Zeit vom Winter 411/410 bis ins Jahr 409/408 passen: Die Fahrt des Dorieus (1,1,2) gehört in den Anfang des Winters, und die vorhergehende Episode des [317]

Thymochares noch in den Herbst des Jahres 411; P. Defosse hat diese Stelle glänzend interpretiert (*Rev. Belge* 46, 1968, 5–24), ohne jedoch das Fehlen der Einleitung erklärt zu haben. Wenn Diodor die Abenteuer des Dorieus erst ins Jahr 410/409 datiert (13,45,1), so ist das damit zu erklären, daß er das Jahr 411/410 mit dem Ende des thukydideischen Werkes abschließt.

Die nächsten Ereignisse, die Fahrt des Thrasyllos nach Athen und die Sammlung von Geldern, gehören noch in den Winter 411/410 (1,1,2–8 = Diodor 13, 45,1–47,2), und die Ankunft des Tissaphernes ist (1,1,9–10) wohl ans Ende des Winters (411/410) zu datieren. Das heißt, daß die Schlacht bei Kyzikos im Sommer des Jahres 410 stattfand (1,1,11–26). Xenophons Bericht schließt mit der Aufforderung von Pharnabazos an die Griechen, eine neue Flotte zu bauen, ebenso groß wie die verlorene (1,1,25). Das muß aber geraume Zeit gedauert haben, und während dieser Zeit geschah eben nichts oder sehr wenig. Das erklärt die Pause vom Sommer des Jahres 410 bis zum Jahre 409. Als die neuen syrakusanischen Feldherren die Flotte in Milet übernahmen (1,1,31), mußte diese die neugebaute gewesen sein, denn im vorhergehenden Jahre hatten die Syrakusaner ihre Flotte verbrannt (1,1,18).

Warum Xenophon nichts von den Friedensverhandlungen in Athen erzählt, die wohl ins Jahr 410/409 gehören (Diodor 13,52–53), ist uns nicht bekannt, muß aber seinen Grund haben. Jedenfalls können wir die Chronologie richtigstellen und annehmen, daß 1,1 die Behandlung der Jahre 411/410, 410/409 und 409/408 enthält, und 1,2 mit 408/407 anfängt, wie der Text von 1,2,1 uns versichert.

Die zweite Schlüsselstelle ist der Zeitpunkt der Rückkehr des Alkibiades nach Athen, die von Xenophon in die Zeit der Plynterien datiert wird (1,4,12), das heißt in den Juni des Jahres 407. Der Anfang dieses Jahres ist in 1,3,1 beschrieben, und in 1,4,1 kehrt Xenophon zum Frühjahr 407 zurück, um über die Schicksale der athenischen Gesandten zu berichten, die sich jahrelang in Kleinasien aufhielten (1,4,7); hier (d.h. in 1,4,1) einen neuen Jahresanfang zu erkennen, liegt kein Grund vor, wird aber von allen Verteidigern der Interpolationshypothese angenommen (s. Baden, a.O.51 Anm. 110 und 54 Nr. 3). Dann nimmt Xenophon den Hauptbericht wieder auf (1,4,8) und erzählt die Heimfahrt des Alkibiades, die im Sommer 407 stattfand, d.h. noch im Archontat des Euktemon.

Jedenfalls blieb Alkibiades nicht lange zu Hause, sondern war schon vor Ende des Jahres 407 wieder in Samos (1,4,21–22).

[318] Mit der Erwähnung der Ausfahrt des Lysander als Nauarch (1,5,1), wohl schon im Frühjahr 407, kommen wir zu dem Bericht der Schlacht von Notion, zu den außerordentlichen Srategenwahlen für 407/406 (1,5,12–16) und zu Konons Fahrt nach Samos, im Winter 407/406 (1,5,18–20). Im nächsten Jahr, nämlich im Frühjahr 406, übernahm Kallikratidas den Befehl von Lysander (1,6,1), und in diesem Jahre fanden auch die Schlacht bei den Arginusen (1,6,24–38) und der Prozeß der Feldherren statt, wohl im Winter 406/405 (1,7,1–35) zur Zeit der Apaturien, also im November 406 und später. Im Frühjahr 405 kehrte Lysander zwar nicht als Nauarch, aber als Epistoleus zu

der Flotte zurück (2,1,7); in diesem Jahre war Alexias Archon (405/404), fand die Schlacht bei Aigos Potamoi statt (2,1,10–32) und betrat Lysander Athen (2,2,26). Im folgenden Jahre 404/403, im Archontat des Pythodoros, wurden die Dreißig gewählt (2,3,1).

Es hat sich also gezeigt, daß der Bericht des Xenophon über das Ende des Peloponnesischen Krieges bedenkenlos angenommen werden kann, einschließlich der Datierung durch Archonten, Ephoren, Olympioniken und Olympiaden.

Wir müssen uns jetzt der Zählung der Kriegsjahre zuwenden, in der Xenophon ganz präzise ist (2,3,9). Am Ende des Sommers von 403 (i.e., im Jahre 404/403) waren $28^{1/2}$ Kriegsjahre vergangen. Diese Angabe als "überhaupt falsch" zu bezeichnen, wie das Lotze tat (*Philologus* 106, 1962, 2), war unberechtigt, und W. P. Henry deutete das Richtige an, als er behauptete (*Greek Historical Writing*, Amsterdam 1967, 43): "auch wenn wir glauben, daß diese Berechnung richtig ist, können wir nicht glauben, daß Xenophon unter dem Einfluß des Thukydides stand, dessen ausdrückliche Angabe (5,26,1) ihr widerspricht."

Erstens einmal läßt sich leicht zeigen, daß Xenophons chronologische Zusammenfassung richtig ist und mit den niemals angezweifelten Angaben des Thukydides übereinstimmt. Die Ephorenliste fängt mit Ainesias an (2,3,9), der Ephor war, als der Krieg ausbrach, im fünfzehnten Jahre nach der Einnahme von Euboia und dem Abschluß des dreißigjährigen Vertrages; all das steht wörtlich bei Thukydides (2,2,1). Der elfte Ephor der xenophontischen Liste (2,3,10), Pleistolas, war vom Herbst 422 bis zum Herbst 421 im Amt, und im Frühjahr dieser Periode wird er von Thukydides als amtierender Ephor dreimal erwähnt (5,19,1; 24,1; 25,1). Alexippides war der 21. Ephor, vom Herbst 412 bis zum Herbst 411, und wird, kurz nachdem er seine Pflichten übernahm, von Thukydides (8,58,1) als amtierender Ephor erwähnt.

Anders liegt die Sache mit der Länge des gesamten Krieges, die Thukydides ganz genau mit siebenundzwanzig oder dreimal neun Jahren angibt (5,26); da|bei wird von ihm das Ende des Krieges durch die Einnahme Athens und die Schleifung der Mauern bestimmt (5,26,1). Auch Xenophon datiert dieses Ereignis ins Jahr 405/404 (2,2,23), sagt aber, daß im folgenden Jahre (404/403), als Endios Ephor und der von den Dreißig gewählte Pythodoros Archon waren (2,3,1), nach der Einnahme von Samos (2,3,6–7) Lysander endlich reichbeladen nach Sparta zurückkehrte (2,3,8), ein Ereignis, das Xenophon an das Ende des Sommers von 403 datiert (2,3,9). Xenophon betont noch einmal (2,3,10), daß das Ende der $28\frac{1}{2}$ Jahre durch die Rückkehr des Lysander bestimmt war. Trotzdem glaube ich, daß seine Angaben nicht ohne Kenntnis der berühmten Stelle des Thukydides geschrieben worden sind und daß sie sich direkt auf sie beziehen. Die Zahlen des Xenophon (2,3,9–10) sind demnach verläßlich und nicht im sachlichen Widerspruch zu denen des Thukydides. Der Ausdruck ... ἔτη τῷ πολέμῳ ἐτελεύτα ist dem jährlich wiederkehrenden des Thukydides (... ἔτος τῷ πολέμῳ ἐτελεύτα) nachgebildet. Xenophons eigene Jahresangaben sind regelmäßig mit denselben Phrasen wiedergegeben (ὁ ἐνιαυτὸς ἔληγεν: 1,1,37; 2,19; 5,21; 2,1,8; 2,24; τῷ δ'ἐπιόντι ἔτει: 1,2,1; 3,1;

[319]

6,1; 2,1,10; 3,1; s. Baden, a.O.49 Anm. 107), wobei in allen fünf Fällen die Ephoren und Archonten, die im Laufe des Jahres ihr Amt antraten und die betreffenden zwei olympischen Jahre (1,2,1 und 2,3,1) mit den eponymen Stadionsiegern genannt werden; in 1,2,1 steht nicht nur die Olympiadenzahl, sondern auch eine Angabe über die Erweiterung der Festordnung. Nicht nur sind die Archonten- und die Ephorennamen richtig wiedergegeben, auch die Olympiaden und die Stadionsieger stimmen mit den von Diodor (13,68,1 und 14,3,1) genannten überein, und die Bemerkung über den Sieg des Euagoras von Elis wird durch andere Quellen bestätigt (s. L. Moretti, *Olympionikai*, Roma 1957, 111 Nr. 350). Es fällt allerdings auf, daß das erste der fünf genau datierten Jahre 408 ist (1,2,1), das heißt, daß die zwei vorausgehenden Jahre (410 und 409), die im ersten Kapitel behandelt wurden, unbezeichnet blieben; vielleicht wurde das in der verlorenen Einleitung erklärt. Noch sonderbarer ist es, daß Xenophon, der Thukydides darin folgt, daß er die Gesamtdauer des Krieges angibt (2,3,9), ansonsten nur dreimal die Summe der vergangenen Kriegsjahre erwähnt (1,3,1; 1,6,1; 2,1,7).

Diese Angaben wurden allgemein verdächtigt, da sie nicht nur falsch zu sein, sondern auch einander zu widersprechen schienen. Um von hinten anzufangen, die letzte der Summen wird anläßlich der Übergabe der Flotte an Lysander spät im Jahre 406 gegeben (2,1,7): damals waren 25 Jahre des Krieges vorbei, das heißt, es war das 26. Kriegsjahr. Nach der thukydideischen Zählung, mit der Xenophon sonst übereinstimmt, läuft das 26. Kriegsjahr vom Frühjahr 406 bis zum Frühjahr 405, und Xenophons Zählung stimmt demnach. Am Anfang desselben Jahres 406 sollen schon 24 Jahre vergangen sein (1,6,1), das heißt, man war in 25. Jahr, das vom Frühjahr 407 bis zum Frühjahr 406 dauerte. Damals, nämlich im Frühjahr von 406 (aber noch im 25. Kriegsjahr, obgleich dieses Kalenderjahr dem Archontat des Kallias zugewiesen wird, 406/405, der erst im Spätsommer sein Amt antrat), legte Lysander seine Stellung als Nauarch nieder, die er während des Jahres 408/407 eingenommen hatte; s. D. Lotze, Lysander und der peloponnesische Krieg 84. Die Angaben des Xenophon bezüglich des 25. und 26. Kriegsjahres (1,6,1 und 2,1,7) scheinen demnach verläßlich zu sein und in Übereinstimmung mit der sonstigen xenophontischen Chronologie zu stehen; bemerkenswert ist, daß sie sich beide auf Lysanders Tätigkeit beziehen und daß sie so umständlich ausgedrückt sind, daß man sie oft für falsch hielt. Anders steht es mit der Angabe (1,3,1), daß am Anfang des Jahres 407 erst 22 Jahre des Krieges vorbei waren, das heißt, daß man sich damals noch im 23. Kriegsjahr befand, das dann im Frühjahr 408 begonnen hätte. Diese Zählung widerspricht nicht nur der xenophontischen Chronologie, nach der dieses Jahr das 24. war, auf dessen Ende Xenophon in 1,6,1 verweist, sondern natürlich auch der thukydideischen.

Ehe wir diesen Widerspruch genauer untersuchen können, müssen wir feststellen, daß keine der drei Summen (1,3,1; 1,6,1; 2,1,7), ob sie nun richtig oder falsch sind, dem unbefangenen Leser überhaupt verständlich sein kann, wenn ihm das Werk des Thukydides nicht gegenwärtig ist. Dieser Einwand hängt eng mit der Frage zusammen, ob uns der ursprüngliche Anfang des xenophontischen Geschichtswerkes erhalten ist, eine Frage, die P. Defosse neulich mit guten Gründen verneint hat (*Rev. Belge* 46, 1968, 5–24), die aber erst vor kurzem wieder ohne Berechtigung von L. Canfora bejaht wurde (*Totalità e selezione nella storiografia classica*, Bari 1972, 102 Anm. 13). Defosse nimmt nicht nur (wie schon vor ihm Hatzfeld) an, daß ein Stückchen der Erzählung fehlt, sondern auch daß die Einleitung verloren gegangen ist. Leider ist er der Frage nicht weiter nachgegangen, wann, von wem und zu welchem Zweck der Anfang der Hellenika entfernt wurde. Das Ergebnis und damit vielleicht auch der Zweck der Verstümmelung war zweifellos der überaus enge Anschluß des xenophontischen Geschichtswerkes an das thukydideische, und demselben Zweck dienen wohl auch die drei Summen (1,3,1; 1,6,1; 2,1,7). Das heißt, daß nach der Abfassung wenigstens der ersten beiden Bücher der *Hellenika* die Einleitung getilgt und drei Summen eingefügt wurden, um einen engen Anschluß an Thukydides herzustellen. Henry behauptet ohne Zögern, daß die ersten beiden Bücher der *Hellenika* ursprünglich nicht als Fortsetzung des Thukydides geschrieben wurden, aber er besteht darauf, daß der Text, wie wir ihn heute haben, erst nach dem ersten Jahrhundert unserer Zeitrechnung zurechtgeschnitten wurde (a.O.44–45 mit Anm. 103). Wir haben dazu schon bemerkt, daß die chronologische Zusammenfassung 2,3,9 mit direktem Bezug [*321*] auf verschiedene Stellen des thukydideischen Werkes geschrieben wurde, und müssen jetzt hinzufügen, daß wenigstens eine der drei Summen (1,6,1) nicht ohne weiteres aus dem Verband, in dem sie steht, herausgelöst werden kann. Xenophon berichtet hier, daß "im folgenden Jahre (406), in dem eine Mondfinsternis eintrat (am 15. April) und der alte Athenatempel in Athen in Brand geriet (s. W. B. Dinsmoor, *AJA* 51, 1947, 111 Anm. 14), im Ephorat des Pityas und im Archontat des Kallias (406/405), die Lakedaimonier Kallikratidas zur Flotte schickten, nachdem die Amtszeit (χρόνος) des Lysander abgelaufen war (παρεληλυθότος) und ebenso vierundzwanzig Jahre des Krieges." Das Participium παρεληλυθότος ist hier von Xenophon auf die Amtszeit des Lysander bezogen (wie in 1,5,1 auf die des Kratesippidas) und dann auf die Dauer des Krieges ausgedehnt. Dasselbe Wort wird auch anläßlich der beiden anderen Summen gebraucht (1,3,1 und 2,1,7), und Thukydides verwendet einmal (2,47,1) in chronologischem Zusammenhang ein ähnliches Wort (διελθόντος), während er sonst immer sagt, daß das Jahr "zu Ende kam" (ἐτελεύτα), ein Ausdruck, den Xenophon in seiner Zusammenfassung wiederholt (2,3,9). Ein Interpolator, dem es daran lag, die xenophontische Darstellung so an die thukydideische anzupassen, daß sie als ihre Fortsetzung gelten konnte, wäre nicht nur dem thukydideischen Sprachgebrauch (ἐτελεύτα) genau gefolgt, sondern hätte ihn gewissenhaft Jahr für Jahr am rechten Platze (1,1,37–1,2,1; 1,2,19–1,3,1; 1,5,21–1,6,1; 2,1,8–2,1,10; 2,2,24–2,3,1) verwendet. Das heißt, daß die Summen zwar nicht zur ursprünglichen Fassung der *Hellenika* gehören, daß man sie aber Xenophon ebensowenig absprechen kann wie die Angabe der Olympiaden, der Ephoren und der Archonten.

Wir haben gesehen, daß zwei der Summen (1,6,1 und 2,1,7) nicht nur richtig sind, sondern auch höchstwahrscheinlich von Xenophon stammen, und wir wenden uns jetzt der ersten Summe zu (1,3,1), die mit den anderen Angaben des Xenophon und des Thukydides im Widerspruch zu stehen scheint. Im Frühjahr des Jahres 407 kam das 24. Kriegsjahr zu Ende und fing das 25. an. Wenn die hier beschriebenen Ereignisse (die Fahrt nach Prokonnesos, 1,3,1) noch in das 24. Kriegsjahr gehören, aber schon im Jahre, in dem Antigenes Archon wurde (407), stattfanden, dann wäre es berechtigt gewesen zu sagen, daß "dreiundzwanzig Jahre des Krieges schon vorbei waren" und man sich am Ende des 24. befand. Im Text des Xenophon steht aber δυοῖν καὶ εἴκοσιν ἐτῶν, und man wird kaum δυοῖν in τριῶν ändern oder annehmen wollen, daß die Summe an den Anfang des vorigen Kapitels (1,2,1) gehört und irrtümlicherweise dem falschen Jahr beigeschrieben wurde. Eher wäre zu erwägen, ob Xenophon nicht annahm, daß er in den ersten beiden Kapiteln Teile von drei Kriegsjahren behandelt hatte (das 21., 22. und 23.), während er tatsächlich auch das 24.

[322] einbe|zogen hatte. Wie schon oben erwähnt (s. S. 316f.), enthält das zweite Kapitel die Ereignisse des Jahres 408, das zum Teil das 23. und zum Teil das 24. Kriegsjahr war, während im ersten Kapitel nicht nur die Ereignisse des Endes des 21. Jahres (1,1,1–10) beschrieben werden, sondern auch die des 22. (1,1,11–26) und ein Teil der Ereignisse des 23. Jahres (1,1,27–31). Dabei wurde bemerkt, daß nach der Schlacht bei Kyzikos, die im Sommer des Jahres 410 stattfand und in der die gesamte peloponnesische Flotte zerstört wurde, ein Jahr lang Frieden herrschte, nämlich während des Flottenbaues der Peloponnesier (1,1,25). Mit anderen Worten, das Jahr, von dessen Ende Xenophon am Ende des ersten Kapitels berichtet, war nicht dasselbe, in dem die Schlacht von Kyzikos stattfand, sondern das der Schlacht folgende. Da die Schlacht selber im 22. Jahre des Krieges (im Sommer 410) geschlagen wurde, konnte Xenophon, wie so viele seiner modernen Leser, angenommen haben, daß das Ende dieses 22. Jahres am Ende des ersten Kapitels beschrieben wurde und daß daher das zweite Kapitel das 23. Jahr beschrieb, das am Ende des Winters (1,3,1) noch nicht zu Ende gekommen war. Der Fehler wurde jedenfalls in der Beschreibung des nächsten Jahres (1,6,1) verbessert, das als 25. bezeichnet wurde.

Als Ergebnis dieser recht umständlichen Untersuchung könnte man die Behauptung aufstellen, daß Xenophon selber die ersten beiden Bücher der *Hellenika* überarbeitete, um sie als Fortsetzung des thukydideischen Werkes herausgeben zu können. Sichere Spuren dieser Überarbeitung sind die Entfernung der Einleitung und die Hinzufügung der Summen der Kriegsjahre, Eingriffe, die bisher einem oder mehreren Interpolatoren zugeschrieben wurden. Ob die Angaben der Jahresanfänge und -enden, der Olympiaden, Ephoren und Archonten, die ja alle auch als Interpolationen bezeichnet wurden, auch der späteren Überarbeitung Xenophons zuzuweisen sind, läßt sich viel schwerer beweisen, ist aber von untergeordneter Bedeutung, solange die Authentizität der Gesamtdarstellung anerkannt wird.

Während wir im Augenblick die Frage außer acht lassen dürfen, zu welchem Zeitpunkt die chronologischen Angaben von Xenophon in den Text gesetzt wurden, müssen wir untersuchen, welchen Zweck diese Angaben verfolgen und zu welcher historiographischen Methode sie gehören. Es ist doch zweifellos falsch, anzunehmen, daß Xenophon Thukydides imitiert, wenn er Olympiaden, Ephoren und Archonten als chronologische Fixpunkte seiner Geschichtsdarstellung gibt, Angaben, die historisch völlig bedeutungslos sind, wie das schon Thukydides in seiner Kritik an Hellanikos (1,97 = *FGrHist* 323 F 8) festgestellt hat, aber für die Chronologie innerhalb eines Jahres, und nicht von Jahr zu Jahr; denn Hellanikos hat nach Archonten datiert (*FGrHist* 323 a F 25 und 26). Tatsächlich hat Xenophon mit seiner Olympiaden- und Beamtenchronologie zwar Timaios vorweggenommen, aber Thukydides ignoriert und mit seiner ungeschickten Zählung der Kriegsjahre die Methode des Thukydides ganz mißverstanden. Augenscheinlich folgt also Xenophon nicht dem Thukydides, noch hat er seine Chronologie selbst entwickelt, denn sonst hätte er sie nicht so schnell wieder aufgegeben; siehe die amüsanten Bemerkungen von E. M. Soulis, *Xenophon and Thucydides*, Athen 1972, 67–69. *[323]*

Diese Überlegungen leiten zu der Frage nach dem Verhältnis zwischen den chronologischen und annalistischen Stellen in den beiden ersten Büchern von Xenophons *Hellenika* und der annalistischen Methode des Hellanikos, dessen Atthis mit dem Ende des Peloponnesischen Krieges abschloß (*FGrHist* IIIb, S. 11). Wir denken hier nicht nur an Olympiaden und Beamtennamen, sondern auch an die historischen Kurzverweise, die Xenophon oft an den Anfang und an das Ende seiner Jahresberichte gestellt hat und die auch bei Thukydides nicht ganz fehlen (6,7).

Baden (a.O.52–54) hat die Behandlung der sizilischen (1,1,37; 5,21; 2,2,24; 3,5) und persischen (1,2,19; 2,1,8–9) Geschichte demselben Verfasser zugeschrieben wie die Datierung nach Olympiaden und Beamten, nämlich demselben Interpolator, hat aber die Kurzverweise auf Mond- und Sonnenfinsternisse und auf Tempelbrände (1,3,1; 6,1; 2,3,4) Xenophon selber belassen, nicht zuletzt weil die Datierungen der Mond- und Sonnenfinsternisse richtig sind. Im Falle des Brandes des alten Athenatempels auf der Akropolis (1,6,1) wird die Richtigkeit der xenophontischen Angabe jetzt auch wieder von Dinsmoor (*AJA* 51, 1947, 111 Anm. 14) angenommen.

Dabei hat Baden vernachlässigt zu erwähnen, daß Xenophon die folgenden Gruppen von Kurzverweisen im Zusammenhang mit dem Jahreswechsel gibt:

a. 408: Karthago (1,1,37–2,1);

b. 407: Koryphasion, Trachis, Persien, Phokaia (1,2,18/19–3,1);

c. 406: Karthago, Mondfinsternis, Brand des alten Athenatempels in Athen (1,5,21–6,1);

d. 405: Persien (2,1,8/9–10);

e. 404: Karthago/Syrakus, Wahl der Dreißig, Sonnenfinsternis, Lykophron, Karthago/Syrakus (2,2,24–3,1,5).

Es zeigt sich somit, daß jeder der fünf besonders betonten Jahreswechsel nicht nur durch die betreffende Olympiade (nur der erste und fünfte natürlich) und die Namen der amtierenden Ephoren und Archonten identifiziert wurde, sondern auch durch eine Reihe von annalistischen Verweisen, von denen jeder in eine längere Geschichte hätte erweitert werden können. In dieser Weise hat Xenophon, vielleicht nicht sehr erfolgreich, versucht, eine allgemeine griechische Geschichte zu schreiben und hat damit spätere Bemühungen größerer Geschichtsschreiber im Prinzip vorweggenommen.

[324]

Wir dürfen daher unsere verschiedenen Überlegungen zusammenfassen und den ursprünglichen Entwurf der ersten beiden Bücher von Xenophons *Hellenika* im allgemeinen als gut erhalten betrachten. Xenophon hat später selber seine Einleitung und den Anfang der Geschichtsdarstellung weggestrichen und an drei Stellen die Summe der verlaufenen Kriegsjahre hinzugefügt; beides tat er, um sein Geschichtswerk als Fortsetzung und Vollendung des unvollständig zurückgelassenen Buches des Thukydides auszugeben, vielleicht herauszugeben. In seiner ursprünglichen Fassung und auch in der uns überlieferten Form verfolgt Xenophon andere Ziele als Thukydides und folgt einer anderen Methode. Das zeigt sich besonders in der Chronologie. Die Genauigkeit, mit der Thukydides jedes Ereignis innerhalb der Jahreszeiten und diese innerhalb der Kriegsjahre zu bestimmen imstande war, erregt noch heute Aufsehen (s. Thukydides 5,20,2–3 und Badens Bemerkungen ad locum, a.O.57), aber die Nachteile dieser Methode zeigen sich darin, daß die meisten Herausgeber diese relativen Angaben für den Leser in absolute übersetzen müssen, seien diese nun attische Archontenjahre, Olympiaden oder die Jahre der christlichen Zeitrechnung. Demgegenüber hat Xenophon an Allgemeinverständlichkeit gewonnen, was er an Genauigkeit oft aufgeben mußte, und seine unsynchronisierten Olympiaden, Ephoren und Archonten, die kaum eine Beziehung zur wirklichen Geschichte hatten, geben uns doch ein klareres Bild der Zeitverhältnisse als die Jahreszeiten des Thukydides, die an Jahreszahlen gebunden sind, die nicht allgemein bekannt waren.

Xenophon war kein Genie, aber seine Neuerungen in literarischen Genres sind wohl bekannt. Bücher wie die Anabasis, die Memorabilien, der Agesilaos, die Kyrupädie, die technischen Schriften über Einkünfte, Ökonomie, Pferdezucht, waren alle epochemachend. Dazu gehört auch seine Griechische Geschichte, besonders wegen ihrer in neuerer Zeit arg kritisierten Chronologie. Xenophon mag hier viel von den bahnbrechenden Arbeiten des Hellanikos gelernt haben, und das Vorhandensein von Olympiaden-, Ephoren- und Archontenlisten muß seine Arbeit sehr erleichtert haben; auch war er, der Athener, sowohl in Sparta wie in Olympia gut zu Hause und konnte die Zeitrechnungen dieser berühmten Orte leicht miteinander in Einklang bringen. Aber letzten Endes war es der gesunde Menschenverstand des Xenophon, der in den manchmal auf Stein veröffentlichten Listen der Archonten, der Ephoren und der numerierten Olympiaden mit ihren Siegerlisten eine allgemein gültige und verständliche geschichtliche Zeitordnung sah.

[325]

Für den Epigraphiker ist die Ernte dieser Untersuchung zwar bescheiden, aber doch beachtlich. Er kann die inschriftlich erhaltenen Datierungen nach Archonten in der Geschichtsschreibung verwendet sehen und kann auch erkennen, daß die Listen der panhellenischen Olympiaden und Olympioniken, deren

Existenz beglaubigt ist, für die Zeitrechnung von den Geschichtsschreibern schon mehr als hundert Jahre vor Timaios nutzbar gemacht wurden.

In der Diskussion bemerkt Herr Schachermeyr: "Wir haben mit der Möglichkeit zu rechnen, daß gewisse Textveränderungen und Interpolationen in dem Zeitpunkt gemacht wurden, als man das erste Mal ein Corpus der xenophontischen Schriften zusammenstellte, was wohl schon in Athen geschah. Aber auch die Herausgabe des Corpus in Alexandria könnte dazu Anlaß geboten haben.

Andererseits möchte ich an der Datierung nach Olympiaden bei Xenophon insofern weniger Anstoß nehmen, als der Autor ja durch viele Jahre zu Skillous in allernächster Nähe von Olympia gelebt hat und mit den olympischen Angelegenheiten bestens vertraut war (vgl. auch *Hell.* 7, 4, 28ff.). Da konnte er wenigstens weit eher als andere Zeitgenossen auf den Gedanken kommen, auch eine Olympiade zu Datierungszwecken heranzuziehen."

39

Theopompos on Thucydides the Son of Melesias

INTRODUCTION

[81] The historian Thucydides was not the only distinguished man of this name in Periclean Athens. In fact, he was probably less well known in his own day than three other men named Thucydides—a poet, a Thessalian who became an honorary citizen of Athens, and the famous opponent of Perikles.[1] Only one of these three men continued to be remembered, mainly in connection with his great adversary, Perikles; this was Thucydides the son of Melesias from the deme Alopeke.[2] It appears that Thucydides was Perikles' opponent from the time of Kimon's death (see below, 4A) until his own ostracism in 443 B.C., which marks the beginning of Perikles' undisputed rule. During this period, Athens gained and lost a land empire, and moved the seat of the Confederacy from Delos to Athens, thus establishing the Athenian Empire. We do not know what part Thucydides played in the deliberations which preceded these actions but he must have been very active since he is singled out in our literary tradition as *the* adversary of Perikles. H. T. Wade-Gery has already given us a fine historical and critical monograph on "Thucydides the son of Melesias,"[3] but he refrained from examining carefully the literary sources on which our knowledge of Thucydides' activity is based. It is the purpose of this paper to reexamine these sources and to reconstruct one of them, the account given by Theopompos in the tenth book of his *History of Philip*. Such an examination is especially justified because neither Thucydides the historian nor Ephoros (to

---

Phoenix 14 (1960) 81–95

1. On the various bearers of the name Thucydides, see my comments in *Hesperia* 24 (1955) 287–288.
2. The information on him has been conveniently assembled by J. Kirchner, *P.A.* no. 7268; Fiehn, *RE* 6A1 (1936) s.v. Thukydides no. 2, cols. 625–627; G. F. Hill, *Sources for Greek History* (revised by R. Meiggs and A. Andrewes) (Oxford 1951) 371.
3. *JHS* 52 (1932) 205–227, reprinted in *Essays in Greek History* (Oxford: Blackwell 1958) 239–270.

judge from Diodoros) even mentions Thucydides the son of Melesias, and most of our knowledge of this man comes from later sources.[4]

THE PRIMARY EVIDENCE

While the name of Thucydides does not occur on any of the stone inscriptions which have been discovered so far, there are fourteen ostraka | preserved bearing his name, mainly in the form Θουκυδίδες Μελεσίο.[5] We know that Stesimbrotos of Thasos wrote about Thucydides (*FGrH* 107 F 10a); there are allusions to him in Aristophanes' *Acharnians* (703–718) and *Wasps* (947–948), and there may be one in Sophocles' *King Oedipus* (879–881).[6] Plato's passing references (*Laches* 179A–180B, and *Meno* 94; see also *Theages* 130A) merely reflect the high regard in which the man was held by the Athenian aristocrats.

[82]

Unfortunately, we do not know what Stesimbrotos said about Thucydides, but Wade-Gery (221, note 74) suggested that some of the later accounts may be derived "ultimately, perhaps, from Stesimbrotos." F. Jacoby (Notes on the Commentary on *FGrH* 328 F 120) made the additional suggestion that Theopompos "may have known the pamphlet" of Stesimbrotos; he also suggested (*CZ* 41 [1957] 8–9 = *Abhandlungen zur Griech. Geschichtsschreibung* [Leiden: Brill 1956] 156) that Plutarch's account of the conversation between Archidamos and Thucydides (*Moralia* 802C = *Pericles* 8.4) may go back to Stesimbrotos, who is mentioned in the same chapter. This paper tends to supplement these suggestions by showing that virtually *all* the later references to Thucydides go back to Theopompos, either directly or through the commentary on Aristophanes by Didymos, who relied directly on Theopompos.[7] The relationship between Stesimbrotos and Theopompos will remain a problem; perhaps one should attribute to the former the anecdotes, to the latter the political analysis and the chronological framework.

THE SECONDARY EVIDENCE

1. Aristotle. The oldest historical statement on Thucydides is in Aristotle's *Constitution of Athens*. After discussing the various Athenian statesmen, among them Thucydides (28.2: Θουκυδίδης δὲ τῶν ἑτέρων, κηδεστὴς ὢν Κίμωνος), Aristotle says (28.5) that the best Athenian politicians of the later period were Nicias, Thucydides, and Theramenes: καὶ περὶ μὲν Νικίου καὶ Θουκυδίδου

---

4. See my general remarks in *Class. et Med.* 19 (1958) 73–77 [*supra*, pp. 81–84].
5. See the references given by E. Vanderpool, *Hesperia* Suppl. 8 (1949) 411 and 412 (Addenda). These ostraka belong, as Vanderpool (following A. Koerte, *Ath. Mitt.* 47 [1922] 2–4) assumed (*Hesperia* 21 [1952] 114), to the famous ostracism of 443 B.C. in which not only Perikles and Thucydides but also Teisandros the son of Epilykos, Andokides the son of Leogoras, and Kleippides the son of Deinias and the father of Kleophon were involved.
6. See G. H. Macurdy, *CP* 37 (1942) 307–310; V. Ehrenberg, *Sophocles and Pericles* (Oxford: Blackwell 1954) 115, note 1.
7. For Didymos's use of Theopompos, see W. Florian, *Studia Didymea Historica* (Diss. Leipzig 1908) 22–31.

[83] πάντες σχεδὸν ὁμολογοῦσιν ἄνδρας γεγονέναι οὐ μόνον | καλοὺς κἀγαθοὺς ἀλλὰ καὶ πολιτικοὺς καὶ τῇ πόλει πάσῃ χρωμένους, περὶ δὲ Θηραμένους...Three details of this statement are also found in Plutarch (*Pericles* 11.1–2), in a passage which evidently does not depend on Aristotle. There, Plutarch calls Thucydides κηδεστὴν Κίμωνος, says that he was πολιτικός, and concerned with the καλοὺς κἀγαθούς. The identity of Plutarch's source shown by a comparison of one of his statements with the closing sentence of an account of Kimon and Perikles (Scholia on Aristides 3.446, lines 24–26, ed. Dindorf) which will be shown (below, 4A) to go back to Theopompos.

Plutarch	Scholia
οἱ δ'ἀριστοκρατικοί...	οἱ δὲ ὀλίγοι...
Θουκυδίδην τὸν 'Αλωπηκῆθεν...	Θουκυδίδην τὸν Μελησίου
ἀντέστησαν ἐναντιωσόμενον....	ἀπεσπάσαντο....

Aristotle's expression that "almost all agree" shows that he is following a literary tradition, perhaps the same as Plutarch and the Scholiast, namely Theopompos; and indeed, I have thought to detect (*Phoenix* 9 [1955] 125 [b][*infra*, p. 323]) an agreement between Theopompos and the immediately preceding sentence of Aristotle's account (28.4) where he describes the deterioration of the Athenian politicians after Kleophon. Another link between Aristotle's and Theopompos' description of Thucydides is provided by the term καλὸς κἀγαθός which is used not only by Aristotle (28.5) but also by Plutarch (*Pericles* 8.4; *Nicias* 2.2). The second of these passages follows immediately upon a reference to Aristotle (28.5) and comes probably from a different source. The identity of this source and of that of the first passage is revealed by the occurrence in them of the term ἀντιπολιτεύομαι which is especially attributed to Theopompos' account of Thucydides (*FGrH* 115 F 91: Θεόπομπος μέντοι ὁ ἱστορικὸς τὸν Πανταίνου φησὶν ἀντιπολιτεύεσθαι Περικλεῖ; F 261: Θεόπομπος δὲ καὶ τοὺς ἐν μιᾷ πόλει φιλοτιμουμένους πρὸς ἀλλήλους ἀντιπολιτεύεσθαι ἔφη).[8]

Considering the nature of Aristotle's brief excursus on the Athenian demagogues (28; the term is used in a neutral sense in 28.1 and 4), one cannot go far wrong in assuming that Aristotle owed a great deal to Theopompos's more detailed treatment of the subject.

2. Androtion–Philochoros–Satyros. The third Scholion on Aristophanes'
[84] *Wasps* 947 tells us that Androtion identified Thucydides the opponent of Perikles as the son of Melesias (*FGrH* 324 F 37), while Philochoros seems to be credited with the statement (*FGrH* 328 F 120; see below 4E) that Thucydides was the political opponent of Perikles. Jacoby assumed (in his commentary) that

---

8. For F 91, see my comments in *Hesperia* 24 (1955) 288, note 11. Thucydides is called ὁ Περικλεῖ ἀντιπολιτευσάμενος in the Scholia on Aristophanes' *Wasps* 947: δημαγωγός, Μελησίου, ὃς καὶ Περικλεῖ διεπολιτεύσατο, by Marcellinus, *Vita Thucydidis* 28, and of Perikles it is said in the Scholia on Aristides (3.477, lines 3–4) πρὸς τοῦτον (sc. Thucydides) ἀντιπολιτευόμενος. All these passages would ultimately go back to Theopompos; for Theopompos's use of the same word in another context, see my comments in *WSt* 71 (1958) 113–114.

Androtion and Philochoros referred to the ostracism of Thucydides or perhaps to his return. It may be considered, however, that the occasion was the trial of Anaxagoras in which Thucydides acted as one of the accusers; see Satyros in Diogenes Laertius 2.12 = H. Diels, *FdV* 59A 1.2 = *FHG* 3.163, no. 14 = Meiggs-Andrewes, *Sources* 261: ὑπὸ Θουκυδίδου εἰσαχθῆναι τὴν δίκην, ἀντιπολιτευομένου τῷ Περικλεῖ. It is known that Satyros used Philochoros (see *FGrH* 328 F 219 and Jacoby's commentary), and that Philochoros dealt extensively with the trial of Pheidias (*FGrH* 328 F 121; see O. Lendle, *Hermes* 83 [1955] 284–303) which belongs historically and politically together with that of Anaxagoras; see D. Kienast, *Gymnasium* 60 (1953) 210–221 on Thucydides' part in the trials of Pheidias, Aspasia, and Anaxagoras: "Der eigentliche Hintermann...ist natürlich...kein anderer als Thukydides, der alte Führer der Oligarchen."

While the information concerning Thucydides as accuser of Anaxagoras is probably part of the documentary tradition of the Atthis (drawing upon the same sources as Krateros, *FGrH* 341), the identification of Thucydides as "the opponent of Perikles" evidently goes back to Theopompos. It appears that ὁ Περικλεῖ ἀντιπολιτευσάμενος sufficiently identified Thucydides the son of Melesias, and that this expression was coined by Theopompos (see also above, note 8).

3. Various Lists of Names. The name of Thucydides occurs also in a number of lists of famous Athenians of the fifth century; some of these lists can still be traced back to authors of the fourth century.

First may be mentioned the introduction to the Perikles anecdote (Plutarch *Moralia* 802 C) which Jacoby attributed to Stesimbrotos (see above, Introduction): ἐπεὶ καὶ Κίμων ἀγαθὸς ἦν καὶ Ἐφιάλτης καὶ Θουκυδίδης. These three, together with Perikles, form a group of contemporaries listed by Aristotle in one paragraph (28.2). It seems reasonable to assume that Plutarch drew here on Aristotle who in turn may have relied upon Theopompos (see above, 1. Aristotle).

The following passage in Dio Chrysostomus 22.1 which contains a list of Athenian rhetors and demagogues from Solon to Perikles seems also to go back to Aristotle or Theopompos: Περικλῆς καὶ Θουκυδίδης | Ἀθήνησι καὶ [85] Θεμιστοκλῆς ἔτι πρότερον καὶ Κλεισθένης, καὶ Πεισίστρατος ἕως ἔτι ῥήτωρ καὶ δημαγωγὸς ἠνείχετο καλούμενος· Ἀριστείδην μὲν γὰρ καὶ Λυκοῦργον καὶ Σόλωνα καὶ Ἐπαμεινώνδαν, καὶ εἴ τις ἕτερος τοιοῦτος φιλοσόφους ἐν πολιτείᾳ θετέον ἢ ῥήτορας κατὰ τὴν γενναῖον τε καὶ ἀληθῆ ῥητορικήν.

Aristotle's list, in a more complete form, may be found in Plutarch's *Pericles* (16.2–3) where Perikles is said to have stood first ἐν Ἐφιάλταις καὶ Λεωκράταις καὶ Μυρωνίδαις καὶ Κίμωσι καὶ Τολμίδαις καὶ Θουκυδίδαις. Kimon, Myronides, Leokrates, and Tolmides are mentioned once again in Plutarch's comparison of *Pericles* and *Fabius Maximus* (1.2). The knowledge of Thucydides as a distinguished contemporary of Perikles surely goes back at least as far as Theopompos; see also, below 4D and 4J.

Next comes a passage in [Plato's] *De virtute* 376 B: Θουκυδίδης καὶ Θεμιστοκλῆς καὶ Ἀριστείδης καὶ Περικλῆς. There can be no doubt that this list is derived from Plato's *Meno* (93B–94D) where the same four Athenians are mentioned, though in chronological order.

Finally, there is a list of Athenian aristocrats who fell victim to ostracism in the Scholia on Aristophanes' *Knights* 855: σχεδὸν δὲ οἱ χαριέστατοι πάντες ὠστρακίσθησαν, Ἀριστείδης, Κίμων, Θεμιστοκλῆς, Θουκυδίδης, Ἀλκιβιάδης. I have attributed (*Class. et Med.* 19 [1958] 79–80 [*supra*, pp. 85–86]) this list to Theophrastos's account of ostracism in his *Laws*, and I have suggested (98) that a trace of the same list can be detected in Plutarch's *Nicias* 11.5: Θουκυδίδῃ καὶ Ἀριστείδῃ καὶ τοῖς ὁμοίοις. It appears that Theophrastos knew that Thucydides was ostracized, and he may have learned this fact from Theopompos.

There is still another mention of Thucydides and Perikles in the *vita* of Sophokles (1): οὐ γὰρ εἰκὸς τὸν ἐκ τοῦ τοιούτου (sc. τέκτων or χαλκεύς) γενομένου στρατηγίας ἀξιωθῆναι σὺν Περικλεῖ καὶ Θουκυδίδῃ, τοῖς πρώτοις τῆς πόλεως. Wade-Gery, who did not mention this passage, questioned (*Essays* 257 and 261) whether Thucydides was ever a general; T. B. L. Webster (*Sophocles*, Oxford 1936, 11–12) offered 444/3 and 428 as alternative years for a joint generalship of Thucydides and Sophokles; while V. Ehrenberg (*Sophocles and Pericles*, Oxford: Blackwell 1954, 117, note 1) questioned whether Thucydides and Sophokles were ever fellow generals; see also H. D. Westlake, *Hermes* 84 (1956) 110, note 1. The biographical tradition of Sophokles preserved the information concerning Sophokles' generalship with Perikles during the Samian War (Webster 11, note 2; Ehrenberg 118 and note 3), and the same may also be true of the connection of Thucydides and Sophokles as generals. It may be noted that the peculiar phrase στρατηγίας ἀξιωθῆναι reappears in the *argumentum* of the *Antigone* by the grammarian Aristophanes (φασὶ δὲ

[86] τὸν Σοφοκλέα ἠξιῶσθαι τῆς ἐν Σάμῳ στρατηγίας) and in the | account of the historian Thucydides in the Scholia on Aristophanes' *Wasps* 947 (στρατηγίας [στρατείας codd.] ἀξιωθέντα).[9] Satyros may have mentioned all this in his biography of Sophokles, and he may have drawn on Androtion who in at least one case gave a complete list of generals (*FGrH* 324 F 38). At any rate, the information concerning the generalship of Thucydides is firmly lodged in the Sophokles biography, and it represents therefore independent evidence and a strong support of the reference to Thucydides' generalship contained in the *vita* of Thucydides the historian (see below, 4C).

4. Theopompos. A. The beginning of Theopompos' account of Thucydides can be found in a Scholion on Aristides (3.446, lines 15–26) which briefly summarizes the conflict between Kimon and Perikles: δύο δὲ ἦσαν Ἀθήνησι πολιτεῖαι· οἱ μὲν γὰρ ἦσαν καλοὶ καὶ ἀγαθοί, οἱ καλούμενοι ὀλιγαρχικοί, οἱ

---

9. See below, 4 D. It may be noted that there is still another link between this Scholion and the Sophokles *vita*: in both documents deductions are drawn from the fact that Sophokles and Thucydides are not attacked in comedy.

δὲ δημοτικοί· καὶ τούτων μὲν προίστατο Κίμων, πολλὰ διανέμων καὶ συγχωρῶν ἐκ τῶν χωρίων ὀπωρίσασθαι τοῖς βουλομένοις, καὶ ἱμάτια διανέμων τοῖς πένησι· τῶν δὲ ὀλιγαρχικῶν προίστατο Περικλῆς· κατηγορηθεὶς δὲ ὁ Κίμων ὑπὸ Περικλέους ἐπὶ Ἐλπινίκῃ τῇ ἀδελφῇ καὶ ἐπὶ Σκύρῳ τῇ νήσῳ, ὡς ὑπ' αὐτοῦ προδιδομένῃ, ἐξεβλήθη. δεδιὼς δὲ ὁ Περικλῆς μὴ ζητηθῇ ὑπὸ τῶν δημοτικῶν, πρὸς αὐτοὺς ἐχώρησεν· οἱ δὲ ὀλίγοι γαμβρὸν ὄντα Θουκυδίδην τὸν Μελησίου τοῦ Κίμωνος ἀπεσπάσαντο, σκυλακώδη ὄντα καὶ ὀλιγαρχικόν.

In the establishment of the text, which differs from the edition of Dindorf, Professor F. W. Lenz has been of great help; see his *Untersuchungen zu den Aristeidesscholien* (Weidmann: Berlin 1934) 29–56. For the old text, see my remarks in *AJP* 80 (1959) 81, note 1.

This is not the place to examine the bearing of this passage on our knowledge of Kimon and Perikles; suffice it to say that Kimon as "democrat" is attested by Nepos (*Cimon* 2.1) and Plutarch (*Cimon* 5.4; 10.1–5), and Perikles as aristocrat likewise by Plutarch (*Pericles* 7; 9.2–3). There can be little doubt that these passages all go back to the account of Theopompos whose statement on the generosity of Kimon (underlined) is still preserved (*FGrH* 115 F 89; add the Scholia on Aristides 3.517–518; compare H. N. Fowler, *HSCP* 12 [1901] 212, note 3, and J. E. Sandys' comments on Aristotle's *Constitution of Athens* 27.3). Theopompos evidently pointed out that Perikles and Kleon, following the example of Kleisthenes, turned away from their aristocratic friends to become popular leaders; see my comments in *AJP* 80 (1959) 85, note 6 [*supra*, p. 164, n. 6].

The career of Thucydides seems to have begun after the ostracism of Kimon and at the time when Perikles turned democrat, but the precise date of this event cannot be established with certainty. Aristotle speaks obscurely (*Constitution* 26.1) of the aristocrats as being without a leader after the fall of the Areopagus (and the ostracism of Kimon), and he dates Perikles' entry as demagogue (*Constitution* 27.1) after 451/0 B.C. Similarly, Plutarch (*Pericles* 10.7–11.1) tells of Thucydides' emergence after the death of Kimon. It may, therefore, be safe to assume that Thucydides came to the fore sometime during the fifties, perhaps not too long before the middle of the century. At any rate, Thucydides was chosen by the aristocrats because of his known oligarchic views, and because he was pugnacious and thus likely to play an active part in internal affairs—different from Kimon who had little interest in political debates.[10]

B. At this point can be added the account given by Plutarch (*Pericles* 11.1–2) which is linked to the statement just quoted by the occurrence in both passages of the following details: Thucydides, the brother-in-law of Kimon, was chosen by the aristocrats as leader against Perikles, because he was an aggressive fighter and an oligarch. Another link between the Scholion and Plutarch consists in the similarity between the general characterization of Athenian democracy with

[87]

---

10. The word σκυλακώδης (for its meaning, see Xenophon, *Cyropaedia* 1.4.4) corresponds to Plutarch's expression (*Pericles* 11.1; see above, 1. Aristotle) συμπλεκόμενος, and to the wrestling metaphors employed by Sophokles, Aristophanes, Plato, and Plutarch; see my comments in *Hesperia* 24 (1955) 288, note 11, above, note 6, and below, note 18.

which the Scholiast begins his comments and with which Plutarch closes his (*Pericles* 11.3): both emphasize the division between oligarchs and democrats.

We are now in a position to consider Plutarch's account (*Pericles* 11.1–2): οἱ δὲ ἀριστοκρατικοὶ ... Θουκυδίδην τὸν Ἀλωπηκῆθεν, ἄνδρα σώφρονα καὶ κηδεστὴν Κίμωνος, ἀντέστησαν ἐναντιωσόμενον, ὃς ἧττον μὲν ὢν πολεμικὸς τοῦ Κίμωνος ἀγοραῖος δὲ καὶ πολιτικὸς μᾶλλον, οἰκουρῶν ἐν ἄστει καὶ περὶ τὸ βῆμα τῷ Περικλεῖ συμπλεκόμενος, ταχὺ τὴν πολιτείαν εἰς ἀντίπαλον κατέστησεν. οὐ γὰρ εἴασε τοὺς καλοὺς καὶ ἀγαθοὺς καλουμένους ἄνδρας ἐνδιεσπάρθαι καὶ συμμεμῖχθαι πρὸς τὸν δῆμον ὡς πρότερον, ὑπὸ πλήθους ἠμαυρωμένους τὸ ἀξίωμα, χωρὶς δὲ διακρίνας καὶ συναγαγὼν εἰς ταὐτὸ τὴν πάντων δύναμιν ἐμβριθῆ γενομένην ὥσπερ ἐπὶ ζυγοῦ ῥοπὴν ἐποίησεν.

Apart from the introductory sentence, which has already been discussed, there are two pieces of significant information in this statement, namely that Thucydides was less a military than a political leader, and that he formed a separate political organization of the aristocrats. The first of these details is connected with the "political wrestling" for which Thucydides was famous; see above, note 10. The second cannot refer to anything but the formation of an [88] *hetaireia*, and it has been so inter|preted by F. Sartori, *La eterie nella vita politica Ateniese* (Rome: L'Erma 1957) 64–65 and 154; see also my comments in *AJP* 80 (1959) 87–88 [*supra*, pp. 165–166]. This interpretation is confirmed by a later passage in Plutarch where the biographer, following the same source, reports that after the ostracism of Thucydides, Perikles dissolved this *hetaireia*; see below, D.

There is, unfortunately, no certain evidence to show that Plutarch used here Theopompos' account. The conviction that he did so rests on the introductory sentence which agrees with the closing phrase of the earlier statement (A) which has been more confidently attributed to Theopompos. Mention should also be made of the phrase συναγαγὼν εἰς ταὐτό which occurs in Plutarch in several other places which, on other grounds, have been associated with Theopompos.[11]

C. The narrative of Thucydides' activities after assuming the leadership of the aristocrats (see above, B) appears to be continued in the anonymous *vita* of Thucydides (6–7), the historical value of which was stressed by Wade-Gery (Appendix A): ἦν δὲ τῶν πάνυ κατὰ γένος δοξαμένων ὁ Θουκυδίδης δεινὸς δὲ δόξας εἶναι ἐν τῷ λέγειν ⟨πρὸ τῆς συγγραφῆς⟩ προέστη τῶν πραγμάτων. πρώτην δὲ τῆς ἐν τῷ λέγειν δεινότητος τήνδε ἐποιήσατο τὴν ἐπίδειξιν. Πυριλάμπης γάρ τις τῶν πολιτῶν ἄνδρα φίλον καὶ ἐρώμενον ἴδιον διά τινα ζηλοτυπήσας ἐφόνευσε, ταύτης δὲ τῆς δίκης ἐν Ἀρείῳ Πάγῳ κρινομένης πολλὰ τῆς ἰδίας σοφίας ἐπεδείξατο, ἀπολογίαν ποιούμενος ὑπὲρ τοῦ Πυριλάμπους, καὶ Περικλέους κατηγοροῦντος ἐνίκα. ὅθεν καὶ στρατηγὸν αὐτὸν ἑλομένων Ἀθηναίων ἄρχων προέστη τοῦ δήμου. μεγαλόφρων δὲ ἐν τοῖς

---

11. See my remarks in *AJP* 80 (1959) 85, note 6 [*supra*, p. 164, n. 6], referring to the *hetaireia* of Kleon which is described by Plutarch (following, I think, Theopompos) *Moralia* 806F–807A: συναγαγὼν εἰς ταὐτὸ διελύσατο τὴν φιλίαν.... The phrase συναγαγὼν εἰς ταὐτό is used by Plutarch also in his account of the ostracism of Hyperbolos (*Aristides* 7.3; *Nicias* 11.4; *Alcibiades* 13.4), passages which may also go back to Theopompos.

πράγμασι γενόμενος, ἅτε φιλοχρηματῶν, οὐκ εἴατο πλείονα χρόνον προστατεῖν τοῦ δήμου. πρῶτον μὲν γὰρ ὑπὸ τοῦ Ξενοκρίτου εἰς (codd. ὡς) Σύβαριν ἀποδημήσας, ὡς ἐπανῆλθε εἰς ᾿Αθήνας, συγχύσεως δικαστηρίου φεύγων ἑάλω· ὕστερον δὲ ἐξοστρακίζεται ἔτη δέκα. φεύγων δὲ ἐν Αἰγίνῃ διέτριβε ⟨κἀκεῖ λέγεται τὰς ἱστορίας αὐτὸν συντάξεσθαι⟩. τότε δὲ τὴν φιλαργυρίαν αὐτοῦ μάλιστα φανερὰν γενέσθαι· ἅπαντας γὰρ Αἰγινήτας κατατοκίζων ἀναστάτους ἐποίησεν.

The first sentence, referring to Thucydides' nobility, links this account with the preceding passage (B) and allows us to assume that we have before us two parts of a consecutive story which began with the rise of Thucydides and ended with his exile on Aigina. The anecdote about the trial of Pyrilampes is told to explain the political prominence of Thucydides; this ancedote belongs, therefore, together with the stories of Sophokles (who became general on account of the *Antigone*; see above, 3) and of Perikles (who distinguished himself in the trial against Kimon). The information that Thucydides was in control of affairs is repeated both before and after the anecdote, and it is also found in the Scholia on Aristides (3.446, lines 29–30, and 447, lines 1–2): ᾧ ποτε ᾿Αθηναῖοι τὰ πολιτικὰ ἐπιτρέψαντες ἐπείθοντο πάντες, ἅτε εὐθυνουμένης τῆς πόλεως...διάστροφον τῆς πόλεως, ᾧ καὶ ὁ δῆμος ἅπας ἐπείθετο.[12] Inasmuch as his prostasia did not last very long and since it came to an end even before the ostracism, it may be presumed that Thucydides was most powerful between 450 and 445 B.C.

Thucydides' part in the founding of Thourioi, his relationship to Xenokritos, and the nature of the trial, all these points are disputed; see H. T. Wade-Gery's positive statement (*Essays* [see note 3] 257 and 261–262) and V. Ehrenberg's outright rejection (*AJP* 69 [1948] 160–161). Wade-Gery may have gone too far in claiming "that Thucydides took more than a casual interest in the Thouria project seems beyond all doubt...," but Ehrenberg surely exaggerates when he says "the whole story is incoherent and mistaken." The text states that Thucydides went to Sybaris (not to Thourioi), and this may have been (*pace* Ehrenberg) during the "first" expedition which is now dated in *ATL* (3.305) "before the crisis in 446."[13] Concerning the trial of Thucydides, we do not know what σύγχυσις δικαστηρίου is, but it may not have had anything to do with the issue which was later settled by Thucydides' ostracism.[14] The exile on Aigina and Thucydides' financial activities there are also mentioned by

[89]

---

12. The strange word διάστροφος is otherwise not attested as a noun; for the political meaning of διαστροφή, see Polybios 2.21.8, for that of διαστρέφω, see Euripides, Frag. 597 (N²). F. W. Lenz kindly called my attention to Arethas's Scholion on Plato's *Gorgias* 455b which mentions διάστροφος πολιτεία, in contrast to ὀρθὴ πολιτεία. The identification by the scholiast (447, lines 5–8) of this Thucydides with the one mentioned by the historian (1.117.2 with Gomme's comments) is wrong and, so far as I can see, of no help.
13. The emendation εἰς Σύβαριν for ὡς Σύβαριν is encouraged by L-S's statement (s.v. C III) "the examples of ὡς with names of places are corrupt." It is tempting to assume that Thucydides went under the command of Xenokritos, and thus to change ὑπὸ τοῦ Ξενοκρίτου to ὑπὸ τῷ Ξενοκρίτῳ.
14. See Wade-Gery 261–262, note 4. Note, however, that Isaios 5.18 (see W. Wyse's notes on pp. 424–425 of his edition) speaks of συγχέειν τὰς ψήφους.

Marcellinus, vita Thucydidis 24: γενόμενος δ' ἐν Αἰγίνῃ μετὰ τὴν φυγὴν ὡς ἂν πλουτῶν ἐδάνεισε τὰ πλεῖστα τῶν χρημάτων.[15] The account of Thucydides' activities in Aigina, presumably between 442 and 433 B.C., should be connected with Thucydides' statement (1.67.2) that the Aiginetans prepared secretly for the war, and it helps to explain the violent action taken by the Athenians (2.27.1); see A. W. Gomme, *Commentary on Thucydides* I (Oxford 1945) 225–226; B. D. Meritt et al. *ATL* 3.203 and 320.

[90]   D. The account of the *vita* appears in a greatly abbreviated form in the second Scholion on Aristophanes' *Wasps* 947: "Ἄλλως. Θουκυδίδης Μελησίου υἱὸς Περικεῖ ἀντιπολιτευσάμενος. τέσσαρες δέ εἰσι Θουκυδίδαι 'Αθηναῖοι, ὁ ἱστοριογράφος καὶ ὁ Γαργήττιος καὶ ὁ Θετταλὸς καὶ οὗτος <u>ῥήτωρ ἄριστος τυγχάνων,</u> ὃς κατηγορηθεὶς ἐν τῷ δικάζειν οὐκ ἐδυνήθη ἀπολογήσασθαι ὑπὲρ ἑαυτοῦ, ἀλλ' ὥσπερ ἐγκατεχομένην ἔσχε τὴν γλῶσσαν, καὶ οὕτω <u>κατεδικάσθη εἶτα ἐξωστρακίσθη.</u> U. v. Wilamowitz-Moellendorff (*Hermes* 12 [1877] 348, note 34) defended the abbreviated reading of the *Venetus* καὶ οὕτω ἐξωστρακίσθη, while Wade-Gery (262) observed that Scholion "says the same" as the *vita*, but that it contains only the statements (which I have underlined) that Thucydides was an excellent orator and that he was convicted and then ostracized. The central part of the Scholion (which mentions Thucydides' inability to defend himself) is independent of the *vita*; it evidently consists of a prose version of the lines of Aristophanes which it is designed to explain. This commentary is, however, not correct, since it confuses the two trials of Thucydides; Wilamowitz confidently asserted that the author of the Scholion was Symmachos.

E. All three Scholia on this line of the *Wasps* (947) start with an identification of the Thucydides mentioned by Aristophanes with the son of Melesias, from Alopeke, the famous adversary of Perikles, who was ostracized: Θουκυδίδην λέγει τὸν Μελησίου 'Αλωπεκῆθεν, τὸν Περικλεῖ ἀντιπολιτευσάμενον. τοῦτον δὲ ἐξωστράκισαν 'Αθηναῖοι τὰ ί ἔτη κατὰ τὸν νόμον (this is a composite text). This may be the oldest part of the commentary, evidently dependent on Theopompos because of the use of ἀντιπολιτεύομαι; see above, note 9.

The author of the first Scholion added a general account of ostracism which is ultimately taken from Didymos's commentary on Demosthenes 23.205; Didymos himself relied on Philochoros (whom he quotes), who in turn used Theophrastos's account of ostracism, as I have tried to show in *Class. et Med.* 19 (1958) 89–92 [supra, pp. 92–95].

The author of the second Scholion (with which we are here concerned) added a list of other bearers of the name Thucydides which appears also in the Scholion on the *Acharnians* 703: γεγόνασι δὲ δ', ὁ ἱστορικός, ὁ Γαργήττιος, ὁ Θετταλός, ὁ Μελησίου υἱός. This list is, according to Wilamowitz (*loc. cit.* 347–352) based on the account of Polemon; see Jacoby's notes on the commentary to *FGrH* 328 F 120 and my remarks in *Hesperia* 24 (1955) 287–288. Accordingly, in a

15. It is assumed (e.g., by W. Schmid, *Gesch. d. griech. Lit.* 5.14, note 2) that this passage refers to Thucydides the son of Melesias, but it should be noted that this would be the only instance in which Marcellinus confused the historian and the statesman.

papyrus containing, among other things, a list of the various bearers of the name Thucydides (*Pap. Ox.* 13. 1611, lines 102–104), a reference to Polemon has been restored: [Πολέμων] ἐν τῷ [περὶ ἀκροπό]λεως; see below F.

F. The third Scholion is, however, the most important, although the most difficult to understand; see Wilamowitz, *loc. cit.* 354, note 39; | F. Jacoby, *FGrH* 115 F 91; 324 F 37; 328 F 120; 350 F 1 (with commentaries and notes). Both Wilamowitz and Jacoby assume that this Scholion is an abbreviated and corrupt version of Didymos's commentary on Aristophanes, but (as stated above, E) it is more likely that the old Scholion consisted only of the identification of Thucydides.

After a reference to Philochoros which can no longer be understood (see above, 2), the Scholiast evidently mentioned the identification of the Thucydides of whom Aristophanes speaks with the historian of this name; see above, note 10. The Scholiast then continues: ἔνιοι δέ, ὧν καὶ Ἀμμώνιος, τοῦ Στεφάνου· καὶ τοῦτο δὲ ὑπίδοι τις, ὥσπερ προείρηται. *Pap. Ox.* 13. 1611 may contain the original statement of Ammonios, since lines 105–113 can be restored as follows: [πῶτον (?)]| τὸν Μελησίου [υἱόν, Στε]|φάνου δὲ τοῦ κω[μωιδου]| μένου πατέρα, [δεύτερον]| δέ τὸν συγγραφ[έα ὄν (?)]| φασιν Ὀλόρου υἱ[όν, τρί]|τον δὲ τὸν Φαρσ[άλιον]. | περὶ μὲν οὖν τοῦ [τοῦ Στε]|φάνου πατρὸς ....
Ammonios took the information about the various bearers of the name Thucydides from Polemon (whom he mentions by name; see above, E), but he himself added that Thucydides was the father of Stephanos whose appearance in comedy caused Ammonios' comments in the first place.[16] Thus, there can be little doubt that the papyrus contains part of Ammonios' work on the κωμῳδούμενοι (*FGrH* 350), and that the Scholiast (on *Wasps* 947) is ambiguous or misleading when he claims that Ammonios called Thucydides the, presumably, "son" but actually "father" of Stephanos.[17] This information has nothing to do with Theopompos's account of Thucydides, or even with Thucydides himself, since it is concerned with Stephanos whose name must have occurred in a comedy, perhaps in the form ὁ Θουκυδίδου.

G. A few words must be added about the remainder of the Scholion (*Wasps* 947): ὁ γενόμενος ὀστρακισμὸς ἐμφαίνει τὸν Μελησίου καὶ τὸν ὀστρακισθέντα. This sentence does not refer to the text of Aristophanes but to the identification of Thucydides with which all three Scholia begin and which probably formed the earliest (Didymean) comment on the passage: τοῦτον δὲ ἐξωστράκισαν Ἀθηναῖοι τὰ ἴ ἔτη κατὰ τὸν νόμον; see above, E. The Scholiast continues his commentary on the older commentary by observing that Theopompos did not call the adversary of Perikles the son of Melesias (as does the Didymean comment; see above, E) but the son of Pantainos, while Androtion did call his

---

16. The restoration of κω[μωιδου]|μένου is new but it has been approved by Colin Roberts after an examination of the papyrus which he kindly undertook.
17. The same mistake appears to have been repeated in Philodemos's *Rhetorica* 1.188, lines 16–17, where there is mention of [τὸν] Στεφάνου Θουκυδί[ίδην] followed by a reference to Thucydides the son of Oloros. I do not see, at the moment, the connection between Philodemos and the Scholiast.

[92] father | Melesias (see above, 2). The name Pantainos must not be emended[18] but it must be understood as a challenge to Thucydides' legitimacy. Theopompos knew, of course, as well as we do, that Thucydides was officially known as the son of Melesias (see, however, Jacoby's note 13 on the commentary to *FGrH* 328 F 120), but he claimed that Thucydides' real father was a certain Pantainos of whom nothing else is known. Similarly, in the case of Hyperbolos, Theopompos insisted (*FGrH* 115 F 95) that Hyperbolos's "real" father was Chremes; see my comments in *Phoenix* 9 (1955) 124–125 [*infra*, pp. 322–324]. It may be presumed, therefore, that Theopompos's comment on Thucydides' father belongs to the beginning of his account of Thucydides; see above, A.

H. There is another point of Thucydides' career which can be fitted into the general account of the *vita* (see above, C), and which may be attributed to Theopompos. Plutarch continues his story of the conflict between Perikles and Thucydides (*Pericles* 11.1–3) with a brief statement on the "public works" of Perikles (11.4–12.1) and on the opposition to them (12.1–2): ... τοῦτο μάλιστα τῶν πολιτευμάτων τοῦ Περικλέους ἐβάσκαινον οἱ ἐχθροὶ καὶ διέβαλον ἐν ταῖς ἐκκλησίαις, βοῶντες ὡς ὁ μὲν δῆμος ἀδοξεῖ καὶ κακῶς ἀκούει τὰ κοινὰ τῶν Ἑλλήνων χρήματα πρὸς αὐτὸν ἐκ Δήλου μεταγαγών.... This attack upon the transfer of the treasury from Delos to Athens and upon the use of the money for the rebuilding of the Akropolis belongs to the years immediately following the middle of the century; see now R. Sealey, *Hermes* 86 (1958) 443–445. After a long digression on the building activities (12.3–13), Plutarch resumes his account of the attacks on Perikles and mentions now the name of the opponent (14): τῶν δὲ περὶ τὸν Θουκυδίδην ῥητόρων καταβοώντων τοῦ Περικλέους ὡς σπαθῶντος τὰ χρήματα καὶ τὰς προσόδους ἀπολλύντος .... Perikles was successful, however, in overcoming this opposition and in having his policy upheld in the assembly. τέλος δὲ πρὸς τὸν Θουκυδίδην εἰς ἀγῶνα περὶ τοῦ ὀστράκου καταστὰς καὶ διακινδυνεύσας ἐκεῖνον μὲν ἐξέβαλεν, κατέλυσε δὲ τὴν ἀντιτεταγμένην ἑταιρείαν. This evidently completes the story of the conflict between Thucydides and Perikles and it fits easily into the account attributed to Theopompos.

I. The circumstances immediately preceding the ostracism of Thucydides are contained in an anecdote which must have come into our biographical tradition from contemporary sources, comedy or Stesimbrotos, and which may very well have been recorded also by Theopompos. This is the famous story (told by Plutarch, *Pericles* 6.2–3) of Lampon's prophecy of the victory of Perikles over Thucydides. Looking upon a one-horned ram, Lampon said: δυεῖν οὐσῶν ἐν
[93] τῇ πόλει δυναστειῶν, τῆς | Θουκυδίδου καὶ Περικλέους, εἰς ἕνα περιστήσεται τὸ κράτος παρ' ᾧ γένοιτο τὸ σημεῖον. Anaxagoras' scientific explanation first won applause, but Lampon's proved to be accurate ὀλίγῳ δ' ὕστερον ... τοῦ μὲν Θουκυδίδου καταλυθέντος .... The two phrases recall the terminology and the political thought of Theopompos. The word δυναστεία is used by him with

---

18. As I tried to do in *Hesperia* 24 (1955) 288, note 11; for the occurrence of this name, see also A. W. Parsons, *Hesperia* Suppl. 8 (1949) 271, note 14.

reference to Kleon (Scholion on Aristophanes' *Peace* 681; see my comments in *Phoenix* 9 [1955] 123–125) [*infra*, pp. 321–324], and to Perikles (see below, J), and the concept of the two opposing factions appears in Theopompos's discussion of Kimon (see above, A) and of Thucydides himself (see above, B). The use of the form καταλυθέντος corresponds to Plutarch's reference (*Pericles* 16.3) to Thucydides' ostracism as κατάλυσις, in a passage which we shall associate with Theopompos (J), and to the use of the same verb in the passage (*Pericles* 14.2; see above, H) telling how Perikles expelled Thucydides and dissolved (κατέλυσε) his *hetaireia*.

J. The last passage to be associated with Theopompos' account of Thucydides pertains to the effect of his ostracism and to the chronology of the "reigns" of the Athenian demagogues. Plutarch tells of Perikles' extended period of power (*Pericles* 16.2–3): τεσσαράκοντα μὲν ἔτη πρωτεύων ἐν Ἐφιάλταις καὶ Λεωκράταις καὶ Μυρωνίδαις καὶ Κίμωσι καὶ Τολμίδαις καὶ Θουκυδίδαις, μετὰ δὲ τὴν Θουκυδίδου κατάλυσιν καὶ τὸν ὀστρακισμὸν οὐκ ἐλάττω τῶν πεντεκαίδεκα ἐτῶν διηνεκῆ καὶ μίαν οὖσαν ἐν ταῖς ἐνιαυσίοις στρατηγίαις ἀρχὴν καὶ δυναστείαν κτησάμενος .... For the list of names of Athenian statesmen from Ephialtes to Thucydides, see above, 3; it may be noted that this list is in some chronological order, probably that of the dates at which the men died.

The forty years of Perikles' reign would be composed of the fifteen years of his supremacy and the twenty-five years preceding Thucydides' ostracism; this means that Perikles was "ruling" from 469/8 until 429/8 and that his "power" lasted from 443/2 until 429/8. It has been noted before (*Phoenix* 9 [1955] 125–126 [*infra*, pp. 323–324]) that Theopompos counted the period of Hyperbolos as six years and that of Kleon as seven, continuing directly upon the forty years of Perikles. This means that we have now a continuous line of demagogues from 469/8 to 416/5: Ephialtes, Leokrates, Myronides, Kimon, Tolmides, Thucydides, Perikles, Kleon, Hyperbolos.

The question arises whether the list can be continued backwards, and what significance is to be attached to the year 469/8, or to the moment when the forty years of Perikles and presumably the "reign" of Ephialtes began. G. Busolt (*Griech. Gesch.* 3.1 [Gotha 1897] 253–254, note 2) has already called our attention to another passage of Plutarch's *Pericles* (7.2) where we read that Perikles entered politics ἐπεὶ δ' Ἀριστείδης | μὲν ἀποτεθνήκει καὶ Θεμιστοκλῆς *[94]* ἐξεπεπτώκει, Κίμωνα δ' αἱ στρατεῖαι τὰ πολλὰ τῆς Ἑλλάδος ἔξω κατεῖχον, and he pointed out that this passage has been attributed either to Stesimbrotos or to Theopompos; it may be attributed to both, of course. Busolt also assumed that the information concerning Perikles' forty years rule goes back to Theopompos, and that the beginning of this period is marked by the death of Aristeides.

This would mean that the political leadership during the period before 469/8 belonged to Aristeides, at least according to Theopompos. And, indeed, Aristotle not only lists (28.2) Themistokles and Aristeides before Ephialtes as leaders of the people, but he discusses (24.3) the democratic reforms of the period between the battle of Salamis (23.5) and the "reign" of Ephialtes (25.1) as due to the policy of Aristeides: ὥσπερ 'Αριστείδης εἰσηγήσατο. In this context belongs the closing sentence of Nepos' biography of Aristeides (3.2): *decessit autem fere post annum quartum quam Themistocles Athenis erat expulsus*, especially if this chronological indication should go back to Theopompos' account of Aristeides; see R. J. Lenardon, *Historia* 8 (1959) 27, who rightly suggested that Nepos refers here to Themistokles' flight and not to his ostracism.

SUMMARY

It appears that most of our evidence concerning Thucydides, except for the contemporary sources, can be combined to form a consistent and consecutive account which begins with the death of Kimon and ends with Thucydides' own exile on Aigina. During this period (the duration of which is not given in our sources) Thucydides, who was the brother-in-law of Kimon, attained prominence by defending successfully Pyrilampes against an accusation by Perikles, was elected general, and became the leader of the people. During his period of power, he organized the conservative elements into what may be called a political party, and he attacked strongly the foreign and domestic economic policy of Perikles, and especially his building program. His own handling of financial matters, however, brought him into ill repute, and after his return from Sybaris he was accused and condemned, probably to pay a fine. In the end, ostracism decided the conflict between Perikles and Thucydides. The latter was exiled and his political organization destroyed. He went to Aigina and continued his financial operations, contributing to the dissatisfaction and ultimately to the revolt of the Aiginetans.

The tradition which has been reconstructed here does not tell us that Thucydides returned from exile and that he was one of the accusers of Perikles and his friends, just before the outbreak of the Peloponnesian War. Nor does [95] this tradition record that Thucydides was in his old age once more subjected to a court trial, and that he was once more convicted. Thus, the biographical account of Thucydides covers only the man's public career, especially the period of his greatest power and prominence.

It is assumed that this tradition goes ultimately back to Theopompos' account of Thucydides as the last of the statesmen who opposed Perikles during the twenty-five years of his ascendancy. Plutarch and the Scholiast on Aristeides evidently used Theopompos, and so did the author of the *vita* of Thucydides (the historian), although he confused the politician and the historian. Finally, the reference to Thucydides in the *Wasps* of Aristophanes was interpreted, probably by Didymos, as referring to Thucydides the son of Melesias, who was identified, following Theopompos, as "the opponent of Perikles." Later scholiasts elaborated this identification by adding details on Thucydides which were taken from Theopompos but which had nothing to do with the allusion to Thucydides in the *Wasps*.

It is hoped that it will now be possible to reexamine the biographical evidence concerning some of the other Athenian statesmen (especially Perikles) in order to reconstruct Theopompos' treatment of them.[19]

19. For some preliminary efforts, see E. Ruschenbusch, *Historia* 7 (1958) 422–423 (on Solon), and my comments on Themistokles (*Hermes* 84 [1957] 500–501; *WSt* 72 [1958] 112–115), on Kleon (*AJP* 80 [1959] 85, note 6 [*supra*, p. 164, n. 6]), and on Hyperbolos (*Phoenix* 9 [1955] 122–126 [*infra*, pp. 320–324]).

40

Theopompos on Hyperbolos

[122] The story of the ostracism of Hyperbolos is told by Plutarch three times (*Aristides* 7 = A; *Nicias* 11 = B; *Alcibiades* 13 = C).[1] In the first passage (A), Plutarch gives a general characterization of ostracism, which is found not only in B and C but also in his Themistocles (22),[2] and closes his remarks with the statement that Hyperbolos was the last man to be ostracized. He then continues, "It is said that Hyperbolos was ostracized for the following reason. Alkibiades and Nikias, who were the most powerful men in the city, were opposed to each other. When the demos was about to hold an ostracism, and when it was clear that the people would proscribe one or the other, they came to an understanding with each other, united their respective factions, and saw to it that Hyperbolos was ostracized." The second passage (B) agrees substantially with the first, but, being part of the biography of Nikias, it goes into greater detail. At the end, Plutarch adds, however, "I am not unaware that Theophrastos says that Hyperbolos was ostracized when Phaiax, not Nikias, was opposing Alkibiades."

---

1. C. Stolz, *Zur relativen Chronologie der Parallelbiographien des Plutarch* (Lund-Leipzig 1929) has shown (*a*) that Plutarch was already working on the *Alcibiades* when he composed the *Nicias* (16–19), (*b*) that the composition of the *Alcibiades* is, however, later than that of the *Nicias* (21–22; 100–101), (*c*) that the *Aristides* was not necessarily (but, I think, probably) composed before the *Nicias* and *Alcibiades* (119–122); compare also K. Ziegler, *RE*, s.v. Plutarchos 902, line 52–903, line 19.

2. W. Graf Uxkull-Gyllenband, *Plutarch und die griechische Biographie* (Stuttgart 1927) 22–23, assumed that all these passages and Diodoros 11.54–55 depend ultimately, though not directly, upon Philochoros; he could have added Diodoros 11.87; 19.1; Plutarch *Aristides* 1; Pollux 8.20; Bekker, *An. Gr.* 1.285. I suspect that the intermediate source of Diodoros was Timaios (see R. Laqueur, *RE*, s.v. Timaios 1093, lines 42–63) and not Ephoros as Laqueur (*Hermes* 46 [1911] 205) and F. Jacoby (commentary on *F. Gr. Hist.*, 328 F 30, p. 315) suggested. Whether or not Plutarch used a book entitled κωμῳδούμενοι (as Uxkull maintains, followed by W. Peek, *Kerameikos* 3 [Berlin 1941] 79, n. 3) is less important than the fact that his characterization of ostracism can be traced back beyond Philochoros to Theophrastos (H. Bloch, *HSCP*, Suppl. 1 [1940] 357–361, on Schol. Aristophanes *Equites* 855), Demetrios of Phaleron (*F. Gr. Hist.*, 228 F 43 = Frag. 95 Wehrli, from Plutarch *Aristides* 1) and ultimately to Aristotle (*Politics* 3, 1284, 17–23 and 36–38).

The third passage (C) introduces Phaiax *and* Nikias as Alkibiades' opponents and gives a general description of Hyperbolos. Plutarch then continues, "persuaded at that time by him (namely by Hyperbolos), they were about to hold an ostracism..." (Plutarch uses here the same language as in A, and continues the sentence with a general characterization of ostracism). "When it was clear that they would ostracize one of the three, Alkibiades united the factions into one, came to an understanding with Nikias, and | turned the [123] ostracism against Hyperbolos. Some say, however, that he came to an understanding not with Nikias but with Phaiax, added this man's faction (to his own), and drove out Hyperbolos who was taken completely by surprise."

It is clear that one can distinguish here three different accounts of the same story: the first (*a*) mentioned only Alkibiades and Nikias (A, B); the second (*b*) mentioned Alkibiades, Nikias, and Phaiax (C); the third (*c*), attributed specifically to Theophrastos in B, insisted that Nikias was not involved in the ostracism at all. The ultimate source of the first account (*a*) is unknown; it cannot have been Ephoros since Diodoros does not even mention the story, but it may have been Theopompos (see below). The second account (*b*), with its emphasis on Phaiax, is connected with (if not largely derived from) the fourth oration of Andokides which Plutarch used (directly or indirectly) in his *Alcibiades* and especially in the account of the ostracism (which is but an expansion of the first account with the addition of Phaiax's name).[3] The significance of this oration, whether it is genuine or not, lies in the fact that its "dramatic date" is early in 415 B.C. which would then be also the date of Hyperbolos' ostracism; I have argued this date again in *Hesperia* 23 (1954) 68, n. 2. It is generally assumed, however, that Hyperbolos was ostracized early in 417 B.C. (see n. 5), but the sole evidence for this date is a fragment of Theopompos (*F. Gr. Hist.*, 115 F 96). I may be permitted to examine this evidence once more, especially since I have myself failed to understand its significance (*TAPA* 79 [1948] 192–193, 208 [*supra*, pp. 117–118, 129–130]).

Little is known of the excursus which Theopompos devoted to the Athenian demagogues of the fifth and fourth centuries (*F. Gr. Hist.*, 115 F 85–100), and any addition to our knowledge of this essay is welcome for our better understanding not only of Athenian history but also of the political views of Theopompos himself.[4] It is my contention that we possess in the Scholion on Aristophanes *Pax* 681, a faithful summary, if not a quotation, of Theopompos' treatment of Hyperbolos.

Χρέμητος δὲ υἱὸς ἦν Ὑπέρβολος, ἀδελφὸς δὲ Χάρωνος, λυχνοπώλης, φαῦλος τοὺς τρόπους. οὗτος μετὰ τὴν τοῦ Κλέωνος δυναστείαν διεδέξατο τὴν

---

3. I can not agree with A. R. Burn who recently suggested (*CZ* 4 [1954] 138–142) that Plutarch did not use (even indirectly) Andokides 4, although he admitted (141 and 142) that Plutarch may have mentioned the speech in his *Alcibiades* 13; for this passage, see my comments in *TAPA* 79 (1948) 210 [*supra*, pp. 130–131].

4. K. V. Fritz has presented a good account of Theopompos in *Am. Hist. Rev.* 46 (1941) 765–787, without taking Theopompos's treatment of Hyperbolos into consideration.

δημαγωγίαν. ἀπ' αὐτοῦ δὲ πρώτου ἤρξαντο 'Αθηναῖοι φαύλοις παραδιδόναι τὴν πόλιν καὶ τὴν δημαγωγίαν, πρότερον δημαγωγούντων πάνυ λαμπρῶν πολιτῶν. προείλετο δὲ τοὺς τοιούτους ὁ δῆμος, ἀπιστῶν διὰ πόλεμον τὸν πρὸς Λακεδαιμονίους τοῖς ἐνδόξοις τῶν πολιτῶν, μὴ τὴν δημοκρατίαν καταλύσαιεν. ἐξωστρακίσθη δὲ οὗτος, οὐ διὰ δυνάμεως φόβον καὶ ἀξιώματος, ἀλλὰ διὰ
[124]   πονηρίαν καὶ αἰσχύνην τῆς | πόλεως. 'Εν Σάμῳ δὲ διατρίβων, ὑπὸ τῶν 'Αθηναίων ἐχθρῶν ἐπιβουλευθεὶς ἀπέθανε, καὶ τὸν νεκρὸν αὐτοῦ εἰς σάκκον βαλόντες ἔρριψαν εἰς τὸ πέλαγος.

F. Jacoby, in his commentary on Theopompos (*F. Gr. Hist.*, 115 F 95–96, p. 371) called attention to this passage, and indeed it contains two statements which are specifically attributed to the historian in the following two Scholia.

Schol. Aristophanes *Vespae* 1007: ὑπὲρ τῆς πονηρίας δὲ 'Υπερβόλου εἴρηται. καὶ νῦν δὲ ὀλίγα παραγράψομεν. 'Ανδοκίδης φησὶ τοίνυν· περὶ 'Υπερβόλου λέγειν αἰσχύνομαι· οὗ ὁ μὲν πατὴρ ἐστιγμένος ἔτι καὶ νῦν ἐν τῷ ἀργυροκοπείῳ δουλεύει τῷ δημοσίῳ, ὡς δὲ ξένος ὢν καὶ βάρβαρος λυχνοποιεῖ. Θεόπομπος δέ φησι καὶ τὸν νεκρὸν αὐτοῦ καταποντωθῆναι, γράφων ὅτι ἐξωστράκισαν τὸν 'Υπέρβολον ἓξ ἔτη· ὁ δὲ καταπλεύσας εἰς Σάμον καὶ τὴν οἴκησιν αὐτοῦ ποιησάμενος, ἀπέθανε. καὶ τούτου τὸν νεκρὸν εἰς ἀσκὸν ἀγαγόντες εἰς τὸ πέλαγος κατεπόντωσαν.

Schol. Lucian *Timon* 30, p. 114, line 21—p. 115, line 12 (ed. H. Rabe): 'Υπερβόλῳ: 'Υπέρβολος οὗτος, ὡς 'Ανδροτίων φησίν, 'Αντιφάνους ἦν Περιθοίδης, ὃν καὶ ὠστρακίσθαι διὰ φαυλότητα. ὁ δὲ αὐτὸς καὶ λυχνοποιὸς ἦν καὶ ἐλυχνοπώλει, ὡς 'Ανδοκίδης ἱστορεῖ, ὃς καὶ ξένον αὐτὸν εἶναι καὶ βάρβαρον βούλεται. ἐπὶ τούτου δὲ καὶ τὸ ἔθος τοῦ ὀστρακισμοῦ κατελύθη, ὡς Θεόφραστος ἐν τῷ περὶ νόμων λέγει. Πολύζηλος δὲ ἐν Δημοτυνδάρεῳ Φρύγα αὐτὸν εἶναί φησιν εἰς τὸ βάρβαρον σκώπτων. Πλάτων δὲ ὁ κωμικὸς ἐν 'Υπερβόλῳ Λυδὸν αὐτόν φησιν εἶναι Μίδα γένος, καὶ ἄλλοι ἄλλως. ἔστι δὲ τῇ ἀληθείᾳ Χρέμητος, ὡς Θεόπομπος ἐν τῷ περὶ δημαγωγῶν. Κρατῖνος δὲ ἐν "Ωραις ὡς παρελθόντος νέου τῷ βήματι μέμνηται καὶ παρ' ἡλικίαν καὶ 'Αριστοφάνης Σφηξὶ καὶ Εὔπολις Πόλεσι. Θεόπομπος δὲ πάλιν ἐν δεκάτῳ Φιλιππικῶν ἐν Σάμῳ φησὶν ἐπιβουλευθέντα ὑπὸ τῶν 'Αθήνηθεν ἐχθρῶν ἀναιρεθῆναι, τὸ δὲ νεκρὸν αὐτοῦ εἰς σάκκον βληθὲν ῥιφθῆναι εἰς τὸ πέλαγος.

The whole scholion (*Pax* 681) consists of three parts two of which have already been known as fragments of Theopompos: (*a*) the name of Hyperbolos' father was Chremes (F 95), (*b*) the character of the Athenian demagogues, (*c*) the ostracism of Hyperbolos (F 96). Each of these three parts may be interpreted separately.

(*a*) There can be no doubt that Hyperbolos was officially known as the son of Antiphanes, of the deme Perithoidai. Androtion gives his full name (*F. Gr. Hist.*, 324 F 42), Plutarch mentions his demotic (*Nicias* 11.3; *Alcibiades* 13.3; was he following here Androtion?), and two ostraka with his name and patronymic were found in the Agora (*Hesperia* 8 [1939] 246, fig. 47: P 12494; *Hesperia* 17 [1948] 186, fig. 8 and Plate 66/3: P 18495). And yet, Theopompos asserted that "in truth" he was the son of Chremes and the brother of Charon. Evidently rumor had it that Chremes and not Antiphanes was the father of Hyperbolos, and it was to this Chremes that Andokides referred when he said (frag. 5, ed. Blass) that Hyperbolos's father was a slave working in the public mint. | Surely, Andokides could not mean Antiphanes, since as son of Antiphanes Hyperbolos was a citizen and held public office. Jacoby suggested that Theopompos got his information from Comedy (Notes on the Commentary on *F. Gr. Hist.*, 324 F 42, p. 137), and it is in Comedy that we find Hyperbolos called a foreigner. We are dealing here with two different people, the Athenian businessman Antiphanes, Hyperbolos's official father, and the foreign slave Chremes who had a son called Charon (and was suspected of being the real father of Hyperbolos).

[125]

(*b*) We know from Aristophanes (*Pax* 680–692) that Hyperbolos took over the leadership of the demos after Kleon's death, and that this change brought a bad man to power. Theopompos adds that up to this time the leaders were very distinguished men, but under the impact of the Peloponnesian War the demos distrusted the nobles and were afraid that they might overthrow the democratic government. Hence they turned to people like Kleon and Hyperbolos. The sentiments expressed here agree remarkably well with the views of Thucydides (2.65.10), of Isokrates (9.121–132) and of Aristotle ('*Aθ.Πολ*.28), and the agreement with Isokrates extends even to the use of words. This means that Theopompos's account of the demagogues of Athens was sympathetic for the period before Kleon, but critical for the later period. He showed himself in this respect a faithful pupil of Isokrates.

(*c*) The statement of Theopompos on the ostracism of Hyperbolos (F 96) has been, I think, entirely misunderstood. C. G. Cobet's rendering of the sentence (*Observationes criticae in Platonis comici reliquias* [Amsterdam 1840] 143) has been generally accepted: *ostracismo eiecerunt hominem, in quo exilio sex annos vixit.*[5] Only H. Neumann (*Klio* 29 [1936] 37) insisted that Cobet's interpretation failed to remove the grammatical difficulties of this passage,[6] but he unfortunately discarded the evidence of Theopompos altogether. It has not been noted that Theopompos speaks in the following sentence of Hyperbolos' departure for Samos and of his subsequent assassination. The six

---

5. See, for instance, H. Müller-Strübing, *Aristophanes und die historische Kritik* (Leipzig 1873) 411; G. Gilbert, *Beiträge zur innern Geschichte Athens* (Leipzig 1877) 231, n. 7; G. Busolt, *Gr. Gesch.* III/2 (Gotha 1904) 1257, n. 1; H. Swoboda, *RE*, s.v. Hyperbolos 257, lines 27–30; J. Carcopino, *L'ostracisme Athénien* (Paris 1935) 194–195; O. W. Reinmuth, *RE*, s.v. Ostrakismos 1683, lines 38–47; W. Peek, *Kerameikos* 3 (Berlin 1941) 101; A. Calderini, *L'ostracismo* (Como 1945) 71; A. E. Raubitschek, *TAPA* 79 (1948) 191–192, 208 [*supra*, pp. 116–117, 129–130]; A. G. Woodhead, *Hesperia* 18 (1949) 82–83; C. Hignett, *A History of the Athenian Constitution* (Oxford 1952) 395–396.

6. The Greek means "they ostracized Hyperbolos (for) six years" which does not make sense—the sentence of ostracism was for a ten year period. Cobet's Latin version makes sense but it is not a translation of the Greek.

years must, I think, refer to the period *before* not after the ostracism. Theopompos himself says of Kleon that he stood at the head of the demos for seven years (F 92: Κλέων δημαγωγὸς ἦν Ἀθηναίων προστὰς αὐτῶν ἑπτὰ ἔτη), and Perikles was in power for forty years (Plutarch *Pericles* 16.2: τεσσαράκοντα μὲν ἔτη | πρωτεύων; from Theopompos?). The six years of which Theopompos speaks in connection with Hyperbolos refer, I think, to the period during which he was the leader of the demos, from the death of Kleon, late in the summer of 422 B.C. (Thucydides 5.12), to his own ostracism, in the spring of 415 B.C., little more than six years. The text as we have it (of F 96) is incomplete and can be corrected by the addition of one word or two: ἐξωστράκισαν τὸν Ὑπέρβολον ἓξ ἔτη ⟨δημαγωγήσαντα vel πρωτεύσαντα vel προστάντα αὐτῶν⟩.[7] This statement should precede rather than follow the characterization which Theopompos gave of Hyperbolos's ostracism, following the words of Thucydides (8.73.3): ἐξωστρακίσθη δὲ οὗτος, οὐ διὰ δυνάμεως φόβον καὶ ἀξιώματος, ἀλλὰ διὰ πονηρίαν καὶ αἰσχύνην τῆς πόλεως. It is possible that Theopompos told the story of Hyperbolos's ostracism which we read in Plutarch (*Aristides* 7; *Nicias* 11) after he characterized the demagogues (Schol. Aristophanes *Pax* 681) and before he indicated the length of Hyperbolos' "rule" and described his end (F 96, as amended here). In this way, all of Theopompos' statements on Hyperbolos can be combined to form a consistent account, and the ostracism of Hyperbolos may now be confidently dated in 415 B.C.

7. Compare the similar construction in Thucydides 1.110.1: οὕτω μὲν τὰ τῶν Ἑλλήνων πράγματα ἐφθάρη ἓξ ἔτη πολεμήσαντα.

41

Review of L. Pearson, *The Local Historians of Attica*

Lionel Pearson. *The Local Historians of Attica*. Philadelphia, Lancaster Press, 1942. Pp. x + 167. $2.25 (to members $1.50). (Philological Monographs published by the American Philological Association, No. XI). [244]

The Philological Monographs can look back on a tradition of more than ten years, which is characterized by good and careful printing, by reasonable prices, and by sound judgment in the selection of manuscripts. The volume under discussion is no exception to this rule. It contains a well-written study, accepted as a dissertation by Yale, and its author is already well known through his Early Ionian Historians (Oxford, 1939). Although Pearson does not give us an up-to-date list of the fragments of the Atthidographers ("since the days of fragment collecting are now almost over"), his accounts of the individual historians are good English substitutes for (and additions to) the German articles to be found in *R.-E.* and elsewhere (see pp. 26, 69, 76, 86, 135f., 144); his final chapter, moreover, contains an original contribution to the problem of the Atthis. Yet, one may wonder whether the work will not have to be done all over again once Jacoby's critical collection of the fragments has become available to all.

Pearson's book does not encourage detailed criticism since he carefully avoids any discussion of controversial material and since he passes with firmness and determination over side issues inviting a more careful study; see A. A. Boyce, *C.W.*, 37 (1943), p. 91. The following outline of the contents may therefore suffice.

The first chapter is devoted to Hellanicus and especially to his Atthis which gave the title to several subsequent Attic histories. "The first and most striking feature of the Atthis is that it sets out to cover the whole of Attic history *ab urbe condita* up to the author's own time." The only two fragments which refer to events of the Peloponnesian War "may be taken as certain evidence that a portion of his Atthis was devoted to an annalistic record, with the | events of [245] each year grouped under the name of its eponymous archon."

The second chapter deals with Thucydides (Herodotus's account of Attic history is not treated separately; see H. Bloch, *A.H.R.*, 48 [1943], pp. 765-6). It is largely a demonstration of this author's dependence upon the tradition of the Atthides. Yet, aside from Hellanicus's Atthis (to which Thucydides refers but which must have been published long after Thucydides began his own work; see T. W. Swain, *Class. Phil.*, 38 [1943], p. 202), there is no evidence for such a tradition in the fifth century. It would have been more appropriate to show how much Thucydides on his part influenced the Attic historiography of the fourth and third centuries.

The third chapter contains mainly a discussion of Xenophon's *Hellenica* and of the *Hellenica of Oxyrhynchus* under the heading "The successors of Thucydides." Pearson asserts that "Xenophon played a small part, much smaller than Thucydides, in keeping alive the traditions of Attic local history." One may wonder whether these "traditions" already existed in Xenophon's time.

In the fourth chapter, Pearson presents documented accounts (similar to those contained in his *Early Ionian Historians*) of the earlier Atthidographers: Cleidemus, Phanodemus, Androtion, Melesagoras, Demon, and Melanthius. The Atthis of Cleidemus who was "the earliest of those who wrote on Athenian local history" (Pausanias) "probably antedates Aristotle's *Constitution of Athens* by not more than thirty years." Of Phanodemus only little is known, but he can be identified with the Athenian politician who was honored on various occasions for his interest in religious ceremonies. Androtion is better known mainly because, as a pupil of Isocrates and an adversary of Demosthenes, he played an important part in Athenian public life; see W. Jaeger, *Paideia*, III, pp. 117ff. Androtion apparently followed Thucydides and Isocrates (and possibly influenced Aristotle) in his admiration for Theramenes and the "moderates." It seems that the conservative elements (Plato among them) dominated the literary scene while the progressive (or radical) democrats (Lysias, Eubulus, Demosthenes, and others) were more actively engaged in politics and administration.

Chapter V: Ephorus, Theopompus, and Aristotle. None of these historians wrote a local history of Attica, but they probably knew all the Atthides published up to their own time. This is particularly true for Aristotle whose *Constitution of Athens* has fortunately been preserved to us. Being a history and description of the Athenian constitution, Aristotle's work contains few of the "familiar features of an Atthis": religious and mythological matters. On the whole it is a critical rather than an original account. Pearson cautiously mentions twice the "political pamphlets" (pp. 57 and 104) which may have been the source of some of Aristotle's obviously biased stories. If only Pearson had found the courage to question the existence of these pamphlets, and to assume, if only for argument's sake, that some of the (pro-Theramenean and conservative) bias came from the Atthides themselves!

With Chapter VI: Philochorus and Ister, Pearson passes into the third century. His account of Philochorus is a welcome addition to | R. Laqueur's article in *R.-E.* (see Pearson's own article in *Greece and Rome*, 12 [1943], pp. 51-56, and A. M. Woodward, *J.H.S.*, 62 [1943], p. 87). The eclectic character of

Philochorus's Atthis is apparent, while Ister's work must have been merely a mythological handbook rather than a complete account of Attic history.

In the seventh chapter, Pearson summarizes the results of his observations under the heading "The Atthis Tradition"; this is by far the most important part of the book. Pearson asserts "the fragments of the Atthidographers ... have given no ground for believing in any such traditions of historical opinions .... The Atthis tradition, which forms the subject of the present chapter, is not an historical but a literary tradition." "The Atthidographers ... restricted themselves ... in dealing with mythology." They "did not merely hand on the old myths as they found them. Each one added something of his own, some new interpretation or some new incident." As Pearson draws with caution the outlines of this development, it becomes apparent that one of the chief tendencies was an ever increasing "rationalistic interpretation." This tendency is cleverly exposed in a comparative study of the accounts of Theseus (and Eumolpus) as given by the various Atthidographers.

In spite of all differences in the treatment of individual myths, there seems to have been a great uniformity not only in the general composition (since the material was arranged chronologically, the chronology having been established by Hellanicus) but also in the relative space devoted to the individual periods. Here, too, a certain development is noticeable. "Since the later writers were in a position to apply detailed, annalistic treatment to a very much longer period than Hellanicus, it cannot be expected that they should devote so large a proportion of their work to mythical times as he did." The annalistic method itself (and the point at which it sets in) is a traditional element of the local histories of Attica. Pearson suggests tentatively and unconvincingly that only the events from the middle of the fifth century onward were treated year by year (following Hellanicus rather than Thucydides). The period between Theseus and Pericles was passed over quickly in most accounts (if we are to trust the small number of fragments referring to this period). Most of the fragments show antiquarian rather than historical interest and refer to "such constitutional questions as would arise in the treatment of Solon and Cleisthenes." Pearson emphasizes that "no Atthis is quoted even as authority for the conspiracy of Cylon or the code of Draco." Aristotle's detailed account of these incidents is therefore thought to be based on a political pamphlet rather than upon serious historical study. Unfortunately, Pearson does not dwell upon this point, as he generally avoids all controversial issues. Yet, since the conservative tendencies of Aristotle's *Constitution* are not confined to the discussion of Draco's code, it may seem unnecessary to assume that this particular chapter is based on sources so entirely different from those used for the rest of the work.

In summing up, Pearson repeats the common characteristics of all the Atthidographers: "their concern with religious ritual and the mythological explanations of religious customs, with constitutional antiquities and the development of Athenian democratic institutions; their interest in the topography of Athens and Attica and the sacred associations of different Attic sites; and their interest in anecdote and literary figures of Attic history."

42

Review of F. Jacoby, *Atthis: The Local Chronicles of Ancient Athens*

[135] *Atthis: The Local Chronicles of Ancient Athens.* By Felix Jacoby. Oxford: At the Clarendon Press, 1949. Pp vii, 431. $9.00.

This book "is now being presented to the public without the edition of and the commentary on the texts with which it deals" and to which it "was originally intended to form an introduction." The following remarks are therefore little more than a preliminary announcement of a most excellent piece of scholarship.

No modern student has contributed more to our knowledge of Greek historical writing in general, and (except for Thucydides) of any one Greek historian in particular than Felix Jacoby. His *Atthis* is perhaps the most original of his contributions; his main thesis that Hellanicus "created" the history of Athens must be accepted, his method is exemplary, and the wealth of subsidiary observations deserves the respect of all serious students.

Jacoby demonstrates that the history of Athens "does not derive from an old and semi-official chronicle kept by the priestly board of Exegetai, but was created in the lifetime of Thukydides by a learned man, the foreigner Hellanikos of Lesbos." Atthidography is a literary, not an historical tradition.

Jacoby emphasizes the political and historical character of the Atthis. The various authors pursued different aims, and the most outstanding of them, Androtion and Philochorus, were critical of radical democracy. They had in common the annalistic arrangement of the material (a possible invention of Hellanicus), and a basically historical interest which permitted digressions on matters of cult, institutions, and antiquities. The sudden start of the Attic series (ca. 350 B.C.), as well as its end (ca. 263 B.C.), coincide, the one with the beginning of Athens' struggle with Macedon, the other with her loss of independence. The Atthidographers were historians, not pamphleteers; their material was the history of Athens, and their method the same critical rationalism which is characteristic of the Great Historiography of Herodotus and Thucydides. Although their own investigations increased the available material from the

Classical Weekly 44 (1951) 135

earlier period (especially in matters of cult and institutions), they devoted, as time went on, more and more space to contemporary affairs, i.e. the fourth century.

The third and perhaps the best chapter is devoted to the sources of the Atthis, especially those of Hellanicus. Jacoby stresses the unique importance of Herodotus's account of Athenian history from ca. 560 to 479 B.C., and he shows in a "test case" that the tradition of the liberation of Athens in 510 B.C. goes back to the story told by Herodotus. After a thorough treatment of the list of archons as a source of historical information (see now T. J. Cadoux, *JHS* 68 [1948] 70–123), Jacoby denies the existence of any preliterary chronicles either in Athens or elsewhere. He suggests, moreover, that the various local chronicles followed and corrected the historical accounts of Hecataeus and Herodotus and did not precede them. In another "test case" dealing with the chronology of Peisistratos, Jacoby is able to show how the oral tradition, presented in part by Herodotus, was transformed by Hellanicus and his successors into a strictly chronological account with the use of the list of archons and the relative chronology preserved by the oral tradition. In a final section, Jacoby rejects the assumption that the early historians and Atthidographers made extensive use of documents, and he calls attention to the fact that only the first, historical, part of Aristotle's *Constitution* is based on the Atthis (probably Androtion).

Students of Athenian history of the sixth and fifth centuries B.C., and of Greek historical writing in general, will find Jacoby's book a rich mine of information and a good example of sound historical criticism.

43

Die schamlose Ehefrau
(Herodot 1, 8, 3)

[139] Theano, die Gattin oder Schülerin des Pythagoras, hat, wenn man der Überlieferung trauen darf,[1] einige erstaunliche Ansichten ausgesprochen, die sie als eine schöne, vernünftige und überaus züchtige Frau erweisen. Der bemerkenswerteste dieser Aussprüche ist uns von Diogenes Laertius (8, 43) überliefert, der sagt, daß Theano einer verheirateten Frau riet, mit ihrem Gewande ihre Scham abzulegen, wenn sie zu ihrem Gatten ginge, aber nach dem Besuch beide wieder anzuziehen: τῇ δὲ πρὸς τὸν ἴδιον ἄνδρα μελλούσῃ πορεύεσθαι παρῄνει ἅμα τοῖς ἐνδύμασι καὶ τὴν αἰσχύνην ἀποτίθεσθαι, ἀνισταμένην τε πάλιν ἅμ' αὐτοῖσιν ἀναλαμβάνειν. Theano muß wohl gemeint haben, daß in der Gegenwart ihres Mannes, aber nur in seiner Gegenwart, eine Frau keine Zurückhaltung, und Scham wäre eine Zurückhaltung, zeigen soll. Offensichtlich steht hinter dieser Ansicht eine sehr hohe und klare Auffassung der Ehe, wie man sie im Kreise des Pythagoras zu finden erwarten würde.

Richard Harder hat auf diese Stelle hingewiesen,[2] aber gemeint, daß sie sich auf einen Ausspruch des Herodot beziehe, den der Geschichtsschreiber dem
[140] Gyges in den Mund legte | (1,8,3): ἅμα δὲ κιθῶνι ἐκδυομένῳ συνεκδύεται καὶ τὴν αἰδῶ γυνή. Tatsächlich muß das Verhältnis der beiden Stellen umgekehrt sein, und Herodot, oder seine Quelle, weist augenscheinlich auf den Ausspruch der Theano hin.[3] Nur so kann man die Worte des Gyges verstehen: "Eine Ehefrau legt mit dem Kleide ihre Scham ab (wenn sie ihren Gatten im Schlafzimmer besucht; daher ist es unerlaubt für mich, dem Fernstehenden, solch einer Begegnung beizuwohnen)." Die Richtigkeit dieser Erklärung ist von

---

Rheinisches Museum 100 (1957) 139–140

1. Zusammengestellt bei K. v. Fritz, *R.E.*, s.v. Theano, no. 5, Sp. 139–1381.
2. *Studies Presented to David M. Robinson*, II, S. 446–449; siehe auch C. E. Frhr. von Erffa, *Philologus*, Suppl. 30/2, 1937, S. 180, fn. 160.
3. So schon W. Aly, *Volksmärchen, Sage und Novelle bei Herodot und seinen Zeitgenossen*, S. 34. Vgl. auch E. Bickels Bemerkungen in *N. Jb.*, 24, 1921, S. 343.

Plutarch bestätigt, der gegen Herodot Stellung nimmt (*Conj. praec.*, 10, 139 C): οὐκ ὀρθῶς Ἡρόδοτος εἶπεν ὅτι ἡ γυνὴ ἅμα τῷ χιτῶνι ἐκδύεται καὶ τὴν αἰδῶ· τοὐναντίον γὰρ ἡ σώφρων ἀντενδύεται τὴν αἰδῶ, καὶ τοῦ μάλιστα φιλεῖν τῷ μάλιστα αἰδεῖσθαι συμβόλῳ χρῶνται πρὸς ἀλλήλους. Plutarchs abschließende Bemerkung ("als Beweis größter Liebe zeigen die Gatten einander höchste Scham") zeigt, daß seiner Ansicht nach Herodot nicht von irgendeiner Frau im allgemeinen spricht, sondern von einer Ehefrau in ihrem Betragen gegenüber ihrem Gatten.

Ich habe an anderen Stellen darauf hingewiesen, daß Herodot hier wohl einer Tragödie folgt, von der wir noch ein Bruchstück besitzen und die dem Ion von Chios zugeschrieben werden kann (*Cl. Weekly*, 48, 1955, S. 48–50),[4] und daß die unmittelbar folgende Gnome (1,8,4: σκοπέειν τινὰ τὰ ἑωυτοῦ) eine wohlbekanntes Pittakeion ist (*W.S.*, 70, 1957). Diese neue Erkenntnis, daß Herodot in einander folgenden Sätzen gnomische Aussprüche der Theano und des Pittakos vereinigte, mag die Vermutung bestärken, daß der Geschichtsschreiber hier einer dichterischen Quelle, nämlich einer Tragödie, folgt. Die Behauptung ist daher kaum zu gewagt, daß auch andere Stellen des Herodot auf dichterische Vorlagen zurückgeführt werden können.*

---

4. Den dort gegebenen Verweisen möchte ich A. Lesky, *Österr. Anz.* 7, 1954, Sp. 150 hinzufügen.

* Ernst Bickel: Rekonstruktions-Versuch einer hell. Gyges-Nysia-Trag.
"Die schamlose Ehefrau," wovon Raubitschek im obigen kurzen, aber ungemein anregenden Beitrag spricht, ist ein Ausdruck, der nicht mißverstanden werden darf. Die Sache, die in Wahrheit gemeint ist, läßt sich die Antike nicht nehmen. Wenn Theano für die Ehefrau im erotischen Verkehr mit dem Gatten die Nacktheit—kurz gesagt—will, so steht | diese Auffassung der Ehe in deutlichem [141] Gegensatz zu dem von R. gleichfalls angeführten Standpunkt der synkretistischen Ethik bei Plutarch, wo als Symbol der Gattenliebe die größte Scham bei dem Eheverkehr angesetzt wird. Im Altertum tritt dieser Standpunkt in Erscheinung in der *pudicitia* der altrömischen Matrone. Aber im hellenistischen Rom war dies anders. Eine "schamlose Ehefrau" war in der Antike jedenfalls die von Martial als Vorbild keuscher und dauerhafter Monogamie gepriesene vornehme Römerin Sulpicia, die in ihrer Lyrik die letzten Geheimnisse ihres Ehelebens preisgab: *si me cadurci dissolutis fasciis nudam Caleno concubantem proferat*. Kein Unterschied besteht zwischen der Cynthia des Properz: *seu nuda erepto mecum luctatur amictu* und der Gattin des Calenus: *o quae proelia, quas utrimque pugnas felix lectulus et lucerna vidit*. Derartige Lyrik einer Römerin der frühen Kaiserzeit läßt sich mit dem Hinweis auf Juvenals Weibersatire abtun, sondern sie erinnert an den Spruch der Theano, die nicht zur Hetäre die Gattin macht, wenn sie für diese in der Ehe durch Ablegung der Scham mit dem Gewande freie Bahn zu erotischer Initiative will.

"Eine sehr hohe und klare Auffassung der Ehe" sieht R. hinter Theanos Spruch. Dies läßt sich nur so verstehen, daß hier jener archaischen Gesittung gegenüber, in welcher die Ehe nicht auf ein persönliches Verhältnis des Brautpaares hin geschlossen ward, ein Zug von Seele zu Seele und zugleich praxitelische Sinnenfreude am nackten Körper den ehelichen Verkehr aesthetisieren soll. Ist dem so, dann erhebt sich beim Einblick in den Kulturgang der Hellenen die Frage, ob den Primat in dieser Entwicklung die Pythagoreerin hat, oder Voraussetzung für Theano auch das freie Liebeslied aus dem Frauenmund der Sappho war. Über die Erotik in Sapphos Ehe gehen die Meinungen auseinander. Die aus einer schwerlich antiken Gesamtauffassung der Sappho resultierende Kritik von Wilamowitz an Welcker in dieser Beziehung ist behandelt *Rh. Mus.* 89 (1940) S. 202f. Siehe auch Merkelbach, "Sappho und ihr Kreis" (*Philol.* 101, 1957, S. 1ff.).

E.B.

44

Damon

[78] Die ausgezeichneten Beiträge von H. Ryffel (*Mus. Helv.* 4 [1947] S. 23–38) und von C. del Grande (*Giorn. It. di Fil.* 1 [1948] S. 3–16) haben so viel zu unserem Verständnis von Damons Werk beigetragen, dass es jetzt möglich ist seine geschichtliche Stellung zu bestimmen.

Auf das älteste Zeugnis hat erst U. v. Wilamowitz-Moellendorff, hingewiesen (*Griech. Verskunst*, S. 59–60, fn. 2): Libanius, *Apologia Socratis* 157. Da diese Stelle, wie überhaupt die ganze Schrift des Libanius, sich gegen die von Polykrates verfasste Anklage des Sokrates richtet (siehe *R.E.*, s.v. Libanios, Sp. 2509; s.v. Polykrates, Sp. 1741, 23–24, und 1746, 15–18), muss man annehmen, dass Polykrates, und vielleicht vor ihm Anytos,[1] Damons Verbannung als Beispiel für die geforderte Verurteilung des Sokrates verwendet hat.[2] Damit erledigen sich alle Zweifel an der Geschichtlichkeit von Damons Ostrakisierung wie sie J. Carcopino ausgesprochen (*R.E.G.* 18, 1905, S. 415–429) und wiederholt hat (*L'ostracisme Athénien*[2], S. 125–142). Man könnte sogar annehmen, dass Plutarchs Angaben über Damons Verbannung (Plut., *Aristides* 1, 7; *Nicias* 6, 1; und besonders *Pericles* 4, 1–2) mittelbar auf die Anklage des Polykrates [79] zurückgehen. In allen diesen Stellen (wie auch in Isocrates, 15, 235) ist die Beziehung zwischen Damon und Perikles betont, und Polykrates hat sicherlich auf diese Beziehung auch hingewiesen. Die sprachliche Aehnlichkeit von Plutarch, *Pericles* 4, 1–2, und Plato, *Protagoras* 316 D–E (wo Damon in den ἄλλοι πολλοί versteckt ist; siehe H. Ryffel, Μεταβολὴ πολιτειῶν, S. 29, fn.

---

1. Ich glaube, dass Polykrates die ursprüngliche Anklage verfasste, aber sie erst einige Jahre später, vielleicht als Antwort auf Platons *Apologie*, mit anachronistischen Zusätzen veröffentlichte; diese Annahme muss jedoch erst bewiesen werden.
2. In ähnlicher Weise hat Andocides, IV 32–34, solche Beispiele vorgebracht; siehe meine Bemerkungen in *T.A.Ph.A.* 79 (1948) S. 203–205 [*supra*, pp. 125–127]. Unsere Kenntnis von Menons Ostrakisierung (Hesychius, s.v. Μενωνίδαι) mag aus einer ähnlichen Quelle stammen; zu Menon, siehe meine Bemerkungen in *Hesperia* XXIV (1955).

88) kann man am besten mit der Annahme erklären, dass beide Schriftsteller denselben Gedankengang wiedergeben.

Die meisten anderen biographischen Zeugnisse über Damon stehen entweder in Platos Schriften oder gehen auf diese zurück. Aus ihnen kann man die Wirksamkeit des Damon über eine Spanne von etwa dreissig Jahren verfolgen. Der früheste Verweis steht im Alcibiades 1 (118 C, zitiert von *Aristides*, 47, p. 427 Dind.), wo Alkibiades die Erziehung des Perikles schildert. Plate dachte sich den alten Perikles in der Gesellschaft des Damon, der selber damals viel jünger gewesen sein mag.[3] Die hier erwähnte Verbindung von Damon und Perikles ermöglicht die Zeitbestimmung des von Plutarch berichteten Verhältnisses der beiden Männer: es muss in die letzten Jahre des Perikles gehören (wie auch W. Schmid, *Gesch. d. Griech. Lit.*, I/2 S. 731, fn. 9, betont hat). Daher sollte man die Verbannung des Damon in dieselbe Zeit setzen, wie es ja auch A. Martin (*Notes sur l'ostracisme*, S. 31), J. Beloch (*Gr. Gesch.* II/1 S. 313, fn. 1) und O. W. Reinmuth (*R.E.*, s.v. Ostrakismos, Sp. 1682–1683) getan haben. Beloch weist auch darauf hin, dass der von Andocides (I 16) erwähnte Damon, mit dem Agariste in erster Ehe verheiratet war, mit dem Freund des Perikles gleichzusetzen ist; nur darf man nicht behaupten, dass Damon damals (415 v. Chr.) schon tot war.

Plato spricht von Damon auch im Laches, der um 420 v. Chr. gespielt haben mag. Dort nennt Sokrates ihn seinen Genossen (197 D), und die späteren (Diogenes Laertius, 2, 5, 3) haben ihn daher seinen Schüler genannt.[4] In einer zweiten Stelle (180 C–D) erfah|ren wir, dass Sokrates Nikias überredete seinen Sohn Nikeratos dem Damon zur Erziehung zu übergeben, und dass Nikias den Damon für einen ausgezeichneten Lehrer, nicht allein der Musik, hielt. Diese Verbindung fand kurz vor der Zeit das Gespräches statt (ἔναγχος), was auch auf die Zeit um 420 v. Chr. hinweist, da der Sohn des Nikeratos, wenn sein Vater in 403 v. Chr. ermordet wurde, noch ein Kind war (Lysias, 18, 10). Schliesslich erwähnt Plato den Damon noch einmal im *Laches* (200 A–B) in einem Zusammenhang, der klar macht, dass Damon als noch am Leben gedacht ist.[5]

[80]

Plato hat den Damon auch in der Republik wiederholt angeführt (424 C; 400 B–C; siehe auch Proclus, *Comm.* 1, S. 42, 54, 56, 61, 62 Kroll), in Stellen die der Annahme nicht widersprechen dass Damon auch damals (um 412 v. Chr.) als noch am Leben gedacht war; diese Stellen sind von Ryffel ausführlich erklärt worden. Damon ist auch im Axiochos un 405 v. Chr. als anwesend

---

3. Das Scholion zu der Stelle nennt Damon als Urenkelschüler des Pythokleides, und Plato hat Damon einen Schüler des Agathokles genannt (*Laches* 180 C–D).

4. Diese Annahme bestätigt Aristophanes, der in den Wolken Sokrates die Lehre des Damon vortragen lässt, wie del Grande (siehe oben) wiederum schön gezeigt hat; siehe auch W. Jäger, *Paideia* 2, S. 404, fn. 110. Die Erwähnung des Damon in den Briefen der Sokratiker (14) ist nicht ganz klar, obgleich J. Sykoutris die Stelle einigermassen verbessert hat (*Die Briefe des Sokrates und der Sokratiker*, S. 59, fn. 4).

5. Die tadelnden Worte, die Sokrates hier an den *Laches* richtet (οὗ σύ που οἴει καταγελᾶν, καὶ ταῦτα οὐδ' ἰδὼν πώποτε τὸν Δάμωνα) erklärt ein Vers des Aristophanes (*Vespae* 959) der von Laches sagt: κιθαρίζειν γὰρ οὐκ ἐπίσταται.

gedacht (364 A), in einem Gespräch, das zwar dem Plato abgesprochen wird, das aber daher vielleicht geschichtlich treu sein wird (siehe 368 D). Man kann also mit Sicherheit annehmen, dass die Wirksamkeit des Damon ins letzte Drittel des fünften Jahrhunderts gehört.[6] In dieser Zeit wurde er auch von Plato in einer Komödie als Cheiron des Perikles verspottet (*Frog.* 191 Kock), und sein Areopagiticus mag auch gegen das Ende des Jahrhunderts verfasst worden sein.[7] Dieser Ansatz widerspricht in keiner Weise Ryffels und del Grandes | Bemerkungen über Damons Werk (siehe oben), und bestätigt die engere Beziehung zwischen Damons Schrift und der Kritik seiner Theorie die man schon seit langem in einem Papyrus erkannt hat (siehe W. Crönert, *Hermes* 44 [1909] S. 510 und 520; A. Busse, *Rh. Mus.* 77 [1928] S. 38, fn. 2.[8] Man könnte sogar die kurze Wiederherstellung der Macht des Areopagus von der Aristoteles spricht (*Ath. Pol.*, 35, 2) mit Damons *Areopagiticus* zusammenstellen.

Es muss jetzt versucht werden den Zeitpunkt von Damons Ostrakisierung festzulegen. Platos Zeugnis kann hier nicht benützt werden, da er auf Damons Verbannung nicht einmal anspielt; doch wäre es möglich, sie zwischen die Erwähnung im Alcibiades und im Laches zu setzen und anzunehmen (siehe oben), dass Damon vor 430 v. Chr. ostrakisiert wurde und vor 420 v. Chr. nach Athen zurückkehrte. Wir besitzen auch eine Angabe in der pseudo-Xenophontischen *Verfassung der Athener* (1, 13), die in dieselbe Zeit passt und mit Damons Verbannung in Zusammenhang gebracht werden darf.[9]

Man hat allgemein angenommen, dass der Verfasser hier von zwei Gruppen von Leuten spricht, deren eine sich der Gymnastik, die andere sich der Musik widmeten;[10] doch ist es besser τὴν μουσικήν als Objekt von γυμναζομένους und ἐπιτηδεύοντας zu verstehen, die ja einen gemeinsamen Artikel τούς haben. Wir lesen hier, dass das Volk den Leuten, die sich der Musik widmeten und sie betrieben, ein Ende gemacht hat, aus Gründen, die wir, wenn man an Damons Erziehungsweise denkt, verstehen können. Die Verbannung des

---

6. Ob Plato selber den Damon gekannt hat, wissen wir nicht, doch ist es sehr wahrscheinlich. Jedenfalls hat Damons Schüler Drakon Plato in der Musik unterrichtet, wie schon Aristoxenos betonte (H. Well und Th. Reinach, *Plutarque, de la musique*, S. 65–69; F. Wehrli, *Aristoxenos*, S. 30, Frag. 82) und Olympiodoros wiederholte (*Vita Platonis*, II; *Prolegomena*, II).
7. Zu den zuletzt von W. Kranz gesammelten Fragmenten (*Fragmente der Vorsokratiker*, I⁶ S. 382–384; siehe auch S. 501–502) könnte man eines der Testimonia (No. 8) hinzufügen. Auf diese von Galen und Martianus Capella erzählte Anekdote mag schon Aristoteles (*Politik* 8, 1340 b) Bezug nehmen; siehe auch F. Buecheler, *Rh. Mus.*, 40, 1885, S. 310. Zu den Testimonia könnte man Cicero, *De oratore*, 3, 132, Nepos, *Espaminondas* 2, 1, und (wenn man E. L. Highbarger, *Cl. Phil.*, 40 [1945] S. 43–44, Glauben schenkt) auch Vergil, *Bucolica* 3 und 8 hinzufügen
8. Für den Text, siehe K. Jander, *Oratorum et rhetorum ... fragmenta*, S. 18–20, No. 37; M. Untersteiner, *Sofisti* 3, S. 208–211, No. 12. Der Satz κακῶς εἰδότες, ὅτι οὔτε χρῶμα δειλοὺς οὔτε ἁρμονία ἂν ἀνδρείους ποιήσειεν τοὺς αὐτῇ χρωμένους kann mit Aristoteles, *Probleme* 19, 29, verglichen werden.
9. Siehe besonders F. Buecheler, *Rh. Mus.* 40 (1885) S. 312; E. Kalinka, *Die pseudo-xenophontische* Ἀθηναίων πολιτεία, S. 138–140; H. Frisch, *The Constitution of the Athenians*, S. 211–213; G. Grossmann, *Politische Schlagwörter aus der Zeit des Peloponnesischen Krieges*, S. 177, fn. 197.
10. Für die Verwendung des Wortes ἐπιτηδεύω, siehe die Bemerkungen von H. Ryffel, *Mus. Helv.* 4 (1947) S. 35, fn. 47, und 36.

Lehrers hat natürlich das Ende der Schule bedeutet; auf beide weist das Wort καταλέλυκεν hin. Leider kann auch diese Stelle nicht für eine genauere [82] Datierung von Damons Ostrakisierung verwendet werden, da die Schrift, in der sie steht, selbst nicht datiert ist.

Wir besitzen auch ein Ostrakon auf dem Δάμον Δαμονίδο steht und das wegen seiner Verziehrung in die Zeit nach 450 v. Chr. gehört.[11] Der vollständige Name des Damon ist in einer Notiz des Stephanus von Byzanz, s.v. Ὄα, erhalten: Damon, der Sohn des Damonides, von Oa.[12]

Es ist daher verwunderlich, dass so viele Forscher die Tätigkeit des Damon un 462 v. Chr. beginnen lassen und seine Verbannung in die vierziger Jahre des fünften jahrhunderts setzen.[13] Sie sind nämlich U. v. Wilamowitz-Moellendorff gefolgt, der (*Hermes* 14 [1879] S. 318–320) behauptete, dass der Musiker Damon, der Freund des Nikias und des alten Perikles, mit Damonides, dem Ratgeber des Perikles in 462 v. Chr., gleichzusetzen ist. Dieser Damonides ist uns von Aristoteles bekannt (*Ath. Pol.* 27, 3), den Plutarch (Pericles 9, 1) benutzt hat. Nach der Feststellung, dass Perikles den Gerichtssold eingeführt hat (vergleiche Aristoteles, *Politik*, 1274 a 9), erzählt Aristoteles eine Anekdote über Kimons Freigebigkeit, die er in Theopompus gelesen haben kann (*F. Gr. Hist.*, 115 F 89), und fügt dann die Bemerkung über Damonides hinzu. Plutarch gibt nicht nur dieselbe Anekdote wieder, er fügt ihr auch denselben Schluss über Damonides an. Ich glaube mit H. T. Wade-Gery (*A.J.P.* 59 [1938] S. 133–134) dass der Verweis auf Damonides schon im zehnten Buche des Theopomp stand.

Vor der Entdeckung des Ostrakon mit dem Namen von Damons Vater [83] Damonides mag es bestechlich erschienen sein, Damon und Damonides gleichzusetzen[14] und die Angabe des Stephanus ausser Acht zu lassen, die uns Damons vollen Namen gibt. Wir haben aber jetzt ein urkundliches Zeugnis für den Namen von Damons Vater, und nichts steht im Wege sich Kenyon, Gomperz, und Sir John E. Sandys anzuschliessen (*Aristotle's Constitution of Athens*², S. 116), die zwischen Damon und Damonides unterscheiden wollten. Natürlich sind sie nicht zwei ganz verschiedene Leute[15] sondern Vater und

---

11. A. Brueckner, *Ath. Mitt.* 60 (1915 S. 20–21, No. 50 (Tafel IV); *I.G.* I² 912; M. N. Tod, *A Selection of Greek Hist. Inscr.*, I² S. 92–93; Die Scherbe war ein Einzelfund und gehört nicht zur Gruppe der Thukydides-Kleippides Ostraka, wie J. Carcopino, *L'ostracisme Athénien*², S. 127, behauptet. Der Name Damon weist nicht auf fremde Abkunft hin (U. v. Wilamowitz. Moellendorff, *Platon* II² 1920, S. 15, fn. 1), wie W. Crönert richtig betonte (*Hermes* 44 [1009] S. 510, fn. 1).
12. Die Richtigkeit des hier überlieferten Demennames haben nur G. Kaibel (*Stil und Geschichte der πολιτεία Ἀθηναίων des Aristoteles*, S. 183–184) und J. Carcopino (*L'ostracisme*², S. 135–136) bestritten, aber für eine Entscheidung muss man auf neue Funde warten.
13. Siehe, sum Beispiel, A. Rosenberg, *Neue Jahrbücher* 36 (1915) S. 211; G. de Sanctis, ᾿Ατθίς² S. 478–479, fn. 2; W. Kroll. *R.E.* Suppl. 3, Sp. 324–325; W. Peek, *Kerameikos*³, S. 77; P. Goosens, *Chronique d'Égypte* 39/40, 1945, S. 132; A, Calderini, *L'ostracismo* S. 54–56; W. Nestle, *Vom Mythos zum Logos*², S. 435; V. Ehrenberg, *Sophocles and Pericles*, S. 92–93.
14. Siehe auch G. Busolt, *Gr. Gesch.* III/1 S. 248–249, fn. 1; J. Kirchner, *P.A.*, No. 3133; U. v. Wilamowitz-Moellendorff, *Aristoteles und Athen* I S. 134–135.
15. Fur die Verwechslung der Demennamen Ὄαθεν (Stephanus) und Οἴηθεν (Aristoteles-Plutarch) verwies schon J. Carcopino, *L'ostracisme*², S. 134–135, auf Harpokration, s.v. Οἴηθεν; siehe auch F. Jacoby's Bemerkungen zu *F. Gr. Hist.*, 328 F 28 (und 372 F 8).

Sohn. Gegen die Gleichsetzung spricht nicht nur die zeitliche Verschiedenheit der Belege aber hauptsächlich auch die verschiedene politische Einstellung der beiden Männer, auf die erst V. Ehrenberg, *Sophocles and Pericles*, S. 93, hingewiesen hat. Für die Gleichsetzung spricht allein die erklärende Notiz in Aristoteles (*Ath. Pol.* 27, 3): ὃς ἐδόκει τῶν πολλῶν εἰσηγητὴς εἶναι τῷ Περικλεῖ· διὸ καὶ ὠστράκισαν αὐτὸν ὕστερον. J. Carcopino hat schon betont (*L'ostracisme*,[2] S. 139–141), dass dieser Satz inhaltlich mit dem übereinstimmt, was Plutarch über Damon gesagt hat, und er hat ihn daher als eine Interpolation bezeichnet. Wie dem auch sei, soviel ist sicher, dass er sich auf Damon und nicht auf Damonides bezieht, und dass Aristoteles hier einem Irrtum unterfallen ist; die Verwendung des Wortes ὕστερον zeigt, dass er sich des späteren Zeitpunktes von Damons Ostrakisierung wohl bewusst war.

Die vorliegenden Ausführungen haben gezeigt, dass die Wirksamkeit des Musikers Damon ins letzte Drittel des fünften Jahrhunderts gehört, dass er während der dreissiger Jahre dem Perikles nahestand und dem Scherbengericht verfiel, zu einer Zeit als Perikles selber und seine anderen Freunde unter Angriff kamen, und dass sein Vater Damonides etwa dreissig Jahre früher dem Perikles vorschlug, den Richtersold einzuführen.

45

Phaidros and His Roman Pupils

Cicero's hostility toward Epicurean philosophy did not extend to the representatives and champions of this school both in Rome and in Athens. In fact, many of his most initimate friends were Epicureans.[1] Among these were the three men in charge of the Epicurean school in Athens during Cicero's life time, Zenon, Phaidros, and Patron. Little is known about Zenon and even less about Patron, but our knowledge of Phaidros can be considerably augmented by the study of certain Attic inscriptions.[2] It has not been known, until now, whether Phaidros was an Athenian by birth, but his son Lysiades, was an Athenian citizen since he was a member of the Areopagus in 43 B.C.[3] J. Sundwall in fact rightly suggested that Phaidros's son was the Athenian archon Lysiades who held office in 51/0 B.C.[4] It has not been noted, however, that Lysiades, son of Phaidros, was also πυθόχρηστος ἐξηγητής before the last quarter of the first century.[5] The inscription honoring him in this capacity (*I.G.*, II², 3513) contains not only his demotic (Βερενικίδης), but also the name of his sister Chrysothemis. It is now possible to recognize in Phaidros son of Lysiades from Berenikidai (*I.G.*, II², 3897–3899) the Epicurean philosopher Phaidros.[6] Moreover, it becomes evident that Phaidros belonged to a distinguished

[96]

Hesperia 18 (1949) 96–103

1. See C. M. Hall, *Class. Weekly*, 28, 1935, pp. 113–115; H. M. Poteat, *Class. Weekly*, 38, 1945, p. 155; N. W. DeWitt, *Transact. Royal Soc. of Canada*, 39, 1945, Sect. II, pp. 34–35.
2. For Zenon, see E. Zeller, *Philosophie der Griechen*, 3rd edition, 3, 1, pp. 373–374, note 2. The literary evidence concerning Phaidros has been conveniently assembled by K. Philippson (I suspect that the author is Robert Philippson), *R.E.*, s.v. Phaedrus no. 8; see also R. Philippson, *Symbolae Osloenses*, 19, 1939, p. 15. I wish to thank Professors J. F. Gilliam and H. M. Hubbell for their kind help and advice in the preparation of this article.
3. Cicero, *Philippica*, 5, 5, 13–14; see also 8, 27.
4. *Klio*, 6, 1906, pp. 330–331; J. Kirchner and F. Münzer, *R.E.*, s.v. Lysiades nos. 3 and 5; see also *I.G.*, II², 1046, lone 25; 1713, line 21.
5. See *I.G.*, II², 3513, lines 8–11; the date of this document can be determined from the restoration of lines 1–7 proposed below, and from the fact that Lysiades' successor as πυθόχρηστος ἐξηγητής, Polykritos, held office as early as ca. 30 B.C.; see *Hesperia*, 12, 1943, p. 59, note 134.
6. New restorations of *I.G.*, II², 3897 and 3899 will be given below.

337

Athenian family which can be traced back to the end of the third century before Christ.[7]

[97] Ἀγαθοκλῆς Λυσιάδου Βερενικί[δ]ης, epimeletes in 186/5 (*I.G.*, II², 896, lines 42-43)
|
Λυσιάδης, archon in 148/7 (*I.G.*, II², 1938, line 1; *Inscriptions de Délos*, no. 1505, line 34) Λυ[σι]άδης Βερε[νικίδης], dedicator ca. 140 B.C. in Delos (*I. de Délos*, no. 1445A, line 8) Λυσ[ι]ά[δ]ης [Ἀγα]θοκλέους Βερενικ[ί]δης, epimeletes in Delos in 136/5 (*I. de Délos*, no. 1922, lines 2-3)

Καλλίθεος Λυσιάδου Βερενικίδης epimeletes ca. 130 B.C. (*I.G.*, II², 1939, line 52) Καλλίθεος Λυσιάδου, victor in δίαυλον at Delphi in 128/7 B.C. (*Fouilles de Delphes*, III, 2, no. 38) Καλλίθεος (Πτολεμαιίδος), hippeus in 128/7 B.C. (*F. de Delphes*, III, 2, no. 27, line 39) [Κα]λλίθ[εος] Λυσιάδου Βερενικίδης, in list ca. 125/4 B.C. (*I.G.*, II², 2452, line 20)	[Φαῖδρ]ος Λυσιάδου Βερ⟨ε⟩νικίδ[ης] ephebe in 119/8 (*I.G.*, II², 1008, line 125) Φαῖδρος [Λυ]σιά[δ]ου Βε[ρενικίδης], honored by statue ca. 78 B.C. (*I.G.*, II², 3897) [Φαῖ]δρος Λυ[σιάδου] Βερενικί[δης], honored by statue ca. 78 B.C. (*I.G.*, II², 3899) Φαῖδρος Λυσιάδου Βερενικίδης, dedicator of statue ca. 78 B.C. (*I.G.*, II², 3898)
Λυσιάδης Φαίδρου Βερενικίδης exegetes ca. 55 B.C. (*I.G.*, II², 3513, line 9) Λυσιάδης, archon in 51/0 B.C. (*I.G.*, II², 1046, line 25; 1713, line 21)	Χρυσόθεμις Φαίδρου Βε[ρενικίδου] honored by statue ca. 55 B.C. (*I.G.*, II², 3513, line 13) [Χρυσόθεμις Φαίδρου Βερ]ενικίδου, dedicator of statue ca. 55 B.C. (*I.G.*, II², 3513, line 2)

Phaidros was born ca. 138 B.C. since he was an ephebe in 119/8 B.C. This information is gained from a new restoration of *I.G.*, II², 1008, line 125; as seen from the restored tracing (Figure 45.1), the old restoration [Καλλίθε]ος is impossible.[8]

---

7. Since the deme Berenikidai to which the family belonged was created at the same time as the tribe Ptolemais (see W. K. Pritchett, *The Five Attic Tribes after Kleisthenes*, pp. 13-23), ca. 225/4 B.C., one cannot recognize earlier members of the family unless their previous deme affiliation is known. F. O. Bates (*The Five Post-Kleisthenean Tribes*, p. 43) suggested that "some preëxisting deme was re-named, for it seems hardly reasonable to suppose that a new deme was created outright." Neither Bates, however, nor Pritchett (*op. cit.*, p. 30) examined the evidence in order to find members of Berenikidai families, who belong to the period prior to the creation of the deme. One such family may have been that of the herald Eukles; see *Hesperia Index*, p. 61.
The first to draw up a stemma of the family to which Phaidros belonged was J. Kirchner (*P.A.*, no. 7910; he accepted a wrong date for *I.G.*, II², 1939, and a wrong restoration for *I.G.*, II², 1008, line 125); he was followed by P. Roussel who made a few additions and improvements (*B.C.H.*, 32, 1908, p. 347, no. 377; but he retained the wrong restoration for *I.G.*, II², 1008, line 125). J. Sundwall (*Nachträge*, p. 121) accepted Roussel's scheme with one small addition, and Roussel himself, finally, enlarged the stemma in his book *Délos colonie Athénienne*, pp. 103-104.
8. One may now restore also line 124 of *I.G.*, II², 1008 to [Πλούταρ]χος Σωσιβίου Θημα[κεύς]; see *I.G.*, II², 6207, 6208.

Figure 45.1. *I.G.*, II², 1008, lines 123–27 (tracing from squeeze).

It appears, therefore, that Phaidros was an Athenian by birth, and that he was less than seventy years of age when he died as head of the Epicurean school at Athens in 70 B.C., being succeeded by Patron.[9] It is likely that Phaidros stayed in Athens until shortly before 88 B.C. when most pro-Romans left the city in order to escape the new "democratic" regime.[10] At that time, he may have gone to Rome as a teacher of philosophy, and Cicero made his acquaintance there.[11] We know that Phaidros was in Athens before he went to Rome, not only from the inscriptions but also from the report of Atticus (Cicero, *De legibus*, I, 53) that Phaidros remembered the visit at Athens of Gellius which must have taken place shortly after 94 B.C.[12] It is reasonable to assume that Phaidros returned to Athens soon after Sulla restored "peace and order." At any rate, Phaidros was active as a teacher of Epicurean philosophy when Cicero came to Athens in 79 B.C.[13] Cicero probably did not see Phaidros again since the philosopher died in 70 B.C. (see not 9), but Cicero as well as Atticus retained a deep affection for the great teacher. It must have been during these years, while Phaidros was alive and active as head of the Epicurean school, that one of his Roman pupils erected the statue of Phaidros on the Akropolis. What is left of this monument (*I.G.*, II², 3899 = Plate 25) may be restored as follows:

[98]

[Τίτος] Π[ομπών-]
[ιος Τ]ίτου υ[ἱὸς]
[Φαῖ]δρον Λυ[σιά-]
[δου] Βερενικί[δην]
[- - - - - - - - - - -]

[99]

9. See Phlegon of Tralles, *F.H.G.*, III, p. 606, frag. 12 = *F. Gr. Hist.*, II, p. 1164, no. 257, frag. 12, §8.
10. See W. S. Ferguson, *Hellenistic Athens*, pp. 444–445.
11. *Ad familiares*, 13, 1, 2.
12. See E. Groag, *Röm. Reichsbeamten von Achaia bis auf Diokletian*, p. 8, note 29.
13. Cicero, *De finibus*, I, 5, 16.

It is, of course, not certain that Titus Pomponius Atticus was the dedicator of this statue, but it is made likely by the great devotion of the Roman gentleman philosopher for his Athenian teacher.[14] At about the same time a larger monument was erected on the Akropolis of which a considerable part remains. The monument consisted of three statues set up on a large base which was composed of two slabs of marble. The inscriptions on the right slab are almost completely preserved (*I.G.*, II², 3897, lines 1–6 = Plate 25) while only one fragment is left of the inscription on the left slab (*I.G.*, II², 3897, lines 7–9 = Plate 25). Yet the whole inscription can be restored with certainty:

'Ο δ[ῆμος]
[Λεύ]κιον Σωφήιο[ν 'Αππίου υὸν]
[ἀρε]τῆς ἕνεκ[α ἀνέθηκεν].

[ὁ δῆ|μο]ς
5 ["Αππιον Σωφ]|ήιον 'Αππίου υὸν
[ἀρετῆς ἕν]|εκα ἀνέθηκεν.

Λ[ε]ύκιος Σωφήιος 'Απ[ίου υὸς]
Φαῖδρον [Λυ]σιά[δ]ου Βε[ρενεκίδην]
τὸν ἑαυτ[οῦ κα]θηγη[τὴν ἀνέθηκεν].

The new restoration of line 9 was made possible by the addition of a new fragment containing the letters ΘΗΓΗ and allowing as the only reasonable restoration some form of the noun καθηγητής. N. W. DeWitt has repeatedly called our attention to the fact that in the Epicurean school, and originally only there, the teachers were called καθηγηταί.[15] To the scant literary evidence available to him may now be added the epigraphical evidence which is by no means plentiful. In addition to *I.G.*, II², 3897 (see above), and the Agora inscription published below, one may mention *I. de Délos*, no. 1801 honoring the καθηγητής Dionysios, perhaps the third head of the Epicurean school (*Diog. Laertius*, 10, 25). Even more interesting is *I.G.*, II², 3793 honoring the καθηγητής Alexander, son of Maro, from Phaleron. One would not hesitate to identify this man with the Epicurean philosopher Alexander mentioned by Plutarch (*Quaest. conviv.*, 2, 3, 1), were it not for the fact that another Attic inscription honoring apparently the same man (*I.G.*, II², 3819) calls him διδάσκαλος and records that his statue was set up ἄλσει μέσσωι which has been understood as a reference to the Academy. Yet ἄλσος may have been used in the poem for κῆπος and the term διδάσκαλος (which fitted metrically better than καθηγητής) may have become more acceptable to the Epicureans by the time of Plutarch. Finally, one may consider the statue base of C. Sulpicius Galba (*I.G.*, II², 4157) as restored by J. H. Oliver, *A.J.A.*, 46, 1942, p. 382; the perplexing word in the third line may be restored as καθ[ηγητήν] referring possibly to some philosophic teaching activity of the scholar which has otherwise

---

14. Cicero, *Ad Familiares*, 13, 1, 4; see A. H. Byrne, *Titus Pomponius Atticus*, pp. 25–26.
15. *Cl. Ph.*, 31, 1936, p. 206; *Cl.J.*, 42, 1947, p. 197.

remained unknown to us; see also *T.A.P.A.*, 77, 1946, p. 149, note 10; *A.J.P.*, 69, 1948, p. 436.

The Romans mentioned in *I.G.*, II², 3897 were evidently two brothers Lucius and Appius, sons of Appius Saufeius. It is surprising that their identity has not been recognized before. Lucius Saufeius, son of Appius, is, of course, the Epicurean friend of Cicero and Atticus, mentioned so often in Cicero's correspondence.[16] Lucius's brother Appius is mentioned in one of Cicero's letters to Atticus (6, 1) written from Laodicea early in 50 B.C. Speaking of his daughter Tullia's suitors, Cicero wrote: *Quare adiunges Saufeium nostrum, hominem semper amantem mei, nunc, credo, eo magis, quod debet etiam fratris Appi amorem erga me cum reliqua hereditate crevisse; qui declaravit, quanti me faceret, cum saepe tum in Bursa.* From this one may deduce that *Saufeius noster* is Lucius Saufeius, and that Lucius had a brother Appius who had died not long before the letter was written, that is ca. 51 B.C. Appius Saufeius is said to have been particularly fond of Cicero on account of the Bursa affair. It seems reasonable to assume that Bursa was Titus Munatius Plancus Bursa who had been active in the Milo trial as one of the tribunes and who was shortly afterwards, early in 51 B.C., accused by Cicero (*de vi*) and convicted. In order to understand why Appius Saufeius should have been so delighted by Bursa's conviction and subsequent exile, one may make reference to a passage in Asconius (*In Milonianam*, 48–49 Clark) recording the trial and acquittal of one M. Saufeius *M. f. qui dux fuerat in expugnanda taberna Bovillis et Clodio occidendo*. Saufeius was defended by Cicero and by M. Caelius, and F. Münzer has called attention (*R.E.*, s.v. Saufeius no. 6) to an inscription from Tusculum, Caelius's hometown (?) (*C.I.L.*, XIV, 2624):

> Caelia P. f. municipio suo
> donum dedit imaginem
> L. Saufei
> Ap. f. ex se natei.

This Lucius Saufeius, whose statue was erected in Tusculum, was undoubtedly Cicero's friend and the distinguished Epicurean who wrote on the history of civilization.[17] This is clearly shown by the name of Lucius's father, Appius, a praenomen which occurs in Republican times almost exclusively in the families of the Claudii and Saufeii.[18] If it were possible to change in Asconius's text M. Saufeius M. f. to Ap. Saufeius Ap. f., one could assume that Cicero defended Appius Saufeius and secured his acquittal.

[101]

The statue base on the Akropolis (*I.G.*, II², 3897) accordingly supported

---

16. The evidence has been conveniently assembled by F. Münzer, *R.E.*, s.v. Saufeius no. 5; see also N. W. Dewitt, *loc. cit.* (see note 1), pp. 34–35. Some members of the family did business in Delos; see J. Hatzfeld, *B.C.H.*, 36, 1912, pp. 74–75 (referring to *I. de Délos*, nos. 1754, lines 2 and 10; 1755, lines 8–9).

17. See F. Münzer, *Rh. Mus.*, 69, 1914, pp. 625–629; compare also G. Vlastos, *A.J.P.*, 67, 1946, p. 55, note 20. The place of birth of Caelius is discussed by R. G. Austin, *Pro M. Caelio*, pp. 111–112.

18. It may be suggested, therefore, that the Appius Saufeius mentioned by Pliny (*Nat. Hist.*, 7, 183) was either the father or the brother of Lucius.

three statues, one of Lucius Saufeius, erected by the people of Athens, one of Lucius's brother Appius, also erected by the people of Athens, and one of the philosopher Phaidros erected by his pupil Lucius Saufeius. It might be suspected that Appius Saufeius, too, erected a statue of Phaidros, and a substantial fragment of the base of this statue has been found in the Agora Excavations (Plate 25).

1. Inscribed base of Pentelic marble, found on June 1st, 1938, in late wall beneath the church, in Section II. Left side hacked off and reworked. On top, a rectangular cutting: Length, 0.27 m.; width, 0.22 m.; depth, 0.065 m. Front and right faces are smoothly dressed, back is carefully picked with a narrow smooth-dressed band at the top. On the front and right sides a shallow rebate along the bottom.
Height, 0.23 m.; width, 0.46 m.; thickness, 0.44 m.
Height of letters, 0.06 m.
Inv. no. I 5485.

["Ἄππιος Σωφήιος 'Ἀπ]πίο[υ]
[υἱὸς Φαῖδρον Λυσιά]δου
[Βερενι]κίδην τὸν ἑαυτοῦ
[καθηγ]ητὴν ταῖν θεαῖν.

The text of this inscription and its lettering agree in every respect with those of *I.G.*, II², 3897, but the Agora monument was evidently set up in the Eleusinion located near the place of the discovery of the stone. The cutting on top of the base shows that it received a pillar, and this means that Appius Saufeius erected a herm of the Epicurean philosopher Phaidros.

The most remarkable aspect of this herm dedication found in the Agora is the fact that it was dedicated to the Eleusinian deities and that it was probably set up in their sanctuary which was located between the Agora and the Akropolis. To erect the statue or herm of an ordinary mortal in the sanctuary of the Eleusinian goddesses would have been unusual, and I have found no other example, but to erect the herm | of an Epicurean philosopher in a sanctuary and to dedicate this herm to goddesses whose activity and perhaps very existence was questioned by the man thus honored, require some explanation.[19] Yet not only Cicero, but also Atticus was initiated in the Eleusinian Mysteries,[20] and Phaidros's own son (of whose philosophic beliefs we know nothing) was chosen πυθόχρηστος ἐξηγητής and his statue (or herm) and that of his sister Chrysothemis were erected in Eleusis and dedicated to the Eleusinian Goddesses (*I.G.*, II², 3513).[21] Titus Pomponius Atticus was initiated in the Mysteries (see

---

19. The only specific reference to Epicurus's attitude toward the Eleusinian deities which I was able to find is in Plutarch, *Adv. Coloten*, 22, p. 1119d = H. Usener, *Epicurea*, pp. 259–260, no. 392: cf. C. Jensen, *Ein neuer Brief Epikurs*, pp. 78–83.
20. See P. Graindor, *Athènes sous Auguste*, p. 138, note 3.
21. Line 15 should be restored Δήμητρι καὶ Κόρηι ἀν[έθηκεν].

above, note 20), and his biographer reports that statues of him and of his wife Pilia were erected in the most sacred places.[22] It is possible to identify one of these monuments, because lines 1–7 of *I.G.*, II^2, 3513 (Plate 25) may be restored as follows:

[Χρυσόθεμις Φαίδρου | Βε]ρενεικίδου
[θυγάτηρ Κόιντον Και | κ]ίλιον Πο̣υ̣-
[πωνιανὸν Ἀττικὸν τὸν | ἀ]κουστὴν
[τοῦ Φαίδρου τοῦ πατρ]|ὸς αὐτῆς
[τὸν ἑαυτῆς φίλον καὶ ε]||ὐεργέτην
[Δήμητρι καὶ Κόρηι ἀνέ]||θηκεν.

The restoration of the name of Atticus as Quintus Caecilius Pomponianus Atticus shows that the document belongs to the period after 58 B.C., when Atticus had been adopted by his uncle who made this provision in his will.[23] Atticus married Pilia in 56 B.C., and if the monument to which *I.G.*, II^2, 3513 belongs contained also a statue of Pilia, it must belong to the years following 56 B.C.

In 51 B.C., Lysiades was chosen archon of Athens, and after that year he served as member of the Areopagus. It seems likely, therefore, that Lysiades should have been πυθόχρηστος ἐξηγητής before he became archon. In that case, *I.G.*, II^2, 3513 should be dated ca. 55 B.C. The term ἀκουστής for pupil corresponds exactly to the Latin *auditor* used by Cicero, *De officiis*, 1, 26, 90.

The combination of literary and epigraphical evidence has once again produced satisfactory results. Several individuals known only from the writings of Cicero have become better known to us. Phaidros, the Epicurean philosopher, was an Athenian from a very distinguished family; he participated in the life of his city, and was beloved by his students. His son Lysiades well deserved to be put in a position of official responsibility by Marc Antony. He had been chosen as a religious interpreter and later as an archon; when Cicero attacked Antony for putting men like Lysiades | on the panel of judges, Lysiades was nearly sixty years of age and must have been a distinguished member of the Areopagus.

Among the Romans with whom we have become better acquainted, one may mention in the first place the two brothers Appius and Lucius Saufeius, sons of Appius Saufeius. They must have played a role in Athens similar to that of Atticus, for the people of Athens erected statues (or herms) of them at public expense. The inscriptions provided us with their full names, and we are thus in a position to combine certain other evidence. Appius Saufeius seems to have been involved in the murder of Clodius and owed his acquittal to Cicero. Yet he died only a year after the trial and left his brother Lucius as his heir. Lucius was more of a philosopher, but his fortune was almost confiscated in 43 B.C. It was his friend Atticus who intervened on his behalf.

---

22. Cornelius Nepos, *Atticus*, 3, 2.
23. See the address of Cicero, *Ad Atticum*, 3, 20.

Finally, we have now two documents mentioning Titus Pomponius Atticus. One is the base of a herm of Phaidros, Atticus's teacher, the other is the pedestal of a statue of Atticus himself. These identifications give added significance to the Attic inscriptions of the first century before Christ. Renewed study of the documents of this period will undoubtedly produce identifications which will ultimately add to our knowledge of the period.

46

Zu einigen Wiederholungen bei Lukrez

Die versus iterati im Gedicht des Lukrez wurden vor kurzem in einer Leipziger *[218]*
Dissertation von Christoph Lenz (*Die wiederholten Verse bei Lukrez*, 1937) eingehend behandelt und ein Vergleich mit meiner Zusammenstellung zeigt mir, daß Lenz das Material erschöpfend behandelt hat.[1]

Die Disposition, an die sich Lenz hielt (15f.) und der Umstand, daß er Quellenfragen im allgemeinen (62f.) nicht in den Kreis seiner Betrachtungen zog, bringen es mit sich, daß er die Wiederholungen nicht inhaltlich ordnete. Deshalb möchte ich hier ganz kurz auf eine Reihe von Wiederholungen hinweisen, in denen Lukrez Gedanken von Epikur wiedergibt, die oft Wort für Wort mit Stellen in den erhaltenen Schriften Epikurs übereinstimmen.

1

Zu der Wiederholung 1, 146ff. = 2, 55ff.; 3, 87ff.; 6, 35ff. (Lenz, 40ff.) sei auf Epikur *r.s.* 12 (*ratae sententiae*) und Hb. §78f. (*Herodotbrief*) hingewiesen (auch die Wiederholung der Anfangsverse dieser Iteration bei Seneca, *ep.* 110, 6 ist für die Bedeutung dieser Verse kennzeichnend).

2

Nicht aufgenommen wurde von Lenz die Wiederholung: 1, 150 = 156f., 205, 237, 248, 262, 265f., 543f., die dadurch an Interesse gewinnt, daß diese eindrucksvolle Formulierung des ersten Hauptsatzes der epikurischen Physik

---

1. Außer einer, gleich zu erwähnenden Iteration, die Lenz nicht berücksichtigt hat, fiel mir unter anderem das Fehlen folgender Stellen auf: 2, 998 = 5, 795, 821 (s. Kranz, *Herm.* 64, 499); 4, 189 = 5, 283; 5, 67f. = 416f. und 76 = 774 (Zeichen strenger Komposition des V Buches; s.u.S.222); 5, 866 = 6, 1245 (im VI Buch ein dazugeschriebener Vers, der das Elend der Pestkranken charakterisieren soll).

eine wörtliche Übertragung aus dem griechischen Original ist (*Hb.* § 38). Es entsprechen: *nulla res* = οὐδέν, *gigni, creari, fieri* = γίγνεται, *e nihilo, de nihilo* = ἐκ τοῦ μὴ ὄντος, *reverti, redit, pereunt, revocari* = ἐφθείρετο, *ad nihilum* = εἰς τὸ μὴ ὄν. Erwähnt sei auch der Hinweis auf fr. B 12 des Empedokles (Diels), das bei Philo mit den Worten Epikurs eingekleidet wird.

[219]

3

Ebenso liegt in der Wiederholung 1, 510 = 538, 548, 574, 609, 612; 2, 157, die als Inhalt eine "feste Formulierung einer grundlegenden Aussage über die *primordia*" (Lenz, S. 31) enthält, eine genaue Übertragung eines epikurischen Satzes (Hb. § 41) vor. Es entsprechen: *corpora prima* = σώματα, *solida* = πλήρη, *simplex* = ἄτομος.

4

Zu Vers 3, 519f. (= 1, 670/4, 757, 790/3, 797; 2, 753/6, 864) hat schon Heinze im Kommentar bemerkt, daß hier "ein Hauptsatz der epikurischen Physik" vorliegt, doch glaubte er, die originale Sentenz sei nicht erhalten (Lenz, 46; Anm. 85). Bei genauerem Zusehen findet man aber folgende Entsprechungen:

1, 672: *proinde aliquid superare necesse est incolume ollis* = Hb. § 54: ἐπειδή περ δεῖ ὑπομένειν (ἐν ταῖς διαλύσεσι τῶν συγκρίσεων) στερεὸν καὶ ἀδιάλυτον.

1, 673: *ne tibi res redeant ad nihilum funditus omnes* = Hb. § 41: μὴ μέλλει πάντα ἐς τὸ μὴ ὂν φθαρήσεσθαι.

1, 673: *ne tibi res redeant ad nihilum...de nihilo renata...* = Hb. § 54: ὁ τὰς μεταβολὰς οὐκ εἰς τὸ μὴ ὂν ποιήσεται οὐθ' ἐκ τοῦ μὴ ὄντος....

5

Kurz sei auf die Wiederholung von 2, 336ff. in 2, 692ff., 723f. hingewiesen, die Lenz nur in der Statistik (9) anführt, die aber auch einen epikurischen Gedanken enthält (*Hb.* § 42). Es entsprechen:

2, 722: *dissimili figura* = διαφοραῖς σχημάτων.

2, 724: *omnia omnibus non paria constant* = καθ' ἑκάστην σχημάτισιν ἁπλῶς ἄπειροί εἰσιν αἱ ἄτομοι.

6

Ein schönes Beispiel, das diese ganze Gruppe von Wiederholungen charakterisiert, ist 2, 1128 = 4, 860. Lenz (33) weist ausdrücklich darauf hin, daß die Formulierung: *fluere atque recedere* sonst nur noch 4, 695 vorkommt.
[220] Der von | Lukrez ausgesprochene Gedanke findet sich nun auch bei Epikur (*Hb.* § 48) und es entsprechen: fluere = ῥεῦσις, recedere = ἀνταναπλήρωσις. Auf die

Analyse des Geruchsinnes geht Epikur nicht ausführlich genug ein (§ 53), um, wie Lukrez 4, 695, dieselben Worte wie anläßlich des Gesichtsinnes zu gebrauchen.

7

An das Ende dieser Reihe, die noch vervollständigt werden könnte, möchte ich eine Wiederholung stellen, deren Inhalt besonderes Interesse verdient: 1, 817ff. = 907ff.; 2, 688ff., 1007ff. Ehe ich auf das entsprechende griechische Vorbild für diese Wiederholung eingehe, möchte ich doch der Meinung Ausdruck geben, daß Lenz dem Dichter Unrecht tut, wenn er (48) meint, Lukrez hätte, aus "Vorliebe für den Buchstabenvergleich," diesen "auch da angewendet, wo er nicht paßte und Unstimmigkeiten herbeiführte." Meiner Meinung nach spricht Lukrez an den vier Stellen folgende, von einander abweichende, Gedanken aus:

  I. So wie dieselben Atome verschiedene Dinge, so bilden dieselben Buchstaben verschiedene Worte.

  II. So wie es bei den Dingen auf die Anordnung der Atome ankommt, so bei den Worten auf die der Buchstaben.

  III. So wie die Atome, so sind such die Buchstaben untereinander verschieden.

  IV. Die Eigenschaften der Dinge werden durch die Atome selbst ebenso wenig beeinflußt, wie Form, Klang und Sinn der Worte durch die einzelnen Buchstaben.

Wenn Lenz (47), allerdings unter Hinweis auf Diels, meint, der Buchstabenvergleich sei "für Epikur nicht zu belegen," so kann ich ihm nicht so ohne weiteres zustimmen. Epikur spricht (*Hb.* § 48) davon daß die εἴδωλα bei der Loslösung von den Körpern θέσιν καὶ τάξιν bewahren und die gleichen Worte finden sich im Zusammenhang mit dem Buchstabengleichnis in dem Bericht über Demokrits Lehre bei Aristoteles (*Metaph.* 985b, 13ff.). Ein ähnliches Verhältnis, wie zwischen den Worten bei Epikur und Aristoteles (oder stammen die Ausdrücke θέσις καὶ τάξις von Demokrit selbst?) besteht zwischen zwei Stellen bei Lukrez und Laktanz (*Div. inst.* 3, 17, 24), der berichtet, Leukipp, Demokrit [221] und Epikur hätten die Verschiedenheit der Dinge *vario ordine ac positione* erklärt und dann zur Erläuterung auch den Buchstabenvergleich anführt. Wir können demnach folgende Entsprechungen feststellen:

Aristoteles (Demokrit)	σχῆμα	=	θέσις	=	τάξις
Epikur	σχῆμα	=	θέσις	=	τάξις
Lukrez	figura	=	positura	=	ordo (cum quibus)
Laktanz	—	=	positio	=	ordo

Daß das Buchstabengleichnis im *Hb.* nicht vorkommt, darf niemanden wundern, da sich der *Hb.* nicht an ein weiteres Publikum wendet und daher solche anschauliche Darstellungen im Interesse der Kürze vermeidet.

Lenz weist (47) mit Recht auf das gelungene Wortspiel: *ignes–lignum* (1, 912) hin, nur meine ich, daß diese Anwendung des Buchstabenvergleichs, an die

Lukrez 1, 871 noch nicht dachte, bereits im Vers I, 901 entstanden ist. Diese Stelle (1, 897ff.) ist überhaupt sehr interessant. Wir haben hier nämlich das unmittelbare Vorbild für jene Darstellung vom *Entstehen des Feuers* (5, 1096ff.), die Jelenko (*WS.* 54, 59ff.) als spätere Lage der Kulturgeschichte zu erweisen versucht hat. Es ist nun sicher kein Zufall, daß dieselbe Geschichte vom Feuer, das sich durch im Wind aneinandergeriebene Äste entzündet, auch bei Thukydides (2, 77, 4) zu lesen ist und daß die beiden Stellen fast Wort für Wort übereinstimmen: *in montibus* = ἐν ὄρεσι, *terantur* = τριθεῖσα, *donec flammai fulserunt flore coorto* = πῦρ καὶ φλόγα ἀπ' αὐτοῦ ἀνῆκεν.

Die Darstellung wird bei Lukrez mit "*inquis*" eingeleitet (ähnlich: 2, 931ff.; 3, 350ff., 533ff., 698ff., 894ff.; 4, 409; 5, 338ff., 1041ff., 1091ff. u.a.) und könnte daher als Stellungnahme zu einem von Anaxagoras angeführten Beispiel aufgefaßt werden. Der Hinweis auf Anaxagoras fr. A 89 und A 98 (Diels) genügt, um zu zeigen, daß solche Beispiele Anaxagoras zuzutrauen sind. Daß die Geschichte vom Entstehen des Feuers bei Thukydides vorkommt, darf als Argument für die Zurückführung dieser Stelle auf Anaxagoras angeführt werden; die Thuk.-Stelle gehört zu den wenigen Sätzen dieses Autors, die gleichsam zwischen Gedankenstrichen geschrieben sind, und eine kurze Durchsicht zeigte, daß sich noch zwei weitere Hinweise auf Anaxagoras in dem Werk des Thuk. finden. 2, 28 und 7, | 50, 4 sind mit Anaxagoras fr. A 42 und A 77 (Diels) zu vergleichen. Wenn Anaxagoras (fr. A 75) als erster das Verhältnis zwischen Erde, Sonne, Mond erkannt hat und wir bei Thuk. die Kenntnis dieser Erkenntnis voraussetzen müssen, so sind wir berechtigt, die Thuk.-Stellen auf Anaxagoras zurückzuführen.

Durch diese Zuweisung fällt einiges Licht auf jene, von Jelenko festgestellte, zweite Lage der Kulturgeschichte des Lukrez.

Auch hier kann uns die Wiederholung einer größeren Anzahl von Versen weiterführen: 3, 784ff. = 5, 128ff. Nach den Ausführungen von Lenz (54–60) kann die Priorität der Verse im V Buch als erwiesen gelten. Da sich Lenz aber mit der Frage nach der Ursache der Wiederholung nicht weiter beschäftigt, so kann darüber einiges gesagt werden. Wenn Lukrez das III Buch nicht später eingeschoben hätte, so wäre die Frage nach der ψυχή im V Buch behandelt worden. Schon 1, 112ff. und 130ff. wird die Problemstellung angedeutet (Stellen, die sicher vor Abfassung des III Buches geschrieben sind; s. Lenz, 25ff.), dann folgt, entsprechend der Disposition des *Hb.* (wohl auch der Schrift περὶ φύσεως), nach der Darstellung der εἴδωλα–Lehre, also im V Buch, die Behandlung des Seelenproblems. Jelenko wies schon (*WS.* 54, 67f.) auf das Verhältnis zwischen Ankündigung des Themas im Prooemium und seiner Ausführung hin (es sei hier auf die von Lenz nicht aufgenommene Wiederholung 5, 67f. = 416f.; 76 = 774 hingewiesen) und wir finden nun auch im Prooemium des V Buches Vers 59ff. die Ankündigung der Behandlung des Seelenproblems, dessen Durchführung wir dann vermissen. Ein Rest der Darstellung ist uns in 5, 128ff. erhalten, wo zwei Gedanken ausgesprochen werden:

1. Die Seele ist an den Körper gebunden und mit ihm sterblich.

II. Die körperlichen Gestirne, deren Materie ähnlich der von Erde, Feuer, Wasser oder Luft (142ff.) ist, sind leblos und können nicht göttlich sein. Da der zweite Gedanke in der Darstellung des V Buches nicht fehlen durfte, so blieb die ganze Stelle dort stehen, wurde aber ins III Buch mit manchem anderen übernommen (erhalten blieb aus den gleichen Gründen 5, 351ff. = 3, 806ff.; s. Lenz, 58ff.).

An Stelle dieser Partien über die Seele, die das V Buch an das neu eingeschobene III Buch abgab, wurde von Lukrez die zweite Lage der Kulturgeschichte geschrieben.

1.(7)A [= No. 78 Fragment a, p. 290]

2.(7)B [= No. 7B Fragment b, p. 291]

3.(7)C [= No. 7B Fragment c, p. 292]

4.(7)D [= No. 7B Fragment d, p. 293]

5.(16)A [= Figure 1, p. 93]
Inscription from Attica
(Getty Museum, 78.AA.377).

6.(16)B [= Figure 2, pp. 94]
Inscription in 5.(16)A under different lighting.

7.(18)A [= Tafel 33, 3]   8.(18)B [= Tafel 33, 4]

9.(25) [= Figure 1, p. 162]
The Leagros Cup in Baltimore.

10.(26)A [= Figure 2A]
Back of *I.G.*, I², 763.

11.(26)B [= Figure 2B]
Top of *I.G.*, I², 763.

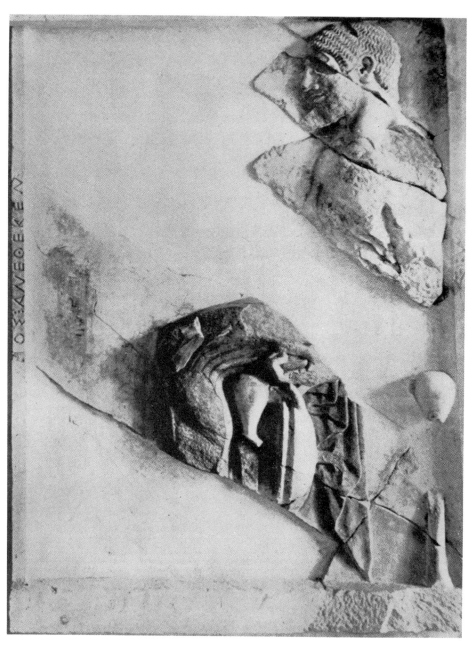

12.(27)A [= Figure 1, p. 246]
The Potter Relief.

13.(27)B [= Figure 4, p. 248]
Fragment of Potter Relief.

14.(27)C [= Figure 6, p. 248]
Top: I.G., I², 978. *Bottom*: New fragment of Potter Relief.

15.(27)D [= Figure 5, p. 249]
Dedicatory inscription of Potter Relief.

16.(27)E [= Figure 7, p. 248]
End of signature of Potter Relief.

17.(29)A [= Plate 86, a]
I.G., II², 4676.

18.(29)B [= Plate 86, b]
I.G., II², 4992.

19.(29)C [= Plate 86, c]
I.G., II², 5029a.

20.(29)D [= Plate 86, d]
E.M. 3913.

21.(30)A [= Plate 15, a]
Volute-crater (Stanford University Museum of Art, 70.12: Side A).

22.(30)B [= Plate 15, b]
Detail of Side A and figure of Pan.

23.(30)C [= Plate 16, a]
Volute-crater (Stanford University Museum of Art, 70.12: Side B).

24.(30)D [= Plate 16, b]
Figure of satyr.

25.(45) [= Plate 3]
Dedications by Phaidros and his Roman pupils.

I.G., II², 3899

I.G., II², 3897, 1–3

I.G., II², 3897, 4–6

I.G., II², 3897, 7–9

Agora I 5485

I.G., II², 3513, 1–7

Bibliography of A. E. Raubitschek

1935

"*Epikureische Untersuchungen.*" Dissertation, Wien, 1935.

1936

"Bericht über Zusammensetzungen archaischer Inschriftensteine von der Akropolis in Athen." *Anzeiger der Akademie der Wissenschaften in Wien* 73 (1936) 29–30.

1937

"Oiagros," "Oianthe," "Oinobios," "Oinophilos," "Oionokles." In *Paulys Realencyclopädie der classischen Altertumswissenchaft*, vol. 17.2 (1937).

1938

"Phainarete," "Phainippos," "Phanagores," "Phanas," "Phanion," "Phano," "Phanomachos," "Phanope," "Phanosthenes," "Phanostrate," "Phanostratos," "Pharnapates," "Pharnaspes," "Phayllos," "Pheidippos," "Philoktemon." In *Paulys Realencyclopädie der classischen Altertumswissenchaft*, vol. 19–2 (1938).

"Zur Technik und Form der altatischen Statuenbasen." *Bulletin de l'Institut Archéologique Bulgare* 12 (1938) 132–181.

"Zu einigen Wiederholungen bei Lukrez." *American Journal of Philology* 59:2 (1938) 218–223.

"Zu altattischen Weihinschriften." *Jahreshefte des Oesterreichischen Archäologischen Instituts in Wien* 31 (1938) 21–68.

1939

"Onetor." In *Paulys Realencyclopädie der classischen Altertumswissenchaft*, vol. 18 (1939).

"*Erga megala te kai thomasta.*" *Revue des Études Anciennes* 41:3 (1939) 217:222.

"Leagros." *Hesperia* 8 (1939) 155–164.

"Early Attic Votive Monuments." *Annual of the British School at Athens* 40 (1939) 17–37.

Review of E. Loewy, *Zur Datierung attischer Inschriften,* and *Der Beginn der rotfigurigen Vasenmalerei. American Journal of Archaeology* 43:4 (1939) 710–713.

1940

"The Inscription on the Base of the Athena Promachos Statue." *American Journal of Archaeology* 44:1 (1940) 109.

"Two Monuments erected after the Victory of Marathon." *American Journal of Archaeology* 44:1 (1940) 53–59.

"Some Notes on Early Attic Stoichedon Inscription." *Journal of Hellenic Studies* 60 (1940) 50–59.

"A New Fragment of A.T.L., D8." *American Journal of Philology* 61:4 (1940) 475–479.

1941

"Philon," "Philonides," "Philytas," "Phokides," "Phryne," "Phrygia," "Phrynichos," "Phrynis," "Phrynoi." In *Paulys Realencyclopädie der classischen Altertumswissenchaft*, vol. 20. 1 (1941).

"Note on a Study of the Acropolis Dedications." *American Journal of Archaeology* 45:1 (1941) 70.

"The Heroes of Phyle." *Hesperia* 10:3 (1941) 284–295.

"A Possible Signature of Kalamis." *American Journal of Archaeology* 45:1 (1941) 90.

"Two Notes on Isocrates." *Transactions of the American Philological Association* 72 (1941) 356–364.

1942

"Orthobulos," "Oulios." In *Paulys Realencyclopädie der classischen Altertumswissenchaft*, vol. 18.2 (1942).

"Notes on Attic Prosopography." *Hesperia* 11:3 (1942) 304–313.

"The Potter Relief from the Akropolis." *American Journal of Archaeology* 46:1 (1942) 123.

"An Original Work of Endoios." *American Journal of Archaeology* 46:2 (1942) 245–253.

"*I. G.* II^2, 2839 and 2844." *Classical Philology* 37:3 (1942) 317–319.

Review of W. K. Pritchett and B. D. Meritt, *The Chronology of Hellenistic Athens*. *American Journal of Archaeology* 46:4 (1942) 574–575.

1943

"Greek Inscriptions." *Hesperia* 12 (1943) 12–96.

"The Inscriptions." In *Small Objects from the Pnyx* I, ed. G. R. Davidson and D. B. Thompson, *Hesperia: Supplement* 7 (1943) 1–11.

"Heinrich Gomperz." *American Journal of Archaeology* 47:2 (1943) 227–228.

1944

"A Note on *I.G.* I^2, 945." *Hesperia* 13:4 (1944) 352.

"Athens and Halikyai." *Transactions of the American Philological Association* 75 (1944) 10–14.

Review of L. Pearson, *The Local Historians of Attica*. *American Journal of Philology* 45:3 (1944) 294–297.

Review of W. A. McDonald, *The Political Meeting Places of the Greeks*. *American Historical Review* 49:3 (1944) 692–694.

Review of C. E. Black, *The Establishment of Constitutional Government in Bulgaria*. *Russian Review* 4:1 (1944) 118–120.

1945

"A Greek Folksong Copied for Lord Byron." *Hesperia* 14:1 (1945) 33–35 (with C. M. Dawson).

"*Kynebion-Kys.*" *Record of the Museum of Historic Art*, Princeton University 4:1 (1945) 9–10.

"Hadrian as Son of Zeus Eleutherios." *American Journal of Archaeology* 49:2 (1945) 128–133.

"Two Notes on Athenian Epigrams." *Hesperia* 14:4 (1945) 367–368.

"The Pyloroi of the Akropolis." *Transactions of the American Philological Association* 76 (1945) 104–107.

"The Priestess of Pandrosos." *American Journal of Archaeology* 49:4 (1945) 434–435.

Review of S. D. Markman, *The Horse in Greek Art*. *American Journal of Philology* 46:2 (1945) 220–222.

Review of B. Newman, *Balkan Background*. *The Philhellene* 4:4–5 (1945) 4.

Review of A. W. Parsons, *Klepsydra and the Paved Court of the Pythion*. *American Journal of Archaeology* 49:2 (1945) 187–189.

Review of *Transactions and Proceedings of the American Philological Association* 74 (1943). *American Journal of Philology* 46:3 (1945) 330–333.

1946

Hesperia: Index, Volumes I-X, Supplements I-IV: Epigraphical Indexes. American School of Classical Studies at Athens 1946.

"The Pedestal of the Athena Promachos." *Hesperia* 15:2 (1946) 107–114.

"Octavia's Deification in Athens." *Transactions of the American Philological Association* 77 (1946) 146–150.

Review of H. Bloesch, *Agalma, Kleinod, Weihgeschenk, Götterbild: Ein Beitrag zur frühgriechischen Kultur- und Religionsgeschichte*. *American Journal of Archaeology* 50:1 (1946) 196–197.

Review of E. Langlotz, *Die Darstellung des Menschen in der griechischen Kunst*. *American Journal of Archaeology* 50:1 (1946) 197–198.

Review of E. Langlotz, *Uber das Interpretieren griechischer Plastik*. *American Journal of Archaeology* 50:1 (1946) 198.

Review of E. Langlotz, *Griechische Klassik*. *American Journal of Archaeology* 50:1 (1946) 199.

1947

"The Ostracism of Xanthippos." *American Journal of Archaeology* 51:3 (1947) 257–262.

"Three Attic Proxeny Decrees." *Hesperia* 16:2 (1947) 78–81 (with C. P. Loughran).

"Early Christian Epitaphs from Athens." *Hesperia* 16:1 (1947) 1–54 (with J. S. Creaghan).

"Jean Hatzfeld." *American Journal of Archaeology* 51:3 (1947) 305.

Early Christian Epitaphs from Athens. Woodstock, M. 1947 (with J. S. Creaghan).

1948

"The Case against Alcibiades (Andocides IV)." *Transactions of the American Philological Association* 79 (1948) 191–210.

"Sophocles of Sunion." *Jahresheft des Oesterreichischen Archäologischen Instituts in Wien* 37 (1948) 35–40.

"Ostracism." *Archaeology* 1:1 (1948) 79–82.

Selected Works of Cicero. New York, 1948 (with I. K. Raubitschek and L. R. Loomis).

Review of A. P. Dorjahn, *Political Forgiveness in Old Athens: The Amnesty of 403 B.C. American Journal of Philology* 69:1 (1948) 126–127.

Review of E. Buschor, *Vom Sinn der griechischen Standbilder. American Journal of Archaeology* 52:2 (1948) 414–415.

1949

Supplementum Epigraphicum Graecum, vol. 10 ed. A. E. Raubitschek and J. J. E. Hondius, Lugduni Batavorum, 1949.

"Commodus and Athens." In *Commemorative Studies in Honor of Theodore Leslie Shear, Hesperia*: Supplement 8 (1949) 279–290.

"Phaidros and his Roman Pupils." *Hesperia* 18:1 (1949) 96–103.

Dedications from the Athenian Akropolis. Cambridge, Mass., 1949 (with L. H. Jeffery). Pages 433–453 reprinted in German translation as "Einige technische Bemerkungen zu den frühen attischen Weihungen." In *Das Alphabet*, Wege der Forschung 88, 339–434. Darmstadt 1968.

Review of H. A. Bauer, *Das antike Athen in zwanzig Farbaufnahmen. American Journal of Archaeology* 53: 1–2 (1949) 84 and 219–220.

Review of H. Bogner, *Der tragische Gegensatz. Classical Philology* 44:2 (1949) 131–132.

Review of F. R. Cowell, *Cicero and the Roman Republic. American Historical Review* 55:1 (1949) 105–106.

1950

Excavations at Gözlü Kule, Tarsus, I: The Hellenistic and Roman Periods, The Inscriptions, 384–387. Princeton, 1950.

"Toynbee and the Classics." *Classical Weekly* 43:7 (1950) 99–100.

"Another Drachma Dedication." *Yale Classical Studies* 11 (1950) 295–296.

"The Crisis of the Athenian Democracy." *Transactions of the American Philological Association* 80 (1950) 434–435.

"The Origin of Ostracism." *American Journal of Archaeology* 54:3 (1950) 258–259.

Review of A. Lesky, *Thalatta: Der Weg der Griechen zum Meer. American Journal of Archaeology* 55:1 (1950) 92–93.

Review of J. A. Notopoulos, *The Platonism of Shelley. Thought* 25 (1950) 253–255.

Review of G. M. A. Richter, *Archaic Greek Sculpture. Saturday Review of Literature* 23:1 (1950) 37–38.

Review of W. H. D. Rouse, *Homer, The Iliad and the Odyssey. Princeton Alumni Weekly* 50 (1950), no. 28 (*Good Reading* 1:2).

1951

"The Mechanical Engraving of Circular Letters." *American Journal of Archaeology* 55:3 (1951) 343–344. Reprinted in *Festschrift A. Rumpf zum 60. Geburtstag*, 125–126. Krefeld 1951.

"Sylleia." In *Studies in Honor of A. C. Johnson*, 49–57. Princeton, 1951.

"The Origin of Ostracism." *American Journal of Archaeology* 55:2 (1951) 221–229.

"The Key to the Classical Tradition." *Vanderbilt Alumnus* 37:2 (1951) 12–16.

Review of C. Bonner, *Studies in Magical Amulets*. *American Journal of Archaeology* 55:4 (1951) 419–420.

Review of F. Jacoby, *Atthis: The Local Chronicles of Ancient Athens*. *Classical Weekly* 44:9 (1951) 135.

Review of J. H. Oliver, *The Athenian Expounders of the Sacred and Ancestral Law*. *Classical Weekly* 44:9 (1951) 135–136.

Review of G. M. A. Richter, *Archaic Greek Sculpture*. *Princeton Alumni Weekly* 52 (1951), no. 8 (*Good Reading* 3:1).

Review of K. Schefold, *Orient, Hellas, und Rom in die archäologischen Forschung seit 1939*. *Classical Philology* 46:1 (1951) 62–63.

Review of W. Schmalenbach, *Griechische Vasenbilder*, and K. Schefold, *Griechische Plastik, I: Die grossen Bildhauer des archaiischen Athen*. *American Journal of Philology* 72:2 (1951) 213–14.

Review of F. Solmsen, *Hesiod and Aeschylus*. *Classical Weekly* 45:5 (1951), 70–72.

1952

"Plato's College." *Classical Weekly* 45:13 (1952) 193–196.

"International Epigraphy: Report on the Second International Congress of Greek and Latin Epigraphy, Paris, April 15–19, 1952." *Archaeology* 5:2 (1952) 119–120.

"When History was Young," *Classical Bulletin* 28:5 (1952) 49–52, 62–65, 68.

Review of A. Calderini, *L'Ostracismo*. *Classical Philology* 47 (1952) 203–204.

Review of D. Grene, *Man in His Pride: A Study in the Political Philosophy of Thucydides and Plato*. *Classical Weekly* 46:3 (1952) 40.

Review of B. D. Meritt, H. T. Wade–Gery, and M. F. McGregor, *The Athenian Tribute Lists*. *Classical Weekly* 45:15 (1952), 230–231.

Review of W. T. Stace, *Religion and the Modern Mind*. *Daily Princetonian* 76 (1952), no. 138.

1953

"Notes on the Post-Hadrianic Boule," In *Geras Antoniou Keramopoullou*, 242–255. Athens, 1953.

"Two Notes on the Fasti of Achaia." In *Studies Presented to D. M. Robinson*, 2: 330–333. St. Louis 1953.

"Ostracism: The Athenian Ostraca." In *Actes du 2e Congrès International d'Épigraphie Grecque*, 59–74. Paris, 1953.

"Education and the Classics: Unaided Reason." *Folia* 7:2 (1953) 86–107.

"Athenian Ostracism." *Classical Journal* 48:4 (1953) 113–122.

Review of H. R. Breitenbach, *Historiographische Anschauungsformen Xenophons.* *Classical Philology* 48:1 (1953) 36–37.

Review of E. Buschor, *Frühgriechische Jünglinge.* *Archaeology* 6:3 (1953) 187–188.

Review of M. Grant, *Roman Anniversary Issues: An Exploratory Study of the Numismatic and Medallic Commemoration of Anniversary Years, 49B.C.–A.D. 375.* *Classical Journal* 48:6 (1953) 271–272.

Review of R. Hampe, *Die Gleichnisse Homers und die Bildkunst seiner Zeit.* *Archaeology* 6:1 (1953) 62.

Review of *Studies Presented to D. M. Robinson,* ed. G. E. Mylonas. *Archaeology* 6:4 (1953) 254.

Review of C. J. Radcliffe, *The Problem of Power.* *Princeton Alumni Weekly* 53 (1953), no. 26 (*Good Reading* 4:3).

1954

"The New Homer." *Hesperia* 23:4 (1954) 317–319.

"Philinos," *Hesperia* 23:4 (1954) 68–71.

"Epigraphical Notes on Julius Caesar." *Journal of Roman Studies* 44 (1954) 65–75.

"The Dates of Caesar's Second and Third Dictatorship." *American Journal of Archaeology* 58.2 (1954) 148.

"Der Ostrakismos des Theseus." *Bulletin d'archélogie et d'histoire Dalmate* 56–59:2 (1954) 50–51 (Mélanges Abramiĉ II).

"The Classics and the Bible." *Classical Bulletin* 30:4 (1954) 37–39.

"Values and Value Judgement." *Princeton Alumni Weekly* 54 (1954), no. 18.

Euripides' *Trojan Women* (translation). In *Anthology of Greek Drama,* 199–239. New York 1954.

Review of J. A. Oliver, *The Ruling Power: A Study of the Roman Empire in the Second Century after Christ through the Roman Oration of Aelius Aristides.* *Phoenix* 8 (1954) 165–166.

Review of M. Rambaud, *L'Art de la déformation historique dans les Commentaires de César.* *Latomus* 13 (1954) 615–616.

Review of A. Rumpf, *Archäologie.* *Archaeology* 7:3 (1954) 187.

Review of A. Schmitt, *Der Buchstabe H im Grieschischen.* *Gnomon* 26:2 (1954) 121–122.

Review of W. Steidle, *Sueton und die antike Biographie.* *Classical Philology* 49:1 (1954) 62–63.

Review of A. E. Taylor, *Socrates.* *Princeton Alumni Weekly* 54 (1954), no. 18 (*Good Reading* 5:2).

1955

"Zur attischen Genealogie." *Rheinisches Museum* 98:3 (1955) 258–262.

"Damon." *Classica et Mediaevalia* 16: 1–2 (1955) 78–83.

"Philochoros *Frag.* 30 (Jacoby)." *Hermes* 83:1 (1955) 119–120.

"Theopompos on Hyperbolos." *Phoenix* 9:3 (1955) 122–126.

"Gyges in Herodotus." *Classical Weekly* 48:4 (1955) 48–50.

"Menon, Son of Menekleides." *Hesperia* 24:4 (1955) 286–289.

"Aeschylus: *The Oresteia*." *Alumnae Bulletin*, Randolf-Macon Woman's College 49:1 (1955) 21–33.

Review of N. DeWitt, *St. Paul and Epicurus*. *Princeton Alumni Weekly* 56 (1955), no. 8 (*Good Reading* 7:1).

Review of *The Acts of the Pagan Martyrs, Acta Alexandrinorum*, ed. H. A. Musurillo. *Thought* 30 (1955), 446–447.

Review of *Heinrich Gomperz, Philosophical Studies*, ed, D. S. Robinson. *Classical Weekly* 48:13 (1955), 184.

1956

"The Gates in the Agora." *American Journal of Archaeology* 60:3 (1956) 279–282.

"(H)abronichos." *Classical Review* 6:3–4 (1956) 199–200.

Review of *Miss Mabel's Oresteia*. *Alumnae Bulletin*, Randolf-Macon Woman's College 49:2 (1956) 14.

Review of A. Aymard and J. Auboyer, *L'Orient et la Grèce antique*. *Erasmus* 9:9–10 (1956) coll. 302–303.

Review of J. Pieper, *Justice*. *Princeton Alumni Weekly* 56 (1956), no. 3 (*Good Reading* 7:3).

1957

"Die Verstossung des Themistokles," *Hermes* 84:4 (1957) 500–501.

"Brutus in Athens." *Phoenix* 11:1 (1957) 1–11.

"Das Datislied." *Charites E. Langlotz: Studien zur Altertumswissenchaft* 234–242. Bonn, 1957.

"Die schamlose Ehefrau (Herodot 1, 8, 3)." *Rheinisches Museum* 100:2 (1957) 139–140.

Review of Z. Gansiniec, *Geneza tropaionu*. *Archaeology* 10:4 (1957) 295–296.

Review of G. Klaffenbach, *Griechische Epigraphik*. *Deutsche Literaturzeitung* 78:8 (1957) coll. 685–689.

Review of C. S. Lewis, *Till We Have Faces: A Myth Retold*. *Princeton Alumni Weekly* 57 (1957), no. 26 (*Good Reading* 8:3).

Review of A. Rumpf, *Archäologie* II. *Archaeology* 10:4 (1957) 292–293.

1958

"Meeresnähe und Volksherrschaft." *Wiener Studien* 71 (1958) 112–115.

"Theophrastos on Ostracism." *Classica et Mediaevalia* 19:1–2 (1958) 73–109.

"Ein neues Pittakeion." *Wiener Studien* 71 (1958) 170–172.

Review of *The Christian Idea of Education*, ed. E. Fuller. *Princeton Alumni Weekly* 59 (1958), no, 9 (*Good Reading* 10:1).

Review of M. P. Nilsson, *Die hellenistische Schule*. *Phoenix* 12:1 (1958) 40–42.

Review of *Ovid's Art of Love*. *Princeton Alumni Weekly* 58 (1958), no. 26 (*Good Reading* 9:3).

1959

"The Brutus Statue in Athens." *Atti del III Congresso Internazionale di Epigrafia Greca e Latina*, 15–21. Rome, 1959.

"Arae August." *Hesperia* 28:1 (1959) 65–85 (with Anna Benjamin).
"A Note on the Inscription on the Plaster Cast." *Record of the Art Museum, Princeton University* 18:2 (1959) 60.
"Die Rükkehr des Aristeides." *Historia* 8 (1959) 127–128.
Review of F. E. Adcock, *Roman Political Ideas and Practice*. *Classical World* 52:8 (1959) 256.
Review of *Monumenta Asiae Minoris Antiqua* VII, ed. W. M. Calder (II). *Journal of Biblical Literature* 78:4 (1959) 359–361.
Review of R. Flacelière, *Inscriptions de la terrasse du temple et de la région nord du sanctuaire* (Fouilles de Delphes 3, 4), and J. Marcadé, *Recueil des signatures de sculpteurs grecs*, I. *Gnomon* 31:3 (1959) 266–268.
Review of *C. Julii Caesaris Commentarii De bello civili*, ed. F. Kraner and F. Hoffmann. *Classical World* 53:2 (1959) 59 (with I. K. Raubitschek).
Review of N. A. Maschkin, *Zwischen Republik und Kaiserreich: Ursprung und sozialer Character des Augusteischen Prinzipats*, transl. M. Brandt. *Erasmus* 12: 11–12 (1959), coll. 361–362.
Review of E. Mireaux, *Daily Life in the Time of Homer*, transl. I. Sells. *Classical World* 53:1 (1959) 13–14.
Review of H. W. Pleket, *The Greek Inscriptions in the "Rijksmuseum van Ouheden" at Leyden*. *American Journal of Archaeology* 63:1 (1959) 99.
Review of F. Sartori, *La eterie nelle vita politica Ateniese del VI e V secolo A.C.* *American Journal of Philology* 80:1 (1959) 81–88.

1960

"Theopompos in Thucydides the Son of Melesias." *Phoenix* 14:2 (1960) 81–95.
"The Covenant of Plataea." *Transactions of the American Philological Association* 91 (1960) 178–183.
Review of S. P. Bovie, *Horace's Satires and Epistles*. *Yearbook of Comparative and General Literature* 9 (1960), 114–115.

1961

"Herodotus and the Inscriptions." *Bulletin of the Institute of Classical Studies, University of London* 8 (1961), 59–61.
Review of H. H. Scullard, *From the Gracchi to Nero: A History of Rome from 133 B.C. to A.D. 68*. *Erasmus* 14: 11–12 (1961), coll. 365–366.
Review of A. Severyns, *Grèce et Proche-Orient avant Homère*. *Classical World* 55:1 (1961) 16.

1962

"Demokratia. *Hesperia* 31 (1962) 238–243.
Review of L. H. Jeffery, *The Local Scripts of Archaic Greece*. *Gnomon* 34:3 (1962) 225–231. Reprinted in *Das Alphabet*. Wege der Forschung 88, 435–444, Darmstadt, 1968.
Review of H. Meyer, *Die Aussenpolitik des Augustus und die angusteische Dichtung*. *Classical World* 56:3 (1962) 86.

Review of F. Ollier, *Xenophon, Banquet, Apologie de Socrate*. *Classical World* 55:6 (1962) 170.

Review of J. N. Sevenster, *Paul and Seneca*. *Classical World* 55:8 (1962) 260.

Review of C. G. Starr, *The Origins of Greek Civilization 1100–650 B.C. Gnomon* 34:2 (1962) 201–202.

1963

"War Melos tributpflichtig?" *Historia* 12:1 (1963) 78–83.

"The Marble Prohedria in the Theater of Dionysus." *American Journal of Archaeology* 67:2 (1963) 216.

"Ernst Buschor." *American Journal of Archaeology* 67:4 (1963) 421.

Review of U. Albini, *Andocide, L'orazione De Reditu*. *American Journal of Philology* 84:3 (1963) 332–334.

Review of C. M. Eliot. *Coastal Demes of Attica: A Study in the Policy of Kleishenes*. *American Journal of Archaeology* 67:3 (1963) 313.

Review of A. Mannzmann, *Griechische Stiftungsurkunden: Studie'ze Inhalt und Rechtsformen*. *Classical Journal* 58:4 (1963) 185–186.

Review of A. Masaracchia, *Solone*. *Classical Philology* 58:2 (1963) 137–140.

Review of B. D. Meritt, *The Athenian Year*. *Phoenix* 17:2 (1963) 137–142.

1964

"Iamblichos at Athens." *Hesperia* 33:1 (1964) 63–68.

"Demokratia." In *Akte des IV. Internationalen Kongresses für Griechische und Lateinische Epigraphik*, 332–337 Wien, 1964.

"Die Inschrift als Denkmal: Bermerkungen zur Methodologie der Inschrift kunde." *Studium Generale* 17:4 (1964) 219–228.

"The Treaties between Persia and Athens." *Greek, Roman, and Byzantine Studies* 5:3 (1964) 151–159.

Review of E. Badian, *Studies in Greek and Roman History*. *Classical World* 58:2 (1964) 57.

Review of G. Gottlieb, *Das Verhältnis der ausserherodoteischen Überlieferung in Herodot*. *Gnomon* 36:8 (1964) 829–830.

Review of R. Syme, *Sallust*. *Classical World* 58:1 (1964) 21.

1965

"A Note on the Themistocles Decree." In *Studi in onore di Luisa Banti*, 285–287. Rome 1965.

"Die Inschrift als geschichtliches Denkmal." *Gymnasium* 72:6 (1965) 511–522.

"Hai Athenai tou Perikleous." *Epeteris tes philosophikes Scholes tou Panepistimou Athenon* (1965), 101–124.

Review of B. Snell, *Scenes from Greek Drama*. *Classical World* 58:8 (1965) 255–256.

1966

"The Peace Policy of Pericles." *American Journal of Archaeology* 70:1 (1966) 37–41.

"Otto Walter." *American Journal of Archaeology* 70:1 (1966) 74.

"Early Boeotian Potters." *Hesperia* 35:2 (1966) 154.

"Greek Inscriptions." *Hesperia* 35:3 (1966) 241–257.

"Zetemata Epigraphikes." *Epeteris tes philosophikes Scholes tou Panepistemiou Athenon* (1966) 148–170.

Review of G. E. Bean and T. B. Mitford, *Journeys in Rough Cilicia in 1962 and 1963*. *Anzeiger für die Altertumswissenschaft* 19:1 (1966) 70–71.

Review of L. Bieler, *The Grammarian's Craft: An Introduction to Textual Criticism*. *Thought* 61 (1966) 455–458.

Review of R. Hutmacher, *Das Ehrendekret für den Strategen Kallimachos*. *Gnomon* 38:8 (1966) 837–838.

Review of M. A. Levi, *Political Power in the Ancient World*. *American Historical Review* 72:1 (1966) 137–138.

Review of H. Montgommery, *Gedanke und Tat: Zur Erzählungstechnik bei Herodot, Thucydides, Xenophon, und Arrian*. *Classical World* 59:6 (1966) 197.

Review of *L. Annaei Senecae ad Lucilium Epistulae Morales*, ed. L. D. Reynolds. *Classical World* 59:9 (1966) 321.

Review of F. Semi, *Il Semento di Cesare*. *Classical World* 60:4 (1966) 166–167.

1967

Review of G. Pfohl, *Griechische Inschriften als Zeugnisse des privaten und öffentlichen Lebens*. *Anzeiger für die Altertumswissenschaft* 20:4 (1967) 222–223.

1968

"Prokrisis." *Palingenesia* 4 (1968) 89–90.

Review of *The Classical Tradition: Literary and Historical Studies in Honor of Harry Caplan*, ed. L. Wallach. *American Historical Review* 73:4 (1968) 1111–1112.

Review of R. M. Haywood, *Ancient Rome*. *Classical World* 62:1 (1968) 24.

1969

"Das Denkmal-Epigramm." In *L'Epigramme Grecque, Entretiens sur l'antiquité classique*, Fondation Hardt XIV, 32/36. Vandoevres-Genève, 1969. 19/3.

"Drei Ostraka in Heidelberg." *Archäologischer Anzeiger* (1969) 107–108.

"Die attische Zwölfgötter." In *Opus Nobile: Festschrift zum 60. Geburtstag von U. Jantzen*, 129–130. Wiesbaden, 1969.

Review of K. Buechner, *Sallustinterpretationen: In Auseinandersetzung mit dem Sallustbuch von Ronald Syme*. *Classical World* 62:6 (1969) 228.

1970

"IG II 2314 + 13122 + *Hesperia* Suppl. VII Nr. 8." *Klio* 52 (1970) 379–381.

"The Cretan Inscription MB 1969, 4-2,1: A Supplementary Note." *Kadmos* 9:2 (1970) 155–156.

1971

"Inschriften als Hilfsmittel der Geschichtsforschung." *Rivista Storica dell' Antichità* 1:1–2 (1971) 177–195.

Review of F. Eckstein, *Anathemata: Studien zu den Weihgeschenken strengen Stils im Heiligtum von Olympia. Erasmus* 23:6 (1971) 292–296.

1972

"A Late Byzantine Account of Ostracism." *American Journal of Philology* 93:1 (1972) 87–91 (with J. J. Keaney).

Early Cretan Armorers. Mainz, 1972 (with H. Hoffmann).

1973

"The Speech of the Athenians at Sparta." In *The Speeches in Thucydides*, ed. P. A. Stadter, 32–48. Chapel Hill, 1973.

"Die sogenannten Interpolation in den ersten beiden Büchern von Xenophons griechischer Geschichte." *Vestigia* 17 (1973) 315–325.

"Kolieis." In *Phoros: Tribute to B. D. Meritt*, ed. D. W. Bradeen and M. F. McGregor, 137–138. Locust Valley, N.Y., 1974.

"Der korinthische Sänger Pyrrhias." *Zeitschrift für Papyrologie und Epigraphik* 12:1 (1973) 99–100.

1974

"Eine Bermerkung zu Aristoteles, Verfassung von Athen 29, 2." *Chiron* 4 (1974) 101–102.

"Zu Periklesstatue des Kresilas." *Archeologia Classica* 25–26 (1974) 620–621 (Volume in onore di M. Guarducci).

"Zu den zwei attischen Marathondenkmälern in Delphi." In *Mélanges Helléniques offerts à Georges Daux*, 315–316. Paris, 1974.

1975

"Nomos and Ethos." In *Classica et Iberica: A Festschrift in Honor of J. M.-F. Marique.*

"Der kretische Gürtel." In *Wandlungen: Studien zur antiken und neueren Kunst, E. Homann-Wedeking gewidmet*, 49–52 Waldhassen-Bayern, 1975. (with Isabelle Raubitschek).

Review of E. F. Bloedow, *Alcibiades Reexamined. Classical World* 68:6 (1975) 391.

1976

"Plato and Minos." *Quaderni di Storia* 3 (1976) 233–238.

"Epilogue." In *La Paz de Calias* by Carlos Schrader, 215–217. Barcelona, 1976.

"Me quoque excellentior (Boethii Consolatio 4.6.27)." In *Latin Script and Letters A.D. 400–900: Festschrift Presented to L. Bieler*, 62. Leiden, 1976.

1977

"Bermerkungen zu den Buchstabenformen der griechischen Inscriften des fünften Jahrhunderts." In *Das Studium der griechischen Epigraphik*, ed. G. Pfohl, 62–72. Darmstadt, 1977.

"Bildhauerinschriften." In *Das Studium der griechischen Epigraphik*, ed. G. Pfohl, 116–120. Darmstadt, 1977.

"Corinth and Athens before the Peloponnesian War." In *Greece and the Eastern Mediterranean in Ancient and Prehistory: Studies Presented to F. Schachermeyer*, 266–269. Berlin, 1977.

Review of C. Mossé, *Athens in Decline 404-86 B.C.*, transl. J. Stewart. *Classical World* 70:5 (1977) 341.

1979

"Attic Black-Figure Eye-Cup (Nikosthenic Workshop)." In *Greek Vase-Painting in Midwestern Collections*, 88-89. Chicago, 1979.

1980

"Das Schwertband des Herakles." *Tainia: Roland Hampe zum 70. Geburtstag*, 65-67. Mainz, 1980.

"Zum Ursprung und Wesen der Agonistik." In *Studien zur antiken Sozialgeschichte: Feschrift F. Vittinghoff*, 1-5. Wien, 1980.

"Zu *Pap. Köln* 38 = Arch. Fr. 196A West." *Zeitschrift für Papyrologie und Epigraphik* 39 (1980) 48.

"Miscellanea: Theognis 313-314 and *Philoctetes* 1050-51." *Archaiognosia* 1:1 (1980) 188.

"Ein Peltast." *Grazer Beiträge* 9 (1980) 21-22.

1981

Inscriptiones Graecae: Inscriptiones Atticae Euciidis Anno Anteriores, vol. I, Editio Tertia, Fasc. I: *Decreta et Tabulae Magistratum*, ed. D. Lewis, with M. H. Jameson, A. E. Raubitschek, B. D. Meritt, M. F. McGregor, D. W. Bradeen, W. E. Thompson, A. G. Woodhead. Berlin, 1981.

"Andocides and Thucydides." In *Classical Contributions: Studies in Honor of M. F. McGregor*, 121-123. Locust Valley, N.Y., 1981.

"A New Attic Club (Eranos)." *J. Paul Getty Museum Journal* 9 (1981) 93-98.

1982

"The Dedication of Aristokrates." In *Studies in Attic Epigraphy, History and Topography Presented to Eugene Vanderpool, Hesperia: Supplement* 19 (1982) 130-132.

"The Mission of Triptolemos." In *Studies in Athenian Architecture, Sculpture and Topography Presented to Homer A. Thompson, Hesperia: Supplement* 20 (1982) 109-117 (with Isabelle Raubitschek).

1983

"The Agonistic Spirit in Greek Culture." *Ancient World* 7 (1983) 3-7.

1984

"Die historisch-politische Bedeutung des Parthenon und seines Skulpturenschmuckes." In *Parthenon-Kongress Basel: Referate und Berichte 4. bis 8. April 1982*, 19. Mainz, 1984.

"Zur Uberschrift von *IG* I^3 259." In *Studia in honorem C. M. Danvov*, Ann. de l'univ. de Sofia 77/2, 372-373. Sofia, 1984.

1985

"*Philokaloumen met' Euteleias'*, Praktika tou XII Diethnous Sunedriou Klasikēs Archaiologias, Athēna 4-10 Sept. 1983, 1: 248-249. Athens, 1985.

"Die Gründungsorakel der Dionysien." In *Pro Arte Antiqua. Festschrift für Hedwig Kenner*, 2: 300–301. Österreichische Archaeologische Institut Sonderschriften 18. Berlin, 1985.

"Zur Fruhgeschichte der Olympischen Spiele." In *Lebendige Altertumswissenschaft. Festgabe zur Vollendung des 70. Lebenjahres von Hermann Vetters*, 64–65. [A. Hakkert], 1985.

"Des Buch als Waffe," *Zaberndruck 200 Jahrfeier*, 212–214.

"Athen unter Perikles." In *Studien zur Alten Geschichte. Siegfried Lauffer zum 70. Geburtstag am 4. August 1981*, 2: 757–761. Rome, 1985.

1986

"Tyche zum Geleit." *Tyche. Beiträge zur Alten Geschichte Papyrologie und Epigraphik* 1 (1986) 1.

"Aristotles über den Ostrakismos." *Tyche. Beiträge zur Alten Geschichte Papyrologie und Epigraphik* 1 (1986) 169–174.

"Theseus at the Isthmia." In *Corinthiaca: Studies in honor of Darell A. Amyx*, 1–2. Columbia, M., 1986.

"A Note on the *Philoctetes* (1402)." In *Studies in honour of T. B. L. Webster*, 198–199. Bristol, 1986.

"The Eleusinian Spondai (*IG* I^3, 6, lines 8–47)." In *Philia Epē*, 2; 263–265. Bibliothêkê tês en Athênais Archaiologikês Hetaireias 103. Athens, 1986. (with Mariko Sakurai).

Review of T. Corsten, *Die Inschriften von Kios*, Inschriften griechischer Städte aus Kleinasien 29. *American Journal of Philology* 107:4 (1986) 601–602.

1988

"The Panhellenic idea and the Olympic Games." *Archaeology of the Ancient Olympics* (1988) 35–37.

"Homonoia kai Eirene dia ton Olymiakon Agonon." *Horos* 6 (1988) 9–12.

1989

"What the Greeks Thought of Their Early History." *Ancient World* 20 (1989) 39–45.

"Kultur und Fortschritt in der Blütezeit der griechischen Polis." *Klio* 71:1 (1989) 285–290.

1990

"Ideologische und methodologische Grundprobleme der antiken Geschichtsschreibung." *Klio* 72:2 (1990) 552–559.

Subject Index

Academy, 340
Acharnai, Attic deme, 42
Acheron River, 103
Achilles, representations on pottery, 268
Acropolis. *See* Akropolis
Adeimantus, 49
Aegean Sea, 6
Aegeis, tribe, 70
Aegina. *See* Aigina
Aegisthos, 123
Aegospotami, battle of. *See* Aigospotamoi, battle of
Aelian
   De Naturibus Animalium, 106
   Varia Historia, on ostracism, 78, 79, 80
   Varia Historia, 221
Aemilii, Lepidus and Juncus, 142
Aeschines, potter, 214n.6
Aeschines the orator, 38, 43, 36, 83, 89n.5, 98, 116, 160
   on decrees of Miltiades and Themistokles, 277
   as an historical source, 174
Aeschylus, 269
   Agamemnon, 46
   Eumenides, 269
   Persians, 151, 275
   Prometheus Bound, 268
Aetolian War, 114, 115. *See also* Dorian War
Agariste, daughter of Hippokrates, 111, 112
Agathokles, 97
Aglauros, priesthood of, 158
Agonothetes, 221
Agora, Athenian, 36, 38, 43, 55, 58, 60, 61, 78, 90, 92
Agoratos, son of Eumares, 39
Aigeis, Attic tribe, 7, 43
Aigeus, 150

Aigina, 23–25, 271, 273, 274, 313, 314, 318
   and the battle of Salamis, 276
Aigospotamoi, battle of, 73, 164, 299
Ailios Dionysios, on famine at Athens, 114
Aisimos, 41
Aiskhines the orator. *See* Aeschines the orator
Aithalidai, Attic deme, 158, 159
Aitolians, $Κοινόν$ of, 240
Akropolis
   dedications, 59, 60
   excavations on the North slope, 55. *See also* Metroön
   museum of, 274
   rebuilding of, 316
Al Mina. *See* Posideion-Al Mina
Alcmeonidae. *See* Alkmeonidae
Alcmeonides. *See* Alkmeonides
Alcmeonids. *See* Alkmeonidae
Alexander the Great, 174, 228
Alexander, son of Maro, 340
Alexandria, schools, 169. *See also* Education, Greek
Alexandrus, Claudius, 87
Alexias, Athenian archon 405/4, 299
Alexippides, 299
Alkibiades, 8, 86, 298
   accused of adultery, 121, 123–24
   and Damon, 333
   and Hyperbolos, 320, 321
   and Melos, 31
   and Nikias and Hyperbolos, 129
   and political clubs, 121, 160, 163, 164
   and Sicilian expedition, 117
   and $στάσις$, 163
   compared to Peisistratos, 118
   compared to Perikles, 291
   debate with Nikias. *See* Nikias: debate with Alkibiades

Subject Index

marriage of, 122–23
Olympic victories, 119, 117, 125
ostracism, 5, 103, 117, 125
relation to Megakles, 125
trial for impiety, 74n.29
Alkibiades, the Elder, 86
Alkman, use of ἀλιτηρός, 111n.12
Alkmeon, 5
Alkmeonidae, 4, 55, 61, 67n.6, 105, 111, 152, 271
attack on Leipsydrion, 162
and battle of Marathon, 111n.10
curse of, 114, 115. See also Kylon, conspiracy of
temple at Delphi, 273
Alkmeonides, 111n.11
Alopeke, Attic deme, 306, 314
Alphabets. See Writing
Altamura Painter, 234–25
Alyattes, 272
tomb of, 270, 272
Amasis, 46, 272
Ambrosios Painter, 230n.6
Ammon, oracle of, 18
Ammonios, on Thoukydides son of Melesias, 315
Amnesty
granted by Athens after Aigospotamoi. See Amnesty law of 405
granted by Athens before Salamis. See Amnesty law of 481
Solonian, 73, 73n.23
Amnesty law of 405, 75n.30
Amnesty law of 481, 28, 73, 75n.30
Amorges, 6, 7, 8
Amphictyonic Council, 21
Amphictyonic League, 20
Amphierastai, 138
Amphitrite, representations on pottery, 233
Amynias, 199
year of archonship, 200n.24
Amyrtaios, 18
Anagyrous, Attic deme, 39, 56
Anakreon, 46
Anaxagoras, 348
and Perikles, 316
trial of, 309
Anaxileos, 111n.11
Andokides, 7, 8
use of Hetaireiai, and Synomosiai, 162
fourth oration, on Hyperbolos, 321
as an historical source, 174
on law against tyrants, 72
on Melos, 31
on Miltiades, 56
on ostracism, 55, 57, 58, 83, 97, 103
on ostracism of Megakles, 106
on the mutilation of the Hermai, 165
on the Peace of Kallias, 277

on political clubs, 164n.7
on the speech against Alkibiades, 116—31
Androtion, 47, 224, 326
admiration for Theramenes, 326
on Hipparkhos, 105, 106
as an historical source, 174
on ostracism, 65–67, 69, 71
on ostracism of Hyperbolos, 129
as source for Aristotle's Constitution of Athens, 329
on Thoukydides son of Melesias, 308–9
Ankhimolius, Spartan, 210
funerary monument of, 271
Antalkidas, Peace of, 3, 9, 21
Anthemocritus, Athenian herald, 280
Antiokhis, Attic tribe, 42
Antiphon, 74n.2
attack on Alkibiades, 127
Antigonos, 267
Antony, Marc, 343
Anytos, 41
Aphrodite, 258, 261
Hegemone, 226
statue of, 196
Apollo, 144, 168, 255
Delian, 17
Delphic, 19
statue of, 271
temples of, 247, 255
Apollodoros, 89n.5
Appius Saufeius, 341, 342, 343
Aratos, court poet to Antigonos, 267
Archaic style in art, 207
Archeranistes. See Arkheranistes
Archermos. See Arkhermos
Archibacchos. See Arkhibakkhos
Archidamian War. See Arkhidamian War
Archidamus. See Arkhidamos
Archilochus. See Arkhilokhos
Archinus. See Arkhinos from Koile
Archon epynomos, 67, 69, 138, 200n.24, 211
Archon, basileus, 67
Archon, King. See Archon, basileus
Archon, polemarkhos, 67, 69, 70, 71, 204
Archons. See also Alexias; Amynias; Aristion; Damasias; Euktemon; Herodes; Hypsikhides; Kallias; Kharias; Nikodemos; Nikokrates; Pythodoros; Sophokles; Telesinos; Titus Flavius; Tyranny
election of, 67, 69, 74n.28
law forbidding their mockery in Comedy, 200
and their relationship to tyrants, 67–69
role in ostracism, 121
Areius Didymus, philosopher, 221
Areopagos, 67, 68n.10, 73n.24, 109n.3, 311, 334, 343
jurisdiction in cases of tyranny, 72–75

Arethas, scholiast of Plato, 89n.6
Arginousai, battle of, 298
Argive alliance, 118
Argos, 85, 231, 249
  and epigrams, 261
Arion, offering of, 271
Ariphron, 49
Aristeides, Aelius, 85n.3
  on the Covenant of Plataia, 277
  on ostracism, 100–101
  scholia, 85n.3, 94, 101
Aristeides, 12, 14n.7, 15, 161
  death of, 317
  and fighting at Psyttaleia, 275
  ostracism, 26–28, 59, 61, 91, 98, 104, 106, 107, 109n.3, 113, 152, 161
  and tribute assessment, 122
Aristion, Athenian archon (421/0), 16
  ostracism, 67n.8
Aristophanes
  Acharnians, 6, 7
    on Thoukydides son of Melesias, 307
    scholia, on Thoukydides son of Melesias, 314
  and Amynias, 200
  and aristocracy, 199
  Birds
    on Melos, 31, 33
    on political clubs, 165
  Clouds, 33
    scholia, 196
  Frogs, on political clubs, 166
  Knights, 111
    on political clubs, 164, 164n.7
    scholia on ostracism, 73, 78, 85, 86, 87, 94, 95
    scholia to, 114
    scholia, list of ostracisms, 310
  Lysistrata, 74n.29, 165
    and Kleomenes, 165
    on panhellenism, 145
  Peace, 274
    scholia, on Kleon, 317
  use of *Hetaireiai*, and *Synomosiai*, 162
  Wasps, 72n.22, 93, 95
    on Athenian law courts, 288
    on political clubs, 164
    on Thoukydides son of Melesias, 307
    scholia on Hyperbolos, 322
    scholia on Thoukydides son of Melesias, 308–310, 315
    scholion to, 117
  Wealth, 39
    scholia, 85n.3
Aristophanes, grammarian, *argumentum* to Sophokles' *Antigone*, 310
Aristotle, 326
  as source for Timaios, 96
  chronology for Solon's carrer, 48

Constitution of Athens, 14n.7, 15, 37n.9, 46, 47, 54, 329
  on Aristeides, 26–28
  chronology for the election of generals, 72
  on Damasias, 73
  on generals, 69, 70, 70n.15, 71n.16, 71, 72
  on Hipparkhos, 104–6
  on Hippias, 73n.24
  on Isagoras, 74n.29
  on Kleophon, 308
  on law against tyrants, 66n.4, 68, 72
  on Miltiades, 110
  on Nikias, 307–8
  on ostracism, 65–67, 69, 71, 72n.20, 76, 78, 80, 88, 100, 101, 103
  on ostracism of Megakles, 106
  on Perikles, 311
  on Solon's electoral reforms, 68n.10
  and Theophrastus' *Laws*, 84
  on Theramenes, 307–8
  on Thoukydides son of Melesias, 307–8
  on Xanthippos, 110
Ethics, on political clubs, 134
  on Damon, 334–36
  use of *Hetaireiai*, and *Synomosiai*, 162
  on Kimon, 83
  on ostracism, 58
  on Peisistratus, 48
Politics, 68n.10, 92, 99
  on ostracism, 84, 85, 86n.4, 97, 98
Rhetoric, 85
  on Solon, 45–48
Arkheranistes, 135–7, 140–42
Arkhermos, 207, 216
Arkhibakkhos, 139
Arkhidamian War, 7, 122
Arkhidamos, 307
  speech at Sparta, 281
Arkhilokhos, and epigrams, 250, 252–53
Arkhinos from Koile, 37, 39, 40, 43
  decree of, 37, 37n.7
Arniades, 250
Arrian, 228
Artabanus, 46
Artaphernes, 4, 149n.1
Artaxerxes, 3, 7, 9, 14, 33
Artemis *Agrotera*, 207
Asklepieia, 225
Asopos, river, 25
Aspasia, trial of, 309
Assessment decrees, Athenian. *See* Tribute, assessment of
Assessors, board of. *See* Tribute, assessment of
Athena, 16, 17, 20, 177, 204
  dedications to, 197
  Demokratia, 225
  Polias, 17

Subject Index

sanctuary at Plataia, 12
Skiras, sanctuary of, 158
statue of, 216
treasury of, 273. *See also* Delian Treasury, transfer to Athens
Athenaeus, and Alkibiades, 127
Athenian Agora, excavations, 109
Athenian alliance. *See* Athenian *arkhe*; Delian League
Athenian amphictyony. *See* Athenian *arkhe*; Delian League
Athenian *arkhe*, 16, 20, 21, 30, 31, 54, 103, 176, 229, 238, 274, 280, 282, 284, 286, 291, 306. *See also* Delian League
divisions of, 20
tribute to, 12, 16, 17, 20, 21. *See also* Athena, treasury of; Delian Treasury, transfer to Athens; *Tribute Lists, Athenian*
Athenian League. *See* Athenian *arkhe*; Delian League
Athenian Tribute Lists. *See Tribute Lists, Athenian*
Athens
date of treaty with Plataia, 25
famine at, 114
and Macedon, 328
as mother city of all Ionians, 231
treaty with Thebes, 25n.3
Athos
mountain, 87
peninsula of, 270
Atrometos, father of Aischines, 40
Atthidographers, 275. *See* Androtion, Aristotle, Demon, Ephoros, Hellanicos, Ister, Kleidemos, Melanthius, Melasagoras, Phanodemos, Philochoros, Theopompos
Attic Amphictyony. *See* Athenian *arkhe*; Delian League
Attic festivals. *See* Dionysia; Eleutheria; Panathenaia
Atticus, 339–44
Augustan Games, 221
Augustus, 220–22, 276
Autolykos, pancratiast, 203
Auxesia, statue of, 271
Azenia, Attic deme, 159

Babylon, 270, 272
Bacchylides, and epigrams, 262
Battus, 46
Beldam Painter, 230n.15
Berenike, priest of, 226
Berenikidai, Attic deme, 337, 338n.7
Berlin Painter, 215n.15, 232, 233n.21, 233n.22
Biton, statue of, 271
Black-figure pottery, 61, 229

Boedromion, month, 224, 225
Boeotia, 21, 25, 26, 247, 263
Athenian victory over, 198, 271
and epigrams, 259, 261
and writing, 188
Bosphoros, 271
Boule, 78, 79, 110, 227–28
of 400, 80
of 500, 80
Bouleutérion, 78, 79
of Pergamum, 228
Boutalion, 57
Bouzygai, priestly family, 112
Brasidas, 32, 200
Brygos workshop, 233
Bursa, Titus Munatius Plancus, 341
Byzantine period, scholia and lexika, 82

Calderini, Aristide, 55
Calendar, Athenian, 54
Callias. *See* Kallias
Callimachus. *See* Kallimakhos of Aphidnae
Callisthenes. *See* Kallisthenes
Callixenos. *See* Kallixenos
Camarina, 282, 286, 291
Capua, 234
Carcopino, Jérôme, 54
Caria. *See* Karia
Carthage. *See* Karthage
Casilo, Claudius, 87
Cassius Dio, 222
Chaironeia. *See* Khaironea, battle of
Chalcidians
Athenians Victory over, 198, 271
at the battle of Salamis, 276
Charias. *See* Kharias
Charites
priest of, 225, 226
statue of, 219n.40
Charmus. *See* Kharmos
Cheiron. *See* Kheiron
Chersonesos, 73, 103
Chilon. *See* Khilon
Chios. *See* Khios
Chiusi, the near Chiusi Painter, 230n.3
Choirs, boys, 168
Cicero, 339, 341, 342
and Epicurean philosophy, 337
Cimon. *See* Kimon
City Dionysia. *See* Dionysia
Cleisthenes. *See* Kleisthenes
Cleomedes. *See* Kleomedes
Cleon. *See* Kleon
Cleophon. *See* Kleophon
Clodius, murder of, 343
Clubs, Athenian, 134–42. *See also* Eranos, political club
Clubs, oligarchic, 160

Clubs, political, 120, 121, 312. *See also* Alkibiades; Andokides; Aristophanes, *Frogs*; Aristophanes, *Knights*; Aristophanes, *Wasps*; Aristotle, *Ethics*; Clubs, oligarchic; Eranistai; Eranos, political club; Euripides, *Phoenician Women*; Heraklistai club; Hetairies; Isagoras; Kleisthenes; Kleon; Lysias; Nikias; Old Oligarch; Peisander, and political clubs; Perikles; Plato; Plutarch; Sophronistai; Soteriastai; Themistokles; Theramenes; Thoukydides; Thoukydides son of Melesias; Xenophon
Cohen, Robert, 55
Colonies
  Athenian, 17
  Ionic, 20
Colonization, Pericles' policy, 20
Common Peace. *See* κοινὴ εἰρήνη Index of Greek words
Congresses, panhellenic, 178
Corcyra. *See* Korkyra
Corfu. *See* Korfu
Corinth. *See* Korinth
Cos. *See* Kos
Crete, 198
  and writing, 188–89
Critias. *See* Kritias
Croesus. *See* Kroisos
Ctesiphon. *See* Ktesiphon
Cyclades, 20, 29
Cylon. *See* Kylon, Conspiracy of
Cyprus, 48
  epigrams, 260
Cypselids. *See* Kypselides
Cyrene, *See* Kyrene
Cythera. *See* Kythera
Cyzicus. *See* Kyzikos, battle of

Dadouchos, Kallias, son of Hipponikos, 237
Daidalos, 273
Damasias, Athenian archon (582/1?), 47, 48, 73
Damasias, Athenian archon, (639/8), 48
Damasis, 47
Damia, statue of, 271
Damon, 332–36
  ostracism, 54, 55, 57, 58, 334–36
  and Perikles, 86n.4, 332, 333, 334, 336
  and Sokrates, 332, 333
Dareios I, 4, 9, 24n.2, 146, 211
Dareios II, 4–9
Darius. *See* Dareios
Datis, 146–55, 149n.1, 210
  Hymn, 146–55
Deinarchos, speech against Polyeuktos, 89n.5
Deinias, Athenian general (428), 56

Delian League, 18, 19, 21, 22, 177, 240, 277, 306. *See also* Athenian *arkhe*
  reorganization of, 20
Delian Treasury, transfer to Athens, 16, 17, 22, 177, 180, 287, 306, 316
Delion, 271
Delion, battle of, 123
Delos, 16, 17, 19, 158, 240, 263
Delphi, 12, 19, 144, 145, 156, 157, 184, 218
  and epigrams, 259
  temple dedicated by the Alkmeonidai, 273
  votive offerings at, 271–72
Delphic Amphictyony, 19, 21, 22, 178, 184, 239
Delphic Oracle, 24
Demades, and Alexander, 228
Demagogues, Athenian, list of, 309–10, 317. *See also* Dio Chrysostomus; Plato, *Meno*; Plato, pseudo; Plutarch
Demaretos, 24
Deme lists, 89n.5
Demes, Attic
  Acharnai, 42
  Aithalidai, 158, 159
  Alopeke, 306, 314
  Anagyrous, 39, 56
  Azenia, 159
  Berenikidai, 337, 338n.7
  Gargettos, 56
  Koile, 39
  Kothokidai, 40
  Marathon, 57
  Oineis, 41–43
  Paiania, 39, 43, 135, 138
  Perithoidai, 322
  Phyle, 35, 37, 38, 42–44
  Sounion, 137
  Xypete, 55, 57
Demesthenes, on Aeschines, 277
Demeter, 136, 137, 140, 177, 232
  and Battle of Salamis, 240
  cult of, 231
  Hymn to, 231
  representations on pottery, 230–37
Demetrios of Phaleron, 49, 84, 86, 98, 99, 100
  Laws, 86n.4
  on Sokrates, 86n.4
  as used by Plutarch, 100
Demetrios Poliorketes, 226, 228
Demokhares, agoranomos, 158–59
Demokhares, demotic, 158
Demokratia, 223–24. *See also* Athena *Demokratia*
  cult of, 225–26
  painting of, 223
  statue of, 226–28
Demon, 326
Demos
  painting of, 223

priest of, 225, 226
Demosthenes, 3, 8, 10, 14n.6, 14n.7, 15, 182, 326
  on duration of ostracism, 103
  on ostracism of Themistokles, 94
  speech against Aristokrates, 87
  speech against Neaira, 89n.5, 224
Demotic, 104, 106, 158
Demotics See Names, Athenian
Diagoras, on Melos, 33
Didymos
  commentary on Demosthenes, 82, 83, 87, 89n.5, 93, 94, 95, 97, 99
  commentary on Demosthenes, on ostracism, 314
  and Diodorus Siculus, 97
  on Kimon, 94
Digamma, 249
Dikaios, 260
Dikast tickets, 63
Dio Chrysostomus, 221, 222
  Athenian rhetors and demagogues, list of, 309
  on Peisistratus, 49
  on Periander, 49
  on Solon, 49
Diodorus Siculus, 3, 9, 12, 13, 14, 17, 18, 21
  and Didymus, 97
  and Ephorus, 95
  as an historical source, 174
  on Melos, 33
  on ostracism, 78, 83, 95–98, 99, 102
  on Peisistratus, 49
  use of Philochorus, 84
  problems of chronology, 17
  on Thoukydides son of Melesias, 307
  on Solon, 49
Diogenes Laertius, 14
  on Solon, 45, 48
  sources, 49
Diomedes, and Alkibiades, 125
Dionysia, 17, 19
  donations at, 20
Dionysios of Halicarnassos
  on Hipparkhos, 105
  on Triptolemos, 232
Dionysios, tyrant, 27n.6
Dionysos, 177, 238
  cart of, 230
  cult of (157), 231
  Gründungsorakel of, 156–57
  representation on pottery, 235–36
  sanctuary of, 138
  Theater of, 225, 226
Dioskuroi, 258
Dipylon pottery, 188, 192, 250
Dodona, 156
Dorian invasion, 262
Dorian War, 114–15. *See also* Aetolian War

Draco, law-code, 327
Droop Cups, 215n.16
Drusus, consul, 135, 136, 138

Ecclesia. *See* Ekklesia
Edinburgh Painter, 230
Education, Greek, 167–69, 334
Egina. *See* Aigina
Egypt, 6, 47
  monuments of, 270
Egyptian Expedition, Athenian. *See* Kimon, expedition to Cyprus
Ekklesia, 110
Ekphyllophoria, 79, 89, 89n.5, 91
Eleans, 239
Electoral law of 487, 66n.5, 68
Eleusinia, 17, 19, 224
  donations at, 20
Eleusinian cult, 231–32
Eleusinian deities, 238, 342
Eleusinian Mysteries, 240, 342
Eleusinian scepter, 236
Eleusinian Spondai, 239–41
Eleusinion, 231, 236, 342
Eleusis, 18, 24, 74n.29, 195, 221, 231, 240
  grain distribution, 177
  mother of Triptolemos, 233
  sanctuary in, 232
  violation by Megara, 280
Eleutherai, 42
Eleutheria, 12, 13, 15
Elis, 144
Elpinike, sister of Kimon, 127
Emperors, Roman, cult of, 168
Emporium Painter, 230n.17
Endoios, sculptor, 216, 218
Ephebe inscriptions, 226
Ephebes, 168, 222, 228
  institution of, 167, 168
Ephebic oath, 232n.19, 277
Ephesos, 7, 207, 211
Ephialtes, 217, 269
Ephoros, 3, 12, 14, 17, 326
  as historical source, 174
  on Melos, 31, 32
  as source for Diodoros, 95
  on Thoukydides son of Melesias, 306–7
Epicharinos, statue of, 196
Epicurean school at Athens, 339, 340
Epicurus, 345–49
Epidamnos, 14
Epigrams, funerary, 245–65
Epigraphical studies, history of, 53–54
Epigraphy, 173–86
Epikrates, 41
Epilykos, Treaty of, 7, 8
Epimenides, 14, 114, 115
Epistatai, 121
Epitaphioi, 237

Eranistai. *See* Eranos, political club
Eranos, political club, 134–42
Erasistratus, father of Phaiax, 131
Erekhtheis, Attic tribe, 43
Eretria, 149
Ergokles, 41
Erythrai, 72n.22, 219n.40
Etymologicum Magnum, 89n.5, 92, 99
Euboia, 103
  revolt of, 24, 286
  and writing, 188
Eubotas of Kyrene, 297
Eubulus, 326
Eucharistos, Marcus Aemelius, 135, 136, 138, 142
Euchidas, 12, 19
Euenor, sculptor, 207
Euergides Painter, 232
Eugammon of Kyrene, 262
Eukleides, Athenian archon (403/2), 38, 39
Eukles, herald, 338n.7
Euktemon, Athenian archon (408/7), 297
Eumelos, 259
Eumolpidai, 239
Eumolpos, father of Triptolemos, 233, 327
Eupalinos, the canal of, 270, 282, 283, 285, 291
Euphranor, painter, 223
Eupolis
  use of ἀλιτηρός, 112
  on Alkibiades, 131
  Demoi, 165
Euribiades, and Themistokles, 284
Euripides
  Bacchae, 237
  Helen, 165
  and Hesiod, 267
  Ion, 20
  Medea, 14n.7, 289
  Orestes, 153
  Phoenician Women, on political clubs, 166
  Suppliants, 14n.7
Eurykleides, 225, 226
Eusebius, *Chronicles*, 80, 85n.3
Eusthatius, 85n.3
Euthydikos, 206
Euthymides, painter, 207
Exegetai, priestly board, 328
Exekias, 230n.8

Famine, in Athens, 114
Financial accounts, 63
Four Hundred, *coup d'état* of. *See* Oligarchic Coup of 411
Francois Vase, 268
Fruits, 'First Fruits'. *See* ἀπαρχαί Index of Greek Words
Funeral lists, 63

Funeral Orations. *See* Epitaphioi
Funerals, Athenian public, 12

Galba, C. Sulpicius, statue of, 340
Gargettos, Attic deme, 56
Gellius, 339
Generals, 163
  date for the founding of the office, 69, 70, 71
  election of, 66n.4
  institution of, 76
  and ostracism, 69–72
  sacrifices to Demokratia, 224, 228
Geraistos, 103
Glauke, priestess, 159n.6
Gomme, Arthur, 55
Gorgias, promoter of panhellenism, 239
Graffiti, 191, 193
Granikos, battle of, 228
Great Dionysia. *See* Dionysia
Great King. *See* Artaxerxes; Dareios I; Dareios II, Xerxes, Alexander the Great
Great Panathenaia. *See* Panathenaia
Greek History, Sources of, 63
Gyges, 272
Gymnasiarch, 168

Habron, son of Patrokles, 57
Hades, representations on pottery, 233, 234
Hadrian, 138, 143
Hagias of Troizen, 262
Harmodius and Aristogeiton
  cult of. *See* Tyrannicides, cult of
  statues of, 211n.24, 271
Harpocration, 65n.1, 89n.5
Hearth, sacred on the Akropolis, 111
Hecataeus, 329
Hekate, 235n.35
  representation on pottery, 236
Hektor, 247–49, 263–64
Hellanicus, 275, 325–29
  on ἀλιτήριοι, 115
Hellanodikai, 144
Hellenic League, 21
Hellenic Oath, 8
Hellenica of Oxyrhynchus, 326
Hellespont, 20, 247
Hephaistos, 227, 230
Hera, sanctuary of, 270–71
Herakleides the Clazomenian, 7, 8
Herakleides Ponticus
  περὶ ἀρχῆς, 49
  on Solon, 48–49
  summary of Aristotle, 110
Herakleion, in the Kynosarges, 209, 210
Herakleitos, on Hesiod, 268
Herakleon, 271
Herakles, 136, 137, 140, 142–45, 168

representations on pottery, 235
sanctuary of, 272
shrine at Kynosarges, 138
Herakliastai, club, 135, 136, 138, 140–42
Hermai, 43, 342, 344
  as an Athenian innovation, 272
  mutilation of, 117, 164, 165
Hermes, 168
  statue of, 204
Hermokopidai, *See* Hermai, mutilation of
Herms. *See* Hermai
Herodes, Athenian archon (59/8), 225
Herodotos, 4, 11, 12, 14, 24, 329
  on Athens and Corinth, 24
  on battle of Phaleron, 209
  on battle of Salamis, 276, 283
  on Eleusinian deities, 232
  on generals, 69
  as historical source, 174, 328
  the incident with Kandaules, his wife, and Gyges, 330–31
  on Miltiades, 73, 73n.28, 109
  on monuments, 270–74
  on Oath of Plataia, 277
  on Olympic games, 145
  on ostracism, 105
  use of πρυτανηίη, 110
  on Peisistratus, 73
  on Solon, 45–47
  on Troizen, 103
  on tyrants, 68
Hesiod, 262
  and Aeschylus, 266–69
Hesychius, 14n.7
Hetaireiai, vs. *Synomosiai,* 162–66
Hetairies. *See* Clubs, oligarchic
Hierapolis, 221
Himerius, 46
Hipparkhos, 5, 69, 70, 74n.29, 104, 232
  ostracism, 28, 65, 66, 104, 106, 112, 152, 161, 196
  statue of, 112, 196
  subject of first ostracism, 58
Hippias, 4, 6, 25, 67, 68, 74n.29, 75n.30, 105, 145, 152, 231
  and battle of Marathon, 69
  expulsion of, 68, 69, 73, 73n.24, 76, 79, 162, 163
  orator, promoter of panhellenism, 239
Hippokrates
  brother of Kleisthenes, father of Agariste, 111
  general at Delium, 123
  son of Alkmeonides, 57, 111
  son of Alkmeonides, ostracism of, 57, 59, 61, 152, 109n.3
  son of Anaxileos, 57, 111
Hipponicus, father-in-law to Alkibiades, 123

Homer
  electoral procedures in, 67, 69
  and epigrams, 247, 251, 253, 259, 261, 264
  and Hesiod, 266–67
  Iliad, 69
    and epigrams, 246, 248, 255, 263
    personification of, 220, 222
  Odyssey, personification of, 220n.1
Homer, 'The New', 220–22
Hoplite General, 221
Horai, statue of, 219n.40
Hyperbolos, 129, 317, 320–24
  assassination of, 232
  ostracism, 56, 57, 75, 84n.3, 86, 89, 96, 98, 103–6, 117, 118, 121, 130, 161, 163, 312, 320
Hyperides
  For Euxenippus, 74n.29
  Plataicus, 15
Hypsikhides, Athenian archon, 104

Idomeneus, 13, 15, 95
Ilias Latina, author of, 222
Inscriptiones Graecae, 63
Iobakkhai, 139
Iobakkheion, 139
Ionia, 20
Ionian Revolt, 4, 6, 9, 10, 70, 211
Ionian War, 124, 289
Iphikrates the Younger, 228
Isagoras, 4, 74n.29
  conflict with Kleisthenes, 162
  and political clubs, 163
Ismenion, inscriptions of, 271
Isocrates. *See* Isokrates
Isodike, granddaughter of Megakles, 111n.13
Isokrates, 89n.5, 160
  on Alkibiades, 124
  defence of Alkibiades the younger, 125
  on Hyperbolos, 323
  on Melos, 30
  Panathenaicus, 224
  Panegyricus, 15, 145
  Panegyricus
    on Ephebic Oath, 232n.19
    on Triptolemos, 237
    on panhellenism, 144
  Philippus, 84n.2
  Plataicus, 15, 277
  promoter of panhellenism, 239
  on Solon, 47
  teacher of Androtion, 326
Ister, 326–27
Isthmian Games, 197–98
Ithome, 25

Jacoby, F., *Atthis: The Local Chronicles of Ancient Athens,* 328

Juncus Aemilius, 142

Kalamis, sculptor, 196
Kallaischros, 197, 198
Kallias, 3, 8, 178, 182, 196
  ostracism, 55, 57, 58, 126
Kallias, Athenian archon, (406/5), 300, 301
Kallias, Peace of, 3, 5, 7–10, 14, 14n.6, 15–20, 103, 181, 183, 184, 277
Kallias, son of Didymias, statue of, 197
Kallias, son of Hipponikos, statue of, 196
Kallias, son of Kallaischros, 200
Kallichoron, well, 236
Kallidike, 236
Kallimakhos of Aphidnae, polemarch, 204
  dedication at Marathon, 204–7
Kallisthenes, on the Peace of Kallias, 278
Kallixenos, son of Aristonymos, 152
  ostracism, 55, 57, 59, 61
Kalos-names, on pottery. *See* Love-name, on pottery
Karia, 20
Karkinos, son of Xenotimos, 147
Karthage, 303
Kekropis, Attic tribe, 42, 43
Kephalos, potter, 214n.6
Kephisophon from Paiania, 39, 43
Kerameikos, 56, 210, 211, 219, 226
Kerykes, 239
Khaironeia, battle of, 184
Kharias, Athenian archon (415/4), 34
Kharias, son of Kharias, 159
Kharinos, Decree of, 280
Kharmos, father of Hipparkhos, 105
Kheiredemos, son of Euangelos, 198
Kheiron, 334
Khilon, 13, 14, 14n.5, 15
Khios, 29, 228
Khoinix, public measure, 136, 137, 141
Khremes, father of Hyperbolos?, 322
Khrisothemis, sister of Phaidros, statue of, 342
Khryses, 255
Kimon, 3, 5, 6, 17, 20, 22, 25, 112, 161, 229, 269, 317
  campaign in Cyprus, 286
  death of, 306, 311, 318
  expedition to Cyprus, 5, 17, 180, 286
  Olympic victory, 127
  ostracism, 56, 58, 59, 91, 96, 103, 107, 126, 161, 311
  Peace of, 18
  and Perikles, 310–11
  pro-Spartan attitude, 58
  son of Militides, 111n.13
  trial of, 313
Kimonian herms, 43
King's Peace, 14n.6,
Kiss Painter, 202

Kleidemos, 105, 326
Kleinias, Decree of, 21
Kleippides, son of Deinias, 56, 59
Kleisthenes, 4, 5, 55, 67, 68, 110, 162, 327
  conflict with Isagoras, 162
  death of, 67n.6, 110
  exile of, 111
  and ἰσονομία, 168
  ostracism, 56
  and political clubs, 163
  ostracism legislation, 65, 66, 66n.3, 4, 67n.7, 69, 76, 78, 80, 104, 106
  reforms, 59, 110n.8
Kleobis, statue of, 271
Kleomedes, 30
Kleomenes I, 4, 24, 74n.29
Kleomenes, and Aristophanes' *Lysistrata*, 165
Kleon, 317
  and Amynias, 199
  death of, 323
  and political clubs, 163–64
  and tribute assessment, 121
Kleophon Painter, 235
Kleophon, ostracism, 56
Kleophrades Painter, 233
Kleosthenes, 144
Koile, Attic deme, 39
Konon, Titus Flavius, Athenian archon, 135, 136
Kore, 136, 137, 140
  dedicated by Euthydikos, 206
  representations on pottery, 231, 233–36
Korfu, 253
  and epigrams, 248
Korinth, 263
  and Sparta, 287
  and writing, 188
Korkyra, 263
Koroneia epigrams, 199
Koryphasion, 303
Kos, 198
Kosmetes, 168
Kothokidai, Attic deme, 40
Kouroi, 196
Kourotrophos, priesthood of, 158
Krateros, 199
  on Thoukydides son of Melesias, 309
Kratessipidas, 297
Krissa, 254
Kritian Boy, 196
Kritias, 274
  in Plato's *Timaeus*, 46
Kritias, tyrant, 46
  tomb of, 224–25
Kritios, sculptor, 196, 197
Kroisos, 5, 45, 46, 48, 271, 272
Kronos, 267, 268
Kroton, 197
  dedication in Delphi, 197

Subject Index

Krotonians, 103
Ktesiophon, 36
Kydrokles, son of Timokrates, 57
Kylix, 212, 215
Kylon, conspiracy of, 73, 111, 112, 115n.27, 162, 327
Kyme, 274
Kynosarges, 91, 138, 142, 209, 210
Kypselides, 257
Kyrene, 18
Kythera, 14, 32
Kyzikos, battle of, 298, 302

Laches, 333
Ladike, 272
Lakedaimonios, 111n.13, 199
Lampon, prophecy concerning Perikles, 316
Laurion, 106
Law courts, 110n.7, 120
Leagros Period, 230–31
Leagros, son of Glaukon, 200, 201
Leagros, statue of, 201, 202
Leobotes, 199
Leodikea, 341
Leokrates, 183, 309, 317
Lepidus Aemilius, 142
Lepreon, 239
Lesbos, 328
Lexicon Rhetoricum Cantabrigiense, 87, 88
Lexika, 81, 82, 84
Libraries, 168
Linear B, 191
Liopesi, village in Attika, 134, 138
Literacy, 52–63
Locris, 21
Lokroi, 234n.32
Love-names, on pottery, 197, 201
Lucian
Lucius Saufeius, 341, 342, 343
Lucretius, 345–49
Lycurgus. *See* Lykourgos
Lydia, 5, 48
Lykophron, 303
Lykourgos, 198
 In Leocratem, 74n.29, 112, 203
 on Hipparkhos, 105, 106
 on votive offerings, 201
 on the Oath of Plataia, 277
 and στάσις, 163
Lykourgos, law-giver, 143–44
Lysander, 199, 298, 300
Lysiades, Athenian archon (51), 343
Lysiades, son of Phaidros, 337
Lysias, 326
 on Melos, 34
 on ostracism of Megakles, 106
 on panhellenism, 143–44
 and Phaiax, 125
 on political clubs, 166
 promoter of panhellenism, 239

Macedonia, 10, 174, 184
 and Athens, 328
Maecenas, 221
Makron, painter, 233–34
Mandrokles, 271
Marathon, Attic deme, 57
Marathon epigrams, 275
Marathon, battle of, 4, 24, 27, 58, 67, 70, 71, 74–76, 79, 104, 110, 146, 149–52, 163, 183, 186, 204–11, 232, 276, 282
 shield signal, 4, 69, 80, 109
 stele listing the names of the dead, 208–11
Marcellinus, on Thoucydides and Thoucydides son of Melesias, 314n.15
Mardonios, 149n.1
Marmor Parium, 47
Marsyas Painter, 238n.46
Massarachia, A., *Solone*, 45–49
Medea, 150
Medea-epigram, 258
Megakles, 5, 48, 69, 111n.11
 ostracism, 27, 28, 54, 61, 104, 106, 107, 111, 125, 152
 relation to Alkibiades, 126
 and στάσις, 163
Megara, 25, 85
 and epigrams, 259
 revolt and acquisition by Athens, 24
 violation of Eleusis, 280
Megarian Decree, 280
Melanthius, 70, 211, 326
Melesagoras, 326
Melian Dialogue, 30, 33, 282, 283, 287, 288, 291
Melian Expedition, 30
Melian woman, and Alkibiades. *See* Alkibiades, accused of adultery
Melians, enslavement of, 117, 118
Melos, 29–34
 and writing, 188
Menekrates, 250
Menon, son of Menekleides, ostracism, 56–60, 62
Messene, 18
Metageitnion, month, 224
Meter, poetic, 109n.4
Metics, 36
Metroön, 36, 38
Midas, 272
 tomb of, 14n.7
Midas-epigram, 252, 253, 264
Mikion, 225, 226
Mikon, sculptor, 197
Miletos, 85, 272, 298
Milo, trial of, 341
Miltiades, 68, 70, 110, 127, 149n.1, 185, 186
 ostracism, 56

Parian expedition. *See* Parian expedition
speech to Callimachus, 282
trial of, 73n.28, 109
Mnesiphilus, 294
Months. *See* Boedromion; Metageitnion;
Mounichion
Mounichia, 14, 36
Mounichion, month, 138
Mousaios, 231
Muses, 168
Mycenae. *See* Mykenai
Mykale, battle of, 112
Mykenai, 191
Mylonas, 240
Myronides, 112, 309, 317
Myrrhine, daughter of Kallias, wife of
Hippias, 105
Mysteries. *See* Eleusinian Mysteries
Mystery-Mockers. *See* Eleusinian Mysteries,
profanation of
Mytilene, revolt of, 6

Names. *See also* Demokhares; Demotic; Love-
names; Patronymics; Philistion;
Pronapes; Prosopography
Athenian, 58–60, 63
Naxos, 29
Nemean Games, 197–98
Nemesis, statue of, 151
Neoi, 168
Neokleides, 7
Nepos
Aristides, 17, 27
on Salamis, 276
Cimon, 107, 311
Thrasybulus, on trial of Miltiades, 109
Nesiotes, sculptor, 196, 197
Nestor, 250
Nicias. *See* Nikias
Nikanor, Gaius Iulius. *See* Homer, 'The New'
Nike
archaic statues of, 206
as ἄγγελος ἀθανάτων, 204
of Marathon, 204–7
temple of, 178
Nikias
debate with Alkibiades, 117–19, 130
and Hyperbolos, 320, 321
and Melos, 30, 32, 33
ostracism, 56
and political clubs, 121, 164
Peace of, 33, 122, 176, 287
Nikodemos, Athenian archon, 104
Nikokrates, Athenian archon, (333/332), 228
Nikomedes, Athenian archon (483/482), 27
Nikosthenes, painter, 215n.16
Niobid Painter, 234–35
Notion, battle of, 298

Oaths vs. treaties, 8–9
Oineis, Attic tribe, 41–43
Oinophyta, battle of, 21
Old Oligarch, 46, 119, 127
  on political clubs, 164
Oligarchia, 224
Oligarchic Coup of 411, 72n.22, 75, 165
Oligarchy, 72
Oltos, painter, 232
Olympia, 201
  and epigrams, 255
Olympic Divinities, statues of, 273
Olympic Ekecheiria, 239
Olympic Games, 143–45, 196, 240
Olympiodoros
  commentary on Plato's *First Alkibiades*,
    89n.6
  commentary on Plato's *Gorgias*, 89n.6
Onomakritos, 231
Orestes, 153
Orontes River, 188
Oropos, 30, 37n.7
Ostracism, 5, 18, 53–64, 65–107. *See also*
  Aelian; Alkibiades; Andokides;
  Androtion; Archons; Aristeides;
  Aristeides, Aelius; Aristion;
  Aristophanes, *Knights*; Aristotle;
  Aristotle, *Constitution of Athens*;
  Aristotle, *Politics*; Damon; Didymos;
  Diodorus Siculus; Generals;
  Herodotos; Hipparkhos; Hyperbolos;
  Kallias; Kimon; Kleisthenes;
  Kleophon; Lysias; Megakles; Miltiades;
  Nepos; Nikias; Ostraka; Peisistratos;
  Petalism; Phaiax; Philochoros; Plato;
  Plutarch; Sokrates; Suidas;
  Themistokles; Theodoros Metochites;
  Theophrastos; Theopompos;
  Thoukydides; Thoukydides son of
  Melesias; Timaios; Xanthippos
  and the archons, 121
  in Argos, 85
  bibliography, 82n.1
  cause of, 100, 106–7
  chronological limits, 58
  dating specific ostracisms, 104
  discussion in Plutarch, 57
  duration of the institution, 102
  and election of generals, 69–72
  5th and 4th c. B.C. scholarship, 83
  institution of, 120
  late antique sources on, 77–80
  law of, 72, 73
  legislation, 72, 76
  length of exile, 58
  in Megara, 85
  in Miletos, 85
  number of votes necessary, 78, 79, 80, 88,
    91, 92

ostrakon from Athenian Agora, Inv. no. P. 16873, 108, 109n.3
residence requirement, 102–3
role of political clubs, 161
table of dates and sources, 28
the first, 74, 76
the last, 74, 104
and tyranny legislation, 72–75
voting practices, 57–58, 120–21, 132
Ostraka
as evidence for Greek pottery, 61–62
as evidence for literacy and letter forms, 62–63
Athenian, 53–64
sherds from small cups, 62
Oulios, son of Kimon, 111n.13
Ovid
Amores, 14n.7
Ars Amatoria, 222

Paiania, Attic deme, 39, 43, 135, 138
Paides, 168
Pamphaios, potter, 215n.16
Pamphylia, 5
Pan Painter, 234n.28
Pan, 237, 272
Pan, sanctuary of, 271
Panakton, 25n.3
Panathenaic Games. *See* Panathenaia
Panathenaia, 17, 19, 196–98, 201
donations at, 20, 21
Pancratia, 197, 203
Pandionis, Attic tribe, 39, 43
Pandrosos, cult of, 158–59
Panhellenic Games. *See* Panathenaia, Isthmian Games; Nemean Games; Olympic Games
Pantainos, library of, 220n.3
Papasilenos, 237
Parian expedition, 70, 74n.29
Paros, 207
Parthenion, well, 236
Parthenon, 178, 198, 273, 303
Parthenos, 236
Patron, 337, 339
Patronymics. *See* Names, Athenian
Pausanias, 12, 25n.3, 43, 289
on dedications at Marathon, 207
description of the Akropolis, 216
on Endoios, 219n.40
on Euphenor, 223
on Olympic games, 143
on statue of Xanthippos, 112n.20
on Themistokles Decree, 276
on votive offerings, 196
Pearson, L., *Local Historians of Athens, the*, 325–27
Peiraius, 221

Peisander, and political clubs, 165–66
Peisistratidai, 11, 25, 101, 105, 118
Peisistratos, 11, 46–49, 65, 67, 75, 105, 193, 224
agricultural policy, 231
buildings of, 203
chronology of, 329
death of, 68, 75
expulsion of, 67n.8, 73
son of Hippias, 203, 211
and στάσις, 163
Peison, 143–44
Pelasgoi, 272
Peloponnesian influence of Athenian art, 195–96
Pelops 143–44
Penestai, 200
Pentathlon, 202
Pentecontaetia, 285–86
Peplos kore, 218
Perachora, 254
and epigrams, 259
Pergamum, 228
Periander, 49
Perikles, 5, 6, 18, 56, 106, 110n.8, 111, 179, 180, 197, 229, 269, 309, 317
and Alkbiades, 127
and Athenians' speech at Sparta, 282
blamed for plague, 114
building program, 19, 178, 273, 316
citizenship legislation, 59
colonization program, 20
Congress Decree of, 3, 16, 18–22, 287
conviction of, 86n.4
and curse of the Alkmeonidai, 111n.13
and Damon, 86n.4, 332–36
death of, 163, 237
and δύναμις, 284
entry as demagogue, 311
funeral oration, 289
and Kimon, 310–11, 313
and Megarian Decree, 280
Papyrus Decree, 18
and political clubs, 160, 161, 164, 312
Pontic expedition, 6
and Thoukydides, son of Melesias, 306
years in power, 234, 317
Perithoidai, Attic deme, 322
Persephone, representations on pottery, 233, 236
Persian Wars, 3–15, 17–20, 24–29, 36, 58, 146, 150, 153, 174, 176, 177, 180, 182, 186, 202, 204–11, 232, 232n.19, 240, 274–76, 281–83, 285, 286, 289. *See also*, Marathon, battle of; Plataia, battle of; Psyttaleia; Salamis, battle of; Thermopylai
destruction of Sanctuaries, 12, 19, 178
Perugia, 234n.32

Petalism, 96, 102, 190
Phaiax, 117–19, 125–26, 129–31, 320, 321
  ostracism, 56, 58
  and political clubs, 163
Phaidros, 344
  genealogy of, 338
  statue of, 339
Phaleron, 340
  battle of, 208–11
Phanodemos, 326
Pharsalos, 21, 199
  statue of, 197
Pheidias, trial of, 309
Pheidostrate, daughter of Eteokles, priestess of Aglauros, 158
Pherephatta, 233
Phidias, sculptor, 151
Philia, wife of Phaidros, 343
Philip II, 186
Philistion, as a woman's name, 159
Philistion, daughter of Demokhares, 159
Philochoros, 71, 86, 93, 94, 275, 326
  Aristides, Aelius, as used by, 101
  Atthis, 83, 85, 89n.5, 224
  dates of individual ostracisms, 104
  on Hipparkhos, 104, 105, 106
  as historical source, 174
  on ostracism, 78, 80, 83, 87, 88, 102–3, 132–33
  and ostracism of Hyperbolos, 129
  on Thoukydides son of Melesias, 308–9
Philocyprus, 46
Philodemos, *Rhetorica* I, on Thoukydides son of Melesias, 315n.17
Philokrates, 30
Phlegon, 143–45
Phoenicia, and writing, 190
Phoenicians, 6
Phokaia, 14n.7, 240, 303
Phokaians, ancestral images of, 273
Phokis, 21
Phormisios, 41
Photion, Lexikon of, 257
Photios, 92
Phrynikhos
  Fall of Miletus, 74n.29
  Phoenissai, 112n.18
Phyle
  Attic deme, 35, 37, 38, 42–44
  inscription honoring heroes, 40–41
Pindar
  and epigrams, 262
  Nemean 5, 201
  Nemea 8, 107
  Pythia 7,
  on φθόνος, 106, scholia, 107
  on Olympic games, 143, 145
Pisistratus. *See* Peisistratos

Pissouthnes, 6
Plataia, 13, 24, 25
  battle of, 3, 9, 12, 19, 178, 271, 282
  Covenant of, 11, 12, 13, 15, 19, 178
  date of treaty with Athens, 25
  debate concerning, 281
  Oath of, 8, 11, 15, 19, 183, 184, 277
  siege of, 176
Plataians, at the battle of Salamis, 276
Plato
  Apology, 127
  Axiochus, 91
  comic poet, 129
  on Damon, 333
  Gorgias
    on Militades, 109 on ostracism of Kimon, 91
    scholia, on διάστοφος πολιτεία, 313n.12
  Laches, on Thoukydides son of Melesias, 307
  Menexenos, on Triptolemos, 237
  Menexenus, 85n.3
    Meno list of Athenian demagogues, 310
    on Thoukydides son of Melesias, 307
  on political clubs, 166
  Phaedrus, 14n.7
    and epigrams, 251
  pseudo, *De virtute*, list of Athenian demagogues, 310
  Republic,
    on Damon, 333, scholia to, 114
  on Solon, 46–47
  Theages, on Thoukydides son of Melesias, 307
  Theaetetus, 160
  Timaeus, 46, 47
  use of *Hetaireiai*, and *Synomosiai*, 162
Pliny, *Natural History*, on Euphranor, 223
Plutarch, 17
  Alkibiades, 74n.29
    on Alkibiades' relationships with Agatharchos, Taeureas, and woman from Melos, 123
    and Andocides, 127, 128
    and ostracism of Hyperbolos, 129, 322
  Apopth. regum et imp., 49
  Aristeides, 9, 12, 14n.7, 15, 19, 26, 78, 86n.4, 98, 100
    on fighting at Psyttaleia, 275
    on Hyperbolos, 320
    on Salamis, 276
  De curiositate, on famine at Athens, 114
  De Gloria Atheniensium, 15
  on Dorian (or Aetolian) War, 115
  Fabius Maximus, 309
  as historical source, 174
  for later historians, 100
  Kimon, 3, 18

on curse of the Alkmeonidai, 111n.13
on Perikles, 311
Lysander, 30
Moralia
  on Arkhidamos, 307
  on Thoukydides son of Melesias, 307, 309
Nikias, 84n.3, 86n.4, 100
  on Hipparkhos, 104
  on Hyperbolos, 320, 322
  on ostracism of Hyperbolos, 129
  on ostracism, 57, 58, 83, 91, 86n.4, 100
  on Olympic games, 143
Perikles, 3, 6, 18, 20, 25n.4, 86n.4
  on Athenian *arkhe*, 287
  on curse of the Alkmeonidai, 111n.13
  list of Athenian demagogues, 309–10
  on Perikles and Sparta, 285
  on Perikles' career, 234
  on Thoukydides son of Melesias, 308, 317
Praecepta gerendae rei publicae, on political clubs, 164n.6
pseudo, biography of Andocides, 117
references to Philochoros, 84
Solon, 14, 45, 47, 49
  on amnesty, 73
sources, 49
Themistokles, 21, 66n.5
  on Eleusinian deities, 232
  on Hyperbolos, 320
  on Xanthippos, 112
Theseus, 84
use of Philochoros, 98
X Orat. Vit., 37n.9
Poetry, archaic, 14
Polemarch. *See* Archon, polemarkhos
Polemon, on Thoukydides son of Melesias, 315
Pollux, 89n.5, 89n.6, 91, 92, 100
  on ostracism, 83, 84, 89, 90, 97
Polyaenus, *Strategemata*, 49
Polycrates, *Tyche*, 46
Polygnotos, painter, 235
Pompos, name for Papasilenos, 237
Pontic expedition, of Perikles, 6
Poseidippos, comic poet, 13
Poseidon, representations on pottery, 233
Posideion-Al Mina, 188
Potter Relief, from the Akropolis, 212–19
Pottery. *See* Achilles; Aeschines; Agatharkhos; Altamura Painter; Ambrosis Painter; Amphitrite; Archaic style; Beldam Painter; Berlin Painter; Black-figure pottery; Brygos workshop; Chiusi; Demeter; *Demokratia*, painting of; *Demos*, painting of; Dionysos; Dipylon pottery; Droop Cups; Edinburgh Painter; Emporium Painter; Euergides Painter; Euphranor; Euthymides; Exekias; Francois Vase; Hekate; Hades; Herakles; Kephalos; Kiss Painter; Kleophon Painter; Kore; Leagros Period; Love-names; Makron; Marsyas Painter; Nikosthenes; Niobid Painter; Oltos; Ostraka; Pamphaios; Pan Painter; Persephone; Polygnotos; Potter Relief; Red-figure pottery; Severe style; Smikros; Swing Painter; Telos Painter; Theseus Painter; Theseus; Triptolemos; Xenokles; Zeus
Princes, Hellenistic, cult of, 168
Prometheus, 268–69
Pronapes, 198, 199
  name, 198–99
Propylaia, 199
Prosodion, 259
Prosopography, Athenian, 56
Protagoras, 112, 269
Prytaneion. *See* Prytanies
Prytanies, 73n.28, 74n.28, 88, 110, 110n.8, 112n.17, 114n.24, 203
Psyttaleia, 275
Ptolemaios Chennos, 80
Ptolemais, tribe, 338n.7
Ptolemy Euergetes, priest of, 226
Pylades, 153
Pylos, 33
Pyrilampes, trial of, 313, 318
Pythia, 144
Pythodoros, 37n.9
  Atheninas archon (404/3), 299

Rampin Horseman, 218
Red-figure pottery, 61, 215, 231
Reforms of 487, 110, 318
Reggio, 234n.32
Reinmuth, Oscar, 55
Rhetors, Athenian, list of, 309–10
Rhodes, and writing, 188, 189
Rhodopis, 272
Rhombos, statue of, 195

Sabaziastai, 138
Sages, Seven, 13, 264, 271
Salaminioi, genos, 158
Salamis, 8, 197, 198, 200, 275
  battle of, 15, 26, 58, 73, 102, 105, 111, 112, 183, 185, 197, 221–22, 232, 240, 271, 283, 318
  oracle concerning, 185
  reënactement of the battle staged by Augustus, 222
Samos, 6, 23, 24, 38, 56, 103, 271, 298, 299, 323
  revolt of, 6, 23, 287

Sardis, 4, 5, 6, 211
Sartori, F., *Le eterie nelle vita politica Ateniese del VI e V secolo A.C.*, review of, 160–66
Satyr, 237
Satyros, on Thoukydides son of Melesias, 308–9
Scholia on an individual work. *See under work's title*
Scholia, 81, 82, 84
Scribes, professional, 62
Sculpture, Eastern, 207
Scythia, 272
Sea-league. *See* Delian League.
Sedition act, Athenian of 410, 75
Segesta, 128
Seleucid Empire, 169
Serapiastai, 138
Severe style, 196, 207
Sicilian expedition, 7, 56, 58, 117, 118, 119, 124, 128, 129, 164, 165, 182, 282, 285
Sicily, 7, 8
Simonides, and epigrams, 245, 251, 252, 262
Sinope, 6
Siphnian treasury, 216, 218, 273
Skyllaion, 103
Smikros, potter, 214n.6
Sokrates, 33
Sokrates of Anagyrous, Athenian general, 56
Solmsen, F.,*Hesiod and Aeschylus,* 266–269
Solon Logos. See Herodotus, on Solon
Solon, 45, 46, 48, 190, 193, 224, 274, 309, 327
  apodemia, 46, 47
  in Comedy, 46, 47
  electoral reforms, 67, 68, 68n.10
  and εὐνομία, 268
  and Hesiod, 266–67
  laws, 46, 47, 66. *See also* Solon, *seisachtheia*
  legislastion on tyranny, 72–74
  papyrus evidence, 48–49
  poetry, 45–47
  seisachtheia, 46, 47, 48
Sophokles
  Ajax, 14n.5, 107
  Antigone, 289
  generalship, 310, 313
  Oedipus the King, 46, 109n.5, 165
  use of ἀλιτηρός, 111n.12
  on Thoukydides son of Melesias, 307
  Triptolemos, 232
  vita of, on Perikles and Thoukydides, son of Melesias, 310
Sophokles, Titus Flavius, Athenian archon, AD (121/2), 138
Sophronistai, 70n.16
Soros, 210
Soteriastai, 138
Sounion, Attic deme, 137
Sphinx, 253

Stasinos of Cyprus, 262
Stephanos, son of Thoukydides son of Melesias, 315
Stephanus Byzantius, 221
Stesimbrotos of Thasos, 111n.13
  as historical source, 174
  on Themistokles, Thoukydides, Perikles, Kimon, 83
  on Thoukydides, son of Melesias, 307, 309, 316
Sthenelaïdas, Spartan ephor, 281, 289
Stobaios, 97
Strabo, 46
  on Melos, 33
  on Olympic games, 143
Strategoi. *See* Generals
Suetonius, *Divi Augusti Vita*, 221
Suidas, 89n.5, 92–94
  on famine at Athens, 114
  on ostracism, 85n.3, 95
Sulla, 39
Sunium, 209
Supplementum Epigraphicum Graecum, 63
Susa, 7
Swing Painter, 230
Sybaris, 313, 318
Symmachos, on Melos, 33
Synodos, 138–39
Synomosiai, vs. *hetaireiai*, 162
Syrakuse, 96, 303

Tanagra, battle of, 17
Taureas, choregos, 123
Teireisias, 237
Teisias, 30, 125
Telesinos, Athenian archon, 104
Telos Painter, 238
Tenos, 29
Thasos, revolt of, 286
Thebes, 12, 21, 24, 25, 25n.3, 35, 43, 255
  and the Oath of Plataia, 277, 281
  and writing, 188
  siege of, 183
  treaty with Athens, 25n.3
Themistokles, 21, 26, 149n.1, 185, 186, 199
  alleged treason, 58
  building of the fleet, 104
  as choregos for Phrynikhos' *Phoenissai*, 112n.18
  and Eurybiades, 294
  hoard of ostraka found by Broneer, 57
  naval program, 66n.5
  ostracism, 54, 55, 58, 60, 61, 74n.29, 94, 95, 96, 103, 107, 109n.3, 113, 152, 161
  and political clubs, 160
  trial of, 203
Themistokles Decree, 11, 183, 185, 275, 283
Themistokles, 'The New', 221, 222
Theodoros of Samos, sculptor, 273

Theodoros Metochites, on ostracism, 100–101
Theophrastos
  as used by Aristides, Aelius, 101
  Characters, 85n.3
  on Theseus, 224
  dates of ostracisms, 104
  on Hipparkhos, 105, 106
  Laws, 49, 84–87, 89, 89n.5, 93
  on number of votes necessary for an ostracism, 88
  on ostracism, 310
  on ostracism, of Hyperbolos, 89
  on Nikias and Alkibiades, 130
  Nomoi. See Theophrastos, *Laws*
  on ostracism of Hipparkhos, 104
  as used by Plutarch, 100
  Πολιτικὰ τὰ πρὸς τοὺς καιρούς, meaning of title, 84n.3
  Πολιτικὰ τὰ πρὸς τοὺς καιρούς, on ostracism, 80, 84n.3
  summary of evidence regarding his account of ostracism, 102
Theopompos, 3, 8, 9, 14n.6, 17, 47, 326
  as historical source, 174
  History of Philip, on Thoukydides son of Melesias, 306
  on Hyperbolos, 98, 117, 129, 316, 320–24
  on Kimon, 308, 317
  on Perikles, 308
  on Solon, 48
  on the Oath of Plataia, 277
  on the Peace of Kallias, 277
  on Thoukydides son of Melesias, 309–18
  as source on ostracism, 83
Thera, and writing, 188
Theramenes, 35, 326
  and political clubs, 166
Thermopylai, 145
  battle of, 185, 276
  inscription honoring the dead, 271
Theseus Painter, 230n.3
Theseus, 224
  accounts of compared, 327
  painting of, 223
  representations on pottery, 268
Thesmothetai, 67
Thesprotians, 103
Thessalos, son of Kimon, 111n.13
Thessaly, 21, 26, 199, 200
Thirty Tyrants, 35, 36, 164
Thirty Years' Peace, 23, 176, 286
Thisbe, 247
Thomas Magister
  Anecdota Graeca, 99
  Ecloga Vocum Atticarum, 92, 97, 101
Thornax, 271
Thoukydides, 6, 7, 8, 13, 14, 16, 17, 21, 107, 326, 328
  account of the Corinthians' speech at Athens, 23, 25
  account of the Corinthians' speech at Sparta, 25
  account of the destruction of Plataea, 24
  on Alkibiades' relationships with Agatharchos, Taeureas, and woman from Melos, 124
  amendation due to epigraphical evidence, 54
  Athenian treasurer (424/3), 7
  on Athenians' speech at Sparta, 279–91
  on curse of Alkmeonidai, 111
  on dedications at Marathon, 207
  historian, source for Ephorus, 96
  as historical source, 174, 328
  knowledge of Herodotus, 25
  on Δωριακὸς πόλεμος, 114
  on Melos, 29
  on Nikias/Alkibiades debate. See Nikias, debate with Alkibiades
  on τὸ παράλογον of war, 289
  on oligarchic clubs, 160
  on Olympic games, 145
  ostracism, 75, 96, 103, 126
  on Peace of Kallias, 3
  Perikles' funeral oration, 289
  on Solon, 46
  speeches, 119. See also individual listings
  on τὰ δίκαια, 102
  on tyrants, 68
  use of *Hetaireiai*, and *Synomosiai*, 162
Thoukydides son of Melesias, 95, 309–18
  exile to Aigina, 313, 314, 318
  generalship, 310
  ostracism, 58, 59, 62, 161, 306, 309, 310, 312, 313, 316
  and political clubs, 166
  and Thourioi, 313
  vita, 312
  and Xenokritos, 313
Thoukydides son of Oloros, 315n.17
Thourioi, founding of, 313
Thrace, 20, 200
Thrasyboulos, 35–43, 48
  stele in honor of, 36
Thrasyllos, 298
Thucydides. See Thoukydides
Timaios, 92
Timaios, on ostracism, 78, 90, 96
Timonassa, wife of Peisistratos, 231
Timotheos, from Melite, 159
Tissaphernes, 8
Tolmides, 309, 317
Tomb inscriptions, 63
Trachis, 303
Trajan, 134, 221n.3
Tribute lists, Athenian, 13, 16, 17, 19, 20, 21, 54, 72n.22, 176, 177
Tribute, assessment of, 54, 121, 122, 124, 130

Triptolemos
  Mission of, 229–38
  representations on pottery, 230, 231, 232
Troizen, 103, 185, 276
  and the battle of Salamis, 275
Tullia, daughter of Cicero, 341
Tusculum, 341
Twelve Gods, 200, 202
  alter on the Acropolis, 203
  cult of, 201
  dedications to, 197, 198, 227
  sanctuaries of, 198
Tyche, 200n.27
Tyrannicides, cult of, 67n.6
Tyranny legislation, 67, 152, 223, 226
  and ostracism, 27, 65, 67, 71, 72–75
Tyranny, and the archons, 67–69
Tyrants. *See also* Thirty Tyrants
  expulsion of, 97
Tyre, 272
Tzetzes, *Chiliades*, 79, 89n.5, 90, 91, 93

Uranos, 267–68

Valerius Maximus, 49
Vanderpool, Eugene, 53, 60, 61
Vase Painting, Attic, 212. *See also* Pottery
Vergil, 267
  Aeneid, 14n.7
Victor statues. *See* Votive offerings
Victories, statues of, 273
Votive offerings, 195–203

Wells. See Kallichoron, Parthenion
Writing, 187–93, 198, 207, 215. *See also*
      Crete; Digamma; Euboia; Korinth;
      Linear B; Melos; Rhodes and writing;
      Scribes, professional; Thebes; Thera
  Athenian, 63–63, 189

Boustrophedon, 190
Double consonants, used in Attic Greek, 62
  and epigrams, 246
  Ionic, 62, 197, 199, 216, 218n.34
  Phoenician, 188, 256

Xanthippos, 5, 109
  date of leadership, 110
  ostracism, 27, 28, 54, 59, 104, 106, 108–15
  ostrakon, 74n.28
  statue of, 112
Xenainetos, Athenian archon (401/0), 37
Xenokles, 148
Xenokles, potter, 214n.6
Xenokritos, and Thoukydides son of Melesias, 313
Xenophon, 37, 37n.9
  Anabasis, 103
  Anabasis, on τὰ δίκαια, 102
  on Artemis Agrotera, 207
  Athenian general, 199
  Cyropedia, on σκυλακώδης, 311n.10
  Hellenika, 103, 326
  Hellenika, chronology of, 296–305
  Hellenika, on Marathon, 204n.1
  as historical source, 174
  Memorabilia, 70n.15
  on political clubs, 166
  pseudo, *Constitution of Athens*, on Athenian law-courts, 288
Xenotimos, 89n.5
Xerxes, 13, 26, 28, 103, 154, 270, 276, 283
Xypete, Attic deme, 55, 57

Zenon, 337
Zeus, 267, 268
  Eleutherios, cult of at Plataia, 12, 19
  in Hesiod and Homer, 266
  representations on pottery, 233
Zonaras, 92

Index of Greek Words

ἀβουλία, 289
ἀδικεῖν, 109, 281
ἀγαθός, 119. See also καλὸς κἀγαθος
ἄγαλμα, 254
ἄγγελος ἀθανάτων, 204
ἀγέλαστος πέτρα, 236, 237
ἀγορά, 140
ἀγῶνα, 143
ἀγῶνα γυμνικόν, 144
αἰτία, 143
Αἰτωλικὸς πόλεμος, 114
ἀλεῖν, 114
ἀλειτηρός, 105, 109, 110, 111, 113, 114, 112, 152
ἄλευρα, 114
ἀληθής, 149
ἄλσει μέσσῳ, 340
ἄλσος, 142
ἀνάθημα, 37n.7
ἀναρχία, 37n.9
ἀνέθηκε, 260
ἀνεπίφθονος, 283
ἀνταναπλήρωσις, 346
ἀντιπολιτεύομαι, 308, 314
ἄνω ποταμός, 14n.7
ἄξιος, 282, 284, 285
ἀξίως, 282
ἀπαρχαί, 16, 17, 20, 237, 240
ἀπεικότως, 282
ἀπιστήσαντες, 144
ἀποδημία, 46, 47
ἀποικίαι, 17
ἀρετή, 107
Ἀρρίφρονος, 109n.3
ἀρχή, 16
ἀσέβεια, 112
ἀτιμία, 103, 106
ἄτομος, 346
ἀφίκετο, 149

βῆμα, 201
Βία, 267
βιάζω, 285, 288
βουλεύομαι, 289
βουλευτήριον, 78
βουλή, 79
βύβλος, 191

γίγνεται, 346
γνώμη, 14n.5
γυμνάζομαι, 334

δειπνοῦσιν, 140
δεκάτη, 214.n.11
δεξιότης, 284
δέοντα, τά, 281
διάστροφος, 313n.12
διδάσκαλος, 340
διελθόντος, 301
δικάζω, 288
δίκαια, τά, 103
Δίκη, 267, 268
δογματίσαι, 138
δόξης, 107
δούριος ἵππος, 198
δραϜεος, 249
δύναμις, 100, 284
δυναστεία, 316
δυσχεραινότων, 144
Δωριακὸς πόλεμος, 114

ἐγλογισταί, 140
ἔδοξε, 140
εἴδωλα, 347, 348
εἰκότως, 282
Εἰρήνη, 267, 268
εἰρήνη, 144
εἰς τὸ μὴ ὄν, 346

ἐκ τοῦ μὴ ὄντος, 346
ἔκδοσις, 141
ἐκεχειρία, 144
ἑκόντων, 285
ἐκφυλλοφορία, 79, 89, 89n.5, 91, 190
ἐλασσέομαι, 288
ἐλευθέρια, 12
ἔναγχος, 333
ἐξεργασμένα, 270
ἐξήλασαν, 95
ἐξοστρακισμός, 92
ἑορτή, 144
ἐπὶ τύμβῳ, 253
ἐπιστάτης, 7
ἐπιτάφια, 12
ἐπιτηδεύομαι, 334
ἐπίφθονος, 283
ἐπόπται, 240
ἔρανος, 141
ἔργα μέγαλα τε καὶ θωμάστα, 270–74
ἐργολαβοῦντες, 140
ἔργον, 270
ἔστε, 252
ἐτελεύτα, 301
εὐβουλία, 289, 290
Εὐνομίη, 267, 268
ἐφθείρετο, 346

ζημία, 281

ἡβηδὸν ἀποσφαγῆναι, 31

θέσις, 347
θέσιν καὶ τάξιν, 347
θόρυβοι, 139
θωμάσια, 270
θώματα, 270

ἰσονομία, 268
ἰσοτέλεια, 37n.9

καθηγητής, 340
κακίας, 100
καλλιστεφάνου 'Αφροδίτης, 261
καλὸς κἀγαθός, 308
κατάλυσις, 317
κατάστασις, 66n.4
κατάφημι, 109, 109n.5
κατέγνωσαν, 95
κεῖμαι, 252, 254
κεχειροτονημένον, 221
κηδηστής, 308
κοινή, 262
κοινὴ εἰρήνη, 239
κόλασις, 100
κολούειν τοὺς ὑπερέχοντας, 107
Κράτος, 267
Κυλώνειον ἄγος, 114

λίμναις, ἐν, 138, 139
λιμός vs. λοιμός, 114, 115
λόγος, 149, 287
λοιμός vs. λιμός, 114, 115

Μάρκῳ, 138
μάχαι, 139
μεγαλούχων, 151
μένουσα, 252
μετριάζω, 284, 287
Μήτηρ, 138
μνῆμα, 251
μνημόσυνα λιπέσθαι, 270
μουσική, 334
μοχθηρίας, 100
μύδρος, 14n.7
μύσται, 240

νόθοι, 91
νόμος, 288
νόμος ἐρανιστῶν, 138, 141

ξυγγενεῖς, 144
ξύλα, 141, 142
ξυνετώτατος, 284

ὀλυμπιονίκης, 197
ὁμόνοια, 144
ὀνόματα, καλά, 282
ὀρθὴ πολιτεία, 313n.12
ὅπῃ, 149
ὀστρακίζω, 85n.3
ὄστρακον, 109
οὐδέν, 346
ὄφρα, 252

πάλαι κατατεθνηῶτος, 263
πανηγύρεις, 143, 144
παννυχισταί, 140
πάππον, 126
παράλογος, 289
πατρόθεν καὶ τοῦ δήμου, 60. See also Names; Athenian (Subject Index)
πληγαί, 139
πλήρη, 346
πολιτεία, ὀρθή, 313n.12
πολιτικός, 308
πονηρίας, 100
πότε, 264
πράκτορες, 140
προδότης, 105, 152
προεδίδοσαν, 105
προνοία, 290
πρόξενϜος, 261
πρόφασις, ἀληθεστάτη, 279n.1
πρυανείη, 110
πρυτανεῖον, 109, 111
πυθόχρηστος ἐξηγητής, 337, 343

ῥεῦσις, 346

σεισάχθεια, 46, 47, 48
σῆμα, 246, 247, 248
σιμίδαλις, 141
σκυλακώδης, 311n.10
σκυτάλη, 191
σοφρονώτατος, 284
σπονδὰς εἶναι, 240
σπονδοφόροι, 239, 240
στάσις, 14, 143, 162–63
στήλη, 246
στήλη ἀτιμίας, 106
στρατηγία, 203
στρατηγίας ἀξιωθῆναι, 310
στρεπτοί, 140, 141
συμπλεκόμενος, 311n.10
συναγαγὼν εἰς ταὐτό, 312
σύλλογον, 144
συνερανιστῶν, 139
συνθήκη, 3
σύνοδος, 141
συνχέειν τὰς ψήφους, 313n.14
σύγχυσις δικαστηρίου, 313

σχῆμα, 347
σώματα, 346

ταμίας, 140, 141
τάξιν καὶ θέσιν, 347
τάξις, 347
ταπείνωσις, 100
τηρεῖν, 114
τόπος ἀπεδίδοτο, 103
τύμβος, 246, 253. See also ἐπὶ τύμβῳ
τυραννίδος γραφή, 72n.21

φιλοδικέω, 288
φοραί, 140, 141
φόρος, 17, 20, 21
φθόνος, 100, 106, 107

χαριέστατοι, 86, 98
Χσάνθιππος, 109

ψηφίσμα, 178
ψυχή, 248

ὠφελία, 286, 287

Index of Latin Words

ab urbe condita, 325
ad nihilum, 346

corpora prima, 346
creari, 346

de nihilo, 346
dissimili figura, 346

e nihilo, 346

fieri, 346
figura, 347
fluere, 346

gigni, 346

ignes, 347
imitatio cum variatione, 263
inquis, 348

lignum, 347

Magna Mater, 138

Neos Homeros, 220
nulla res, 346

ordo, 347

pereunt, 346
positura, 347

ratae sententiae, 345
recedere, 346
redit, 346
reverti, 346
revocari, 346

simplex, 346
solida, 346